T0261066

Pontiac Firebird Automotive Repair Manual

by John B. Raffa and John H Haynes
Member of the Guild of Motoring Writers

Models covered:
Pontiac Firebird, Firebird Trans Am, Firebird S/E
1982 through 1992

(79019-5AA16)

J H Haynes & Co. Ltd.
Haynes North America, Inc.
www.haynes.com

Acknowledgements

We are grateful for the help and cooperation of Tomco Industries, 1435 Woodson Road, St. Louis, Missouri 63132, for their assistance with technical information and certain illustrations. Wiring diagrams and certain illustrations originated exclusively for Haynes North America Inc. by Valley Forge Technical Information Services.

A book in the Haynes Automotive Repair Manual Series

ISBN-13: 978-1-56392-065-3

ISBN-10: 1-56392-065-4

Library of Congress Catalog Card Number 93-77337

Disclaimer

There are risks associated with automotive repairs. The ability to make repairs depends on individual skill, experience and proper tools. Individuals should act with due care and acknowledge and assume the risk of making automotive repairs. While every attempt is made to ensure that the information in this manual is correct, no liability can be accepted by the authors or publishers for loss, damage or injury caused by any errors in, or omissions from, the information given.

Contents

Introductory pages

About this manual	0-5
Introduction to the Pontiac Firebird	0-5
Vehicle identification numbers	0-6
Buying parts	0-8
Maintenance techniques, tools and working facilities	0-8
Booster battery (jump) starting	0-16
Jacking and towing	0-17
Automotive chemicals and lubricants	0-18
Conversion factors	0-19
Safety first!	0-20
Troubleshooting	0-21

Chapter 1
Tune-up and routine maintenance   **1-1**

Chapter 2   Part A
L4 (four-cylinder engine)   **2A-1**

Chapter 2   Part B
V6 engine   **2B-1**

Chapter 2   Part C
V8 engine   **2C-1**

Chapter 2   Part D
General engine overhaul procedures   **2D-1**

Chapter 3
Cooling, heating and air conditioning systems   **3-1**

Chapter 4   Part A
Fuel and exhaust systems: carbureted models   **4A-1**

Chapter 4   Part B
Fuel and exhaust systems: fuel injected models   **4B-1**

Chapter 5
Engine electrical systems   **5-1**

Chapter 6
Emissions control systems   **6-1**

Chapter 7   Part A
Manual transmission   **7A-1**

Chapter 7   Part B
Automatic transmission   **7B-1**

Chapter 8
Clutch and driveline   **8-1**

Chapter 9
Brakes   **9-1**

Chapter 10
Chassis electrical system and wiring diagrams   **10-1**

Chapter 11
Suspension and steering systems   **11-1**

Chapter 12
Body   **12-1**

Index   **IND-1**

Pontiac Firebird

About this manual

Its purpose

The purpose of this manual is to provide comprehensive, useful and accessible automotive repair information, to help you get the best value from your vehicle. It can do so in several ways. It can help you decide what work must be done, even if you choose to have it done by a dealer service department or a repair shop; it provides information and procedures for routine maintenance and servicing; and it offers diagnostic and repair procedures to follow when trouble occurs.

We hope you use the manual to tackle the work yourself. For many simpler jobs, doing it yourself may be quicker than arranging an appointment to get the vehicle into a shop and making the trips to leave it and pick it up. More importantly, a lot of money can be saved by avoiding the expense the shop must pass on to you to cover its labor and overhead costs. An added benefit is the sense of satisfaction and accomplishment that you feel after doing the job yourself. However, this manual is not a substitute for a professional certified technician or mechanic. There are risks associated with automotive repairs. The ability to make repairs on a vehicle depends on individual skill, experience and proper tools. Individuals should act with due care and acknowledge and assume the risk of performing automotive repairs.

Using the manual

The manual is divided into Chapters. Each Chapter is divided into numbered Sections, which are headed in bold type between horizontal lines. Each Section consists of consecutively numbered paragraphs.

The reference numbers used in illustration captions pinpoint the pertinent Section and the Step within that Section. That is, illustration 3.2 means the illustration refers to Section 3 and Step (or paragraph) 2 within that Section.

Procedures, once described in the text, are not normally repeated. When it's necessary to refer to another Chapter, the reference will be given as Chapter and Section number. Cross references given without use of the word "Chapter" apply to Sections and/or paragraphs in the same Chapter. For example, "see Section 8" means in the same Chapter. References to the left or right side of the vehicle assume you are sitting in the driver's seat, facing forward.

This repair manual is produced by a third party and is not associated with an individual car manufacturer. If there is any doubt or discrepancy between this manual and the owner's manual or the factory service manual, please refer to factory service manual or seek assistance from a professional certified technician or mechanic. Even though we have prepared this manual with extreme care, neither the publisher nor the author can accept responsibility for any errors in, or omissions from, the information given.

NOTE

A **Note** provides information necessary to properly complete a procedure or information which will make the procedure easier to understand.

CAUTION

A **Caution** provides a special procedure or special steps which must be taken while completing the procedure where the Caution is found. Not heeding a Caution can result in damage to the assembly being worked on.

WARNING

A **Warning** provides a special procedure or special steps which must be taken while completing the procedure where the Warning is found. Not heeding a Warning can result in personal injury.

Introduction to the Pontiac Firebird

For the years covered in this manual, the Firebird is available in three models: Firebird, Firebird Trans Am, and Firebird S/E.

Engines available include the 2.5 liter in-line four-cylinder engine (L4); 2.8 liter and 3.1 liter V6 engines; and 5.0 liter and 5.7 liter V8 engines; available as standard or optional equipment, depending on year and model.

Four and five-speed manual transmissions and three and four-speed automatic transmissions are available as standard or optional equipment, depending on year, model and engine.

The Front suspension is an independent modified McPherson strut type with a coil spring mounted between the frame crossmember and the lower control arm. The rear suspension consists of the torque arm, track bar, coil springs, shock absorbers and lower control arms.

The hydraulic brake system consists of disc brakes at the front and either drum-type or disc brakes at the rear depending on model. Power brakes are standard equipment.

Vehicle identification numbers

Modifications are a continuing and unpublicized process in vehicle manufacturing. Since spare parts manuals and lists are compiled on a numerical basis, the individual vehicle numbers are essential to correctly identify the component required.

Vehicle identification number (VIN)

This very important identification number is located on a plate attached to the top left corner of the dashboard and can easily be seen while looking through the windshield from the outside of the vehicle **(see illustration)**. The VIN also appears on the Vehicle Certificate of Title and Registration. It contains valuable information such as where and when the vehicle was manufactured, the model year and the body style.

Body identification plate

This metal plate is located on the top left side of the radiator support **(see illustration)**. Like the VIN, it contains valuable information concerning the production of the vehicle, as well as information about the way in which the vehicle is equipped. This plate is especially useful for matching the color and type of paint during repair work.

Engine identification numbers

The ID number on the L4 engine is found at the left rear side of the engine, on the casting to the rear of the exhaust manifold.

The ID number on the V6 engine is found either below the left cylinder head or below the right cylinder head, to the rear of the timing cover **(see illustration)**.

The Vehicle Identification Number (VIN) is located on the driver's side of the dashboard and is visible through the windshield

The body identification plate is located at the top right side of the radiator support (A), and the Emissions Control Information label is located on the fan shroud (B)

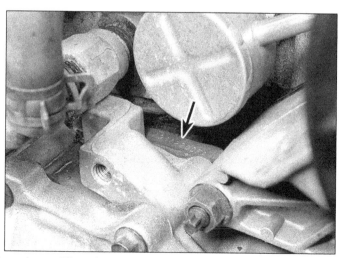

Location of the engine ID number on V6 engine

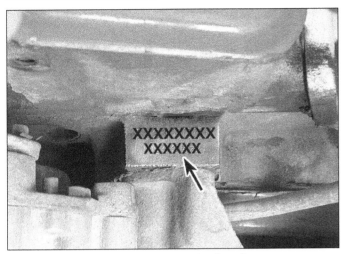

Typical V8 engine number locations

Location of the automatic transmission identification number

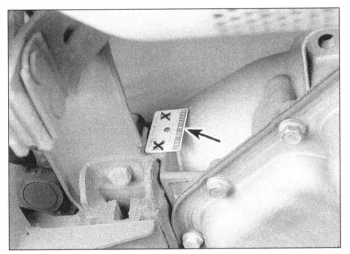

Location of the automatic transmission identification number

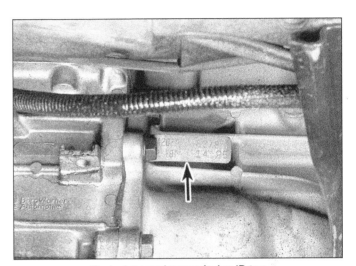

Typical manual transmission ID tag

The ID number on the V8 engine is found just below the right cylinder head, to the rear of the timing cover **(see illustration)**.

Automatic transmission numbers

Automatic transmission ID numbers may be stamped in a variety of locations on pads on either side of the transmission **(see illustration)**.

Manual transmission numbers

Manual transmission ID numbers are stamped on the driver's side, adjacent to the rear of the cover (four-speed 83 mm and five-speed 70 mm), or on the driver's side, below the side cover (four-speed 76 mm) **(see illustration)**.

Rear axle numbers

The rear axle ID number is located on the right axle tube, adjacent to the carrier.

Alternator numbers

The alternator ID number is located on top of the drive end frame.

Starter numbers

The starter ID number is stamped on the outer case, toward the rear.

Battery numbers

The battery ID number is located on the cell cover segment on top of the battery.

Emissions Control Information label

The Emissions Control Information label is attached to the radiator fan shroud.

Buying parts

Replacement parts are available from many sources, which generally fall into one of two categories - authorized dealer parts departments and independent retail auto parts stores. Our advice concerning these parts is as follows:

Retail auto parts stores: Good auto parts stores will stock frequently needed components which wear out relatively fast, such as clutch components, exhaust systems, brake parts, tune-up parts, etc. These stores often supply new or reconditioned parts on an exchange basis, which can save a considerable amount of money. Discount auto parts stores are often very good places to buy materials and parts needed for general vehicle maintenance such as oil, grease, filters, spark plugs, belts, touch-up paint, bulbs, etc. They also usually sell tools and general accessories, have convenient hours, charge lower prices and can often be found not far from home.

Authorized dealer parts department: This is the best source for parts which are unique to the vehicle and not generally available elsewhere (such as major engine parts, transmission parts, trim pieces, etc.).

Warranty information: If the vehicle is still covered under warranty, be sure that any replacement parts purchased - regardless of the source - do not invalidate the warranty!

To be sure of obtaining the correct parts, have engine and chassis numbers available and, if possible, take the old parts along for positive identification.

Maintenance techniques, tools and working facilities

Maintenance techniques

There are a number of techniques involved in maintenance and repair that will be referred to throughout this manual. Application of these techniques will enable the home mechanic to be more efficient, better organized and capable of performing the various tasks properly, which will ensure that the repair job is thorough and complete.

Fasteners

Fasteners are nuts, bolts, studs and screws used to hold two or more parts together. There are a few things to keep in mind when working with fasteners. Almost all of them use a locking device of some type, either a lockwasher, locknut, locking tab or thread adhesive. All threaded fasteners should be clean and straight, with undamaged threads and undamaged corners on the hex head where the wrench fits. Develop the habit of replacing all damaged nuts and bolts with new ones. Special locknuts with nylon or fiber inserts can only be used once. If they are removed, they lose their locking ability and must be replaced with new ones.

Rusted nuts and bolts should be treated with a penetrating fluid to ease removal and prevent breakage. Some mechanics use turpentine in a spout-type oil can, which works quite well. After applying the rust penetrant, let it work for a few minutes before trying to loosen the nut or bolt. Badly rusted fasteners may have to be chiseled or sawed off or removed with a special nut breaker, available at tool stores.

If a bolt or stud breaks off in an assembly, it can be drilled and removed with a special tool commonly available for this purpose. Most automotive machine shops can perform this task, as well as other repair procedures, such as the repair of threaded holes that have been stripped out.

Flat washers and lockwashers, when removed from an assembly, should always be replaced exactly as removed. Replace any damaged washers with new ones. Never use a lockwasher on any soft metal surface (such as aluminum), thin sheet metal or plastic.

Fastener sizes

For a number of reasons, automobile manufacturers are making wider and wider use of metric fasteners. Therefore, it is important to be able to tell the difference between standard (sometimes called U.S. or SAE) and metric hardware, since they cannot be interchanged.

All bolts, whether standard or metric, are sized according to diameter, thread pitch and length. For example, a standard 1/2 - 13 x 1 bolt is 1/2 inch in diameter, has 13 threads per inch and is 1 inch long. An M12 - 1.75 x 25 metric bolt is 12 mm in diameter, has a thread pitch of 1.75 mm (the distance between threads) and is 25 mm long. The two bolts are nearly identical, and easily confused, but they are not interchangeable.

In addition to the differences in diameter, thread pitch and length, metric and standard bolts can also be distinguished by examining the bolt heads. To begin with, the distance across the flats on a standard bolt head is measured in inches, while the same dimension on a metric bolt is sized in millimeters (the same is true for nuts). As a result, a standard wrench should not be used on a metric bolt and a metric wrench should not be used on a standard bolt. Also, most standard bolts have slashes radiating out from the center of the head to denote the grade or strength of the bolt, which is an indication of the amount of torque that can be applied to it. The greater the number of slashes, the greater the strength of the bolt. Grades 0 through 5 are commonly used on automobiles. Metric bolts have a property class (grade) number, rather than a slash, molded into their heads to indicate bolt strength. In this case, the higher the number, the stronger the bolt. Property class numbers 8.8, 9.8 and 10.9 are commonly used on automobiles.

Strength markings can also be used to distinguish standard hex nuts from metric hex nuts. Many standard nuts have dots stamped into one side, while metric nuts are marked with a number. The greater the number of dots, or the higher the number, the greater the strength of the nut.

Metric studs are also marked on their ends according to property class (grade). Larger studs are numbered (the same as metric bolts), while smaller studs carry a geometric code to denote grade.

It should be noted that many fasteners, especially Grades 0 through 2, have no distinguishing marks on them. When such is the case, the only way to determine whether it is standard or metric is to measure the thread pitch or compare it to a known fastener of the same size.

Standard fasteners are often referred to as SAE, as opposed to metric. However, it should be noted that SAE technically refers to a non-metric fine thread fastener only. Coarse thread non-metric fasteners are referred to as USS sizes.

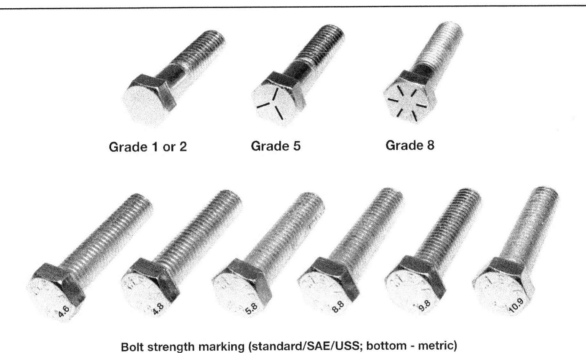

Grade 1 or 2   Grade 5   Grade 8

Bolt strength marking (standard/SAE/USS; bottom - metric)

Grade	Identification
Hex Nut Grade 5	3 Dots
Hex Nut Grade 8	6 Dots

Standard hex nut strength markings

Grade	Identification
Hex Nut Property Class 9	Arabic 9
Hex Nut Property Class 10	Arabic 10

Metric hex nut strength markings

Class 10.9   Class 9.8   Class 8.8

Metric stud strength markings

Since fasteners of the same size (both standard and metric) may have different strength ratings, be sure to reinstall any bolts, studs or nuts removed from your vehicle in their original locations. Also, when replacing a fastener with a new one, make sure that the new one has a strength rating equal to or greater than the original.

Tightening sequences and procedures

Most threaded fasteners should be tightened to a specific torque value (torque is the twisting force applied to a threaded com-ponent such as a nut or bolt). Overtightening the fastener can weaken it and cause it to break, while undertightening can cause it to eventually come loose. Bolts, screws and studs, depending on the material they are made of and their thread diameters, have specific torque values, many of which are noted in the Specifications at the beginning of each Chapter. Be sure to follow the torque recommendations closely. For fasteners not assigned a specific torque, a general torque value chart is presented here as a guide. These torque values are for dry (unlubricated) fasteners threaded into steel or cast iron (not aluminum). As was previously mentioned, the size and grade of a fastener determine the amount of torque that can safely be applied to it. The figures listed here are approximate for Grade 2 and Grade 3 fasteners. Higher grades can tolerate higher torque values.

Fasteners laid out in a pattern, such as cylinder head bolts, oil pan bolts, differential cover bolts, etc., must be loosened or tightened in sequence to avoid warping the component. This sequence will normally be shown in the appropriate Chapter. If a specific pattern is not given, the following procedures can be used to prevent warping.

Metric thread sizes	Ft-lbs	Nm
M-6	6 to 9	9 to 12
M-8	14 to 21	19 to 28
M-10	28 to 40	38 to 54
M-12	50 to 71	68 to 96
M-14	80 to 140	109 to 154
Pipe thread sizes		
1/8	5 to 8	7 to 10
1/4	12 to 18	17 to 24
3/8	22 to 33	30 to 44
1/2	25 to 35	34 to 47
U.S. thread sizes		
1/4 - 20	6 to 9	9 to 12
5/16 - 18	12 to 18	17 to 24
5/16 - 24	14 to 20	19 to 27
3/8 - 16	22 to 32	30 to 43
3/8 - 24	27 to 38	37 to 51
7/16 - 14	40 to 55	55 to 74
7/16 - 20	40 to 60	55 to 81
1/2 - 13	55 to 80	75 to 108

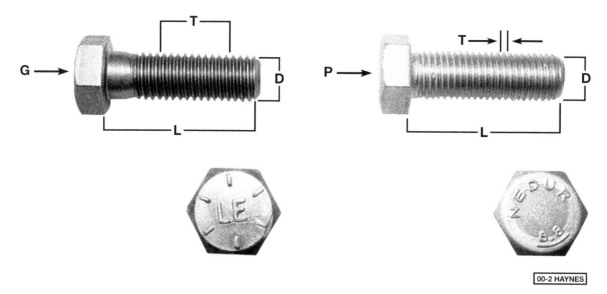

00-2 HAYNES

Standard (SAE and USS) bolt dimensions/grade marks

- G Grade marks (bolt strength)
- L Length (in inches)
- T Thread pitch (number of threads per inch)
- D Nominal diameter (in inches)

Metric bolt dimensions/grade marks

- P Property class (bolt strength)
- L Length (in millimeters)
- T Thread pitch (distance between threads in millimeters)
- D Diameter

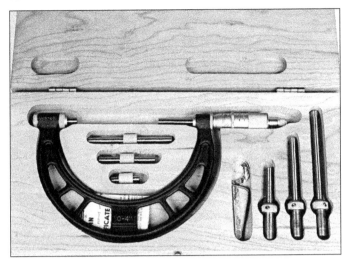

Micrometer set

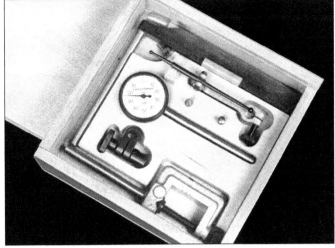

Dial indicator set

Initially, the bolts or nuts should be assembled finger-tight only. Next, they should be tightened one full turn each, in a criss-cross or diagonal pattern. After each one has been tightened one full turn, return to the first one and tighten them all one-half turn, following the same pattern. Finally, tighten each of them one-quarter turn at a time until each fastener has been tightened to the proper torque. To loosen and remove the fasteners, the procedure would be reversed.

Component disassembly

Component disassembly should be done with care and purpose to help ensure that the parts go back together properly. Always keep track of the sequence in which parts are removed. Make note of special characteristics or marks on parts that can be installed more than one way, such as a grooved thrust washer on a shaft. It is a good idea to lay the disassembled parts out on a clean surface in the order that they were removed. It may also be helpful to make sketches or take instant photos of components before removal.

When removing fasteners from a component, keep track of their locations. Sometimes threading a bolt back in a part, or putting the washers and nut back on a stud, can prevent mix-ups later. If nuts and bolts cannot be returned to their original locations, they should be kept in a compartmented box or a series of small boxes. A cupcake or muffin tin is ideal for this purpose, since each cavity can hold the bolts and nuts from a particular area (i.e. oil pan bolts, valve cover bolts, engine mount bolts, etc.). A pan of this type is especially helpful when working on assemblies with very small parts, such as the carburetor, alternator, valve train or interior dash and trim pieces. The cavities can be marked with paint or tape to identify the contents.

Whenever wiring looms, harnesses or connectors are separated, it is a good idea to identify the two halves with numbered pieces of masking tape so they can be easily reconnected.

Gasket sealing surfaces

Throughout any vehicle, gaskets are used to seal the mating surfaces between two parts and keep lubricants, fluids, vacuum or pressure contained in an assembly.

Many times these gaskets are coated with a liquid or paste-type gasket sealing compound before assembly. Age, heat and pressure can sometimes cause the two parts to stick together so tightly that they are very difficult to separate. Often, the assembly can be loosened by striking it with a soft-face hammer near the mating surfaces. A regular hammer can be used if a block of wood is placed between the hammer and the part. Do not hammer on cast parts or parts that could be easily damaged. With any particularly stubborn part, always recheck to make sure that every fastener has been removed.

Avoid using a screwdriver or bar to pry apart an assembly, as they can easily mar the gasket sealing surfaces of the parts, which must remain smooth. If prying is absolutely necessary, use an old broom handle, but keep in mind that extra clean up will be necessary if the wood splinters.

After the parts are separated, the old gasket must be carefully scraped off and the gasket surfaces cleaned. Stubborn gasket material can be soaked with rust penetrant or treated with a special chemical to soften it so it can be easily scraped off. A scraper can be fashioned from a piece of copper tubing by flattening and sharpening one end. Copper is recommended because it is usually softer than the surfaces to be scraped, which reduces the chance of gouging the part. Some gaskets can be removed with a wire brush, but regardless of the method used, the mating surfaces must be left clean and smooth. If for some reason the gasket surface is gouged, then a gasket sealer thick enough to fill scratches will have to be used during reassembly of the components. For most applications, a non-drying (or semi-drying) gasket sealer should be used.

Hose removal tips

Warning: *If the vehicle is equipped with air conditioning, do not disconnect any of the A/C hoses without first having the system depressurized by a dealer service department or a service station.*

Hose removal precautions closely parallel gasket removal precautions. Avoid scratching or gouging the surface that the hose mates against or the connection may leak. This is especially true for radiator hoses. Because of various chemical reactions, the rubber in hoses can bond itself to the metal spigot that the hose fits over. To remove a hose, first loosen the hose clamps that secure it to the spigot. Then, with slip-joint pliers, grab the hose at the clamp and rotate it around the spigot. Work it back and forth until it is completely free, then pull it off. Silicone or other lubricants will ease removal if they can be applied between the hose and the outside of the spigot. Apply the same lubricant to the inside of the hose and the outside of the spigot to simplify installation.

As a last resort (and if the hose is to be replaced with a new one anyway), the rubber can be slit with a knife and the hose peeled from the spigot. If this must be done, be careful that the metal connection is not damaged.

If a hose clamp is broken or damaged, do not reuse it. Wire-type clamps usually weaken with age, so it is a good idea to replace them with screw-type clamps whenever a hose is removed.

Tools

A selection of good tools is a basic requirement for anyone who plans to maintain and repair his or her own vehicle. For the owner who has few tools, the initial investment might seem high, but when compared to the spiraling costs of professional auto maintenance and repair, it is a wise one.

To help the owner decide which tools are needed to perform the tasks detailed in this manual, the following tool lists are offered: *Maintenance and minor repair,*

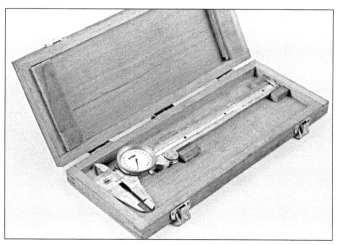

Dial caliper

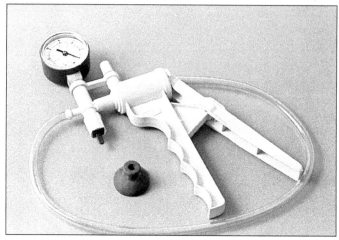

Hand-operated vacuum pump

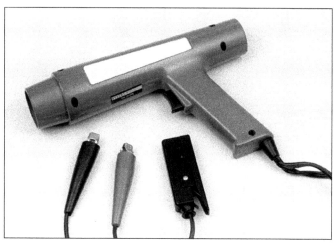

Timing light

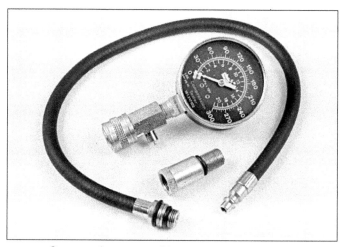

Compression gauge with spark plug hole adapter

Damper/steering wheel puller

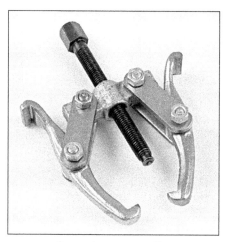

General purpose puller

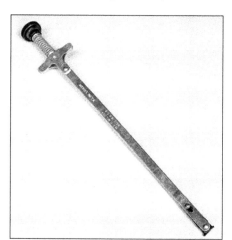

Hydraulic lifter removal tool

Repair/overhaul and *Special.*

The newcomer to practical mechanics should start off with the *maintenance and minor repair* tool kit, which is adequate for the simpler jobs performed on a vehicle. Then, as confidence and experience grow, the owner can tackle more difficult tasks, buying additional tools as they are needed.

Eventually the basic kit will be expanded into the *repair and overhaul* tool set. Over a period of time, the experienced do-it-yourselfer will assemble a tool set complete enough for most repair and overhaul procedures and will add tools from the special category when it is felt that the expense is justified by the frequency of use.

Maintenance and minor repair tool kit

The tools in this list should be considered the minimum required for performance of routine maintenance, servicing and minor repair work. We recommend the purchase of combination wrenches (box-end and open-

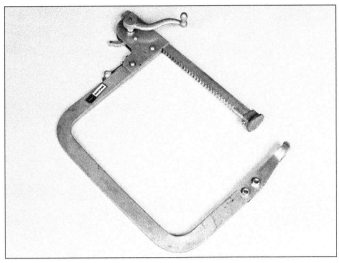

Valve spring compressor

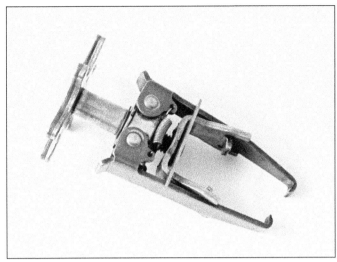

Valve spring compressor

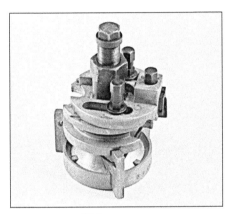

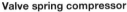

Ridge reamer

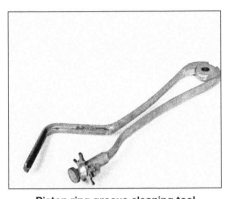

Piston ring groove cleaning tool

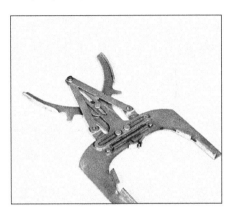

Ring removal/installation tool

end combined in one wrench). While more expensive than open end wrenches, they offer the advantages of both types of wrench.

> *Combination wrench set (1/4-inch to 1 inch or 6 mm to 19 mm)*
> *Adjustable wrench, 8 inch*
> *Spark plug wrench with rubber insert*
> *Spark plug gap adjusting tool*
> *Feeler gauge set*
> *Brake bleeder wrench*
> *Standard screwdriver (5/16-inch x 6 inch)*
> *Phillips screwdriver (No. 2 x 6 inch)*
> *Combination pliers - 6 inch*
> *Hacksaw and assortment of blades*
> *Tire pressure gauge*
> *Grease gun*
> *Oil can*
> *Fine emery cloth*
> *Wire brush*
> *Battery post and cable cleaning tool*
> *Oil filter wrench*
> *Funnel (medium size)*
> *Safety goggles*
> *Jackstands (2)*
> *Drain pan*

Note: *If basic tune-ups are going to be part of routine maintenance, it will be necessary to purchase a good quality stroboscopic timing*

light and combination tachometer/dwell meter. Although they are included in the list of special tools, it is mentioned here because they are absolutely necessary for tuning most vehicles properly.

Repair and overhaul tool set

These tools are essential for anyone who plans to perform major repairs and are in addition to those in the maintenance and minor repair tool kit. Included is a comprehensive set of sockets which, though expensive, are invaluable because of their versatility, especially when various extensions and drives are available. We recommend the 1/2-inch drive over the 3/8-inch drive. Although the larger drive is bulky and more expensive, it has the capacity of accepting a very wide range of large sockets. Ideally, however, the mechanic should have a 3/8-inch drive set and a 1/2-inch drive set.

> *Socket set(s)*
> *Reversible ratchet*
> *Extension - 10 inch*
> *Universal joint*
> *Torque wrench (same size drive as sockets)*
> *Ball peen hammer - 8 ounce*
> *Soft-face hammer (plastic/rubber)*

Ring compressor

> *Standard screwdriver (1/4-inch x 6 inch)*
> *Standard screwdriver (stubby - 5/16-inch)*
> *Phillips screwdriver (No. 3 x 8 inch)*
> *Phillips screwdriver (stubby - No. 2)*
> *Pliers - vise grip*
> *Pliers - lineman's*
> *Pliers - needle nose*
> *Pliers - snap-ring (internal and external)*
> *Cold chisel - 1/2-inch*

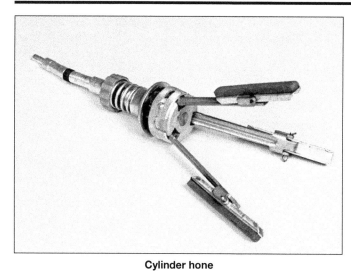

Cylinder hone

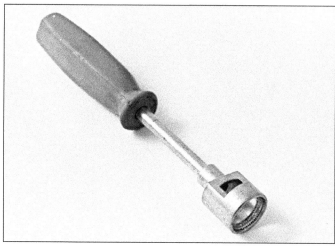

Brake hold-down spring tool

Scribe
Scraper (made from flattened copper
*   tubing)*
Centerpunch
Pin punches (1/16, 1/8, 3/16-inch)
Steel rule/straightedge - 12 inch
Allen wrench set (1/8 to 3/8-inch or
*   4 mm to 10 mm)*
A selection of files

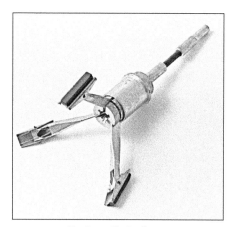

Brake cylinder hone

Wire brush (large)
Jackstands (second set)
Jack (scissor or hydraulic type)
Note: *Another tool which is often useful is an electric drill with a chuck capacity of 3/8-inch and a set of good quality drill bits.*

Special tools

The tools in this list include those which are not used regularly, are expensive to buy, or which need to be used in accordance with their manufacturer's instructions. Unless these tools will be used frequently, it is not very economical to purchase many of them. A consideration would be to split the cost and use between yourself and a friend or friends. In addition, most of these tools can be obtained from a tool rental shop on a temporary basis.

This list primarily contains only those tools and instruments widely available to the public, and not those special tools produced by the vehicle manufacturer for distribution to dealer service departments. Occasionally, references to the manufacturer's special tools are included in the text of this manual. Generally, an alternative method of doing the job without the special tool is offered. How-

ever, sometimes there is no alternative to their use. Where this is the case, and the tool cannot be purchased or borrowed, the work should be turned over to the dealer service department or an automotive repair shop.

Valve spring compressor
Piston ring groove cleaning tool
Piston ring compressor
Piston ring installation tool
Cylinder compression gauge
Cylinder ridge reamer
Cylinder surfacing hone
Cylinder bore gauge
Micrometers and/or dial calipers
Hydraulic lifter removal tool
Balljoint separator
Universal-type puller
Impact screwdriver
Dial indicator set
Stroboscopic timing light (inductive
*   pick-up)*
Hand operated vacuum/pressure pump
Tachometer/dwell meter
Universal electrical multimeter
Cable hoist
Brake spring removal and installation
*   tools*
Floor jack

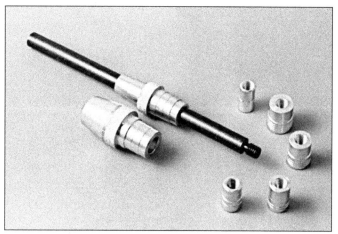

Clutch plate alignment tool

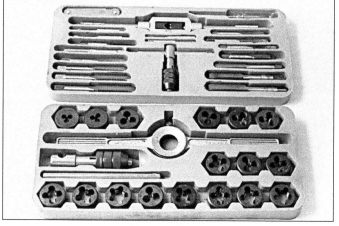

Tap and die set

Buying tools

For the do-it-yourselfer who is just starting to get involved in vehicle maintenance and repair, there are a number of options available when purchasing tools. If maintenance and minor repair is the extent of the work to be done, the purchase of individual tools is satisfactory. If, on the other hand, extensive work is planned, it would be a good idea to purchase a modest tool set from one of the large retail chain stores. A set can usually be bought at a substantial savings over the individual tool prices, and they often come with a tool box. As additional tools are needed, add-on sets, individual tools and a larger tool box can be purchased to expand the tool selection. Building a tool set gradually allows the cost of the tools to be spread over a longer period of time and gives the mechanic the freedom to choose only those tools that will actually be used.

Tool stores will often be the only source of some of the special tools that are needed, but regardless of where tools are bought, try to avoid cheap ones, especially when buying screwdrivers and sockets, because they won't last very long. The expense involved in replacing cheap tools will eventually be greater than the initial cost of quality tools.

Care and maintenance of tools

Good tools are expensive, so it makes sense to treat them with respect. Keep them clean and in usable condition and store them properly when not in use. Always wipe off any dirt, grease or metal chips before putting them away. Never leave tools lying around in the work area. Upon completion of a job, always check closely under the hood for tools that may have been left there so they won't get lost during a test drive.

Some tools, such as screwdrivers, pliers, wrenches and sockets, can be hung on a panel mounted on the garage or workshop wall, while others should be kept in a tool box or tray. Measuring instruments, gauges, meters, etc. must be carefully stored where they cannot be damaged by weather or impact from other tools.

When tools are used with care and stored properly, they will last a very long time. Even with the best of care, though, tools will wear out if used frequently. When a tool is damaged or worn out, replace it. Subsequent jobs will be safer and more enjoyable if you do.

How to repair damaged threads

Sometimes, the internal threads of a nut or bolt hole can become stripped, usually from overtightening. Stripping threads is an all-too-common occurrence, especially when working with aluminum parts, because aluminum is so soft that it easily strips out.

Usually, external or internal threads are only partially stripped. After they've been cleaned up with a tap or die, they'll still work. Sometimes, however, threads are badly damaged. When this happens, you've got three choices:

1) *Drill and tap the hole to the next suitable oversize and install a larger diameter bolt, screw or stud.*

2) *Drill and tap the hole to accept a threaded plug, then drill and tap the plug to the original screw size. You can also buy a plug already threaded to the original size. Then you simply drill a hole to the specified size, then run the threaded plug into the hole with a bolt and jam nut. Once the plug is fully seated, remove the jam nut and bolt.*

3) *The third method uses a patented thread repair kit like Heli-Coil or Slim-sert. These easy-to-use kits are designed to repair damaged threads in straight-through holes and blind holes. Both are available as kits which can handle a variety of sizes and thread patterns. Drill the hole, then tap it with the special included tap. Install the Heli-Coil and the hole is back to its original diameter and thread pitch.*

Regardless of which method you use, be sure to proceed calmly and carefully. A little impatience or carelessness during one of these relatively simple procedures can ruin your whole day's work and cost you a bundle if you wreck an expensive part.

Working facilities

Not to be overlooked when discussing tools is the workshop. If anything more than routine maintenance is to be carried out, some sort of suitable work area is essential.

It is understood, and appreciated, that many home mechanics do not have a good workshop or garage available, and end up removing an engine or doing major repairs outside. It is recommended, however, that the overhaul or repair be completed under the cover of a roof.

A clean, flat workbench or table of comfortable working height is an absolute necessity. The workbench should be equipped with a vise that has a jaw opening of at least four inches.

As mentioned previously, some clean, dry storage space is also required for tools, as well as the lubricants, fluids, cleaning solvents, etc. which soon become necessary.

Sometimes waste oil and fluids, drained from the engine or cooling system during normal maintenance or repairs, present a disposal problem. To avoid pouring them on the ground or into a sewage system, pour the used fluids into large containers, seal them with caps and take them to an authorized disposal site or recycling center. Plastic jugs, such as old antifreeze containers, are ideal for this purpose.

Always keep a supply of old newspapers and clean rags available. Old towels are excellent for mopping up spills. Many mechanics use rolls of paper towels for most work because they are readily available and disposable. To help keep the area under the vehicle clean, a large cardboard box can be cut open and flattened to protect the garage or shop floor.

Whenever working over a painted surface, such as when leaning over a fender to service something under the hood, always cover it with an old blanket or bedspread to protect the finish. Vinyl covered pads, made especially for this purpose, are available at auto parts stores.

Booster battery (jump) starting

Observe the following precautions when using a booster battery to start a vehicle:

a) *Before connecting the booster battery, make sure the ignition switch is in the Off position.*
b) *Turn off the lights, heater and other electrical loads.*
c) *Your eyes should be shielded. Safety goggles are a good idea.*
d) *Make sure the booster battery is the same voltage as the dead one in the vehicle.*
e) *The two vehicles MUST NOT TOUCH each other.*
f) *Make sure the transmission is in Neutral (manual transaxle) or Park (automatic transaxle).*
g) *If the booster battery is not a maintenance-free type, remove the vent caps and lay a cloth over the vent holes.*

Connect the red jumper cable to the positive (+) terminals of each battery.

Connect one end of the black cable to the negative (-) terminal of the booster battery. The other end of this cable should be connected to a good ground on the engine block **(see illustration)**. Make sure the cable will not come into contact with the fan, drivebelts or other moving parts of the engine.

Start the engine using the booster battery, then, with the engine running at idle speed, disconnect the jumper cables in the reverse order of connection.

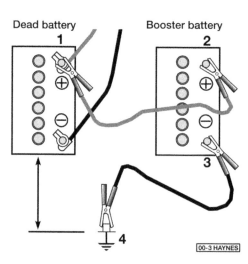

Make the booster battery cable connections in the numerical order shown (note that the negative cable of the booster battery is NOT attached to the negative terminal of the dead battery)

Jacking and towing

Jacking

The jack supplied with the vehicle should only be used for raising the vehicle when changing a tire or placing jackstands under the frame. **Caution:** *Never work under the vehicle or start the engine while this jack is being used as the only means of support.*

The vehicle should be on level ground with the wheels blocked and the transmission in Park (automatic) or Reverse (manual). Pry off the hub cap (if equipped) using the tapered end of the lug wrench. Loosen the wheel nuts one-half turn and leave them in place until the wheel is raised off the ground.

Place the jack under the side of the vehicle in the indicated position and place the jack lever in the "up" position. Raise the jack until the jack head groove fits into the rocker flange notch **(see illustration)**. Operate the jack with a slow, smooth motion, using your hand or foot to pump the handle until the wheel is raised off the ground. Remove the wheel nuts, pull off the wheel and replace it with the spare. (If you have a stowaway spare, refer to the instructions accompanying the supplied inflator).

With the beveled side in, replace the wheel nuts and tighten them until snug. Place the jack lever in the "down" position and lower the vehicle. Remove the jack and tighten the nuts in a crisscross sequence by turning the wrench clockwise. Replace the hub cap (if equipped) by placing it into position and using the heel of your hand or a rubber mallet to seat it.

Towing

The vehicle can be towed with all four wheels on the ground, provided that speeds do not exceed 35 mph and the distance is not over 50 miles, otherwise transmission damage can result.

Towing equipment specifically designed for this purpose should be used and should be attached to the main structural members of the vehicle and not the bumper or brackets.

Safety is a major consideration when towing and all applicable state and local laws must be obeyed. A safety chain system must be used for all towing.

While towing, the parking brake should be released and the transmission should be in Neutral. The steering must be unlocked (ignition switch in the Off position). Remember that power steering and power brakes will not work with the engine off.

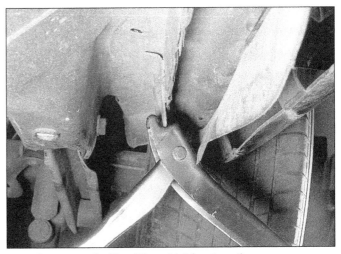

Notches in the rocker panel flange are provided for jack placement (two on each side of the vehicle) - place the groove in the jack head into the notch

Automotive chemicals and lubricants

A number of automotive chemicals and lubricants are available for use during vehicle maintenance and repair. They include a wide variety of products ranging from cleaning solvents and degreasers to lubricants and protective sprays for rubber, plastic and vinyl.

Cleaners

Carburetor cleaner and choke cleaner is a strong solvent for gum, varnish and carbon. Most carburetor cleaners leave a dry-type lubricant film which will not harden or gum up. Because of this film it is not recommended for use on electrical components.

Brake system cleaner is used to remove grease and brake fluid from the brake system, where clean surfaces are absolutely necessary. It leaves no residue and often eliminates brake squeal caused by contaminants.

Electrical cleaner removes oxidation, corrosion and carbon deposits from electrical contacts, restoring full current flow. It can also be used to clean spark plugs, carburetor jets, voltage regulators and other parts where an oil-free surface is desired.

Demoisturants remove water and moisture from electrical components such as alternators, voltage regulators, electrical connectors and fuse blocks. They are non-conductive, non-corrosive and non-flammable.

Degreasers are heavy-duty solvents used to remove grease from the outside of the engine and from chassis components. They can be sprayed or brushed on and, depending on the type, are rinsed off either with water or solvent.

Lubricants

Motor oil is the lubricant formulated for use in engines. It normally contains a wide variety of additives to prevent corrosion and reduce foaming and wear. Motor oil comes in various weights (viscosity ratings) from 0 to 50. The recommended weight of the oil depends on the season, temperature and the demands on the engine. Light oil is used in cold climates and under light load conditions. Heavy oil is used in hot climates and where high loads are encountered. Multi-viscosity oils are designed to have characteristics of both light and heavy oils and are available in a number of weights from 5W-20 to 20W-50.

Gear oil is designed to be used in differentials, manual transmissions and other areas where high-temperature lubrication is required.

Chassis and wheel bearing grease is a heavy grease used where increased loads and friction are encountered, such as for wheel bearings, balljoints, tie-rod ends and universal joints.

High-temperature wheel bearing grease is designed to withstand the extreme temperatures encountered by wheel bearings in disc brake equipped vehicles. It usually contains molybdenum disulfide (moly), which is a dry-type lubricant.

White grease is a heavy grease for metal-to-metal applications where water is a problem. White grease stays soft under both low and high temperatures (usually from -100 to +190-degrees F), and will not wash off or dilute in the presence of water.

Assembly lube is a special extreme pressure lubricant, usually containing moly, used to lubricate high-load parts (such as main and rod bearings and cam lobes) for initial start-up of a new engine. The assembly lube lubricates the parts without being squeezed out or washed away until the engine oiling system begins to function.

Silicone lubricants are used to protect rubber, plastic, vinyl and nylon parts.

Graphite lubricants are used where oils cannot be used due to contamination problems, such as in locks. The dry graphite will lubricate metal parts while remaining uncontaminated by dirt, water, oil or acids. It is electrically conductive and will not foul electrical contacts in locks such as the ignition switch.

Moly penetrants loosen and lubricate frozen, rusted and corroded fasteners and prevent future rusting or freezing.

Heat-sink grease is a special electrically non-conductive grease that is used for mounting electronic ignition modules where it is essential that heat is transferred away from the module.

Sealants

RTV sealant is one of the most widely used gasket compounds. Made from silicone, RTV is air curing, it seals, bonds, waterproofs, fills surface irregularities, remains flexible, doesn't shrink, is relatively easy to remove, and is used as a supplementary sealer with almost all low and medium temperature gaskets.

Anaerobic sealant is much like RTV in that it can be used either to seal gaskets or to form gaskets by itself. It remains flexible, is solvent resistant and fills surface imperfections. The difference between an anaerobic sealant and an RTV-type sealant is in the curing. RTV cures when exposed to air, while an anaerobic sealant cures only in the absence of air. This means that an anaerobic sealant cures only after the assembly of parts, sealing them together.

Thread and pipe sealant is used for sealing hydraulic and pneumatic fittings and vacuum lines. It is usually made from a Teflon compound, and comes in a spray, a paint-on liquid and as a wrap-around tape.

Chemicals

Anti-seize compound prevents seizing, galling, cold welding, rust and corrosion in fasteners. High-temperature anti-seize, usually made with copper and graphite lubricants, is used for exhaust system and exhaust manifold bolts.

Anaerobic locking compounds are used to keep fasteners from vibrating or working loose and cure only after installation, in the absence of air. Medium strength locking compound is used for small nuts, bolts and screws that may be removed later. High-strength locking compound is for large nuts, bolts and studs which aren't removed on a regular basis.

Oil additives range from viscosity index improvers to chemical treatments that claim to reduce internal engine friction. It should be noted that most oil manufacturers caution against using additives with their oils.

Gas additives perform several functions, depending on their chemical makeup. They usually contain solvents that help dissolve gum and varnish that build up on carburetor, fuel injection and intake parts. They also serve to break down carbon deposits that form on the inside surfaces of the combustion chambers. Some additives contain upper cylinder lubricants for valves and piston rings, and others contain chemicals to remove condensation from the gas tank.

Miscellaneous

Brake fluid is specially formulated hydraulic fluid that can withstand the heat and pressure encountered in brake systems. Care must be taken so this fluid does not come in contact with painted surfaces or plastics. An opened container should always be resealed to prevent contamination by water or dirt.

Weatherstrip adhesive is used to bond weatherstripping around doors, windows and trunk lids. It is sometimes used to attach trim pieces.

Undercoating is a petroleum-based, tar-like substance that is designed to protect metal surfaces on the underside of the vehicle from corrosion. It also acts as a sound-deadening agent by insulating the bottom of the vehicle.

Waxes and polishes are used to help protect painted and plated surfaces from the weather. Different types of paint may require the use of different types of wax and polish. Some polishes utilize a chemical or abrasive cleaner to help remove the top layer of oxidized (dull) paint on older vehicles. In recent years many non-wax polishes that contain a wide variety of chemicals such as polymers and silicones have been introduced. These non-wax polishes are usually easier to apply and last longer than conventional waxes and polishes.

Conversion factors

Length (distance)

Inches (in)	X	25.4	= Millimetres (mm)	X	0.0394	= Inches (in)
Feet (ft)	X	0.305	= Metres (m)	X	3.281	= Feet (ft)
Miles	X	1.609	= Kilometres (km)	X	0.621	= Miles

Volume (capacity)

Cubic inches (cu in; in³)	X	16.387	= Cubic centimetres (cc; cm³)	X	0.061	= Cubic inches (cu in; in³)
Imperial pints (Imp pt)	X	0.568	= Litres (l)	X	1.76	= Imperial pints (Imp pt)
Imperial quarts (Imp qt)	X	1.137	= Litres (l)	X	0.88	= Imperial quarts (Imp qt)
Imperial quarts (Imp qt)	X	1.201	= US quarts (US qt)	X	0.833	= Imperial quarts (Imp qt)
US quarts (US qt)	X	0.946	= Litres (l)	X	1.057	= US quarts (US qt)
Imperial gallons (Imp gal)	X	4.546	= Litres (l)	X	0.22	= Imperial gallons (Imp gal)
Imperial gallons (Imp gal)	X	1.201	= US gallons (US gal)	X	0.833	= Imperial gallons (Imp gal)
US gallons (US gal)	X	3.785	= Litres (l)	X	0.264	= US gallons (US gal)

Mass (weight)

Ounces (oz)	X	28.35	= Grams (g)	X	0.035	= Ounces (oz)
Pounds (lb)	X	0.454	= Kilograms (kg)	X	2.205	= Pounds (lb)

Force

Ounces-force (ozf; oz)	X	0.278	= Newtons (N)	X	3.6	= Ounces-force (ozf; oz)
Pounds-force (lbf; lb)	X	4.448	= Newtons (N)	X	0.225	= Pounds-force (lbf; lb)
Newtons (N)	X	0.1	= Kilograms-force (kgf; kg)	X	9.81	= Newtons (N)

Pressure

Pounds-force per square inch (psi; lbf/in²; lb/in²)	X	0.070	= Kilograms-force per square centimetre (kgf/cm²; kg/cm²)	X	14.223	= Pounds-force per square inch (psi; lbf/in²; lb/in²)
Pounds-force per square inch (psi; lbf/in²; lb/in²)	X	0.068	= Atmospheres (atm)	X	14.696	= Pounds-force per square inch (psi; lbf/in²; lb/in²)
Pounds-force per square inch (psi; lbf/in²; lb/in²)	X	0.069	= Bars	X	14.5	= Pounds-force per square inch (psi; lbf/in²; lb/in²)
Pounds-force per square inch (psi; lbf/in²; lb/in²)	X	6.895	= Kilopascals (kPa)	X	0.145	= Pounds-force per square inch (psi; lbf/in²; lb/in²)
Kilopascals (kPa)	X	0.01	= Kilograms-force per square centimetre (kgf/cm²; kg/cm²)	X	98.1	= Kilopascals (kPa)

Torque (moment of force)

Pounds-force inches (lbf in; lb in)	X	1.152	= Kilograms-force centimetre (kgf cm; kg cm)	X	0.868	= Pounds-force inches (lbf in; lb in)
Pounds-force inches (lbf in; lb in)	X	0.113	= Newton metres (Nm)	X	8.85	= Pounds-force inches (lbf in; lb in)
Pounds-force inches (lbf in; lb in)	X	0.083	= Pounds-force feet (lbf ft; lb ft)	X	12	= Pounds-force inches (lbf in; lb in)
Pounds-force feet (lbf ft; lb ft)	X	0.138	= Kilograms-force metres (kgf m; kg m)	X	7.233	= Pounds-force feet (lbf ft; lb ft)
Pounds-force feet (lbf ft; lb ft)	X	1.356	= Newton metres (Nm)	X	0.738	= Pounds-force feet (lbf ft; lb ft)
Newton metres (Nm)	X	0.102	= Kilograms-force metres (kgf m; kg m)	X	9.804	= Newton metres (Nm)

Vacuum

Inches mercury (in. Hg)	X	3.377	= Kilopascals (kPa)	X	0.2961	= Inches mercury
Inches mercury (in. Hg)	X	25.4	= Millimeters mercury (mm Hg)	X	0.0394	= Inches mercury

Power

Horsepower (hp)	X	745.7	= Watts (W)	X	0.0013	= Horsepower (hp)

Velocity (speed)

Miles per hour (miles/hr; mph)	X	1.609	= Kilometres per hour (km/hr; kph)	X	0.621	= Miles per hour (miles/hr; mph)

Fuel consumption*

Miles per gallon, Imperial (mpg)	X	0.354	= Kilometres per litre (km/l)	X	2.825	= Miles per gallon, Imperial (mpg)
Miles per gallon, US (mpg)	X	0.425	= Kilometres per litre (km/l)	X	2.352	= Miles per gallon, US (mpg)

Temperature

Degrees Fahrenheit = (°C x 1.8) + 32   Degrees Celsius (Degrees Centigrade; °C) = (°F - 32) x 0.56

*It is common practice to convert from miles per gallon (mpg) to litres/100 kilometres (l/100km),
where mpg (Imperial) x l/100 km = 282 and mpg (US) x l/100 km = 235*

Safety first!

Regardless of how enthusiastic you may be about getting on with the job at hand, take the time to ensure that your safety is not jeopardized. A moment's lack of attention can result in an accident, as can failure to observe certain simple safety precautions. The possibility of an accident will always exist, and the following points should not be considered a comprehensive list of all dangers. Rather, they are intended to make you aware of the risks and to encourage a safety conscious approach to all work you carry out on your vehicle.

Essential DOs and DON'Ts

DON'T rely on a jack when working under the vehicle. Always use approved jackstands to support the weight of the vehicle and place them under the recommended lift or support points.

DON'T attempt to loosen extremely tight fasteners (i.e. wheel lug nuts) while the vehicle is on a jack - it may fall.

DON'T start the engine without first making sure that the transmission is in Neutral (or Park where applicable) and the parking brake is set.

DON'T remove the radiator cap from a hot cooling system - let it cool or cover it with a cloth and release the pressure gradually.

DON'T attempt to drain the engine oil until you are sure it has cooled to the point that it will not burn you.

DON'T touch any part of the engine or exhaust system until it has cooled sufficiently to avoid burns.

DON'T siphon toxic liquids such as gasoline, antifreeze and brake fluid by mouth, or allow them to remain on your skin.

DON'T inhale brake lining dust - it is potentially hazardous (see *Asbestos* below).

DON'T allow spilled oil or grease to remain on the floor - wipe it up before someone slips on it.

DON'T use loose fitting wrenches or other tools which may slip and cause injury.

DON'T push on wrenches when loosening or tightening nuts or bolts. Always try to pull the wrench toward you. If the situation calls for pushing the wrench away, push with an open hand to avoid scraped knuckles if the wrench should slip.

DON'T attempt to lift a heavy component alone - get someone to help you.

DON'T rush or take unsafe shortcuts to finish a job.

DON'T allow children or animals in or around the vehicle while you are working on it.

DO wear eye protection when using power tools such as a drill, sander, bench grinder, etc. and when working under a vehicle.

DO keep loose clothing and long hair well out of the way of moving parts.

DO make sure that any hoist used has a safe working load rating adequate for the job.

DO get someone to check on you periodically when working alone on a vehicle.

DO carry out work in a logical sequence and make sure that everything is correctly assembled and tightened.

DO keep chemicals and fluids tightly capped and out of the reach of children and pets.

DO remember that your vehicle's safety affects that of yourself and others. If in doubt on any point, get professional advice.

Asbestos

Certain friction, insulating, sealing, and other products - such as brake linings, brake bands, clutch linings, torque converters, gaskets, etc. - may contain asbestos. Extreme care must be taken to avoid inhalation of dust from such products, since it is hazardous to health. If in doubt, assume that they do contain asbestos.

Fire

Remember at all times that gasoline is highly flammable. Never smoke or have any kind of open flame around when working on a vehicle. But the risk does not end there. A spark caused by an electrical short circuit, by two metal surfaces contacting each other, or even by static electricity built up in your body under certain conditions, can ignite gasoline vapors, which in a confined space are highly explosive. Do not, under any circumstances, use gasoline for cleaning parts. Use an approved safety solvent.

Always disconnect the battery ground (-) cable at the battery before working on any part of the fuel system or electrical system. Never risk spilling fuel on a hot engine or exhaust component. It is strongly recommended that a fire extinguisher suitable for use on fuel and electrical fires be kept handy in the garage or workshop at all times. Never try to extinguish a fuel or electrical fire with water.

Fumes

Certain fumes are highly toxic and can quickly cause unconsciousness and even death if inhaled to any extent. Gasoline vapor falls into this category, as do the vapors from some cleaning solvents. Any draining or pouring of such volatile fluids should be done in a well ventilated area.

When using cleaning fluids and solvents, read the instructions on the container carefully. Never use materials from unmarked containers.

Never run the engine in an enclosed space, such as a garage. Exhaust fumes contain carbon monoxide, which is extremely poisonous. If you need to run the engine, always do so in the open air, or at least have the rear of the vehicle outside the work area.

If you are fortunate enough to have the use of an inspection pit, never drain or pour gasoline and never run the engine while the vehicle is over the pit. The fumes, being heavier than air, will concentrate in the pit with possibly lethal results.

The battery

Never create a spark or allow a bare light bulb near a battery. They normally give off a certain amount of hydrogen gas, which is highly explosive.

Always disconnect the battery ground (-) cable at the battery before working on the fuel or electrical systems.

If possible, loosen the filler caps or cover when charging the battery from an external source (this does not apply to sealed or maintenance-free batteries). Do not charge at an excessive rate or the battery may burst.

Take care when adding water to a non maintenance-free battery and when carrying a battery. The electrolyte, even when diluted, is very corrosive and should not be allowed to contact clothing or skin.

Always wear eye protection when cleaning the battery to prevent the caustic deposits from entering your eyes.

Household current

When using an electric power tool, inspection light, etc., which operates on household current, always make sure that the tool is correctly connected to its plug and that, where necessary, it is properly grounded. Do not use such items in damp conditions and, again, do not create a spark or apply excessive heat in the vicinity of fuel or fuel vapor.

Secondary ignition system voltage

A severe electric shock can result from touching certain parts of the ignition system (such as the spark plug wires) when the engine is running or being cranked, particularly if components are damp or the insulation is defective. In the case of an electronic ignition system, the secondary system voltage is much higher and could prove fatal.

Troubleshooting

Contents

Symptom   *Section*

Engine

Engine backfires 13
Engine "diesels" (continues to run) after switching off... 15
Engine hard to start when cold 4
Engine hard to start when hot 5
Engine lacks power 12
Engine "lopes" while idling or idles erratically............... 8
Engine misses at idle speed................................... 9
Engine misses throughout driving speed range.................. 10
Engine rotates but will not start 2
Engine stalls 11
Engine starts but stops immediately.......................... 7
Engine will not rotate when attempting to start 1
Pinging or knocking engine sounds during acceleration
    or uphill 14
Starter motor noisy or excessively rough in engagement............. 6
Starter motor operates without rotating engine 3

Engine electrical system

Battery will not hold a charge. 16
"Check engine" light comes on................................. 19
Ignition light fails to come on when key is turned on......... 18
Ignition light fails to go out 17

Fuel system

Excessive fuel consumption.................................... 20
Fuel leakage and/or fuel odor 21

Cooling system

Coolant loss 26
External coolant leakage 24
Internal coolant leakage 25
Overcooling 23
Overheating 22
Poor coolant circulation 27

Clutch

Clutch slips (engine speed increases with no increase
    in vehicle speed) 29
Clutch pedal stays on floor when disengaged.................. 33
Fails to release (pedal pressed to the floor - shift lever
    does not move freely in and out of Reverse) 28
Grabbing (chattering) as clutch is engaged 30
Squeal or rumble with clutch fully disengaged (pedal
    depressed) 32
Squeal or rumble with clutch fully engaged (pedal released) 31

Manual transmission

Difficulty in engaging gears................................. 38

Symptom   *Section*

Noisy in all gears 35
Noisy in Neutral with engine running 34
Noisy in one particular gear 36
Oil leakage... 39
Slips out of high gear 37

Automatic transmission

Engine will start in gears other than Park or Neutral.................... 42
Fluid leakage 44
General shift mechanism problems 40
Transmission slips, shifts rough, is noisy or has no drive in
    forward or reverse gears..................................... 43
Transmission will not downshift with accelerator pedal
    pressed to the floor... 41

Driveshaft

Knock or clunk when the transmission is under initial load
    (just after transmission is put into gear)................... 46
Metallic grating sound consistent with vehicle speed 47
Oil leak at front of driveshaft 45
Vibration 48

Rear axle

Noise (same when in drive as when vehicle is coasting) 49
Oil leakage... 51
Vibration 50

Brakes

Brake pedal feels spongy when depressed 55
Brake pedal pulsates during brake application................. 58
Excessive brake pedal travel.................................. 54
Excessive effort required to stop vehicle 56
Noise (high-pitched squeal without the brakes applied)............. 53
Pedal travels to the floor with little resistance 57
Vehicle pulls to one side during braking 52

Suspension and steering systems

Excessive pitching and/or rolling around corners or
    during braking... 61
Excessive play in steering 63
Excessive tire wear (not specific to one area)................ 65
Excessive tire wear on inside edge............................ 67
Excessive tire wear on outside edge........................... 66
Excessively stiff steering 62
Lack of power assistance 64
Shimmy, shake or vibration.................................... 60
Tire tread worn in one place.................................. 68
Vehicle pulls to one side 59

This section provides an easy-reference guide to the more common problems which may occur during the operation of your vehicle. These problems and possible causes are grouped under various components or systems i.e. Engine, Cooling system, etc., and also refer to the Chapter and/or Section which deals with the problem.

Remember that successful troubleshooting is not a mysterious "black art' practiced only by professional mechanics; it's simply the result of a bit of knowledge combined with an intelligent, systematic approach to the problem. Always work by a process of elimination, starting with the simplest solution and working through to the most complex - and never overlook the obvious. Anyone can forget to fill the gas tank or leave the lights on overnight, so don't assume that you are above such oversights.

Finally, always get clear in your mind why a problem has occurred and take steps to ensure that it doesn't happen again. If the electrical system fails because of a poor connection, check all other connections in the system to make sure that they don't fail as well; if a particular fuse continues to blow, find out why - don't just go on replacing fuses. Remember, failure of a small component can often be indicative of potential failure or incorrect functioning of a more important component or system.

Engine

1 Engine will not rotate when attempting to start

1   Battery terminal connections loose or corroded. Check the cable terminals at the battery; tighten the cable or remove corrosion as necessary.
2   Battery discharged or faulty. If the cable connections are clean and tight on the battery posts, turn the key to the On position and switch on the headlights and/or windshield wipers. If they fail to function, the battery is discharged.
3   Automatic transmission not completely engaged in Park or clutch not completely depressed.
4   Broken, loose or disconnected wiring in the starting circuit. Inspect all wiring and connectors at the battery, starter solenoid and ignition switch.
5   Starter motor pinion jammed in flywheel ring gear. If manual transmission, place transmission in gear and rock the vehicle to manually turn the engine. Remove starter and inspect pinion and flywheel at earliest convenience.
6   Starter solenoid faulty (Chapter 5).
7   Starter motor faulty (Chapter 5).
8   Ignition switch faulty (Chapter 10).

2 Engine rotates but will not start

1   Fuel tank empty.
2   Battery discharged (engine rotates slowly). Check the operation of electrical components as described in previous Section.
3   Battery terminal connections loose or corroded. See previous Section.
4   Carburetor flooded and/or fuel level in carburetor incorrect. This will usually be accompanied by a strong fuel odor from under the hood. Wait a few minutes, depress the accelerator pedal all the way to the floor and attempt to start the engine.
5   Choke control inoperative (Chapter 1).
6   Fuel not reaching carburetor. With ignition switch in Off position, open hood, remove the top plate of air cleaner assembly and observe the top of the carburetor (manually move choke plate back if necessary). Have an assistant depress accelerator pedal and check that fuel spurts into carburetor. If not, check fuel filter (Chapter 1), fuel lines and fuel pump (Chapter 4).
7   Fuel injector(s), ECM or fuel pump faulty (fuel-injected vehicles) (Chapters 4 and 6).
8   Excessive moisture on, or damage to, ignition components (Chapter 5).
9   Worn, faulty or incorrectly gapped spark plugs (Chapter 1).
10   Broken, loose or disconnected wiring in the starting circuit (see previous Section).
11   Distributor loose, causing ignition timing to change. Turn the distributor as necessary to start the engine, then set ignition timing as soon as possible (Chapter 1).
12   Broken, loose or disconnected wires at the ignition coil or faulty coil (Chapter 5).

3 Starter motor operates without rotating engine

1   Starter pinion sticking. Remove the starter (Chapter 5) and inspect.
2   Starter pinion or flywheel teeth worn or broken. Remove the cover at the rear of the engine and inspect.

4 Engine hard to start when cold

1   Battery discharged or low. Check as described in Section 1.
2   Choke control inoperative or out of adjustment (Chapter 4).
3   Carburetor flooded (see Section 2).
4   Fuel supply not reaching the carburetor or fuel injection system (see Section 2).
5   Carburetor in need of overhaul (Chapter 4).
6   Distributor rotor carbon tracked and/or mechanical advance mechanism rusted (Chapter 5).
7   Defective fuel or emissions systems control components (Chapters 4 and 6).

5 Engine hard to start when hot

1   Choke sticking in the closed position (Chapter 1).
2   Carburetor flooded (see Section 2).
3   Air filter clogged (Chapter 1).
4   Fuel not reaching the carburetor or fuel injection system (see Section 2).
5   Defective fuel or emissions systems control components (Chapters 4 and 6).

6 Starter motor noisy or excessively rough in engagement

1   Pinion or flywheel gear teeth worn or broken. Remove the cover at the rear of the engine (if so equipped) and inspect.
2   Starter motor mounting bolts loose or missing.

7 Engine starts but stops immediately

1   Loose or faulty electrical connections at distributor, coil or alternator.
2   Insufficient fuel reaching the carburetor/fuel injector(s). Disconnect the fuel line at the carburetor/fuel injector(s) and remove the filter (Chapter 1). Place a container under the disconnected fuel line. Observe the flow of fuel from the line. If little or none at all, check for blockage in the lines and/or replace the fuel pump (Chapter 4).
3   Vacuum leak at the gasket surfaces of the intake manifold and/or carburetor/fuel injection unit(s). Make sure that all mounting bolts (nuts) are tightened securely and that all vacuum hoses connected to the carburetor/fuel injection unit(s) and manifold are positioned properly and in good condition.

8 Engine "lopes" while idling or idles erratically

1   Vacuum leakage. Check mounting bolts (nuts) at the carburetor/fuel injection unit(s) and intake manifold for tightness. Make sure that all vacuum hoses are connected and in good condition. Use a stethoscope or a length of fuel hose held against your ear to listen for vacuum leaks while the engine is running. A hissing sound will be heard. A soapy water solution will also detect leaks. Check the carburetor/fuel injector and intake manifold gasket surfaces.
2   Leaking EGR valve or plugged PCV valve (see Chapters 1 and 6).
3   Air filter clogged (Chapter 1).
4   Fuel pump not delivering sufficient fuel to the carburetor/fuel injector (see Section 7).
5   Carburetor out of adjustment (Chapter 4).
6   Leaking head gasket. If this is sus-

pected, take the vehicle to a repair shop or dealer where the engine can be pressure checked.
7   Timing chain and/or gears worn (Chapter 2).
8   Camshaft lobes worn (Chapter 2).
9   Defective fuel or emissions systems control components (Chapters 4 and 6).

9   Engine misses at idle speed

1   Spark plugs worn or not gapped properly (Chapter 1).
2
3   Faulty spark plug wires (Chapter 1).
4   Choke not operating properly (Chapter 1).
5   Defective fuel injector(s) (Chapter 4).
6   Engine mechanical problem (Chapter 2).

10   Engine misses throughout driving speed range

1   Fuel filter clogged and/or impurities in the fuel system (Chapter 1). Also check fuel output at the carburetor/fuel injector (see Section 7).
2   Faulty or incorrectly gapped spark plugs (Chapter 1).
3   Incorrect ignition timing (Chapter 1).
4   Check for cracked distributor cap, disconnected distributor wires and damaged distributor components (Chapter 1).
5   Leaking spark plug wires (Chapter 1).
6   Faulty emissions system components (Chapter 6).
7   Low or uneven cylinder compression pressures. Remove spark plugs and test compression with gauge (Chapter 1).
8   Weak or faulty ignition system (Chapter 5).
9   Vacuum leaks at carburetor/fuel injection unit(s), intake manifold or vacuum hoses (see Section 8).
10   Defective fuel injector(s) (Chapter 4).

11   Engine stalls

1   Idle speed incorrect (Chapter 1).
2   Fuel filter clogged and/or water and impurities in the fuel system (Chapter 1).
3   Choke improperly adjusted or sticking (Chapter 1).
4   Distributor components damp or damaged (Chapter 5).
5   Defective fuel or emissions systems control components (Chapters 4 and 6).
6   Faulty or incorrectly gapped spark plugs (Chapter 1). Also check spark plug wires (Chapter 1).
7   Vacuum leak at the carburetor/fuel injection unit(s), intake manifold or vacuum hoses. Check as described in Section 8.

12   Engine lacks power

1   Incorrect ignition timing (Chapter 1).
2   Excessive play in distributor shaft. At the same time, check for worn rotor, faulty distributor cap, wires, etc. (Chapters 1 and 5).
3   Faulty or incorrectly gapped spark plugs (Chapter 1).
4   Defective fuel or emissions systems control components (Chapters 4 and 6).
5   Carburetor not adjusted properly or excessively worn (Chapter 4).
6   Faulty coil (Chapter 5).
7   Brakes binding (Chapter 1).
8   Automatic transmission fluid level incorrect (Chapter 1).
9   Clutch slipping (Chapter 8).
10   Fuel filter clogged and/or impurities in the fuel system (Chapter 1).
11   Restricted exhaust system.
12   Use of sub-standard fuel. Fill tank with proper octane fuel.
13   Low or uneven cylinder compression pressures. Test with compression tester, which will detect leaking valves and/or blown head gasket (Chapter 1).

13   Engine backfires

1   Defective fuel or emissions systems control components (Chapters 4 and 6).
2   Ignition timing incorrect (Chapter 1).
3   Faulty secondary ignition system (cracked spark plug insulator,
faulty plug wires, distributor cap and/or rotor) (Chapters 1 and 5).
4   Carburetor in need of adjustment or worn excessively (Chapter 4).
5   Vacuum leak at carburetor/fuel injection unit(s), intake manifold or vacuum hoses. Check as described in Section 8.
6   Valve clearances incorrectly set, and/or valves sticking (Chapter 2).

14   Pinging or knocking engine sounds during acceleration or uphill

1   Incorrect grade of fuel. Fill tank with fuel of the proper octane rating.
2   Ignition timing incorrect (Chapter 1).
3   Carburetor in need of adjustment (Chapter 4).
4   Improper spark plugs. Check plug type against Emissions Control Information label located in engine compartment. Also check plugs and wires for damage (Chapter 1).
5   Worn or damaged distributor components (Chapter 5).
6   Faulty emissions system (Chapter 6).
7   Vacuum leak. Check as described in Section 8.

15   Engine "diesels" (continues to run) after switching off

1   Idle speed too high (Chapter 1).
2   Electrical solenoid at side of carburetor not functioning properly (not all models, see Chapter 4).
3   Ignition timing incorrectly adjusted (Chapter 1).
4   Thermo-controlled air cleaner heat valve not operating properly (Chapter 6).
5   Excessive engine operating temperature. Probable causes of this are malfunctioning thermostat, clogged radiator, faulty water pump (Chapter 3).

Engine electrical system

16   Battery will not hold a charge

1   Alternator drivebelt defective or not adjusted properly (Chapter 1).
2   Electrolyte level low or battery discharged (Chapter 1).
3   Battery terminals loose or corroded (Chapter 1).
4   Alternator not charging properly (Chapter 5).
5   Loose, broken or faulty wiring in the charging circuit (Chapter 5).
6   Short in vehicle wiring causing a continual drain on battery.
7   Battery defective internally.

17   Ignition light fails to go out

1   Fault in alternator or charging circuit (Chapter 5).
2   Alternator drivebelt defective or not properly adjusted (Chapter 1).

18   Ignition light fails to come on when key is turned on

1   Warning light bulb defective (Chapter 10).
2   Alternator faulty (Chapter 5).
3   Fault in the printed circuit, dash wiring or bulb holder (Chapter 10).

19   "Check engine" light comes on

1   Problem with the engine/emissions control system, see Chapter 6.

Fuel system

20   Excessive fuel consumption

1   Dirty or clogged air filter element (Chapter 1).
2   Incorrectly set ignition timing (Chapter 1).
3   Choke sticking or improperly adjusted (Chapter 1).
4   Emissions system not functioning properly (Chapter 6).
5   Carburetor idle speed and/or mixture not adjusted properly (Chapter 1).
6   Carburetor/fuel injection internal parts excessively worn or damaged (Chapter 4).
7   Low tire pressure or incorrect tire size (Chapter 1).

21   Fuel leakage and/or fuel odor

1   Leak in a fuel feed or vent line (Chapter 4).
2   Tank overfilled. Fill only to automatic shut-off.
3   Emissions system filter clogged (Chapter 1).
4   Vapor leaks from system lines (Chapter 4).
5   Carburetor leaking (Chapter 4).

Engine cooling system

22   Overheating

1   Insufficient coolant in system (Chapter 1).
2   Water pump drivebelt defective or not adjusted properly (Chapter 1).
3   Radiator core blocked or radiator grille dirty and restricted (Chapter 3).
4   Thermostat faulty (Chapter 3).
5   Fan blades broken or cracked (Chapter 3).
6   Radiator cap not maintaining proper pressure. Have cap pressure tested by gas station or repair shop.
7   Ignition timing incorrect (Chapter 1).

23   Overcooling

1   Thermostat faulty (Chapter 3).
2   Inaccurate temperature gauge (Chapter 10)

24   External coolant leakage

1   Deteriorated or damaged hoses. Loosen clamps at hose connections (Chapter 1).
2   Water pump seals defective. If this is the case, water will drip from the "weep' hole in the water pump body (Chapter 1).
3   Leakage from radiator core or header

tank. This will require the radiator to be professionally repaired (see Chapter 3 for removal procedures).
4   Engine drain plugs or water jacket core plugs leaking (see Chapter 2).

25   Internal coolant leakage

Note: *Internal coolant leaks can usually be detected by examining the oil. Check the dipstick and inside of the rocker arm cover(s) for water deposits and an oil consistency like that of a milkshake.*
1   Leaking cylinder head gasket. Have the cooling system pressure tested.
2   Cracked cylinder bore or cylinder head. Dismantle engine and inspect (Chapter 2).

26   Coolant loss

1   Too much coolant in system (Chapter 1).
2   Coolant boiling away due to overheating (see Section 22).
3   Internal or external leakage (see Sections 24 and 25).
4   Faulty radiator cap. Have the cap pressure-tested.

27   Poor coolant circulation

1   Inoperative water pump. A quick test is to pinch the top radiator hose closed with your hand while the engine is idling, then let it loose. You should feel the surge of coolant if the pump is working properly (Chapter 1).
2   Restriction in cooling system. Drain, flush and refill the system (Chapter 1). If necessary, remove the radiator (Chapter 3) and have it reverse-flushed.
3   Water pump drivebelt defective or not adjusted properly (Chapter 1).
4   Thermostat sticking (Chapter 3).

Clutch

28   Fails to release (pedal pressed to the floor - shift lever does not move freely in and out of Reverse)

1   Improper linkage free play adjustment (Chapter 8).
2   Clutch fork off ball stud. Look under the vehicle, on the left side of transmission.
3   Clutch plate warped or damaged (Chapter 8).

29   Clutch slips (engine speed increases with no increase in vehicle speed)

1   Linkage out of adjustment (Chapter 8).

2   Clutch plate oil soaked or lining worn. Remove clutch (Chapter 8) and inspect.
3   Clutch plate not seated. It may take several normal starts for a new one to seat.

30   Grabbing (chattering) as clutch is engaged

1   Oil on clutch plate lining. Remove (Chapter 8) and inspect. Correct any leakage source.
2   Worn or loose engine or transmission mounts. These units move slightly when clutch is released. Inspect mounts and bolts.
3   Worn splines on clutch plate hub. Remove clutch components (Chapter 8) and inspect.
4   Warped pressure plate or flywheel. Remove clutch components and inspect.

31   Squeal or rumble with clutch fully engaged (pedal released)

1   Improper adjustment; no free play (Chapter 1).
2   Release bearing binding on transmission bearing retainer. Remove clutch components (Chapter 8) and check bearing. Remove any burrs or nicks, clean and relubricate before reinstallation.
3   Weak linkage return spring. Replace the spring.

32   Squeal or rumble with clutch fully disengaged (pedal depressed)

1   Worn, defective or broken release bearing (Chapter 8).
2   Worn or broken pressure plate springs (diaphragm fingers) (Chapter 8).

33   Clutch pedal stays on floor when disengaged

1   Bind in linkage or release bearing. Inspect linkage or remove clutch components as necessary.
2   Linkage springs being over-extended. Adjust linkage for proper free play. Make sure proper pedal stop (bumper) is installed.

Manual transmission

34   Noisy in Neutral with engine running

1   Input shaft bearing worn.
2   Damaged main drive gear bearing.
3   Worn countershaft bearings.
4   Worn or damaged countershaft end play shims.

35 Noisy in all gears

1   Any of the above causes, and/or:
2   Insufficient lubricant (see checking procedures in Chapter 1).

36 Noisy in one particular gear

1   Worn, damaged or chipped gear teeth for that particular gear.
2   Worn or damaged synchronizer for that particular gear.

37 Slips out of high gear

1   Transmission loose on clutch housing (Chapter 7).
2   Shift rods interfering with engine mounts or clutch lever (Chapter 7).
3   Shift rods not working freely (Chapter 7).
4   Damaged mainshaft pilot bearing.
5   Dirt between transmission case and engine or misalignment of transmission (Chapter 7).
6   Worn or improperly adjusted linkage (Chapter 7).

38 Difficulty in engaging gears

1   Clutch not releasing completely (see clutch adjustment in Chapter 1).
2   Loose, damaged or out-of-adjustment shift linkage. Make a thorough inspection, replacing parts as necessary (Chapter 7).

39 Oil leakage

1   Excessive amount of lubricant in transmission (see Chapter 1 for correct checking procedures). Drain lubricant as required.
2   Side cover loose or gasket damaged.
3   Rear oil seal or speedometer oil seal in need of replacement (Chapter 7).

Automatic transmission

Note: *Due to the complexity of the automatic transmission, it is difficult for the home mechanic to properly diagnose and service this component. For problems other than the following, the vehicle should be taken to a dealer or reputable mechanic.*

40 General shift mechanism problems

1   Chapter 7 deals with checking and adjusting the shift linkage on automatic trans-

missions. Common problems which may be attributed to poorly adjusted linkage are:
Engine starting in gears other than Park or Neutral.
Indicator on shifter pointing to a gear other than the one actually being used.
Vehicle moves when in Park.
2   Refer to Chapter 7 to adjust the linkage.

41 Transmission will not downshift with accelerator pedal pressed to the floor

Chapter 7 deals with adjusting the throttle valve (TV) cable to enable the transmission to downshift properly.

42 Engine will start in gears other than Park or Neutral

Chapter 7 deals with adjusting the Neutral start switch used with automatic transmissions.

43 Transmission slips, shifts rough, is noisy or has no drive in forward or reverse gears

1   There are many probable causes for the above problems, but the home mechanic should be concerned with only fluid level and condition.
2   Before taking the vehicle to a repair shop, check the level and condition of the fluid as described in Chapter 1. Correct fluid level as necessary or change the fluid and filter if needed. If the problem persists, have a professional diagnose the probable cause.

44 Fluid leakage

1   Automatic transmission fluid is a deep red color. Fluid leaks should not be confused with engine oil, which can easily be blown by air flow to the transmission.
2   To pinpoint a leak, first remove all built-up dirt and grime from around the transmission. Degreasing agents and/or steam cleaning will achieve this. With the underside clean, drive the vehicle at low speeds so air flow will not blow the leak far from its source. Raise the vehicle and determine where the leak is coming from. Common areas of leakage are:
a)   *Pan: Tighten mounting bolts and/or replace pan gasket as necessary (see Chapters 1 and 7).*
b)   *Rear extension: Tighten bolts and/or replace oil seal as necessary (Chapter 7).*
c)   *Filler pipe: Replace the rubber seal where pipe enters transmission case.*

d)   *Transmission oil lines: Tighten connectors where lines enter transmission case and/or replace lines.*
e)   *Vent pipe: Transmission over-filled and/or water in fluid (see checking procedures, Chapter 1).*
f)   *Speedometer connector: Replace the O-ring where speedometer cable enters transmission case (Chapter 7).*

Driveshaft

45 Oil leak at front of driveshaft

Defective transmission rear oil seal. See Chapter 7 for replacement procedures. While this is done, check the splined yoke for burrs or a rough condition which may be damaging the seal. They can be removed with crocus cloth or a fine whetstone.

46 Knock or clunk when the transmission is under initial load (just after transmission is put into gear)

1   Loose or disconnected rear suspension components. Check all mounting bolts and bushings (Chapter 11).
2   Loose driveshaft bolts. Inspect all bolts and nuts and tighten them to the specified torque (Chapter 8).
3   Worn or damaged universal joint bearings. Check for wear (Chapter 8).

47 Metallic grating sound consistent with vehicle speed

Pronounced wear in the universal joint bearings. Check as described in Chapter 8.

48 Vibration

Note: *Before assuming that the driveshaft is at fault, make sure the tires are perfectly balanced and perform the following test.*
1   Install a tachometer inside the vehicle to monitor engine speed as the vehicle is driven. Drive the vehicle and note the engine speed at which the vibration (roughness) is most pronounced. Now shift the transmission to a different gear and bring the engine speed to the same point.
2   If the vibration occurs at the same engine speed (rpm) regardless of which gear the transmission is in, the driveshaft is NOT at fault since the driveshaft speed varies.
3   If the vibration decreases or is eliminated when the transmission is in a different gear at the same engine speed, refer to the following probable causes.

4   Bent or dented driveshaft. Inspect and replace as necessary (Chapter 8).
5   Undercoating or build-up dirt, etc. on the driveshaft. Clean the shaft thoroughly and recheck.
6   Worn universal joint bearings. Remove and inspect (Chapter 8).
7   Driveshaft and/or companion flange out of balance. Check for missing weights on the shaft. Remove driveshaft (Chapter 8) and reinstall 180-degrees from original position, then retest. Have driveshaft professionally balanced if problem persists.

Rear axle

49   Noise (same when in drive as when vehicle is coasting)

1   Road noise. No corrective procedures available.
2   Tire noise. Inspect tires and check tire pressures (Chapter 1).
3   Front wheel bearings loose, worn or damaged (Chapter 1).

50   Vibration

See probable causes under Driveshaft. Proceed under the guidelines listed for the driveshaft. If the problem persists, check the rear axleshaft bearings by raising the rear of the vehicle and spinning the wheels by hand. Listen for evidence of rough (noisy) bearings. Remove and inspect (Chapter 8).

51   Oil leakage

1   Pinion seal damaged (Chapter 8).
2   Axleshaft oil seals damaged (Chapter 8).
3   Differential cover leaking. Tighten mounting bolts or replace the gasket as required (Chapter 8).

Brakes

Note: *Before assuming that a brake problem exists, make sure that the tires are in good condition and inflated properly (see Chapter 1), that the front end alignment is correct and that the vehicle is not loaded with weight in an unequal manner.*

52   Vehicle pulls to one side during braking

1   Defective, damaged or oil contaminated

disc brake pads on one side. Inspect as described in Chapter 9.
2   Excessive wear of brake pad material or disc on one side. Inspect and correct as necessary.
3   Loose or disconnected front suspension components. Inspect and tighten all bolts to the specified torque (Chapter 11).
4   Defective caliper assembly. Remove caliper and inspect for stuck piston or other damage (Chapter 9).

53   Noise (high-pitched squeal without the brakes applied)

Disc brake pads worn out. The noise comes from the wear sensor rubbing against the disc (does not apply to all vehicles). Replace pads with new ones immediately (Chapter 9).

54   Excessive brake pedal travel

1   Partial brake system failure. Inspect entire system (Chapter 9) and correct as required.
2   Insufficient fluid in master cylinder. Check (Chapter 1), add fluid and bleed system if necessary (Chapter 9).
3   Rear brakes not adjusting properly. Make a series of starts and stops while the vehicle is in Reverse. If this does not correct the situation, remove drums and inspect self-adjusters (Chapter 9).

55   Brake pedal feels spongy when depressed

1   Air in hydraulic lines. Bleed the brake system (Chapter 9).
2   Faulty flexible hoses. Inspect all system hoses and lines. Replace parts as necessary.
3   Master cylinder mounting bolts/nuts loose.
4   Master cylinder defective (Chapter 9).

56   Excessive effort required to stop vehicle

1   Power brake booster not operating properly (Chapter 9).
2   Excessively worn linings or pads. Inspect and replace if necessary (Chapter 9).
3   One or more caliper pistons or wheel cylinders seized or sticking. Inspect and rebuild as required (Chapter 9).
4   Brake linings or pads contaminated with oil or grease. Inspect and replace as required (Chapter 9).
5   New pads or shoes installed and not yet seated. It will take a while for the new material to seat against the drum (or rotor).

57   Pedal travels to the floor with little resistance

Little or no fluid in the master cylinder reservoir caused by leaking wheel cylinder(s), leaking caliper piston(s), loose, damaged or disconnected brake lines. Inspect entire system and correct as necessary.

58   Brake pedal pulsates during brake application

1   Wheel bearings not adjusted properly or in need of replacement (Chapter 1).
2   Caliper not sliding properly due to improper installation or obstructions. Remove and inspect (Chapter 9).
3   Rotor defective. Remove the rotor (Chapter 9) and check for excessive lateral runout and parallelism. Have the rotor resurfaced or replace it with a new one.

Suspension and steering systems

59   Vehicle pulls to one side

1   Tire pressures uneven (Chapter 1).
2   Defective tire (Chapter 1).
3   Excessive wear in suspension or steering components (Chapter 11).
4   Front end in need of alignment.
5   Front brakes dragging. Inspect brakes as described in Chapter 9.

60   Shimmy, shake or vibration

1   Tire or wheel out-of-balance or out-of-round. Have professionally balanced.
2   Loose, worn or out-of-adjustment wheel bearings (Chapters 1 and 8).
3   Shock absorbers and/or suspension components worn or damaged (Chapter 11).

61   Excessive pitching and/or rolling around corners or during braking

1   Defective shock absorbers. Replace as a set (Chapter 11).
2   Broken or weak springs and/or suspension components. Inspect as described in Chapter 11.

62   Excessively stiff steering

1   Lack of fluid in power steering fluid reservoir (Chapter 1).

2   Incorrect tire pressures (Chapter 1).
3   Lack of lubrication at steering joints (Chapter 1).
4   Front end out of alignment.
5   See also section titled Lack of power assistance.

63   Excessive play in steering

1   Loose front wheel bearings (Chapter 1).
2   Excessive wear in suspension or steering components (Chapter 11).
3   Steering gearbox defective (Chapter 11).

64   Lack of power assistance

1   Steering pump drivebelt faulty or not adjusted properly (Chapter 1).
2   Fluid level low (Chapter 1).
3   Hoses or lines restricted. Inspect and replace parts as necessary.
4   Air in power steering system. Bleed system (Chapter 11).

65   Excessive tire wear (not specific to one area)

1   Incorrect tire pressures (Chapter 1).
2   Tires out of balance. Have professionally balanced.
3   Wheels damaged. Inspect and replace as necessary.
4   Suspension or steering components excessively worn (Chapter 11).

66   Excessive tire wear on outside edge

1   Inflation pressures incorrect (Chapter 1).
2   Excessive speed in turns.

3   Front end alignment incorrect (excessive toe-in). Have professionally aligned.
4   Suspension arm bent or twisted (Chapter 11).

67   Excessive tire wear on inside edge

1   Inflation pressures incorrect (Chapter 1).
2   Front end alignment incorrect (toe-out). Have professionally aligned.
3   Loose or damaged steering components (Chapter 11).

68   Tire tread worn in one place

1   Tires out of balance.
2   Damaged or buckled wheel. Inspect and replace if necessary.
3   Defective tire (Chapter 1).

Notes

Chapter 1
Tune-up and routine maintenance

Contents

	Section
Air filter and PCV filter replacement	29
Automatic transmission fluid change	28
Battery check and maintenance	5
Brake check	13
Carburetor choke check	14
Carburetor/throttle body injection (TBI) mounting torque check	21
Chassis lubrication	10
Check engine light	See Chapter 6
Clutch pedal freeplay check and adjustment	24
Cooling system check	7
Cooling system servicing (draining, flushing and refilling)	26
Drivebelt check and adjustment	6
Early Fuel Evaporation (EFE) system check	22
Engine idle speed check and adjustment	15
Engine oil and filter change	16
Evaporative Emissions Control System (EECS) filter replacement	33
Exhaust Gas Recirculation (EGR) valve check	32
Exhaust system check	11

	Section
Fluid level checks	4
Front wheel bearing check, repack and adjustment	27
Fuel filter replacement	18
Fuel system check	17
Ignition timing check and adjustment	34
Introduction and routine maintenance schedule	1
Oxygen sensor replacement	30
Positive Crankcase Ventilation (PCV) valve replacement	31
Rear axle oil change	25
Spark plug replacement	35
Spark plug wires, distributor cap and rotor check and replacement	36
Suspension and steering check	12
Thermo-controlled Air Cleaner (THERMAC) check	20
Throttle linkage check	19
Tire and tire pressure checks	3
Tire rotation	23
Tune-up general information	2
Underhood hose check and replacement	8
Windshield wiper blade inspection and replacement	9

Specifications

Recommended lubricants and fluids

Note: *Listed here are manufacturer recommendations at the time this manual was written. Manufacturers occasionally upgrade their fluid and lubricant specifications, so check with your local auto parts store for the most current recommendations.*

Engine oil
  Type ...... API approved multigrade engine oil
  Viscosity ...... See accompanying chart
Coolant type ...... 50/50 mixture of ethylene glycol based antifreeze and water

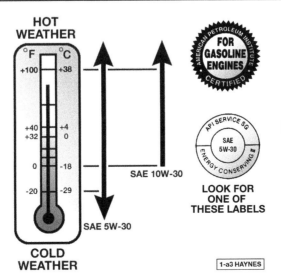

Engine oil viscosity chart - for best fuel economy and cold starting, select the lowest SAE viscosity grade for the expected temperature range

Recommended lubricants and fluids (continued)

Brake fluid type...	DOT-3 brake fluid
Manual transmission lubricant type	
4-speed	
76 mm and 83 mm (cast iron case)	GL-5 gear lubricant, SAE 80W or 80W-90
77 mm (aluminum case).................................	Dexron II or equivalent automatic transmission fluid
5-speed	Dexron II or equivalent automatic transmission fluid
Automatic transmission fluid	Dexron II or equivalent automatic transmission fluid
Rear axle lubricant type..	GL-5 gear lubricant, SAE 80W or 80W-90
with limited-slip	add 4 ounces of limited-slip differential additive when lubricant is changed
Power steering fluid type...	GM power steering fluid or equivalent
Front wheel bearing grease	NLGI No. 2 moly-base wheel bearing grease

Capacities*

Engine oil (with filter change)	
V8	5.0 quarts
V6	4.5 quarts
L4	3.5 quarts
Cooling system	
V8	
1984 and earlier	15 quarts
1985 and later	17.2 quarts
V6	
2.8L	12.5 quarts
3.1L	14.8 quarts
L4	13.0 quarts
Manual transmission lubricant	
4-speed	2.4 quarts
5-speed	3.5 quarts
Automatic transmission fluid (drain and refill)	
200-C	3.5 quarts
700-R4...	4.7 quarts
Rear axle lubricant..	3.5 pints

* Note: *All capacities approximate. Add as necessary to bring to the appropriate level*

Cooling system

Radiator cap opening pressure	15 psi
Thermostat rating	195 degrees F

Fuel system

Engine idle speed	See the Vehicle Emission Control Information label in engine compartment

Ignition system

Distributor direction of rotation...	Clockwise (all)
Firing order	
V8	1-8-4-3-6-5-7-2
V6	1-2-3-4-5-6
L4	1-3-4-2

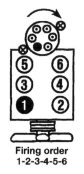

Firing order 1-2-3-4-5-6

Cylinder numbering and direction distributor rotor rotates (arrow) - V6 engine

Firing order 1 - 3 - 4 - 2

Cylinder numbering and direction distributor rotor rotates (arrow) - L4 engine

RIGHT SIDE CYLINDERS

② ④ ⑥ ⑧

FRONT OF ENGINE **DISTRIBUTOR**

① ③ ⑤ ⑦

LEFT (DRIVER'S) SIDE CYLINDERS

Cylinder numbers and distributor spark plug wire terminal locations - V8 engines

The blackened terminal shown on the distributor cap indicates the Number One spark plug wire position

Spark plug type and gap
   L4
      1982 and 1983 .... AC R44TSX (or equivalent) @ 0.060 inch
      1984 on .... AC R43TS6 (or equivalent) @ 0.060 inch
   V6
      1982 through 1984 .... AC R43TS (or equivalent) @ 0.045 inch
      1985 through 1988 .... AC R42TS (or equivalent) @ 0.045 inch
      1989 on .... AC R43TSK (or equivalent) @ 0.045 inch
   V8
      1982 through 1984 .... AC R45TS (or equivalent) @ 0.045 inch
      1985
         Carbureted models VIN G code .... AC R44TS (or equivalent) @ 0.045 inch
         Carbureted models VIN H code .... AC R45TS (or equivalent) @ 0.045 inch
         Fuel injected models .... AC R43TS (or equivalent) @ 0.045 inch
      1986 and 1987
         Carbureted models .... AC R43TS (or equivalent) @ 0.035 inch
         Fuel-injected models
            with cast iron heads .... AC R43TS (or equivalent) @ 0.035 inch
            with aluminum heads .... AC FR3LS (or equivalent) @ 0.035 inch
      1988 through 1991
         with cast iron heads .... AC R45TS (or equivalent) @ 0.035 inch
         with aluminum heads... AC FR5LS (or equivalent) @ 0.035 inch
      1992
         305 cu. in. .... AC CR45TS (or equivalent) @ 0.035 inch
         350 cu.in.
            with cast iron heads .... AC R45TS (or equivalent) @ 0.035 inch
            with aluminum heads .... AC FR5LS (or equivalent) @ 0.035 inch
Ignition timing .... See the Vehicle Emission Control Information label
                                                                                                  in the engine compartment

Clutch
Clutch pedal freeplay (non-hydraulic clutch) .... 0.845 to 0.905 inch

Torque specifications   **Ft-lbs** (unless otherwise indicated)
Oil pan drain plug
   V8 20
   V6 20
   L4 25
Spark plugs
   V8 22
   V6 .... 84 to 180 in-lbs
   L4 15
Carburetor /TBI mounting nuts/bolts
   V8
      Carburetor (long studs/bolts) .... 84 in-lbs
      Carburetor (short studs/bolts) .... 132 in-lbs
      TBI mounting bolts... 15
   V6 .... 132 in-lbs
   L4 15
Distributor hold-down bolt... 20
Manual transmission fill plug
   4-speed 15
   5-speed 20
Automatic transmission pan bolts .... 144 in-lbs
Rear axle filler/inspection plug 20
Rear axle cover bolts 30
Brake caliper mounting bolts.. 35
Wheel lug nuts
   1984 and earlier
      Standard wheels 70
      Cast aluminum wheels 80
   1985 and later
      Standard wheels 80
      Cast aluminum wheels... 100

1 Introduction and routine maintenance schedule

This Chapter was designed to help the home mechanic maintain his/her vehicle for peak performance, economy, safety and long life.

On the following pages you will find a maintenance schedule along with Sections which deal specifically with each item on the schedule. Included are visual checks, adjustments and item replacements.

Servicing your vehicle using the time/mileage maintenance schedule and the sequenced Sections will give you a planned program of maintenance. Keep in mind that it is a full plan, and maintaining only a few items at the specified intervals will not give you the same results.

You will find as you service your vehicle that many of the procedures can, and should, be grouped together, due to the nature of the job at hand. Examples of this are as follows:

If the vehicle is fully raised for a chassis lubrication, for example, this is the ideal time for the following checks: manual transmission oil, rear axle oil, exhaust system, suspension, steering and the fuel system.

If the tires and wheels are removed, as during a routine tire rotation, go ahead and check the brakes and wheel bearings at the same time.

If you must borrow or rent a torque wrench, it is a good idea to service the spark plugs, repack (or replace) the wheel bearings and check the carburetor mounting bolt torque all in the same day to save time and money.

The first step of this or any maintenance plan is to prepare yourself before the actual work begins. Read through the appropriate Sections for all work that is to be performed before you begin. Gather together all necessary parts and tools. If it appears that you could have a problem during a particular job, don't hesitate to seek advice from your local parts counter person. **Caution:** *Some parts on later model vehicles have an electrostatic discharge sensitive sticker applied to them. When installing or servicing any parts with this sticker, the following guidelines should be followed:*

a) *Avoid touching the electrical terminals of the part*

b) *Always ground the package to a known good ground on the vehicle before removing the part from the package*

c) *Always touch a known good ground before handling the part*

Routine maintenance intervals

The following recommendations are given with the assumption that the vehicle owner will be doing the maintenance or service work (as opposed to having a dealer service department do the work). The following are factory maintenance recommendations; however, subject to the preference of the individual owner, in the interest of keeping his or her vehicle in peak condition at all times and with the vehicle's ultimate resale in mind, many of these operations may be performed more often. We encourage such owner initiative.

Every 250 miles or weekly, whichever comes first

Check the engine oil level (Section 4)
Check the engine coolant level (Section 4)
Check the windshield washer fluid level (Section 4)
Check the tires and tire pressures (Section 3)

Every 6000 miles or 6 months, whichever comes first

Check the automatic transmission fluid level (Section 4)
Check the power steering fluid level (Section 4)
Change the engine oil and oil filter (Section 16)
Lubricate the chassis components (Section 10)
Check the cooling system (Section 7)
Check and replace (if necessary) the underhood hoses (Section 8)
Check the exhaust system (Section 11)
Check the steering and suspension components (Section 12)
Check and adjust (if necessary) the engine drivebelts (Section 12)

Check the brake master cylinder fluid level (Section 4)
Check the manual transmission oil level (Section 4)
Check the rear axle oil level (Section 4)
Check the operation of the choke (at this interval, then every 12000 miles or 12 months, whichever comes first) (Section 14)
Check and adjust (if necessary) the engine idle speed (at this interval, then every 12000 miles or 12 months, whichever comes first) (Section 15)
Check the disc brake pads (Section 13)
Check the brake system (Section 13)
Check and service the battery (Section 5)
Check and replace (if necessary) the windshield wiper blades (Section 9)

Every 12000 miles or 12 months, whichever comes first

Check the drum brake linings (Section 13)
Check the parking brake (Section 13)
Check the clutch pedal freeplay (Section 24)
Rotate the tires (Section 23)
Check the Thermo-controlled Air Cleaner (THERMAC) for proper operation (Section 20)
Check the EFE system (Section 22)
Check the fuel system components (Section 17)
Replace the fuel filter (Section 18)
Check the operation of the EGR valve (Section 32)
Change the rear axle oil (if vehicle is used to pull a trailer) (Section 25)
Check the throttle linkage (Section 19)

Every 18000 miles or 18 months, whichever comes first

Check and repack the front wheel bearings (perform this procedure whenever brake pads are replaced, regardless of maintenance interval) (Section 27)
Change the automatic transmission fluid and filter (if driven mainly in heavy city traffic in hot-climate regions, in hilly or mountainous areas, or for frequent trailer pulling) (Section 28)
Check the torque of the carburetor or TBI mounting bolts (Section 21)
Drain, flush and refill the cooling system (Section 26)

Every 24000 miles or 24 months, whichever comes first

Replace the oxygen sensor (if equipped) (Section 30)
Check the EGR system (Section 32)
Replace the PCV valve (Section 31)
Replace the air filter and PCV filter (Section 29)
Replace the spark plugs (Section 35)
Check the spark plug wires, distributor cap and rotor (Section 36)
Check and adjust (if necessary) the ignition timing (Section 34)
Check the EECS emissions system and replace the canister filter (Section 33)
Change the automatic transmission fluid (if driven under severe conditions, see 18000 mile interval) (Section 28)
Change the rear axle oil (if driven under severe conditions, see 12000 mile interval) (Section 25)
Lubricate the clutch cross-shaft (Section 10)
Check the engine compression (Chapter 2D)

2   Tune-up general information

The term "tune-up' is loosely used for any general operation that puts the engine back in its proper running condition. A tune-up is not a specific operation, but rather a combination of individual operations, such as replacing the spark plugs, adjusting the idle speed, setting the ignition timing, etc.

If, from the time the vehicle is new, the routine maintenance schedule (Section 1) is followed closely and frequent checks are made of fluid levels and high wear items, as suggested throughout this manual, the engine will be kept in relatively good running condition and the need for additional tune-ups will be minimized.

More likely than not, however, there will be times when the engine is running poorly due to lack of regular maintenance. This is even more likely if a used vehicle (which has not received regular and frequent maintenance checks) is purchased. In such cases, an engine tune-up will be needed outside of the regular routine maintenance intervals.

The following series of operations are those most often needed to bring a generally poor running engine back into a proper state of tune.

Minor tune-up

Clean, inspect and test battery (Section 5)
Check all engine-related fluids (Section 4)
Check engine compression (Chapter 2D)
Check and adjust drivebelts (Section 6)
Replace spark plugs (Section 35)
Inspect distributor cap and rotor (Section 36)
Inspect spark plug and coil wires (Section 36)

Check and adjust idle speed (Section 15)
Check and adjust timing (Section 34)
Replace fuel filter (Section 18)
Check PCV valve (Section 31)
Check cooling system (Section 7)

Major tune-up

The above operations and those listed below:

Check EGR system (Chapter 6)
Check ignition system (Chapter 5)
Check charging system (Chapter 5)
Check fuel system (Section 17)

3   Tire and tire pressure checks

Refer to illustration 3.6
1   Periodically inspecting the tires may not only prevent you from being stranded with a flat tire, but can also give you clues as to possible problems with the steering and suspension systems before major damage occurs.
2   Proper tire inflation adds miles to the lifespan of the tires, allows the vehicle to achieve maximum miles per gallon figures and contributes to the overall quality of the ride.
3   When inspecting the tires, first check the wear of the tread. Irregularities in the tread pattern (cupping, flat spots, more wear on one side than the other) are indications of front end alignment and/or balance problems. If any of these conditions are noted, take the vehicle to a reputable repair shop to correct the problem.
4   Also check the tread area for cuts and punctures. Many times a nail or tack will embed itself into the tire tread and yet the tire will hold its air pressure for a short time. In most cases, a repair shop or gas station can

repair the punctured tire.
5   It is also important to check the sidewalls of the tires, both inside and outside. Check for deteriorated rubber, cuts, and punctures. Also inspect the inboard side of the tire for signs of brake fluid leakage, indicating that a thorough brake inspection is needed immediately.
6   Incorrect tire pressure cannot be determined merely by looking at the tire. This is especially true for radial tires. A tire pressure gauge must be used **(see illustration)**. If you do not already have a reliable gauge, it is a good idea to purchase one and keep it in the glovebox. Built-in pressure gauges at gas stations are often unreliable.
7   Always check tire inflation when the tires are cold. Cold, in this case, means the vehicle has not been driven more than one mile after sitting for three hours or more. It is normal for the pressure to increase four to eight pounds or more when the tires are hot.

3.6 The use of an accurate tire pressure gauge is essential for long tire life

4.4a After wiping off the engine oil dipstick, make sure it is reinserted all the way before withdrawing it for the oil level check

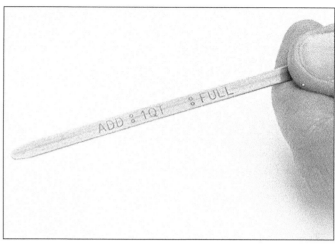

4.4b The oil level should appear between the Add and Full marks; do not overfill the crankcase

8   Unscrew the valve cap protruding from the wheel or hubcap and press the gauge firmly onto the valve stem. Observe the reading on the gauge and compare the figure to the recommended tire pressure listed on the tire placard. The tire placard is usually attached to the driver's door.

9   Check all tires and add air as necessary to bring them up to the recommended pressure levels. Do not forget the spare tire. Be sure to reinstall the valve caps (which will keep dirt and moisture out of the valve stem mechanism).

4   Fluid level checks

1   There are a number of components on a vehicle which rely on the use of fluids to perform their job. During normal operation of the vehicle, these fluids are used up and must be replenished before damage occurs. See Recommended lubricants and fluids at the front of this Chapter for the specific fluid to be used when addition is required. When checking fluid levels, it is important to have the vehicle on a level surface.

Engine oil

Refer to illustrations 4.4a and 4.4b

2   The engine oil level is checked with a dipstick which is located at the side of the engine block. The dipstick travels through a tube and into the oil pan to the bottom of the engine.

3   The oil level should be checked preferably before the vehicle has been driven, or about 15 minutes after the engine has been shut off. If the oil is checked immediately after driving the vehicle, some of the oil will remain in the upper engine components, producing an inaccurate reading on the dipstick.

4   Pull the dipstick from the tube **(see illustration)** and wipe all the oil from the end with a clean rag. Insert the clean dipstick all the way back into the oil pan and pull it out again. Observe the oil at the end of the dipstick **(see illustration)**. At its highest point,

the level should be between the Add and Full marks.

5   It takes approximately one quart of oil to raise the level from the Add mark to the Full mark on the dipstick. Do not allow the level to drop below the Add mark as engine damage due to oil starvation may occur. On the other hand, do not overfill the engine by adding oil above the Full mark since it may result in oil-fouled spark plugs, oil leaks or oil seal failures.

6   Oil is added to the engine after removing a twist-off cap located on the rocker arm cover or through a raised tube near the front of the engine. The cap should be marked "Engine oil' or "Oil.' An oil can spout or funnel will reduce spills as the oil is poured in.

7   Checking the oil level can also be an important preventative maintenance step. If you find the oil level dropping abnormally, it is an indication of oil leakage or internal engine wear which should be corrected. If there are water droplets in the oil, or if it is milky looking, component failure is indicated and the engine should be checked immediately. The condition of the oil can also be checked along with the level. With the dipstick removed from the engine, take your thumb

and index finger and wipe the oil up the dipstick, looking for small dirt or metal particles which will cling to the dipstick. This is an indication that the oil should be drained and fresh oil added (Section 16).

Engine coolant

Refer to illustrations 4.9a and 4.9b

8   All vehicles covered by this manual are equipped with a pressurized coolant recovery system which makes coolant level checks very easy. A clear or white coolant reservoir attached to the inner fender panel is connected by a hose to the radiator cap. As the engine heats up during operation, coolant is forced from the radiator, through the connecting tube and into the reservoir. As the engine cools, the coolant is automatically drawn back into the radiator to keep the level correct.

9   The coolant level should be checked when the engine is hot. Merely observe the level of fluid in the reservoir, which should be at or near the Full Hot mark on the side of the reservoir **(see illustration)**. If the system is completely cool, also check the level in the radiator by removing the cap **(see illustration)**.

4.9a The engine coolant level should appear near the Full Hot mark with the engine at normal operating temperature

4.9b Heed the warning on the radiator cap and never remove it while the engine is hot (arrows align with the vent tube)

4.15 The windshield washer fluid level should be kept an inch or so below the fill line when freezing temperatures are expected

10   **Caution:** *Under no circumstances should either the radiator cap or the coolant recovery reservoir cap be removed when the system is hot, because escaping steam and scalding liquid could cause serious personal injury. In the case of the radiator cap, wait until the system has cooled completely, then wrap a thick cloth around the cap and turn it to the first stop. If any steam escapes, wait until the system has cooled further, then remove the cap. The coolant recovery cap may be removed carefully after it is apparent that no further "boiling' is occurring in the recovery tank.*

11   If only a small amount of coolant is required to bring the system up to the proper level, regular water can be used. However, to maintain the proper antifreeze/water mixture in the system, both should be mixed together to replenish a low level. High-quality antifreeze offering protection to -20~F should be mixed with water in the proportion specified on the container. Do not allow antifreeze to come in contact with your skin or painted surfaces of the vehicle. Flush contacted areas immediately with plenty of water.

12   Coolant should be added to the reservoir until it reaches the Full Cold mark.

13   As the coolant level is checked, note the condition of the coolant.
It should be relatively clear. If it is brown or a rust color, the system should be drained, flushed and refilled (Section 26).

14   If the cooling system requires repeated additions to maintain the proper level, have the radiator cap checked for proper sealing ability. Also check for leaks in the system (cracked hoses, loose hose connections, leaking gaskets, etc.).

Windshield washer fluid

Refer to illustration 4.15

15   Fluid for the windshield washer system is located in a plastic reservoir **(see illustration)**. The level in the reservoir should be maintained at the Full mark, except during periods when freezing temperatures are expected, at which times the fluid level should be maintained no higher than 3/4 full to allow for expansion should the fluid freeze. The use of a windshield washer fluid additive will help lower the freezing point of the fluid and will result in better cleaning of the windshield surface. Do not use antifreeze because it will cause damage to the vehicle's paint.

16   Also, to help prevent icing in cold weather, warm the windshield with the defroster before using the washer.

Battery electrolyte

Refer to illustration 4.17

17   All vehicles with which this manual is concerned are equipped with a maintenance-free battery which is permanently sealed (except for vent holes) and has no filler caps **(see illustration)**. Water does not have to be added to these batteries at any time.

Brake fluid

Refer to illustration 4.19

18   The master cylinder is mounted directly on the firewall (manual brake models) or on the front of the power booster unit (power brake models) in the engine compartment.

19   The master cylinder reservoir incorporates two windows, which allow checking the

brake fluid level without removal of the reservoir cover. The level should be maintained at 1/4-inch above the lowest edge of each reservoir opening **(see illustration)**.

20   If a low level is indicated, be sure to wipe the top of the reservoir cover with a clean rag, to prevent contamination of the brake system, before lifting the cover.

21   When adding fluid, pour it carefully into the reservoir, taking care not to spill any onto surrounding painted surfaces. Be sure the specified fluid is used, since mixing different types of brake fluid can cause damage to the system. See Recommended lubricants and fluids or your owner's manual.

22   At this time the fluid and master cylinder can be inspected for contamination. Normally, the brake system will not need periodic draining and refilling, but if rust deposits, dirt particles or water droplets are seen in the fluid, the system should be dismantled, drained and refilled with fresh fluid.

23   After filling the reservoir to the proper level, make sure the lid is properly seated to prevent fluid leakage and/or system pressure loss.

24   The brake fluid in the master cylinder will drop slightly as the brake shoes or pads at each wheel wear down during normal operation. If the master cylinder requires repeated replenishing to keep it at the proper level, this is an indication of leakage in the brake system, which should be corrected immediately. Check all brake lines and connections, along with the wheel cylinders and booster (see Section 13 for more information).

25   If upon checking the master cylinder fluid level you discover one or both reservoirs empty or nearly empty, the brake system should be bled (Chapter 9).

Manual transmission oil

26   Manual shift transmissions do not have a dipstick. The fluid level is checked with the engine cold by removing a plug from the side of the transmission case. Locate the plug and use a rag to clean the plug and the area around it, then remove it with a wrench.

4.17 This type of battery never requires adding water, but normal maintenance should be performed

4.19 It is essential that dirt and grease be wiped off the master cylinder reservoir cover before it is removed

4.33 Do not confuse the automatic transmission fluid level dipstick (shown here) with the engine oil level dipstick

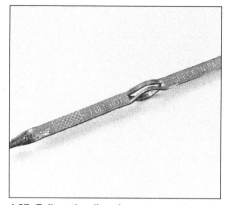

4.37 Follow the directions stamped on the dipstick to get an accurate reading

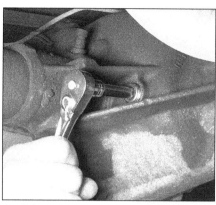

4.43 This rear axle oil filler/check plug is on the differential housing, though on some models the plug may be in the rear cover plate

27   If oil immediately starts leaking out, thread the plug back into the transmission because the level is full. If there is no leakage, completely remove the plug and place your little finger inside the hole. The oil level should be just at the bottom of the plug hole.

28   If the transmission needs more oil, use a syringe to squeeze the appropriate lubricant into the plug hole until the level is correct.

29   Thread the plug back into the transmission and tighten it securely.

30   Drive the vehicle a short distance, then check to make sure the plug is not leaking.

Automatic transmission fluid

Refer to illustrations 4.33 and 4.37

31   The level of the automatic transmission fluid should be carefully maintained. Low fluid level can lead to slipping or loss of drive, while overfilling can cause foaming and loss of fluid.

32   With the parking brake set, start the engine, then move the shift lever through all the gear ranges, ending in Park. The fluid level must be checked with the vehicle level and the engine running at idle. **Note:** *Incorrect fluid level readings will result if the vehicle has just been driven at high speeds for an extended period, in hot weather in city traffic, or if it's been pulling a trailer. If any of these conditions apply, wait until the fluid has cooled (about 30 minutes).*

33   Locate the dipstick at the rear of the engine compartment on the passenger's side and pull it out of the filler tube **(see illustration)**.

34   Carefully touch the end of the dipstick to determine if the fluid is cool (about room temperature), or warm or hot (uncomfortable to the touch).

35   Wipe the fluid from the dipstick and push it back into the filler tube until the cap seats.

36   Pull the dipstick out again and note the fluid level.

37   If the fluid felt cool, the level should be 1/8 to 3/8-inch below the Add mark **(see illustration)**. The two dimples below the Add mark indicate this range.

38   If the fluid felt warm, the level should be

close to the Add mark (just above or below it).

39   If the fluid felt hot, the level should be at the Full mark.

40   Add just enough of the recommended fluid to fill the transmission to the proper level. It takes about one pint to raise the level from the Add mark to the Full mark with a hot transmission, so add the fluid a little at a time and keep checking the level until it is correct.

41   The condition of the fluid should also be checked along with the level. If the fluid at the end of the dipstick is a dark reddish-brown color, or if the fluid has a burnt smell, the transmission fluid should be changed. If you are in doubt about the condition of the fluid, purchase some new fluid and compare the two for color and smell.

Rear axle oil

Refer to illustration 4.43

42   Like the manual transmission, the rear axle has an inspection and fill plug which must be removed to check the oil level.

43   Remove the plug which is located either in the removable cover plate or on the side of the differential carrier **(see illustration)**. Use your little finger to reach inside the rear axle housing to feel the oil level. It should be at the bottom of the plug hole.

44   If this is not the case, add the proper lubricant to the rear axle carrier through the

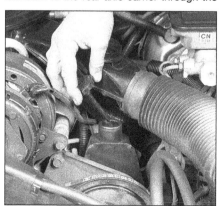

4.52a Make certain no foreign matter gets into the power steering fluid reservoir when the check is made

plug hole. A syringe or a small funnel can be used for this.

45   Make certain the correct lubricant is used, as regular and limited-slip rear axles require different lubricants. You can determine which type of axle you have by reading the stamped number on the axle tube (refer to Chapter 8).

46   Tighten the plug securely and check for leaks after the first few miles of driving.

Power steering fluid

Refer to illustrations 4.52a and 4.52b

47   Unlike manual steering, the power steering system relies on fluid which may, over a period of time, require replenishing.

48   The reservoir for the power steering pump will be located near the front of the engine and can be mounted on either the left or right side.

49   For the check, the front wheels should be pointed straight ahead and the engine should be off.

50   Use a clean rag to wipe off the reservoir cap and the area around the cap. This will help prevent any foreign matter from entering the reservoir during the check.

51   Make sure the engine is at normal operating temperature.

52   Remove the dipstick **(see illustration)**,

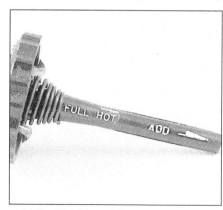

4.52b Checking of the power steering fluid level is done with the engine at normal operating temperature

5.3 Unlike older model batteries, the cable terminals are on the side of these batteries

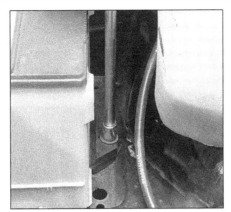

5.6 Use a socket extension to reach the battery hold-down bolt

6.3 The arrow on the tensioner must be within the marked range - if it isn't, the belt is stretched and must be replaced

wipe it off with a clean rag, reinsert it, then withdraw it and read the fluid level **(see illustration)**. The level should be between the Add and Full Hot marks.

53   If additional fluid is required, pour the specified type directly into the reservoir using a funnel to prevent spills.

54   If the reservoir requires frequent fluid additions, all power steering hoses, hose connections, the power steering pump and the steering box should be carefully checked for leaks.

5   Battery check and maintenance

Refer to illustrations 5.3 and 5.6

1   A sealed maintenance-free battery is standard equipment on all vehicles with which this manual is concerned. Although this type of battery has many advantages over the older, capped cell type and never requires the addition of water, it should nevertheless be routinely maintained according to the procedures which follow. **Warning:** *Hydrogen gas in small quantities is present in the area of the two small side vents on sealed batteries, so keep lighted tobacco and open flames or sparks away from them.*

2   The external condition of the battery should be monitored periodically for damage such as a cracked case or cover.

3   Check the tightness of the battery cable clamps to ensure good electrical connections and check the entire length of each cable for cracks and frayed conductors **(see illustration)**.

4   If corrosion (visible as white, fluffy deposits) is evident, remove the cables from the terminals, clean them with a battery brush and reinstall the cables. Corrosion can be kept to a minimum by applying a layer of petroleum jelly or grease to the terminals and cable clamps after they are assembled.

5   Make sure that the rubber protector (if so equipped) over the positive terminal is not torn or missing. It should completely cover the terminal.

6   Make sure that the battery carrier is in good condition and that the hold-down

clamp bolts are tight. If the battery is removed from the carrier, make sure that no parts remain in the bottom of the carrier when the battery is reinstalled **(see illustration)**. When reinstalling the hold-down clamp bolts, do not overtighten them.

7   Corrosion on the hold-down components, battery case and surrounding areas may be removed with a solution of water and baking soda, but take care to prevent any solution from coming in contact with your eyes, skin or clothes, as it contains acid. Protective gloves should be worn. Thoroughly wash all cleaned areas with plain water.

8   Any metal parts of the vehicle damaged by corrosion should be covered with a zinc-based primer, then painted after the affected areas have been cleaned and dried.

9   Further information on the battery, charging and jump-starting can be found in Chapter 5.

6   Drivebelt check and adjustment

Refer to illustrations 6.3, 6.4a, 6.4b, 6.5, 6.7a, 6.7b, 6.9 and 6.10

1   The drivebelts, or V-belts as they are sometimes called, are located at the front of the engine and play an important role in the overall operation of the vehicle and its components. Due to their function and material make-up, the belts are prone to failure after a period of time and should be inspected and adjusted periodically to prevent major engine damage.

2   The number of belts used on a particular vehicle depends on the accessories installed. Drivebelts are used to turn the generator/alternator, air injection (smog) pump, power steering pump, water pump, fan and air-conditioning compressor. Depending on the pulley arrangement, a single belt may be used to drive more than one of these components.

3   On later models, a single serpentine drivebelt is used in place of multiple V-belts. A serpentine belt requires no adjustment, as this is taken care of by a spring-loaded tensioner pulley. The belt should be replaced when the wear indicator on the tensioner

reaches its maximum travel **(see illustration)**. Inspect the belt for missing ribs, fraying and other signs of abnormal wear.

4   With the engine off, open the hood and locate the various belts at the front of the engine. Using your fingers (and a flashlight, if necessary), move along the belts checking for cracks and separation of the belt plies. Also check for fraying and glazing, which gives the belt a shiny appearance **(see illustrations)**. Both sides of the belt should be

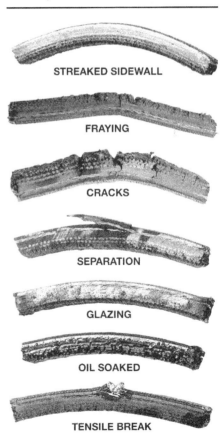

STREAKED SIDEWALL

FRAYING

CRACKS

SEPARATION

GLAZING

OIL SOAKED

TENSILE BREAK

6.4a Here are some of the more common problems associated with the drivebelts (check the belts very carefully to prevent an untimely breakdown)

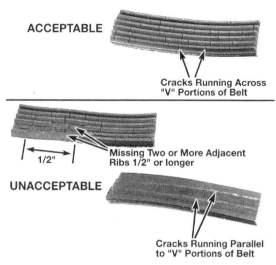

ACCEPTABLE

Cracks Running Across
"V" Portions of Belt

1/2"

Missing Two or More Adjacent
Ribs 1/2" or longer

UNACCEPTABLE

Cracks Running Parallel
to "V" Portions of Belt

6.4b Small cracks in the underside of a serpentine belt are
acceptable - lengthwise cracks or missing pieces
are cause for replacement

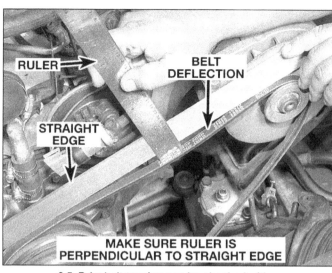

RULER

BELT
DEFLECTION

STRAIGHT
EDGE

MAKE SURE RULER IS
PERPENDICULAR TO STRAIGHT EDGE

6.5 Drivebelt tension can be checked with a
straightedge and ruler

6.7a Typically each component is
mounted with a pivot bolt (1) and
at least one adjustment bolt (2)

inspected, which means you will have to twist the belt to check the underside.

5   The tension of each belt is checked by pushing on the belt at a distance halfway between the pulleys. Push firmly with your thumb and see how much the belt moves down (deflects) **(see illustration)**. A rule of thumb is that if the distance from pulley center-to-pulley center is between 7 and 11 inches, the belt should deflect 1/4-inch. If the belt is longer and travels between pulleys spaced 12 to 16 inches apart, the belt should deflect 1/2-inch.

6   If it is necessary to adjust the belt tension, either to make the belt tighter or looser, it is done by moving the belt-driven accessory on the bracket.

7   For each component there will be an adjustment bolt and a pivot bolt **(see illustrations)**. Both bolts must be loosened slightly to enable you to move the component.

8   After the two bolts have been loosened, move the component away from the engine (to tighten the belt) or toward the engine (to loosen the belt). Hold the accessory in position and check the belt tension. If it is correct, tighten the two bolts until just snug, then recheck the tension. If it is all right, tighten the bolts.

9   It will often be necessary to use some sort of pry bar to move the accessory while the belt is adjusted **(see illustration)**. If this must be done to gain the proper leverage, be very careful not to damage the component being moved or the part being pried on.

10   When replacing the serpentine belt (used on later models), use a socket and breaker bar placed on the pulley center bolt (if a 1/2-inch drive square hole is not provided) to rotate the tensioner pulley away from the belt to release the belt tension **(see illustration)**. Make sure the new belt is routed correctly (refer to the label in the engine compartment) then rotate the pulley away from the belt, slip the belt under the pulley and release the pulley. Also, the belt must completely engage the grooves in the pulleys.

6.7b One of the power steering pump
adjustment bolts is located
behind the pump

6.9 Use a 1/2-inch drive breaker bar in the
square hole provided, to tension the air
conditioning compressor drivebelt

6.10 Use a socket and breaker bar placed
on the tensioner pulley center bolt to
rotate the tensioner away from the belt

Check for a chafed area that could fail prematurely.

Check for a soft area indicating the hose has deteriorated inside.

Overtightening the clamp on a hardened hose will damage the hose and cause a leak.

Check each hose for swelling and oil-soaked ends. Cracks and breaks can be located by squeezing the hose.

7.4a Hoses, like drivebelts, can fail at any time - to prevent the inconvenience of a blown radiator or heater hose, inspect them carefully as shown

7 Cooling system check

Refer to illustrations 7.4a, 7.4b and 7.6

1   Many major engine failures can be attributed to a faulty cooling system. If the vehicle is equipped with an automatic transmission, the cooling system also plays an important role in prolonging its life.

2   The cooling system should be checked with the engine cold. Do this before the vehicle is driven for the day or after it has been shut off for at least three hours.

3   Remove the radiator cap and thoroughly clean the cap (inside and out) with clean water. Also clean the filler neck on the radiator. All traces of corrosion should be removed.

4   Carefully check the upper and lower radiator hoses along with the smaller diameter heater hoses. Inspect each hose along its entire length, replacing any hose which is cracked, swollen or shows signs of deterioration **(see illustrations)**. Cracks may become more apparent if the hose is squeezed.

5   Also make sure that all hose connections are tight. A leak in the cooling system will usually show up as white or rust colored deposits on the areas adjoining the leak.

6   Use compressed air or a soft brush to remove debris from the front of the radiator or air conditioning condenser **(see illustration)**. Be careful not to damage the delicate cooling fins or cut yourself on them.

7   Finally, have the radiator cap and cooling system pressure tested. If you do not have a pressure tester, most gas stations and repair shops will do this for a minimal change.

8 Underhood hose check and replacement

Caution: *Replacement of air conditioning hoses must be left to a dealer or air condition-*

ing specialist who can depressurize the system and perform the work safely.

1   The high temperatures present under the hood can cause deterioration of the numerous rubber and plastic hoses.

2   Periodic inspection should be made for cracks, loose clamps and leaks because some of the hoses are part of the emissions control systems and can affect the engine's performance.

3   Remove the air cleaner if necessary and trace the entire length of each hose. Squeeze each hose to check for cracks and look for swelling, discoloration and leaks.

4   If the vehicle has considerable mileage or if one or more of the hoses is suspect, it is a good idea to replace all of the hoses at one time.

5   Measure the length and inside diameter of each hose and obtain and cut the replacement to size. Since original equipment hose clamps are often good for only one or two uses, it is a good idea to replace them with screw-type clamps.

6   Replace each hose one at a time to eliminate the possibility of confusion. Hoses attached to the heater and radiator contain coolant, so newspapers or rags should be kept handy to catch the spills when they are disconnected.

7   After installation, run the engine until it reaches operating temperature, shut it off and check for leaks. After the engine has cooled, retighten all of the screw-type clamps.

9 Windshield wiper blade inspection and replacement

Refer to illustrations 9.1, 9.4 and 9.5

1   The windshield wiper and blade assembly should be inspected periodically for damage, loose components and cracked or worn blade elements **(see illustration)**.

7.4b Although this radiator hose appears to be in good condition, it should still be periodically checked for cracks (more easily revealed when squeezed)

7.6 Inspect the radiator or air conditioning condenser for debris build-up (view from underneath)

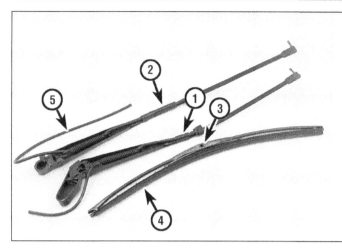

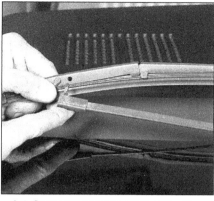

**9.1 Wiper blade
assemblies**

1   *Right wiper arm*
2   *Left wiper arm*
3   *Wiper blade*
4   *Blade element*
5   *Washer hose*

**9.4 Some models are equipped with a
windshield wiper arm assembly
quick release lever**

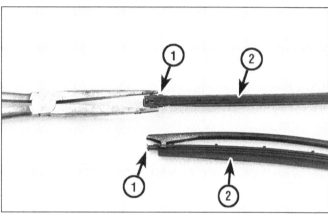

9.5 Wiper blade element installation details

1 *Element*	2 *Housing tabs*

**10.2 Grease fittings (arrows) are located at the steering linkage
and lower balljoints**

2   Road film can build up on the wiper blades and affect their efficiency so they should be washed regularly with a mild detergent solution.
3   The action of the wiping mechanism can loosen the bolts, nuts and fasteners so they should be checked and tightened, as necessary, at the same time the wiper blades are checked.
4   If the wiper blade elements are cracked, worn or warped, they should be replaced with new ones. Some models are equipped with press-type release levers. Push the lever up or down and slide the old wiper blade assembly off the wiper arm pin **(see illustration)**. Install the new blade by positioning it over the wiper arm pin and pressing on it until it snaps into place.
5   The wiper blade element is locked in place at either end of the wiper blade assembly by a spring-loaded retainer and metal tabs. To remove the element, slide the blade under the element near the tabs, then slide the element up, out of the tabs **(see illustration)**.
6   To install, slide the element into the retaining tabs, lining up the slot in the element with the tabs, and snap the element into place.
7   Snap the blade assembly into place on the wiper arm pin.

10   Chassis lubrication

Refer to illustrations 10.2, 10.6, 10.8 and 10.10
1   A grease gun and a cartridge filled with the proper grease (see Recommended lubricants and fluids) are usually the only equipment necessary to lubricate the chassis components.
2   Locate the various grease fittings **(see illustration)**. Inspect the grease fittings for damage. An auto parts store will be able to supply replacement fittings, if necessary. Straight, as well as angled, fittings are available.
3   For easier access under the vehicle, raise it with a jack and place jackstands under the frame. Make sure the vehicle is securely supported by the stands.
4   Before proceeding, force a little of the grease out of the nozzle to remove any dirt from the end of the gun. Wipe the nozzle clean with a rag.
5   With the grease gun and a clean rag, begin lubricating the components.
6   Wipe the grease fitting nipple clean and push the nozzle firmly over the fitting nipple **(see illustration)**. Squeeze the trigger on the grease gun to force grease into the component.

Note: *The lower control arm balljoints (one for each front wheel) should be lubricated until the rubber reservoir is firm to the touch. Do not pump too much grease into these fittings as it could rupture the reservoir. For all other suspension and steering fittings, continue pumping grease into the nipple until grease seeps out of the joint between the two components. If the grease seeps out around the grease gun nozzle, the nipple is clogged*

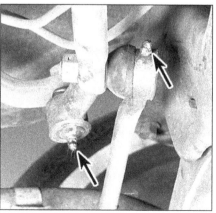

**10.6 Push the grease gun nozzle onto
the fitting and pump grease
into the component**

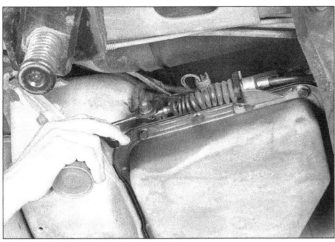

10.8 Lubricate the automatic transmission. shift linkage with a few drops of oil from an oil can

10.10 Multi-purpose grease is used to lubricate the hood latch mechanism

or the nozzle is not seated on the fitting nipple. Resecure the gun nozzle to the fitting and try again. If necessary, replace the fitting.
7   Wipe the excess grease from the components and the grease fitting. Follow the procedures for the remaining fittings.
8   While you are under the vehicle, clean and lubricate the parking brake cable, along with the cable guides and levers. This can be done by smearing some of the chassis grease onto the cable and its related parts with your fingers. Place a few drops of light engine oil on the transmission shift linkage rods and swivels **(see illustration)**.
9   Lower the vehicle to the ground for the remaining body lubrication process.
10   Open the hood and smear a little chassis grease on the hood latch mechanism **(see illustration)**. If the hood has an inside release, have an assistant pull the release knob from inside the vehicle as you lubricate the cable at the latch.
11   Lubricate all the hinges (door, hood, hatch) with a few drops of light engine oil to keep them in proper working order.
12   Finally, the key lock cylinders can be lubricated with spray-on graphite which is available at auto parts stores.

11 Exhaust system check

1   With the engine cold (at least three hours after the vehicle has been driven), check the complete exhaust system from its starting point at the engine to the end of the tailpipe. This should be done on a hoist where unrestricted access is available.
2   Check the pipes and connections for signs of leakage and/or corrosion indicating a potential failure. Make sure that all brackets and hangers are in good condition and tight.
3   At the same time, inspect the underside of the body for holes, corrosion, open seams, etc. which may allow exhaust gases to enter the passenger compartment. Seal all body openings with silicone or body putty.
4   Rattles and other noises can often be traced to the exhaust system, especially the mounts and hangers. Try to move the pipes, muffler and catalytic converter (if so equipped). If the components can come in contact with the body or driveline parts, secure the exhaust system with new mounts.
5   This is also an ideal time to check the running condition of the engine by inspecting inside the very end of the tailpipe. The exhaust deposits here are an indication of engine state-of-tune. If the pipe is black and sooty or coated with white deposits, the engine is in need of a tune-up (including a thorough carburetor inspection and adjustment).

12 Suspension and steering check

1   Whenever the front of the vehicle is raised for service, it is a good idea to visually check the suspension and steering components for wear.
2   Indications of a fault in these systems are excessive play in the steering wheel before the front wheels react, excessive sway around corners, body movement over rough roads or binding at some point as the steering wheel is turned.
3   Before the vehicle is raised for inspection, test the shock absorbers by pushing down to rock the vehicle at each corner. If you push down and the vehicle does not come back to a level position within one or two bounces, the shocks/struts are worn and must be replaced. As this is done, check for squeaks and strange noises coming from the suspension components. Information on suspension components can be found in Chapter 11.
4   Now raise the front end of the vehicle and support it firmly on jackstands placed under the frame rails. Because of the work to be done, make sure the vehicle cannot fall from the stands.
5   Grab the top and bottom of the front tire with your hands and rock the tire/wheel on the spindle. If there is movement of more than 0.005 inch, the wheel bearings should be serviced (see Section 27).
6   Check for loose bolts, broken or disconnected parts and deteriorated rubber bushings on all suspension and steering components. Look for grease or fluid leaking from around the steering box. Check the power steering hoses and connections for leaks. Check the balljoints for wear.
7   Have an assistant turn the steering wheel from side-to-side and check the steering components for free movement, chafing and binding. If the steering does not react with the movement of the steering wheel, try to determine where the slack is located.

13 Brake check

Warning: *The dust created by the brake system may contain asbestos, which is harmful to your health. Never blow it out with compressed air and don't inhale any of it. An approved filtering mask should be worn when inspecting or working on the brakes. Do not, under any circumstances, use petroleum-based solvents to clean brake parts. Use brake system cleaner only.*
Note: *For detailed photographs of the brake system, refer to Chapter 9.*
1   The brakes should be inspected every time the wheels are removed or whenever a defect is suspected. Indications of a potential brake system defect are: the vehicle pulls to one side when the brake pedal is depressed; noises coming from the brakes when they are applied; excessive brake pedal travel; pulsating pedal; and leakage of fluid, usually seen on the inside of the tire or wheel.

Disc brakes

Refer to illustrations 13.5 and 13.6
2   Both front and rear disc brakes (if the rear is so equipped) can be visually checked without removing any parts except the wheels.
3   Raise the vehicle and place it securely on jackstands.

13.5 The inner pads on the front wheels are equipped with a wear sensor

13.6 Disc pad wear inspection hole is visible at the center of caliper

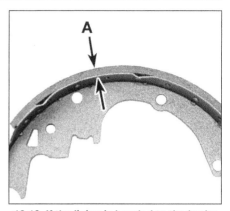

13.13 If the lining is bonded to the brake shoe, measure the lining thickness from the outer surface to the metal shoe, as shown here; if the lining is riveted to the shoe, measure from the lining outer surface to the rivet head

4   The disc brake calipers, which contain the pads, are now visible. There is an outer pad and an inner pad in each caliper. All pads should be inspected.

5   The inner pads on the front wheels are equipped with a wear sensor **(see illustration)**. This is a small, bent piece of metal which is visible from the inboard side of the brake caliper. When the pads wear to the danger limit, the metal sensor rubs against the rotor and makes a screeching sound.

6   Check the pad thickness by looking at each end of the caliper and through the inspection hole in the caliper body **(see illustration)**. If the wear sensor clip is very close to the rotor, or if the lining material is 1/16-inch or less in thickness, the pads should be replaced. Keep in mind that the lining material is riveted or bonded to a metal backing shoe and the metal portion is not included in this measurement.

7   Since it will be difficult, if not impossible, to measure the exact thickness of the remaining lining material, remove the pads for further inspection or replacement if you are in doubt as to the quality of the pad.

8   Before installing the wheels, check for leakage around the brake hose connections leading to the caliper and damage (cracking, splitting, etc.) to the brake hose. Replace the hose or fittings as necessary, referring to Chapter 9.

9   Also check the condition of the rotor. Look for scoring, gouging and burnt spots. If these conditions exist, the hub/rotor assembly should be removed for servicing (Chapter 9).

Drum brakes

Refer to illustration 13.13

10   Using a scribe or chalk, mark the drum and one of the wheel studs so the drum can be reinstalled in the same position.

11   Pull the brake drum off the axle and brake assembly. If this proves difficult, make sure the parking brake is released, then squirt some penetrating oil around the center hub area. Allow the oil to soak in and again try to pull the drum off. Then, if the drum cannot be pulled off, the brake shoes will have to be adjusted in. This is done by first removing the

lanced knock-out in the backing plate with a hammer and chisel. With the lanced area punched in, pull the lever off the sprocket and then use a small screwdriver to turn the adjuster wheel, which will move the shoes away from the drum.

12   With the drum removed, clean the brake assembly with brake system cleaner.

13   Note the thickness of the lining material on both the front and rear brake shoes **(see illustration)**. If the material has worn away to within 1/16-inch of the recessed rivets or metal backing, the shoes should be replaced. If the linings look worn, but you are unable to determine their exact thickness, compare them with a new set at an auto parts store. The shoes should also be replaced if they are cracked, glazed (shiny surface) or contaminated with brake fluid.

14   Check to see that all the brake assembly springs are connected and in good condition.

15   Check the brake components for signs of fluid leakage. With your finger, carefully pry back the rubber cups on the wheel cylinder located at the top of the brake shoes. Any leakage is an indication that the wheel cylinders should be overhauled immediately (Chapter 9). Also check the hoses and connections for signs of leakage.

16   Clean the inside of the drum with brake system cleaner.

17   Check the inside of the drum for cracks, scores, deep scratches and hard spots which will appear as small discolored areas. If imperfections cannot be removed with sandpaper or emery cloth, the drum must be taken to a machine shop for resurfacing.

18   After the inspection process, if all parts are found to be in good condition, reinstall the brake drum (using a rubber plug if the lanced knock-out was removed). Install the wheel and lower the vehicle to the ground.

Parking brake

19   The easiest way to check the operation of the parking brake is to park the vehicle on a steep hill with the parking brake set and the transmission in Neutral (be sure to stay in the vehicle for this check!). If the parking brake cannot prevent the vehicle from rolling, it is in need of adjustment (see Chapter 9).

14 Carburetor choke check

Refer to illustration 14.3

1   The choke only operates when the engine is cold, so this check should be performed before the vehicle has been started for the day.

2   Open the hood and remove the top plate of the air cleaner assembly. It is usually held in place by a wing nut. If any vacuum hoses must be disconnected, make sure you tag them to ensure reinstallation in their original positions. Place the top plate and wing nut aside, out of the way of moving engine components.

3   Look at the top of the carburetor at the center of the air cleaner housing. You will notice a flat plate at the carburetor opening **(see illustration)**.

4   Have an assistant press the accelerator pedal to the floor. The plate should close completely. Start the engine while you observe the plate at the carburetor. **Caution:** *Do not position your face directly over the carburetor, because the engine could backfire and cause serious burns. When the engine starts, the choke plate should open slightly.*

14.3 With the air cleaner top plate removed, the choke plate is easily checked

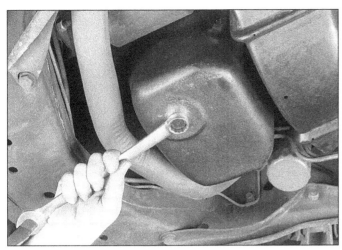

16.9 The oil pan drain plug is located at the bottom of the oil pan and should be removed with a socket or box-end wrench - DO NOT use an open end wrench as the corners of the bolt can be easily rounded off

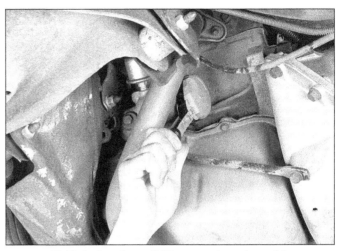

16.14 A strap-type oil filter wrench works well in hard-to-reach locations

5   Allow the engine to continue running at an idle speed. As the engine warms up to operating temperature, the plate should slowly open, allowing more air to enter through the top of the carburetor.

6   After a few minutes, the choke plate should be all the way open to the vertical position.

7   You will notice that the engine speed corresponds with the plate opening. With the plate completely closed, the engine should run at a fast idle. As the plate opens, the engine speed will decrease.

8   If a malfunction is detected during the above checks, refer to Chapter 4 for specific information related to adjusting and servicing the choke components.

15   Engine idle speed check and adjustment

1   The engine idle speed is adjustable on some models and should be checked at the scheduled maintenance interval.

2   On those vehicles with provisions for idle speed adjustment, the specifications for such adjustments are shown on the vehicle Emissions Control Information label; however, the adjustments must be made using calibrated test equipment. The adjustments should therefore be made by a dealer or automotive repair facility.

16   Engine oil and filter change

Refer to illustrations 16.9 and 16.14

1   Frequent oil changes may be the best form of preventative maintenance available for the home mechanic. When engine oil ages, it gets diluted and contaminated, which ultimately leads to premature engine wear.

2   Although some sources recommend oil filter changes every other oil change, we feel

that the minimal cost of an oil filter and the relative ease with which it is installed dictate that a new filter be used whenever the oil is changed.

3   The tools necessary for a normal oil and filter change are a wrench to fit the drain plug at the bottom of the oil pan, an oil filter wrench to remove the old filter, a container with at least a six-quart capacity to drain the old oil into and a funnel or oil can spout to help pour fresh oil into the engine.

4   In addition, you should have plenty of clean rags and newspapers handy to mop up any spills. Access to the underside of the vehicle is greatly improved if the vehicle can be lifted on a hoist, driven onto ramps or supported by jackstands. Do not work under a vehicle which is supported only a bumper, hydraulic or scissors-type jack.

5   If this is your first oil change on the vehicle, it is a good idea to crawl underneath and familiarize yourself with the locations of the oil drain plug and the oil filter. The engine and exhaust components will be warm during the actual work, so it is a good idea to figure out any potential problems before the engine and its accessories are hot.

6   Allow the engine to warm up to normal operating temperature. If the new oil or any tools are needed, use this warm-up time to gather everything necessary for the job. The correct type of oil to buy for your application can be found in Recommended lubricants and fluids near the front of this manual.

7   With the engine oil warm (warm engine oil will drain better and more built-up sludge will be removed with the oil), raise and support the vehicle. Make sure it is firmly supported. If jackstands are used, they should be placed toward the front of the frame rails which run the length of the vehicle.

8   Move all necessary tools, rags and newspapers under the vehicle. Position the drain pan under the drain plug. Keep in mind that the oil will initially flow from the pan with some force, so place the pan accordingly.

9   Being careful not to touch any of the hot exhaust pipe components, use the wrench to remove the drain plug near the bottom of the oil pan **(see illustration)**. Depending on how hot the oil has become, you may want to wear gloves while unscrewing the plug the final few turns.

10   Allow the old oil to drain into the pan. It may be necessary to move the pan farther under the engine as the oil flow slows to a trickle.

11   After all the oil has drained, wipe off the drain plug with a clean rag. Small metal particles may cling to the plug and would immediately contaminate the new oil.

12   Clean the area around the drain plug opening and reinstall the plug. Tighten the plug securely with the wrench. If a torque wrench is available, use it to tighten the plug.

13   Move the drain pan into position under the oil filter.

14   Now use the filter wrench to loosen the oil filter **(see illustration)**. Chain or metal band-type filter wrenches may distort the filter canister, but this is of no concern as the filter will be discarded anyway.

15   Sometimes the oil filter is on so tight it cannot be loosened, or it is positioned in an area which is inaccessible with a filter wrench. As a last resort, you can punch a metal bar or long screwdriver directly through the bottom of the canister and use it as a T-bar to turn the filter. If so, be prepared for oil to spurt out of the canister as it is punctured.

16   Completely unscrew the old filter. Be careful, it is full of oil. Empty the oil inside the filter into the drain pan.

17   Compare the old filter with the new one to make sure they are the same type.

18   Use a clean rag to remove all oil, dirt and sludge from the area where the oil filter mounts to the engine. Check the old filter to make sure the rubber gasket is not stuck to the engine mounting surface. If the gasket is stuck to the engine (use a flashlight if necessary), remove it.

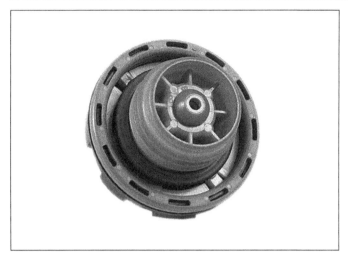

17.5 With today's sophisticated emissions systems, it is essential that seals in the gas tank cap be checked regularly

18.3 Tag all hoses as they are removed from the air cleaner assembly

19   Open one of the cans of new oil and fill the new filter about half full of fresh oil. Also, apply a light coat of oil to the rubber gasket of the new oil filter.

20   Attach the new filter to the engine following the tightening directions printed on the filter canister or packing box. Most filter manufacturers recommend against using a filter wrench due to the possibility of over-tightening and damage to the seal.

21   Remove all tools, rags, etc. from under the vehicle, being careful not to spill the oil in the drain pan, then lower the vehicle.

22   Move to the engine compartment and locate the oil filler cap on the engine. In most cases there will be a screw-off cap on the rocker arm cover (at the side of the engine) or a cap at the end of a fill tube at the front of the engine. In any case, the cap will most likely be labeled "Engine Oil' or "Oil.'

23   If an oil can spout is used, push the spout into the top of the oil can and pour the fresh oil through the filler opening. A funnel may also be used.

24   Pour about three (3) quarts of fresh oil into the engine. Wait a few minutes to allow the oil to drain into the pan, then check the level on the oil dipstick (see Section 4 if necessary). If the oil level is at or near the lower Add mark, start the engine and allow the new oil to circulate.

25   Run the engine for only about a minute and then shut it off. Immediately look under the vehicle and check for leaks at the oil pan drain plug and around the oil filter. If either is leaking, tighten with a bit more force.

26   With the new oil circulated and the filter now completely full, recheck the level on the dipstick and add enough oil to bring the level to the Full mark on the dipstick.

27   During the first few trips after an oil change, make it a point to check frequently for leaks and proper oil level.

28   The old oil drained from the engine cannot be reused in its present state and should be disposed of. Oil reclamation centers, auto repair shops and gas stations will normally accept the oil, which can be refined and used again. After the oil has cooled, it can be drained into a suitable container (capped plastic jugs, topped bottles, milk cartons, etc.) for transport to one of these disposal sites.

17   Fuel system check

Refer to illustration 17.5

Caution: *There are certain precautions to take when inspecting or servicing the fuel system components. Work in a well-venti-lated area and do not allow open flames (cigarettes, appliance pilot lights, etc.) to get near the work area. Mop up spills immediately and do not store fuel-soaked rags where they could ignite.*

1   If your vehicle is fuel injected, refer to the fuel injection pressure relief procedure in Chapter 4 before servicing any component of the fuel system.

2   Before disconnecting any fuel lines, be prepared to catch the fuel as it spills out. Plug all disconnected fuel lines immediately after disconnection.

3   The fuel system is most easily checked with the vehicle raised on a hoist so the components underneath the vehicle are readily visible and accessible.

4   If the smell of gasoline is noticed while driving or after the vehicle has been in the sun, the system should be thoroughly inspected immediately.

5   Remove the gas filler cap and check for damage, corrosion and an unbroken sealing imprint on the gasket **(see illustration)**. Replace the cap with a new one, if necessary.

6   With the vehicle raised, inspect the gas tank and filler neck for punctures, cracks and other damage. The connection between the filler neck and the tank is especially critical. Sometimes a rubber filler neck will leak due to loose clamps or deteriorated rubber, problems a home mechanic can usually rectify.

Caution: *Do not, under any circumstances, try to repair a fuel tank yourself (except rubber components) unless you have had consider-able experience. A welding torch or any open flame can easily cause the fuel vapors to explode if the proper precautions are not taken.*

7   Carefully check all rubber hoses and metal lines leading away from the fuel tank. Check for loose connections, deteriorated hoses, crimped lines and other damage. Follow the lines up to the front of the vehicle, carefully inspecting them all the way. Repair or replace damaged sections as necessary.

8   If a fuel odor is still evident after the inspection, refer to Section 33.

18   Fuel filter replacement

Warning: *Gasoline is extremely flammable, so take extra precautions when you work on any part of the fuel system. Don't smoke or allow open flames or bare light bulbs near the work area, and don't work in a garage where a natural gas-type appliance (such as a water heater or a clothes dryer) with a pilot light is present. Since gasoline is carcinogenic, wear latex gloves when there's a possibility of being exposed to fuel, and, if you spill any fuel on your skin, rinse it off immediately with soap and water. Mop up any spills immedi-ately and do not store fuel-soaked rags where they could ignite. The fuel system on fuel-injected models is under constant pressure, so, if any fuel lines are to be disconnected, the fuel pressure in the system must be relieved first. When you perform any kind of work on the fuel system, wear safety glasses and have a Class B type fire extinguisher on hand.*

Carbureted models

Refer to illustrations 18.3 and 18.6

1   On these models, the fuel filter is located inside the fuel inlet at the carburetor.

18.6 Two wrenches are required to loosen the carburetor fuel inlet fitting - use a flare-nut wrench on the fuel line fitting to prevent rounding-off the nut

It is made of pleated paper and cannot be cleaned or reused.

2   This job should be done with the engine cold (after sitting at least three hours). The necessary tools include open-end wrenches to fit the fuel line nuts. Flare nut wrenches (which wrap around the nut) should be used, if available. In addition, you will have to obtain the replacement filter (make sure it is for your specific vehicle and engine) and some clean rags.

3   Remove the air cleaner assembly. If vacuum hoses must be disconnected, be sure to note their positions and/or tag them to ensure that they are reinstalled correctly **(see illustration)**.

4   Locate the fuel line where it enters the carburetor.

5   Place some rags under the fuel inlet fittings to catch spilled fuel as the fittings are disconnected.

6   Hold the large nut on the carburetor body with the proper size open-end wrench and loosen the fuel line fitting with a flare nut wrench **(see illustration)**.

7   After the fuel line is disconnected, move it aside for better access to the inlet filter nut. Do not crimp the fuel line.

8   Now unscrew the fuel inlet filter nut which was previously held steady. As this fitting is drawn away from the carburetor body, be careful not to lose the thin washer-type gasket or the spring located behind the fuel filter. Also, pay close attention to how the filter is installed.

9   Compare the old filter with the new one to make sure they are the same length and design.

10   Reinstall the spring in the carburetor body.

11   Place the new filter in position behind the spring. It will have a rubber gasket and check valve at one end, which should point away from the carburetor.

12   Install a new washer-type gasket on the fuel inlet nut (a gasket is usually supplied with the new filter) and tighten the nut in the carburetor. Make sure it is not cross-threaded. Tighten it securely (if a torque wrench is available, tighten the nut to 18 ft-lbs). Do not overtighten it, as the hole can strip easily, causing fuel leaks.

13   Hold the fuel inlet nut securely with a wrench while the fuel line is connected. Again, be careful not to cross-thread the connector. Tighten the fitting securely.

14   Plug the vacuum hose which leads to the air cleaner snorkel motor so the engine can be run.

15   Start the engine and check carefully for leaks. If the fuel line connector leaks, disconnect it using the above procedures and check for stripped or damaged threads. If the fuel line connector has stripped threads, remove the entire line and have a repair shop install a new fitting. If the threads look all right, purchase some thread sealing tape and wrap the connector threads with it. Now reinstall and tighten it securely. Inlet repair kits are available at most auto parts stores to overcome leaking at the fuel inlet filter nut.

16   Reinstall the air cleaner assembly, connecting the hoses in their original positions.

Fuel-injected models

Refer to illustration 18.17

Caution: *Relieve the fuel pressure as described in Chapter 4 before performing this procedure.*

17   Fuel-injected engines employ a stainless steel in-line fuel filter. On four-cylinder engines it is located at the left rear of the engine, clamped to the cylinder head. On early V8 fuel-injected models, the filter is bracketed in the position normally occupied by the fuel pump on carbureted models, at the lower right front of the engine. On V6 and later model V8 engines it's located forward of the fuel tank, on the rear crossmember **(see illustration)**.

18   With the engine cold, place a container under the fuel filter.

19   Using wrenches of the proper size, disconnect the fuel lines from each side of the filter. If equipped, inspect the small O-rings on each fuel line and replace them if necessary.

20   Remove the bolts attaching the fuel filter bracket and remove the filter.

21   Install the new filter by reversing the removal procedure. Tighten the fittings securely. turn the ignition key On and check the fittings for leaks.

19   Throttle linkage check

Refer to illustration 19.1

1   The throttle linkage is a cable type and, although there are no adjustments for the linkage itself, periodic maintenance is necessary to assure its proper function **(see illustration)**.

2   Remove the air cleaner (refer to Chapter 2) so the entire linkage is visible.

3   Check the entire length of the cable to make sure that it is not binding.

4   Check all nylon bushings for wear, replacing them with new ones as necessary.

5   Lubricate the cable mechanisms with engine oil at the pivot points.

18.17 On fuel injected models, the fuel filter is located underneath the vehicle, near the fuel tank

19.1 Check the accelerator control cable for binding or damage (carbureted V8 engine shown)

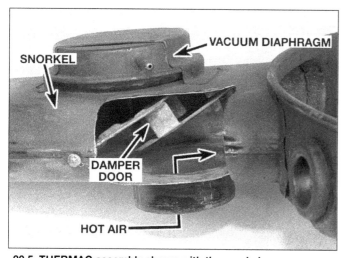

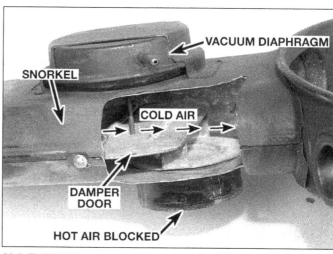

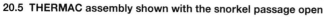

20.5 THERMAC assembly shown with the snorkel passage open   **20.6 THERMAC assembly shown with the snorkel passage closed**

20 Thermo-controlled Air Cleaner (THERMAC) check

Refer to illustrations 20.5 and 20.6

1 Carbureted/TBI engines are equipped with a thermostatically controlled air cleaner which draws air to the carburetor or throttle body from different locations, depending upon engine temperature.

2 This is a simple visual check; however, if access is limited, a small mirror may have to be used.

3 Open the hood and locate the baffle inside the air cleaner assembly. It will be located inside the long snorkel of the metal air cleaner housing. Make sure that the flexible air hose(s) are securely attached and undamaged.

4 If there is a flexible air duct attached to the end of the snorkel, leading to an area behind the grille, disconnect it at the snorkel. This will enable you to look through the end of the snorkel and see the baffle inside.

5 The check should be done when the engine and outside air are cold. Start the engine and look through the snorkel at the baffle, which should move to a closed posi-tion. With the baffle closed, air cannot enter through the end of the snorkel, but instead enters the air cleaner through the flexible duct attached to the exhaust manifold **(see illustration)**.

6 As the engine warms up to operating temperature, the baffle should open to allow air through the snorkel end **(see illustration)**. Depending on ambient temperature, this may take 10 to 15 minutes. To speed up this check to see if you can reconnect the snorkel air duct, drive the vehicle and then check to see if the baffle is completely open.

7 If the thermo-controlled air cleaner is not operating properly, see Chapter 6 for more information.

21 Carburetor/throttle body injection (TBI) mounting torque check

Refer to illustration 21.4

1 The carburetor/TBI is attached to the top of the intake manifold by four nuts or bolts. These fasteners can sometimes work loose from vibration and temperature changes during normal engine operation and cause a vacuum leak.

2 To properly tighten the mounting nuts/bolts, a torque wrench is necessary. If you do not own one, they can usually be rented on a daily basis.

3 Remove the air cleaner assembly, tag-ging each hose to be disconnected with a piece of numbered tape to make reassembly easier.

4 Locate the mounting nuts/bolts at the base of the carburetor/TBI. Decide what spe-cial tools or adapters will be necessary, if any, to tighten the fasteners with a socket and the torque wrench **(see illustration)**.

5 Tighten the nuts/bolts to the specified torque. Do not overtighten them, as the threads could strip.

6 If you suspect that a vacuum leak exists at the bottom of the carburetor/TBI, obtain a length of hose about the diameter of fuel hose. Start the engine and place one end of the hose next to your ear as you probe around the base with the other end. You will hear a hissing sound if a leak exists.

7 If, after the nuts/bolts are properly tight-ened, a vacuum leak still exists, the carbure-tor/TBI must be removed and a new gasket installed. See Chapter 4 for more information.

8 After tightening the fasteners, reinstall the air cleaner and return all hoses to their original positions.

22 Early Fuel Evaporation (EFE) system check

Refer to illustration 22.2

1 The EFE system is designed to recircu-late exhaust gases to help preheat the engine's induction system, thereby improving cold engine driveability and, by reducing the time that the choke is closed, reducing exhaust emissions levels. There are two basic types of EFE systems. The first is a valve in the exhaust system, between the exhaust manifold and the exhaust pipe, and is called a vacuum servo type; the second type is an electrically heated unit that is located between the carburetor base and the intake manifold. The procedures which follow in this Section are concerned only with the first type. If you have the electrically heated type, refer to Chapter 6 for more information.

2 Locate the EFE actuator, which is bolted to a bracket on the right side of the engine. It will have an actuating rod attached to it which will lead down to the valve inside the exhaust pipe. In some cases the entire mech-anism, including actuator, will be located at the exhaust pipe **(see illustration)**.

3 With the engine cold, have an assistant start it. Observe the movement of the actua-tor rod. It should immediately be drawn into the diaphragm, closing the valve.

4 If the valve does not close, it could be seized. Shut off the engine and apply pene-trating oil to the shaft at the pivot points in the valve assembly, allow it to work, then

21.4 A torque check of the carburetor or fuel injection unit mounting bolts will help eliminate vacuum leaks in the induction system

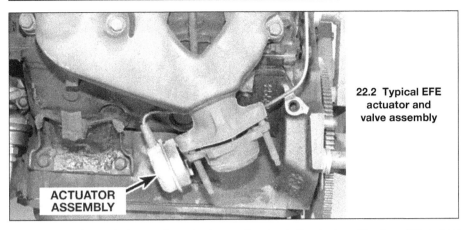

22.2 Typical EFE actuator and valve assembly

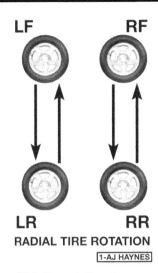

RADIAL TIRE ROTATION

1-AJ HAYNES

23.4 Tire rotation diagram

restart the engine and observe the actuator rod. If it still does not close, disconnect the vacuum hose at the actuator.

5   Hold your thumb over the disconnected hose. If vacuum is felt, test the actuator and valve as follows.

6   Attach a piece of hose to the actuator vacuum outlet, apply vacuum to the hose, then quickly place your thumb over it. If the actuator holds vacuum, it is operating correctly and the valve is defective. Refer to Chapter 6 and replace the valve assembly.

7   If the actuator does not hold vacuum, the actuator diaphragm is defective. Refer to Chapter 6 and replace the actuator.

8   If no vacuum was felt at the disconnected vacuum source hose in Step 5, either the hose is crimped or plugged, or the thermal vacuum switch (TVS) or EFE solenoid is not functioning properly.

9   Check the hose for cracks and restrictions and replace it as necessary. Refer to Chapter 6 to test the TVS. Because of its interrelationship with the ECM, the solenoid must be tested by a dealer service department.

10   With the hose again connected to the EFE actuator, continue to warm up the engine until it reaches normal operating temperature.

11   Make sure that the actuating rod has moved the valve to the open position.

12   If the valve does not open, remove the hose from the actuator. If the valve opens, there is no air bleed for the valve actuator diaphragm, the electrical solenoid plunger or TVS is stuck in the cold mode or the engine is not reaching proper operating temperature. Refer to Chapter 6 to test the TVS or replace the actuator diaphragm. If the engine is not reaching proper operating temperature, check the thermostat (refer to Chapter 3). If the valve does not open with the hose removed from the actuator, it may be stuck closed by corrosion.

23   Tire rotation

Refer to illustration 23.4

1   The tires should be rotated at the specified intervals and whenever uneven wear is noticed. Since the vehicle will be raised and the tires removed anyway, this is a good time

to check the brakes (Section 13) and/or repack the front wheel bearings (Section 27). Read over these Sections if this is to be done at the same time.

2   Refer to the information in *Jacking and towing* at the front of this manual for the proper procedures to follow when raising the vehicle and changing a tire; however, if the brakes are to be checked, do not apply the parking brake as stated. Make sure the tires are blocked to prevent the vehicle from rolling.

3   Preferably, the entire vehicle should be raised at the same time. This can be done on a hoist or by jacking up each corner and then lowering the vehicle onto jackstands placed under the frame rails. Always use four jackstands and make sure the vehicle is firmly supported.

All except models with P245/50VR16 tires

4   Refer to the **accompanying illustration** for the preferred tire rotation pattern. Do not include 'Temporary Use Only' spare tires in the rotation sequence.

5   After rotation, check and adjust the tire pressures as necessary and be sure to check the lug nut tightness.

Models with P245/50VR16 tires

6   If the car is equipped with P245/50VR16 tires, they must be dismounted from the wheels prior to rotation (due to the fact that the front and rear wheels have different offsets and the tires are directional). Directional tires have arrows on both sidewalls that point in the direction the tire must rotate for correct performance. Tire rotation should be done as follows:

7   Dismount the tires from the wheels.

8   Rotate the tires, not the wheels, following an X pattern **(see illustration)**. Make sure the tires will rotate in the direction indicated by the arrows as they're remounted on the wheels.

9   Rebalance the tires.

10   Replace the wheels in their original positions.

11   Recheck the tire pressures and lug nut tightness.

24   Clutch pedal freeplay check and adjustment

Refer to illustration 24.1

Note: *The information which follows applies to 1982 and 1983 vehicles only (non-hydraulic assisted clutch system). A hydraulic clutch operating system is used on all 1984 and later vehicles. The hydraulic system locates the clutch pedal and provides automatic clutch adjustment, so no adjustment of the clutch linkage or pedal position is required.*

1   If your vehicle is equipped with a manual transmission, it is important to have the clutch freeplay properly adjusted. Basically, freeplay is the distance the clutch pedal moves before all play in the linkage is removed and the clutch begins to disengage. It is measured at the pedal pad. Slowly depress the pedal and determine how far it moves before resistance is felt **(see illustration)**.

2   There is only one linkage adjustment to compensate for all normal clutch wear.

3   To check for correct adjustment, apply the parking brake, block the front wheels and start the engine. Hold the clutch pedal approximately 1/2-inch from the floor and move the shift lever between First and

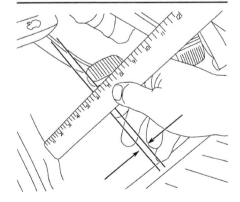

24.1 Clutch pedal freeplay is the distance the pedal moves before resistance is felt

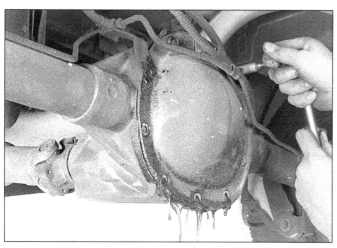

25.3 Remove the lower bolts from the cover, then loosen the upper bolts and allow the oil to drain

26.6 The radiator drain plug (arrow) is located at the bottom of the radiator tank

Reverse gears several times. If the shift is not smooth, clutch adjustment is necessary.

4   Raise the vehicle to permit access to the clutch linkage underneath it. Place the vehicle securely on jackstands.

5   Disconnect the return spring at the clutch fork.

6   Rotate the clutch lever and shaft assembly until the clutch pedal is firmly against the rubber bumper on the dashboard brace. If in doubt about this, have an assistant check it from inside the vehicle.

7   Push the outer end of the clutch fork to the rear until the release bearing lightly contacts the pressure plate fingers.

8   Keep the fork in this position while the pushrod is removed from the operating hole in the lever assembly. Now place the rod in the gauge hole of the lever assembly.

9   With the pushrod in the gauge hole, loosen the locking nut and turn the rod to increase its length. Increase the length of the rod until all lash is removed from the system.

10   Remove the rod from the gauge hole and return it to the lower operating hole in the shaft. Install the retainer and then carefully tighten the nut, being sure the length of the rod is not changed.

11   Connect the return spring and check the pedal freeplay as described previously. Lower the vehicle, then check for correct operation and freeplay. **Note:** *A clutch that has been slipping prior to freeplay adjustment may still slip right after the adjustment has been made due to previous heat damage. If so, allow the clutch to cool for at least 12 hours, then proceed to check the clutch as follows:*

12   While driving the vehicle in High gear at 20 to 25 mph, depress the clutch pedal to the floor, then engage the clutch quickly and momentarily accelerate to full throttle.

13   Engine speed should drop noticeably, then increase as vehicle speed increases. If the clutch is not working properly, the engine speed will increase. **Caution:** *Do not repeat Step 12 more than once or the clutch may overheat.*

14   If the need for further clutch service is indicated, refer to Chapter 8.

25   Rear axle oil change

Refer to illustration 25.3

1   To change the oil in the rear axle it is necessary to remove the cover plate on the differential housing. Because of this, purchase a new gasket at the same time the gear lubricant is bought.

2   Move a drain pan, rags, newspapers and wrenches under the rear of the vehicle. With the drain pan under the differential cover, loosen each of the inspection plate bolts.

3   Remove the bolts on the lower half of the plate, but use the upper bolts to keep the cover loosely attached to the differential **(see illustration)**. Allow the oil to drain into the pan, then completely remove the cover.

4   Using a lint-free rag, clean the inside of the cover and the accessible areas of the differential housing. As this is done, check for chipped gears and metal particles in the oil, indicating that the differential should be thoroughly inspected and repaired (see Chapter 8 for more information).

5   Thoroughly clean the gasket mating surface on the cover and the differential housing. Use a gasket scraper or putty knife to remove all traces of the old gasket.

6   Apply a thin film of gasket sealant to the cover flange and then press a new gasket into position on the cover. Make sure the bolt holes align properly.

7   Place the cover an the differential housing and install the bolts. Tighten the bolts a little at a time, working across the cover in a diagonal fashion until all bolts are tight. If a torque wrench is available, tighten the bolts to the specified torque.

8   Remove the inspection plug on the side of the differential housing (or inspection cover) and fill the housing with the proper lubricant until the level is at the bottom of the plug hole.

9   Securely install the plug.

26   Cooling system servicing (draining, flushing and refilling)

Refer to illustration 26.6

1   Periodically, the cooling system should be drained, flushed and refilled to replenish the antifreeze mixture and prevent formation of rust and corrosion, which can impair the performance of the cooling system and ultimately cause engine damage.

2   At the same time the cooling system is serviced, all hoses and the radiator cap should be inspected and replaced if defective (see Section 8).

3   Since antifreeze is a corrosive and poisonous solution, be careful not to spill any of the coolant mixture on the vehicle's paint or your skin. If this happens, rinse immediately with plenty of clean water. Also, consult your local authorities about the disposal of antifreeze before draining the cooling system. In many areas, reclamation centers have been set up to collect automobile oil and drained antifreeze/water mixtures.

4   With the engine cold, remove the radiator cap.

5   Move a large container under the radiator to catch the coolant as it is drained.

6   Drain the radiator. Most models are equipped with a drain plug at the bottom **(see illustration)**. If this drain has excessive corrosion and cannot be turned easily, or the radiator is not equipped with a drain, disconnect the lower radiator hose to allow the coolant to drain. Be careful that none of the solution is splashed on your skin or into your eyes.

7   If accessible, remove the engine drain plug(s). Four-cylinder models are equipped with one drain plug on the side of the engine block, just below the core plugs. On V6 and V8 models, there are two drain plugs, one plug on each side of the engine about halfway back, on the lower edge near the oil pan rail. These will allow the coolant to drain from the engine itself.

27.6 Remove the wheel bearing dust cap

27.7a Straighten the ends of the cotter pin . . .

27.7b . . . and remove the cotter pin

8   Disconnect the hose from the coolant reservoir and remove the reservoir. Flush it out with clean water.

9   Place a garden hose in the radiator filler neck at the top of the radiator and flush the system until the water runs clear at all drain points.

10   In severe cases of contamination or clogging of the radiator, remove it (see Chapter 3) and reverse flush it. This involves simply inserting the hose in the bottom radiator outlet to allow the clear water to run against the normal flow, draining through the top. A radiator repair shop should be consulted if further cleaning or repair is necessary.

11   When the coolant is regularly drained and the system refilled with 'he correct antifreeze/water mixture, there should be no need to use chemical cleaners or descalers.

12   To refill the system, reconnect the radiator hoses and install the drain plugs securely in the engine. Special thread-sealing tape (available at auto parts stores) should be used on the drain plugs. Install the reservoir and the overflow hose where applicable.

13   Fill the radiator to the base of the filler neck and then add more coolant to the reservoir until it reaches the mark.

14   Run the engine until normal operating temperature is reached and, with the engine idling, add coolant up to the Full Hot level. Install the radiator cap so that the arrows are

in alignment with the overflow hose. Install the reservoir cap.

15   Always refill the system with a mixture of high quality antifreeze and water in the proportion called for on the antifreeze container or in your owner's manual. Chapter 3 also contains information on antifreeze mixtures.

16   Keep a close watch on the coolant level and the various cooling system hoses during the first few miles of driving. Tighten the hose clamps and/or add more coolant as necessary.

27   Front wheel bearing check, repack and adjustment

Refer to illustrations 27.6, 27.7a, 27.7b, 27.8a, 27.8b, 27.9, 27.10, 27.11, 27.15, 27.16, 27.19, 27.20, 27.21a, 27.21b, 27.26 and 27.27

1   In most cases, the front wheel bearings will not need servicing until the brake pads are changed. However, these bearings should be checked whenever the front wheels are raised for any reason.

2   With the vehicle securely supported on jackstands, spin the wheel and check for noise, rolling resistance and freeplay. Now grab the top of the tire with one hand and the bottom of the tire with the other. Move the tire in-and-out on the spindle. If it moves

more than 0.005-inch, the bearings should be checked and then repacked with grease or replaced if necessary.

3   To remove the bearings for replacement or repacking, begin by removing the hub cap and wheel.

4   Remove the caliper (see Chapter 9).

5   Hang the caliper assembly out of the way with a piece of wire. Be careful not to kink or damage the brake hose.

6   Pry the dust cap out of the hub using a screwdriver or hammer and chisel **(see illustration)**. The cap is located at the center of the hub.

7   Use needle-nose pliers or a screwdriver to straighten the bent ends of the cotter pin and then pull the cotter pin out of the locking nut **(see illustrations)**. Discard the cotter pin and use a new one during reassembly.

8   Remove the spindle nut and washer from the end of the spindle **(see illustrations)**.

9   Pull the hub assembly out slightly and then push it back into its original position. This should force the outer bearing off the spindle enough so that it can be removed with your fingers **(see illustration)**. Remove the outer bearing, noting how it is installed on the end of the spindle.

10   Now the hub assembly can be pulled off the spindle **(see illustration)**.

11   On the rear side of the hub, use a screwdriver to pry out the inner bearing lip

27.8a Remove the spindle nut . . .

27.8b . . . the spindle washer . . .

27.9 . . . and the outer wheel bearing

27.10 Remove the hub assembly
from the spindle

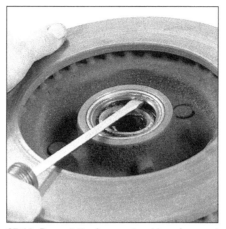

27.11 Pry out the inner wheel bearing seal

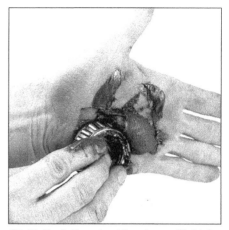

27.15 Pack the wheel bearings with high-
temperature wheel bearing grease
designed especially for disc brakes

27.16 Apply a thin coat of grease to the
spindle surface

27.19 Installing a new seal over the inner
bearing using a soft-faced hammer (tap
around the entire circumference)

machine shop with the facilities to remove the old races and press new ones in.

15   Use an approved high-temperature front wheel bearing grease to pack the bearings. Work the grease completely into the bearings, forcing it between the rollers, cone and cage **(see illustration)**.

16   Apply a thin coat of grease to the spindle at the outer bearing seat, inner bearing seat, shoulder and seal seat **(see illustration)**.

17   Put a small quantity of grease inboard of each bearing race inside the hub. Using your finger, form a dam at these points to provide extra grease availability and to keep thinned grease from flowing out of the bearing.

18   Place the grease-packed inner bearing into the rear of the hub and put a little more grease outboard of the bearing.

19   Place a new seal over the inner bearing and tap the seal with a hammer until it is flush with the hub **(see illustration)**.

20   Carefully place the hub assembly onto the spindle and push the grease-packed outer bearing into position **(see illustration)**.

21   Install the washer and spindle nut. Tighten the nut only slightly (12 ft-lbs of torque) **(see illustrations)**.

seal **(see illustration)**. As this is done, note the direction in which the seal is installed.

12   The inner bearing can now be removed from the hub, again noting how it is installed.

13   Use solvent to remove all traces of the old grease from the bearings, hub and spindle. A small brush may prove useful; however, make sure no bristles from the brush

embed themselves inside the bearing rollers. Allow the parts to air dry.

14   Carefully inspect the bearings for cracks, heat discoloration, bent rollers, etc. Check the bearing races inside the hub for cracks, scoring and uneven surfaces. If the bearing races are defective, the hubs should be taken to a

27.20 Install the outer bearing . . .

27.21a . . . and the spindle washer, aligning the tab with the
groove in the spindle

27.21b Tighten the spindle nut only slightly (no more than 12 ft-lbs)

27.26 Install the cotter pin after aligning up the hole in the spindle with a groove in the spindle nut

27.27 Bend the ends of the cotter pin to secure it

22   Spin the hub in a forward direction to seat the bearings and remove any grease or burrs which could cause excessive bearing play later.

23   Put a little grease outboard of the outer bearing to provide extra grease availability.

24   Now check to see that the tightness of the spindle nut is still 12 ft-lbs.

25   Loosen the spindle nut until it is just loose, no more.

26   Using your hand (not a wrench of any kind), tighten the nut until it is snug. Install a new cotter pin through the hole in the spindle and spindle nut **(see illustration)**. If the nut slits do not line up, loosen the nut slightly until they do. From the hand-tight position, the nut should not be loosened more than one-half flat to install the cotter pin.

27   Bend the ends of the new cotter pin until they are flat against the nut **(see illustration)**. Cut off any extra length which could interfere with the dust cap.

28   Install the dust cap, tapping it into place with a rubber mallet.

29   Install the brake caliper (see Chapter 9).

30   Install the wheel tighten the lug nuts securely.

31   Grab the top and bottom of the tire and check the bearings in the same manner as described at the beginning of this Section.

32   Lower the vehicle to the ground and tighten the lug nuts to the torque listed in this Chapter's Specifications. Install the hub cap, using a rubber mallet to seat it.

28   Automatic transmission fluid change

Refer to illustrations 28.7, 28.8, 28.14, 28.16a, 28.16b, 28.17, 28.20a and 28.20b

1   At the specified time intervals, the transmission fluid should be changed and the filter replaced with a new one. Since there is no drain plug, the transmission oil pan must be removed from the bottom of the transmission to drain the fluid.

2   Before draining, purchase the specified

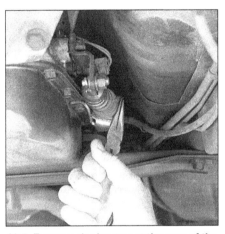

28.7 Remove the keeper at the rear of the selector cable assembly

transmission fluid (see *Recommended lubricants and fluids*) and a new filter. The necessary gaskets should be included with the filter; if not, purchase an oil pan gasket and a strainer-to-valve body gasket.

3   Other tools necessary for this job include jackstands to support the vehicle in a raised position, a wrench to remove the oil pan bolts, a standard screwdriver, a drain pan capable of holding at least eight pints, newspapers and clean rags.

4   The fluid should be drained immediately after the vehicle has been driven. This will remove any built-up sediment better than if the fluid were cold. Because of this, it may be wise to wear protective gloves (fluid temperature can exceed 350-degrees in a hot transmission).

5   After it has been driven to warm up the fluid, raise the vehicle and place it on jackstands for access underneath. Make sure it is firmly supported by the four stands placed under the frame rails.

6   Move the necessary equipment under the vehicle, being careful not to touch any of the hot exhaust components.

7   Remove the keeper at the rear of the

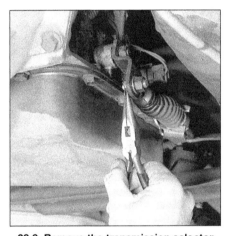

28.8 Remove the transmission selector cable-to-transmission selector lever clip

transmission selector cable assembly **(see illustration)**.

8   Remove the clip that retains the transmission selector cable to the transmission selector lever **(see illustration)**.

9   Remove the cable from the selector lever.

10   Remove the cable housing from the transmission selector cable bracket.

11   Remove the selector cable bracket. To avoid confusion when reinstalling the transmission pan, note that the bolts retaining the cable bracket are a different size than those retaining the pan.

12   Place the drain pan under the transmission oil pan and remove the oil pan bolts along the rear and sides of the pan. Loosen, but do not remove, the bolts at the front of the pan.

13   Carefully pry the pan down at the rear, allowing the hot fluid to drain into the container. If necessary, use a screwdriver to break the gasket seal at the rear of the pan; however, do not damage the pan or transmission in the process.

14   Support the pan and remove the remaining bolts at the front. Lower the pan

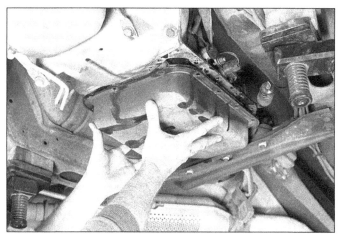

28.14 Carefully support the pan as it is removed, as there is still some fluid in the pan

28.16a On the 200C transmission, the filter/strainer is held in place with two screws

28.16b On the 700-R4 transmission, rotate the filter slightly and pull is straight out of the transmission case

28.17 Some metal particles in the bottom of the pan is normal (if excessive metal is found, you should consult an expert concerning its source and the remedy)

and drain the remaining fluid into the container **(see illustration)**. As this is done, check the fluid for metal particles, which may be an indication of internal failure.

15 Now visible on the bottom of the transmission is the filter/strainer. On models equipped with the 200C transmission, the filter is held in place by two screws. On models equipped with the 700-R4 transmission the filter pick-up tube is pressed into the transmission case.

16 Remove the two screws (if equipped), the filter and the gasket **(see illustrations)**.

17 Thoroughly clean the transmission oil pan with solvent. Inspect It for metal particles and foreign matter **(see illustration)**.

18 Dry the oil pan with compressed air if available. It is important that all remaining gasket material be removed from the oil pan mounting flange. Use a gasket scraper or putty knife for this.

19 Clean the filter mounting surface on the valve body. Again, this surface should be smooth and free of any leftover gasket material.

20 Place the new filter into position, with a new gasket between it and the transmission

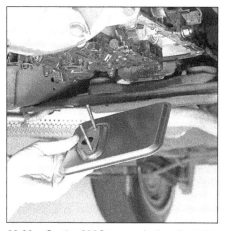

28.20a On the 200C transmission, install a new gasket (shown mounted over the bolts) when attaching the filter to the valve body

valve body or a new O-ring on the pick-up tube **(see illustrations)**. Install the two mounting screws (if equipped) and tighten them securely.

28.20b On the 700-R4 transmission, an O-ring is used on the pick-up tube - if the O-ring did not come out with the pick-up tube, reach up into the bore with your finger and retrieve it

21 Apply a light bead of gasket sealant around the oil pan mounting surface, with the sealant to the inside of the bolt holes. Press

29.2 Remove the wing nut atop the air cleaner assembly

29.4 Remove the air filter element

ones. A thorough program of preventative maintenance would call for the two filters to be inspected between changes.

Carbureted and TBI models

2   The air filter is located inside the air cleaner housing on the top of the engine. The filter is generally replaced by removing the wing nut at the top of the air cleaner assembly and lifting off the top plate (see illustration). If vacuum hoses are connected to the plate, note their positions and disconnect them.

3   While the top plate is off, be careful not to drop anything down into the carburetor.

4   Lift the air filter element out of the housing (see illustration).

5   To check the filter, hold it up to strong sunlight or place a flashlight or droplight on the inside of the filter. If you can see light coming through the paper element, the filter is all right. Check all the way around the filter.

6   Wipe out the inside of the air cleaner housing with a clean rag.

7   Place the old filter (if in good condition) or the new filter (if the specified interval has elapsed) back into the air cleaner housing. Make sure it seats properly in the bottom of the housing.

8   Connect any disconnected vacuum hoses to the top plate and reinstall the plate.

9   The PCV filter is also located inside the air cleaner housing. Remove the top plate and air filter as described previously, then locate the PCV filter on the side of the housing.

10   Remove the retaining clip from the outside of the housing, then remove the PCV filter (see illustrations).

11   Install a new PCV filter, then reinstall the retaining clip, air filter, top plate and any hoses that were disconnected.

TPI and MPFI models

12   The air filter is located in the air cleaner housing, which is mounted to the right of the radiator. When replacing the air filter, be sure to clean the inside of the housing. Also inspect the ducts between the air filter hous-

---

the new gasket into place on the pan, making sure all bolt holes line up.

22   Lift the pan up to the bottom of the transmission and install the mounting bolts. Tighten the bolts in a diagonal fashion, working around the pan. Using a torque wrench, tighten the bolts to the specified torque.

23   Lower the vehicle off the jackstands.

24   Open the hood and remove the transmission fluid dipstick from the guide tube.

25   Since fluid capacities vary between the various transmission types, it is best to add a little fluid at a time, continually checking the level with the dipstick. Allow the fluid time to drain into the pan. Add fluid until the level just registers on the end of the dipstick. In most cases, a good starting point will be four to five pints added to the transmission through the filler tube (use a funnel to prevent spills).

26   With the selector lever in Park, apply the parking brake and start the engine without depressing the accelerator pedal (if possible). Do not race the engine at a high speed; run it at slow idle only.

27   Depress the brake pedal and shift the transmission through each gear. Place the selector back into Park and check the level on the dipstick (with the engine still idling).

Look under the vehicle for leaks around the transmission oil pan mating surface.

28   Add more fluid through the dipstick tube until the level on the dipstick is 1/4-inch below the Add mark on the dipstick. Do not allow the fluid level to go above this point, as the transmission would then be overfull necessitating the removal of the pan to drain the excess fluid.

29   Push the dipstick firmly back into the tube and drive the vehicle to reach normal operating temperature (15 miles of highway driving or its equivalent in the city). Park on a level surface and check the fluid level on the dipstick with the engine idling and the transmission in Park. The level should now be at the Full mark on the dipstick. If not, add more fluid as necessary to bring the level up to this point. Again, do not overfill.

---

29   Air filter and PCV filter replacement

Refer to illustrations 29.2, 29.4, 29.10a and 29.10b

1   At the specified intervals, the air filter and PCV filter should be replaced with new

---

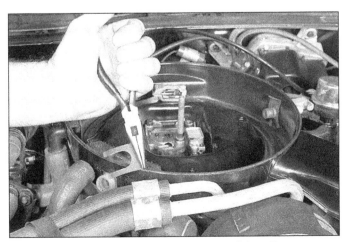

29.10a Remove the PCV filter retaining clip

29.10b Remove the PCV filter

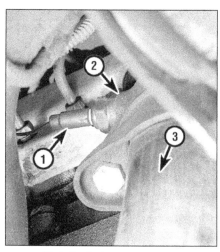

30.1 Typical oxygen sensor installation details

1 Oxygen sensor
2 Exhaust manifold
3 Exhaust pipe

ing and the mass airflow sensor and between the mass airflow sensor and the throttle body for holes and cracks. Replace the ducts if necessary.

30 Oxygen sensor replacement

Refer to illustration 30.1
1   The oxygen sensor is located in the exhaust manifold or exhaust pipe and is accessible from under the hood **(see illustration)**. **Note:** *Special care must be taken when handling the sensitive oxygen sensor.*

 a) *The oxygen sensor has a permanently attached pigtail and connector, which should not be removed from the sensor. Damage or removal of the pigtail or connector can adversely affect its operation.*

 b) *Grease, dirt and other contaminants should be kept away from the electrical connector and the louvered end of the sensor:*
 c) *Do not use cleaning solvents of any kind on the oxygen sensor.*
 d) *Do not drop or roughly handle the sensor.*
 e) *The silicone boot must be installed in the correct position to prevent the boot from being melted and to allow the sensor to operate properly.*

2   Since the oxygen sensor may be difficult to remove with the engine cold, begin by operating the engine until it has warmed to at least 120 degrees F.
3   Disconnect the electrical connector from the oxygen sensor.
4   Note the position of the silicone boot and carefully back out the oxygen sensor from the exhaust manifold. Be advised that excessive force may damage the threads.
5   A special anti-seize compound must be used on the threads of the oxygen sensor to aid in future removal. New or service sensors will have this compound already applied, but if for any reason an oxygen sensor is removed and then reinstalled, the threads must be coated before reinstallation.
6   Install the sensor and tighten it to 30 ft-lbs.
7   Connect the electrical connector.

31 Positive Crankcase Ventilation (PCV) valve replacement

Refer to illustration 31.1
1   The PCV valve is located in the rocker arm cover or intake manifold **(see illustration)**. A hose connected to the valve runs to either the carburetor or intake manifold.
2   When purchasing a replacement PCV valve, make sure it is for your particular vehicle, model year and engine size.

3   Pull the valve (with hose attached) from the rubber grommet in the rocker arm cover or manifold.
4   Loosen the retaining clamp and pull the PCV valve from the end of the hose, noting its installed position and direction.
5   Compare the old valve with the new one to make sure they are the same.
6   Push the new valve into the end of the hose until it is seated and reinstall the clamp.
7   Inspect the rubber grommet for damage and replace it with a new one, if faulty.
8   Push the PCV valve and hose securely into position.
9   More information on the PCV system can be found in Chapter 6.

32 Exhaust Gas Recirculation (EGR) valve check

1   On GM vehicles, the EGR valve is located on the intake manifold, adjacent to the carburetor or TBI unit. Most of the time when a problem develops in this emissions system, it is due to a stuck or corroded EGR valve.
2   With the engine cold to prevent burns, reach under the EGR valve and manually push on the diaphragm. Using moderate pressure, you should be able to press the diaphragm up and down within the housing.
3   If the diaphragm does not move or moves only with much effort, replace the EGR valve with a new one. If in doubt about the quality of the valve, compare the free movement of your EGR valve with a new valve.
4   Refer to Chapter 6 for more information on the EGR system.

33 Evaporative Emissions Control System (EECS) filter replacement

Refer to illustrations 33.3, 33.4 and 33.6
1   The function of the Evaporative Emis-

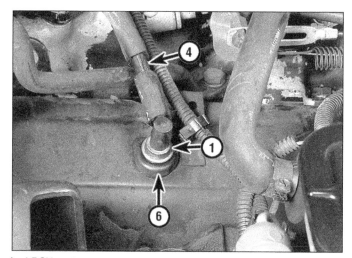

31.1 Details of a typical PCV system

1 PCV valve
2 Air cleaner
3 Crankcase vent tube
4 PCV valve hose
5 Hose to canister
6 Grommet

33.3 The EECS canister is located in the front of the engine compartment, just behind the headlight assembly

33.4 Carefully remove the hoses from the EECS canister

33.6 With the canister retaining bolts removed and the canister inverted, the filter is easily removed

sions Control System is to draw fuel vapors from the tank and carburetor, store them in a charcoal canister and then burn them during normal engine operation.

2   The filter at the bottom of the charcoal canister should be replaced at the specified intervals. If, however, a fuel odor is detected, the canister, filter and system hoses should immediately be inspected.

3   To replace the filter, locate the canister at the front of the engine compartment. It will have between three and six hoses running out of the top **(see illustration)**.

4   Mark the hoses with tape to simplify reinstallation, then disconnect them from the canister **(see illustration)**.

5   Remove the two bolts which secure the bottom of the canister to the body.

6   Turn the canister upside-down and pull the old filter from the bottom of the canister **(see illustration)**.

7   Push the new filter into the bottom of the canister, making sure it is seated all the way around.

8   Place the canister back into position and tighten the two mounting bolts. Connect the various hoses if disconnected.

9   The EECS is explained in more detail in Chapter 6.

34   Ignition timing check and adjustment

Refer to illustrations 34.1, 34.2, 34.3, 34.4 and 34.5

Note: *It is imperative that the procedures included on the Vehicle Emissions Control Information label be followed when adjusting the ignition timing. The label will include all information concerning preliminary steps to be performed before adjusting the timing, as well as the timing specifications.*

1   Locate the VECI label under the hood and read through and perform all preliminary instructions concerning ignition timing. **Note:** *If instructed by the VECI label, place the Electronic Spark Timing (EST) in bypass mode by disconnecting the single wire connector. It is a tan wire with a black stripe that comes out of the wiring harness conduit near the fear of the right-hand rocker arm cover **(see illustration)**. Do not disconnect the four-wire connector to the distributor.*

2   Locate the timing mark pointer plate located beside the crankshaft pulley **(see illustration)**. The 0 mark represents top dead center (TCD). The pointer plat will be marked in either one or two-degree increments and should have the proper timing mark for your

34.1 If instructed by the VECI label, disconnect the EST bypass connector (arrow)

particular engine noted. If not, count back from the 0 mark the correct number of degrees BTDC, as noted on the VECI label, and mark the plate.

3   Locate the notch on the crankshaft balancer or pulley and mark it with chalk or a dab or paint so it will be visible under the timing light **(see illustration)**.

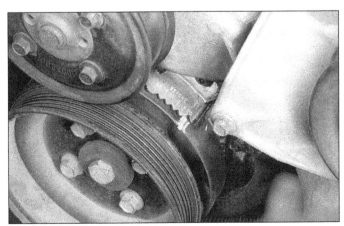

34.2 Location of a typical timing mark indicator

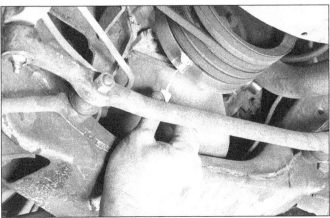

34.3 Mark the crankshaft balancer notch with white paint

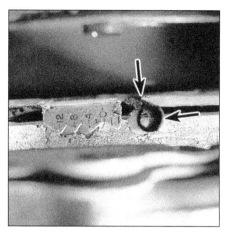

34.4 Location of the magnetic timing probe hole incorporated on some engines

34.5 Aim the timing light at the timing mark

35.9 Remove the spark plugs with a socket, extension and universal-joint, if necessary

4   With the ignition off, connect the pick-up lead of the timing light to the number one spark plug. Use either a jumper lead between the wire and plug or an inductive-type pick-up. *Do not pierce the wire or attempt to insert a wire between the boot and the wire.* Connect the timing light power leads according to the manufacturer's instructions. **Note:** *Some engines incorporate a magnetic timing probe hole for use with special electronic timing equipment* **(see illustration)**. *Consult the manufacturer's instructions for proper use of this equipment.*

5   Start the engine, aim the timing light at the timing mark by the crankshaft pulley and note which timing mark the notch on the pulley is aligning with **(see illustration)**.

6   If the notch is not lining up with the correct mark, loosen the distributor hold-down bolt and rotate the distributor until the notch is lined up with the correct timing mark.

7   Retighten the hold-down bolt and recheck the timing.

8   Turn off the engine and disconnect the timing light. Reconnect the number one spark plug wire, if removed.

35 Spark plug replacement

Refer to illustrations 35.9 and 35.11

1   The spark plugs are located on each side of V6 and V8 engines and on the left side of the four-cylinder engine. They may or may not be easily accessible for removal. If the vehicle is equipped with air-conditioning or power steering, some of the plugs may be tricky to remove. Special extension or swivel tools may be necessary. Make a survey under the hood to determine if special tools will be needed.

2   In most cases, the tools necessary for a spark plug replacement job include a plug wrench or spark plug socket which fits onto a ratchet wrench (this special socket will be padded inside to protect the plug) and a feeler gauge to check and adjust the spark plug gaps. Also, a special spark plug wire removal tool is available for separating the

wires from the spark plugs. To ease installation, obtain a piece of rubber hose, 8 to 12 inches in length, that fits snugly over the porcelain insulator of the spark plug.

3   The best procedure to follow when replacing the spark plugs is to purchase the new spark plugs beforehand, adjust them to the proper gap and then replace each plug one at a time. When buying the new spark plugs, it is important to obtain the correct plugs for your specific engine. This information can be found in the Specifications at the front of this Chapter, but should be checked against the information found on the Emissions Control Information label located under the hood or in the owner's manual. If differences exist between these sources, purchase the spark plug type specified on the Emissions Control label because the information was printed for your specific engine.

4   With the new spark plugs at hand, allow the engine to cool completely before attempting plug removal. During this time, each of the new spark plugs can be inspected for defects and the gaps can be checked.

5   The gap is checked by inserting the proper thickness gauge between the electrodes at the tip of the plug. The gap between the electrodes should be the same as that given in the Specifications or on the Emissions Control label. The wire should just touch each of the electrodes. If the gap is incorrect, use the notched adjuster on the feeler gauge body to bend the curved side electrode slightly until the proper gap is achieved. If the side electrode is not exactly over the center electrode, use the notched adjuster to align the two. Also at this time check for cracks in the porcelain insulator, indicating the spark plug should not be used.

6   Cover the fenders of the vehicle to prevent damage to the paint.

7   With the engine cool, remove the spark plug wire from one spark plug. Do this by grabbing the boot at the end of the wire, not the wire itself. Sometimes it is necessary to use a twisting motion while the boot and plug wire are pulled free. Using a plug wire

removal tool is the easiest and safest method.

8   If compressed air is available, use it to blow any dirt or foreign material away from the spark plug area. A common bicycle pump will also work. The idea here is to eliminate the possibility of material falling into the cylinder as the spark plug is removed.

9   Now place the spark plug wrench or socket over the plug and remove it from the engine by turning in a counterclockwise direction **(see illustration)**.

10   Compare the spark plug with those shown on the inside back cover of this manual to get an indication of the overall running condition of the engine.

11   Due to the angle at which the spark plugs must be installed on most engines, installation will be simplified by inserting the plug wire terminal of the new spark plug into the rubber hose, mentioned previously, before it is installed in the cylinder head **(see illustration)**. This procedure serves two purposes: the rubber hose gives you flexibility for establishing the proper angle of plug insertion in the head and, should the threads be improperly lined up, the rubber hose will merely slip on the spark plug terminal when it meets resistance, preventing damage to the

35.11 A length of rubber hose will save time and prevent damaged threads when installing the spark plugs

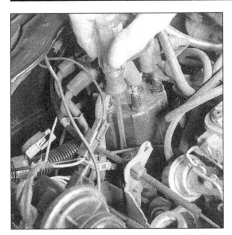

36.3 Release the distributor cap hold-down latches

36.4 Check for deposit build-up on each distributor cap terminal

36.6 Remove the rotor

cylinder head threads.

12   After installing the plug to the limit of the hose grip, tighten it with the socket. It is a good idea to use a torque wrench for this to ensure that the plug is seated correctly. The correct torque figure is included in the Specifications.

13   Before pushing the spark plug wire onto the end of the plug, inspect it following the procedures outlined in Section 36.

14   Attach the plug wire to the new spark plug, again using a twisting motion on the boot until it is firmly seated on the spark plug. Make sure the wire is routed away from the exhaust manifold.

15   Follow the above procedure for the remaining spark plugs, replacing them one at a time to prevent mixing up the spark plug wires.

---

36 Spark plug wires, distributor cap and rotor check and replacement

Refer to illustrations 36.3, 36.4, 36.6, 36.7a, 36.7b, 36.17a and 36.17b

1   Begin this procedure by making a visual check of the spark plug wires while the engine is running. In a darkened garage (make sure there is ventilation) start the engine and observe each plug wire. Be careful not to come into contact with any moving engine parts. If there is a break in the wire, you will see arcing or a small spark at the damaged area. If arcing is noticed, make a note to obtain new wires, then allow the engine to cool and check the distributor cap and rotor.

2   Disconnect the negative cable from the battery. **Caution:** *On models equipped with a Delco Loc II audio system, disable the anti-theft feature before disconnecting the battery.* At the distributor, disconnect the ECM connector and the coil connector (coil-in-cap models) or coil wire (models with separately mounted coil).

3   Remove the distributor cap by placing a screwdriver on the slotted head of each latch. Press down on the latch and turn it 90-degrees to release the hooked end at the

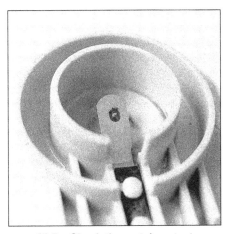

36.7a Check the metal contact on the rotor

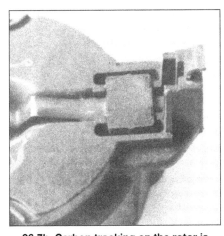

36.7b Carbon tracking on the rotor is caused by a leaking seal between the distributor cap and the ignition coil (always replace both the seal and the rotor when this condition is present)

bottom **(see illustration)**. On some engines, due to restricted working room, a stubby screwdriver will work best. With all latches disengaged, separate the cap from the distributor with the spark plug wires still attached.

4   Inspect the cap for cracks and other damage. Closely examine the contacts on the inside of the cap for excessive corrosion **(see illustration)**. Slight scoring is normal. Deposits on the contacts may be removed with a small file.

5   If the inspection reveals damage to the cap, make a note to obtain a replacement for your particular engine, then examine the rotor.

6   The rotor is visible, with the cap removed, at the top of the distributor shaft. It is held in place by two screws. Remove the screws and the rotor **(see illustration)**.

7   Inspect the rotor for cracks and other damage. Carefully check the condition of the metal contact at the top of the rotor for excessive burning and pitting **(see illustration)**. On coil-in-cap models, also check the top of the rotor for carbon tracks **(see illustration)**. This is a sign of moisture contamination due to a leaking seal between the distributor cap and the coil. The rotor and seal

should be replaced with new ones if carbon tracks are visible.

8   If it is determined that a new rotor is required, make a note to that effect. If the rotor and cap are in good condition, reinstall them at this time. Be sure to apply a small dab of silicone lubricant to the contact inside the cap before installing it. Note that the rotor has two raised pegs on the bottom and that it has a wide slot and a, narrow slot. Make sure that the slots are correctly aligned and that the pegs are firmly seated when the rotor is installed.

9   If the cap must be replaced, do not reinstall it. Leave it off the distributor with the wires still connected.

10   If the spark plug wires are being replaced, now is the time to obtain a new set, along with a new cap and rotor as determined in the checks above. Purchase a wire set for your particular engine, pre-cut to the proper size, with the rubber boots already installed.

11   If the spark plug wires passed the check in Step 1, they should be checked further as follows.

36.17a Release the spark plug wire retaining ring clips from the top of the distributor cap . . .

36.17b . . . and remove the plug wire retainer from the cap

12 Examine the wires one at a time to avoid mixing them up.

13 Disconnect the plug wire from the spark plug. A removal tool can be used for this, or you can grab the rubber boot, twist slightly and then pull the wire free. Do not pull on the wire itself, only on the rubber boot.

14 Inspect inside the boot for corrosion, which will look like a white crusty powder. Some models use a conductive white silicone lubricant which should not be mistaken for corrosion. 1

15 Now push the wire and boot back onto the end of the spark plug. It should be a tight fit on the plug end. If not, remove the wire and use pliers to carefully crimp the metal connector inside the wire boot until the fit is snug.

16 Now, using a clean rag, clean the entire length of the wire. Remove all built-up dirt and grease. As this is done, check for burns, cracks and any other form of damage. Bend the wires in several places to ensure that the conductive wire inside has not hardened.

17 Next, the wires should be checked at the distributor cap in the same manner. On four-cylinder engines, remove the wire from the cap by pulling on the boot, again examining the wires one at a time, and reinstalling each one after examination. Apply new silicone lubricant before reinstallation. On V6 and V8 engines, the distributor boots are connected to a circular retaining ring attached to the distributor cap. Release the locking tabs, turn the ring upside down and check all wire boots at the same time **(see illustrations)**.

18 If the wires appear to be in good condition, reinstall the retaining ring (V6 and V8 engines) and make sure that all wires are secure at both ends. If the cap and rotor are also in good condition, the check is finished. Reconnect the wires at the distributor and at the battery.

19 If it was determined in Steps 12 through 17 that new wires are required, obtain them at this time, along with a new cap and rotor if so determined in the checks above.

20 If a new cap is being installed on a coil-in-cap type distributor, the coil and cover from the cap being replaced should be trans-

ferred to the new cap.

21 Remove the three coil cover attaching screws and lift off the cover.

22 Remove the coil attaching screws, disconnect the leads and separate the coil from the distributor.

23 Attach the coil to the new cap by reversing Steps 21 and 22. Use a new seal between the coil and cap and be sure to lubricate the seal with multi-purpose grease.

24 Attach the rotor to the distributor. Make sure that the carbon brush is properly installed in the cap, as a wide gap between the carbon brush and the rotor will cause rotor burn-through and/or damage to the distributor cap.

25 If new wires are being installed, replace them one at a time. **Note:** *It is important to replace the wires one at a time, noting the routing as each wire is removed and installed, to maintain the correct firing order and to prevent short-circuiting.*

26 Attach the cap to the distributor, reconnecting all wires disconnected in Step 2, then reconnect the battery cable.

Chapter 2 Part A
L4 (four-cylinder) engine

Contents

	Section
Air filter replacement	See Chapter 1
Camshaft - removal and installation	14
Check engine light	See Chapter 6
Compression check	See Chapter 2D
Crankshaft pulley hub and oil seal - removal and installation	12
Cylinder head - removal and installation	7
Drivebelt check and adjustment	See Chapter 1
Engine - removal and installation	17
Engine mounts - replacement	16
Engine oil and filter change	See Chapter 1
Engine oil level check	See Chapter 1
Engine overhaul - general information	See Chapter 2D
Engine removal - methods and precautions	See Chapter 2D
Exhaust manifold - removal and installation	6
Flywheel/driveplate and rear main bearing oil seal - removal and installation	15

	Section
General information	1
Hydraulic lifters - removal, inspection and installation	8
Intake manifold - removal and installation	5
Oil pan - removal and installation	10
Oil pump - removal and installation	11
Oil pump driveshaft - removal and installation	9
Pushrod cover - removal and installation	3
Repair operations possible with the engine in the vehicle	See Chapter 2D
Rocker arm cover - removal and installation	2
Spark plug replacement	See Chapter 1
Timing gear cover - removal and installation	13
Valve train components - replacement	4
Water pump - removal and installation	See Chapter 3

Specifications

Torque specifications

	Ft-lbs (unless otherwise indicated)
Camshaft thrust plate-to-block bolts	84 in-lbs
Crankshaft pulley hub bolt	160
Cylinder head bolts	
1984 and earlier	85
1985 and later	92
Distributor hold-down bolt	22
EFI assembly-to-manifold nuts/bolts	15
EGR valve-to-manifold bolts	120 in-lbs
Engine mount nuts	41
Exhaust manifold-to-cylinder head bolts	37
Fan and pulley-to-water pump bolts	18
Flywheel/driveplate-to-crankshaft bolts	
1984 and earlier	44
1985 and later	55
Intake manifold-to-cylinder head bolts	
All except bolt 7	25
Bolt 7	37
Oil pan bolts	72 in-lbs
Oil pan drain plug	25
Oil pump driveshaft retainer plate bolts	120 in-lbs

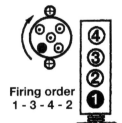

**Firing order
1 - 3 - 4 - 2**

**Cylinder numbering and direction
distributor rotor rotates (arrow)**

*The blackened terminal shown on the
distributor cap indicates the Number One
spark plug wire position*

Torque specifications

Ft-lbs (unless otherwise indicated)

Oil pump-to-block bolts	22
Oil pump cover bolts	10
Oil screen support nut	37
Pushrod cover-to-block bolts...	84 in-lbs
Rocker arm bolts	20
Rocker arm cover bolts	72 in-lbs
Roller lifter retainer studs..	84 in-lbs
Thermostat housing bolts...	20
Timing cover-to-block bolts	84 in-lbs
Water outlet housing bolts..	20
Water pump bracket-to-block bolts	25

Note: *Refer to Chapter 2, Part D, for additional specifications*

1   General information

The forward Sections in this Part of Chapter 2 are devoted to "in-vehicle' repair procedures for the L4 engine. The latter Sections in this Part of Chapter 2 involve the removal and installation procedures for the L4 engine. All information concerning engine block and cylinder head servicing can be found in Part D of this Chapter.

The repair procedures included in this Part are based on the assumption that the engine is still installed in the vehicle. Therefore, if this information is being used during a complete engine overhaul - with the engine already out of the vehicle and on a stand - many of the steps included here will not apply.

The Specifications included in this Part of Chapter 2 apply only to the engine and procedures found here. For Specifications regarding engines other than the L4, see Part B or C, whichever applies. Part D of Chapter 2 contains the Specifications necessary for engine block and cylinder head rebuilding.

2   Rocker arm cover - removal and installation

Refer to illustrations 2.6 and 2.9

1   Remove the air cleaner assembly, tag-ging each hose to be disconnected with a piece of numbered tape to simplify installation.

2   Disconnect the throttle cable from the throttle body assembly, making careful note of the exact locations of the cable components and hardware to ensure correct reinstallation.

3   Remove the PCV valve from the rocker arm cover.

4   Remove the spark plug wires from the spark plugs (refer to the removal technique described in Chapter 1), then remove the wires and retaining clips from the rocker arm cover. Be sure to label each wire before removal to ensure that all wires are reinstalled correctly.

5   Loosen the throttle body mounting nuts and bolts to provide clearance for removal of the EGR valve, then remove the EGR valve.

6   Remove the rocker arm cover bolts (see illustration).

7   Remove the rocker arm cover. **Note:** *If the cover sticks to the cylinder head, use a block of wood and a hammer to dislodge it. If the cover still will not come loose, pry on it carefully, but do not distort the sealing flange surface.*

8   Prior to installation of the cover, clean all dirt, oil and old gasket material from the sealing surfaces of the cover and cylinder head with a scraper and degreaser.

9   Apply a continuous 3/16-inch diameter bead of RTV-type sealant to the sealing flange of the cover. Be sure to apply the sealant inboard of the bolt holes (see illustration).

10   Place the rocker arm cover on the cylinder head while the sealant is still wet and install the mounting bolts. Tighten the bolts a little at a time to the specified torque.

11   Complete the installation by reversing the removal procedure.

3   Pushrod cover - removal and installation

Refer to illustration 3.5

1   Remove the intake manifold as described in Section 5.

2   Remove the distributor (refer to Chapter 5).

3   Remove the pushrod cover bolts and lift off the cover. If the gasket seal is difficult to break, tap on the cover gently with a rubber hammer. Do not pry on the cover.

4   Using a scraper and degreaser, clean the sealing surfaces on the cover and engine block to remove all oil and old gasket material.

5   Prior to installation of the cover, apply a continuous 3/16inch bead of RTV-type sealant to the mounting flange of the pushrod cover (see illustration).

6   With the sealant still wet, place the

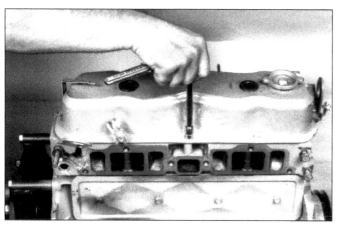

2.6 Remove the rocker arm cover bolts

2.9 Apply a continuous 3/16-inch bead of RTV sealant to the rocker arm cover flange (make sure it is applied to the inside of the bolt holes as shown)

3.5 Apply a bead of RTV sealant to the pushrod cover mounting flange

4.7 A special tool can be used to compress the valve springs and remove the spring keepers

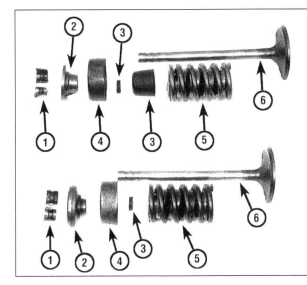

4.8 Valve, valve spring and related components (typical)

1 Keepers
2 Retainer
3 Oil seal
4 Spring shield
5 Spring
6 Valve

4.10 Compress the valve spring and install the O-ring seal squarely in the lower groove before installing the keepers

cover in position on the block and install the cover bolts. Tighten the bolts gradually, from the center out to the ends, to the specified torque.

7   Install the distributor and intake manifold.

4   Valve train components - replacement

Refer to illustrations 4.7, 4.8 and 4.10

1   Remove the rocker arm cover as described in Section 2.

2   If only the pushrod is to be replaced, loosen the rocker bolt enough to allow the rocker arm to be rotated away from the pushrod. Pull the pushrod out of the hole in the cylinder head. If the rocker arm is to be removed, remove the rocker bolt and pivot and lift off the rocker arm.

3   If the valve spring is to be removed, remove the spark plug from the affected cylinder.

4   There are two methods that will allow the valve to remain in place while the valve spring is removed. If you have access to compressed air, install an air hose adapter in the spark plug hole. When air pressure is applied to the adapter, the valves will be held in place by the pressure.

5   If you do not have access to compressed air, bring the piston of the affected cylinder to top dead center (TDC) on the compression stroke. Feed a long piece of 1/4inch nylon cord in through the spark plug hole until it fills the combustion chamber. Be sure to leave the end of the cord hanging out of the spark plug hole so it can be removed easily. Rotate the crankshaft with a wrench (in the normal direction of rotation) until slight resistance is felt.

6   Install the rocker arm bolt (without the rocker arm).

7   Insert the slotted end of a valve spring compressing tool under the bolt head and compress the spring just enough to remove the spring keepers, then release the pressure on the tool **(see illustration)**.

8   Remove the retainer, cup shield, O-ring seal, spring, spring damper and valve stem oil seal **(see illustration)**.

9   Inspect the valve train components for wear or damage. Check for a bent pushrod. Replace any defective parts, as necessary.

10   Installation of the valve train components is the reverse of the removal procedure. Install a new valve stem oil seal after compressing the spring and before installing the spring keepers **(see illustration)**. Prior to installing the rocker arms, coat the bearing surfaces of the arms and rocker arm pivots with moly-based grease or engine assembly lube. The L4 engine valve mechanisms require no special valve lash adjustment. Simply tighten the rocker arm bolts to the specified torque.

5   Intake manifold - removal and installation

Refer to illustration 5.16

1   Disconnect the cable from the negative battery terminal. **Caution:** *On models equipped with a Delco Loc II audio system, disable the anti-theft feature before disconnecting the battery.*

2   Remove the air cleaner assembly, tag-

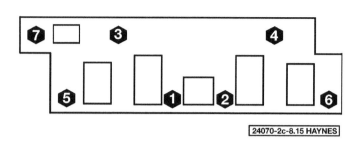

5.16 Intake manifold bolt TIGHTENING sequence

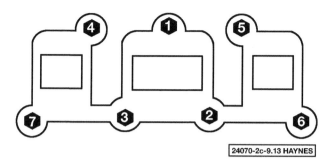

6.13 Exhaust manifold bolt TIGHTENING sequence

ging each hose to be disconnected with a piece of numbered tape to simplify reinstallation.

3   Remove the PCV valve and hose.

4   Drain the cooling system (refer to Chapter 1).

5   Label and disconnect the fuel line, vacuum lines and electrical leads from the throttle body assembly. When disconnecting the fuel line, be prepared to catch some fuel, then plug the fuel line to prevent contamination.

6   Disconnect the throttle linkage, making careful note of how it is installed.

7   Disconnect the cruise control linkage (if equipped).

8   Disconnect the coil wire, remove the two retaining screws and one bolt, then remove the coil.

9   Remove the coolant hoses from the manifold.

10   Remove the manifold retaining bolts and separate the manifold from the cylinder head. Do not pry between the manifold and head, as damage to the gasket sealing surfaces may result.

11   Remove the manifold gasket.

12   If the intake manifold is to be replaced with another, transfer all components still attached to the old manifold to the new one.

13   Before installing the manifold, clean the cylinder head and manifold gasket surfaces. All gasket material and sealing compound must be removed prior to installation.

14   Apply a thin bead of RTV-type sealant to the intake manifold and cylinder head mating surfaces. Make certain that the sealant will not spread into the air or coolant passages when the manifold is installed.

15   Place a new intake manifold gasket on the manifold, place the manifold in position against the cylinder head and install the mounting bolts finger tight.

16   Tighten the mounting bolts a little at a time in the sequence shown, until they are all at the specified torque **(see illustration)**.

17   Install the remaining components in the reverse order of removal.

18   Fill the radiator with coolant, start the engine and check for leaks.

Adjust the ignition timing and idle speed as necessary (Chapter 1).

6   Exhaust manifold - removal and installation

Refer to illustration 6.13

1   If the vehicle is equipped with air-conditioning, carefully examine the routing of the hoses and the mounting of the compressor. You may be able to remove the exhaust manifold without disconnecting the system. If you are in doubt, take the vehicle to a GM dealer or automotive repair shop to have the system depressurized. **Caution:** *Do not, under any circumstances, disconnect any air-conditioning system lines while the system is under pressure.*

2   Remove the cable from the negative battery terminal. **Caution:** *On models equipped with a Delco Loc II audio system, disable the anti-theft feature before disconnecting the battery.*

3   Remove the air cleaner assembly, tagging each hose to be disconnected with a piece of numbered tape to simplify reinstallation.

4   Remove the preheat tube.

5   Remove the engine oil dipstick and dipstick tube.

6   Remove the oxygen sensor (refer to Chapter 1).

7   Label the four spark plug wires, then disconnect them and secure them to the side out of the way.

8   Disconnect the exhaust pipe from the exhaust manifold. You may have to apply penetrating oil to the fastener threads as they are usually corroded. The exhaust pipe can be hung from the frame with a piece of wire.

9   Remove the exhaust manifold end bolts first, then remove the center bolts and separate the exhaust manifold from the engine.

10   Remove the exhaust manifold gasket.

11   Before installing the manifold, clean the gasket mating surfaces on the cylinder head and manifold. All leftover gasket material and carbon deposits must be removed.

12   Place a new exhaust manifold gasket into position on the cylinder head, then place the manifold in position and install the mounting bolts finger tight.

13   Tighten the mounting bolts a little at a time in the sequence shown, until all of the

bolts are at the specified torque **(see illustration)**.

14   Install the remaining components in the reverse order of removal, using new gaskets wherever one has been removed.

15   Start the engine and check for exhaust leaks between the manifold and cylinder head and between the manifold and exhaust pipe.

7   Cylinder head - removal and installation

Refer to illustrations 7.10, 7.11, 7.13, 7.22 and 7.23

1   Remove the intake manifold as described in Section 5.

2   Remove the exhaust manifold as described in Section 6.

3   Remove the bolts that secure the alternator bracket to the cylinder head.

4   If so equipped, unbolt the air conditioner compressor and swing it out of the way for clearance. **Caution:** *Do not disconnect any of the air-conditioning lines unless the system has been depressurized by a dealer or repair shop, because personal injury may occur. Disconnection of the lines should not be necessary in this case.*

5   Disconnect all electrical and vacuum lines from the cylinder head. Be sure to label the lines to simplify reinstallation.

6   Remove the upper radiator hose.

7   Disconnect the spark plug wires and remove the spark plugs. Be sure to label the plug wires to simplify reinstallation.

8   Remove the rocker arm cover. To break the gasket seal it may be necessary to strike the cover with your hand or a rubber hammer. Do not pry between the sealing surfaces. Refer to Section 2 if necessary.

9   When disassembling the valve mechanisms, keep all of the components separate so they can be reinstalled in their original positions. A cardboard box or rack, numbered to correspond to the engine cylinders, can be used for this purpose.

10   Remove each of the rocker arm nuts and separate the rocker arms and pivots from the cylinder head **(see illustration)**.

11   Remove the pushrods **(see illustration)**.

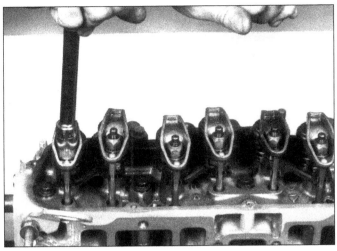

7.10 Remove the rocker arm nuts

7.11 Remove the pushrods and organize them so they can be reinstalled in their original locations

12 If the ignition coil is mounted separately from the distributor, disconnect the wires and remove the coil.

13 Loosen each of the cylinder head mounting bolts one turn at a time, in the reverse of the tightening sequence, until they can be removed **(see illustration)**. Note the length and position of each bolt to ensure correct reinstallation.

14 Lift the head off of the engine. If it is stuck to the engine block, do not attempt to pry it free, as you could damage the sealing surfaces. Instead, use a hammer and block of wood to tap the head and break the gasket seal. Place the head on a block of wood to prevent damage to the gasket surface.

15 Remove the cylinder head gasket.

16 Refer to Chapter 2D for cylinder head disassembly and valve service procedures.

17 If a new cylinder head is being installed, transfer all external parts from the old cylinder head to the new one.

18 If not already done, thoroughly clean the gasket surfaces on the cylinder head and the engine block. Do not gouge or otherwise damage the gasket surfaces.

19 To get the proper torque readings, the threads of the head bolts must be clean. This also applies to the threaded holes in the engine block. Run a tap through the holes to ensure that they are clean.

20 Place the gasket in position over the engine block dowel pins.

21 Carefully lower the cylinder head onto the engine, over the dowel pins and the gasket.

22 Coat the threads of the Number 9 and 10 cylinder head bolts and the point at which the head and the bolt meet with a sealing compound and install the bolts finger tight **(see illustration)**.

23 Tighten each of the bolts a little at a time in the sequence shown **(see illustration)**. Continue tightening in this sequence until the proper torque reading is obtained. As a final check, work around the head in a front-to-rear sequence to make sure none of the bolts have been left out of the sequence.

24 The remaining installation steps are the reverse of removal.

8 Hydraulic lifters - removal, inspection and installation

Refer to illustrations 8.7, 8.8, 8.10a, 8.10b and 8.10c

1 A noisy valve lifter can be isolated when

7.13 Loosen all the cylinder head mounting bolts a little at a time in the reverse of the tightening sequence

the engine is idling. Place a length of hose or tubing near the position of each valve while listening at the other end of the tube.

2 The most likely cause of a noisy valve lifter is excessive wear at the lifter foot and camshaft lobe or wear or contamination of the internal components.

3 Remove the rocker arm cover as described in Section 2.

4 Remove the intake manifold as described

7.22 The Number 9 and 10 cylinder head mounting bolts should be coated with sealant at the points indicated (arrows) before installation

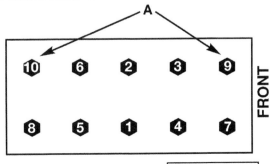

7.23 Cylinder head bolt TIGHTENING sequence - apply sealing compound to the threads and undersides of the heads of bolts 9 and 10

8.7 Roller lifter installation details

1   Pushrod cover studs
2   Lifter guides
3   Lifter guide retainer

8.8 If the lifters are difficult to remove, remove them with a special hydraulic lifter removal tool

8.9 If the lifters will be reused, store them in an organized manner so they can be returned to their original locations

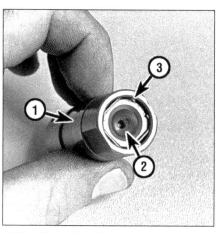

8.10a Check the lifter body, retainer ring and pushrod seat for wear

1   Lifter body
2   Pushrod seat
3   Retainer ring

8.10b If the bottom of any lifter is worn concave, scratched or galled, replace the entire set of lifters and the camshaft

8.10c If equipped with roller lifters, check the roller and bearings for wear

in Section 5.

5   Remove the pushrod cover as described in Section 3.

6   Loosen the rocker arm bolt and rotate the rocker arm away from the pushrod. Remove the pushrod. If more than one pushrod is removed, store them in an organized manner so they can be reinstalled in their original location.

7   On later models, with roller lifters, remove the pushrod cover studs and remove the lifter guide retainer and lifter guide **(see illustration)**.

8   If a lifter is difficult to remove, use a special hydraulic lifter removal tool to withdraw the lifter **(see illustration)**. Do not use pliers or other tools on the outside of the lifter body, because they will damage the finished surface and render the lifter useless.

9   If the lifters are to be reused, they must be kept organized for reinstallation in their original positions **(see illustration)**.

10   Clean and dry the lifters. Examine them for wear and check carefully for flat spots **(see illustrations)**.

11   If the lifters are worn, they must be

replaced with new ones and the camshaft must be replaced as well (see Section 14). If the lifters are in good condition, they can be reinstalled - provided they are installed in their original locations.

12   When installing the lifters, make sure they are replaced in their original bores and coat them with moly-based grease or engine assembly lube.

13   The remaining installation steps are the reverse of removal.

9   Oil pump driveshaft - removal and installation

Refer to illustrations 9.3 and 9.7

1   Remove the air cleaner, marking the hoses as they are disconnected to simplify reinstallation.

2   Remove the alternator (refer to Chapter 5).

3   Remove the oil pump driveshaft retainer plate bolts **(see illustration)**.

4   Remove the bushing.

5   Remove the shaft and gear assembly.

6   Thoroughly clean the sealing surfaces on the engine block and retainer plate.

7   Inspect the gear teeth to see if they are

9.3 The oil pump driveshaft retainer plate is located below the pushrod cover, above the oil filter

1   *Retainer plate*   2   *Mounting bolts*

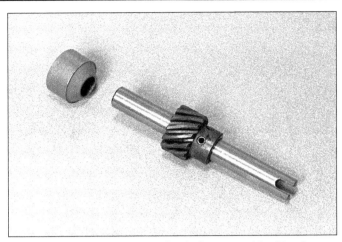

9.7 Inspect the oil pump driveshaft, gear and bushing for wear or damage

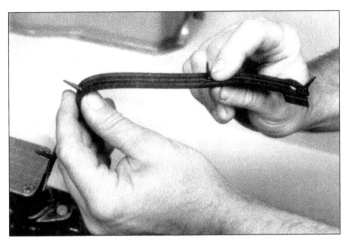

10.9 During installation, the rubber tips on the front oil pan gasket should be pressed into the holes in the timing gear cover

10.10 Secure all four sections of the oil pan gasket in position on the engine block

chipped or cracked (see illustration).

8   Install the oil pump driveshaft in the block and turn it until it engages with the camshaft drive gear and the oil pump. Caution: *Make sure the top of the bushing is flush with the retainer plate mounting surface or the drive shaft is not installed properly.*

9   Apply a 1/16inch diameter bead of RTV sealant to the retainer plate so that it completely seals around the oil pump driveshaft hole in the block.

10   Install the retainer plate mounting bolts and tighten them to the specified torque.

11   Complete the installation by reversing the removal procedure.

10 Oil pan - removal and installation

Refer to illustrations 10.9 and 10.10

1   Disconnect the cable from the negative battery terminal. Caution: *On models equipped with a Delco Loc II audio system, disable the anti-theft feature before discon-*

necting the battery.

2   Raise the vehicle, place it securely on jackstands and drain the engine oil (refer to Chapter 1 if necessary).

3   Disconnect the exhaust pipe at the exhaust manifold and loosen the hanger bracket.

4   Remove the throughbolts from the motor mounts.

5   Using an engine hoist, raise the engine enough so the oil pan can be removed without interference from the steering and suspension components. Block the engine securely in the raised position. Warning: *Do not, under any circumstances, work under the engine when supported only by the hoist.*

6   Remove the oil pan bolts and separate the oil pan from the block.

7   Clean the pan with solvent and remove all old sealant and gasket material from the block and pan sealing surfaces.

8   Install the rear pan oil seal in the groove in the rear main bearing cap and apply a small quantity of RTV sealant to the depressions where the seal meets the block.

9   Install the front pan oil seal on the timing cover, pressing the tips into the holes in the cover (see illustration).

10   Using a light coat of gasket sealant as a retainer, install the pan side gaskets (see illustration). Apply a 1/8inch diameter by 1/4inch long bead of RTV - type sealant at the parting lines of the front seal and side gaskets.

11   Attach the oil pan to the block. The bolts that secure the pan to the timing cover should be installed last.

12   After all bolts are installed, tighten them to the specified torque (use a crisscross pattern and work up to the final torque in three or four steps).

13   The remaining steps are the reverse of the removal procedure.

11 Oil pump - removal and installation

Refer to illustration 11.2

1   Remove the oil pan (refer to Section 10).

11.2 Remove the pick-up tube bracket nut and the oil pump flange mounting bolts

12.3 Remove the crankshaft pulley bolt

12.4 Mark the position of the pulley in relation to the hub

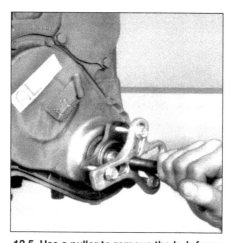

12.5 Use a puller to remove the hub from the crankshaft

12.9 Use the pulley hub bolt to press the hub onto the crankshaft

2   Remove the two oil pump flange mounting bolts and the nut from the main bearing cap bolt **(see illustration)**.

3   Lift out the oil pump and screen as an assembly.

4   To install the pump, align the shaft so it mates with the oil pump driveshaft tang, then install the pump on the block, over the oil pump driveshaft lower bushing. No gasket is used. The oil pump should slide easily into place - if not, remove it and relocate the slot.

5   Install the mounting bolts and nut and tighten them to the specified torque.

6   Reinstall the oil pan (refer to Section 10).

12   Crankshaft pulley hub and front oil seal - removal and installation

Refer to illustrations 12.3, 12.4, 12.5 and 12.9

1   Remove the cable from the negative battery terminal. **Caution:** *On models equipped with a Delco Loc II audio system, disable the anti-theft feature before disconnecting the battery.*

2   Loosen the accessory drivebelt tension

adjusting bolts, as necessary, and remove the drivebelts. Tag each belt as it is removed to simplify reinstallation.

3   With the parking brake applied and the shifter in Park (automatic) or in gear (manual) to prevent the engine from turning over, remove the crankshaft pulley bolt **(see illustration)**. A breaker bar will probably be necessary, since the bolt is very tight.

4   Mark the position of the pulley in relation to the hub **(see illustration)**. Remove the bolts and separate the pulley from the hub.

5   Using a puller, remove the hub from the crankshaft **(see illustration)**.

6   Carefully pry the oil seal out of the front cover with a large screwdriver. Be careful not to distort the cover.

7   Install the new seal with the helical lip toward the rear of the engine. Drive the seal into place using a seal installation tool or a large socket. If there is enough room, a block of wood and hammer can also be used.

8   Apply a thin layer of clean multipurpose grease to the seal contact surface of the hub.

9   Position the pulley hub on the crankshaft and slide it through the seal until it bottoms against the crankshaft gear. Note that the slot

in the hub must be aligned with the Woodruff key in the end of the crankshaft. The hub-to-crankshaft bolt can also be used to press the hub into position **(see illustration)**.

10   Install the crank pulley on the hub, noting the alignment marks made during removal. The pulley-to-hub bolts should be coated with thread-locking compound, whenever they are removed and installed.

11   Tighten the hub-to-crankshaft and pulley-to-hub bolts to the specified torque.

12   The remaining installation steps are the reverse of removal. Tighten the drivebelts to the proper tension (refer to Chapter 1).

13   Timing gear cover - removal and installation

Refer to illustrations 13.3 and 13.9

1   Remove the crankshaft pulley hub as described in Section 12.

2   Remove the oil pan-to-timing gear cover bolts.

3   Remove the cover-to-block bolts **(see illustration)**.

13.3 Remove the timing gear cover mounting bolts

13.9 Use a block of wood to install the oil seal

4   Using a sharp knife, cut the oil pan front gasket flush with the engine block at both sides.
5   Remove the cover and the attached portion of the oil pan gasket.
6   Remove the cover gasket.
7   Using a scraper and degreaser, remove all dirt and old gasket material from the sealing surfaces of the timing gear cover, engine block and oil pan.
8   Replace the front oil seal by carefully prying it out of the timing gear cover with a large screwdriver. Do not distort the cover.
9   Install the new seal with the helical lip toward the inside of the cover. Drive the seal into place using a seal installation tool or a large socket and hammer. A block of wood will also work **(see illustration)**.
10   Prior to installing the cover, install a new front oil pan gasket. Cut the ends off of the gasket and attach it to the cover by pressing the rubber tips into the holes provided.
11   Apply a thin coat of RTV-type gasket sealant to the timing gear cover gasket and place it in position on the cover.
12   Apply a bead of RTV-type sealant to the joint between the oil pan and engine block.
13   Insert the hub through the cover seal and place the cover in position on the block as the hub slides onto the crankshaft.
14   Install the oil pan-to-cover bolts and partially tighten them.
15   Install the bolts that secure the cover to the block, then tighten all of the mounting bolts to the specified torque.
16   Complete the installation by reversing the removal procedure.

---

14 Camshaft - removal and installation

Refer to illustrations 14.16, 14.21 and 14.23
Note: *Before removing the camshaft, refer to Chapter 2, Part D (Section 15), and measure the lobe lift.*

1   If equipped with air conditioning, have the system discharged. **Warning:** *The air*

conditioning system is under high pressure. Do not loosen any hose fittings or remove any components until after the system has been discharged. Air conditioning refrigerant must be properly discharged into an EPA-approved recovery/recycling unit at a dealer service department or an automotive air conditioning repair facility. Always wear eye protection when disconnecting air conditioning system fittings.
2   Remove the cable from the negative battery terminal. **Caution:** *On models equipped with a Delco Loc II audio system, disable the anti-theft feature before disconnecting the battery.*
3   Drain the oil from the crankcase (refer to Chapter 1).
4   Drain the coolant and remove the radiator (refer to Chapter 3).
5   If equipped with air-conditioning, remove the condenser (refer to Chapter 3).
6   Remove the fan and water pump pulley (refer to Chapter 3).
7   Remove the rocker arm cover (refer to Section 2).
8   Loosen the rocker arm bolts and pivot the rocker arms away from the pushrods.
9   Remove the oil pump driveshaft assembly (refer to Section 9).

10   Remove the spark plugs (refer to Chapter 1).
11   Remove the distributor (refer to Chapter 5).
12   Remove the pushrod cover and gasket (refer to Section 3), then remove the pushrods.
13   Remove the valve lifters (refer to Section 8).
14   Remove the crankshaft pulley hub (refer to Section 12).
15   Remove the timing gear cover (refer to Section 13).
16   Working through the holes in the camshaft gear, remove the two camshaft thrust plate screws **(see illustration)**.
17   Supporting the camshaft carefully to prevent damage to the bearings, remove the camshaft and gear assembly by pulling it out through the front of the engine.
18   Refer to Chapter 2, Part D, for the camshaft inspection procedures.
19   If the gear must be removed from the camshaft, it must be pressed off. If you do not have access to a press, take it to your dealer or an automotive machine shop. The thrust plate must be positioned so that the Woodruff key in the shaft does not damage it when the shaft is pressed off.
20   If the gear has been removed, it must be pressed back on prior to installation of the camshaft.

a) *Support the camshaft in an arbor press by placing press plate adapters behind the front journal.*
b) *Place the gear spacer ring and the thrust plate over the end of the shaft.*
c) *Install the Woodruff key in the shaft keyway.*
d) *Position the camshaft gear and press it onto the shaft until it bottoms against the gear spacer ring.*
e) *Use a feeler gauge to check the end clearance of the thrust plate. It should be 0.0015 to 0.0050 inch. If the clearance is less than 0.0015 inch, the spacer ring should be replaced. If the clearance is more than 0.0050 inch the thrust plate should be replaced.*

21   Prior to installing the camshaft, coat

14.16 Remove the camshaft thrust plate screws

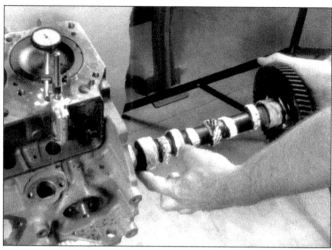

14.21 The camshaft lobes and journals should be lubricated with special camshaft assembly lube prior to installing the camshaft in the block

14.23 The camshaft and crankshaft gears must be positioned so that the timing marks align

15.4 A large screwdriver wedged in the starter ring gear teeth or one of the holes in the driveplate can be used to keep the flywheel/driveplate from turning as the mounting bolts are removed

15.5 Carefully pry the oil seal out with a screwdriver - don't nick or scratch the crankshaft or leaks will develop

each of the lobes and journals with camshaft assembly lube **(see illustration)**.

22 Slide the camshaft into the engine block. Again, be extra careful not to damage the bearings.

23 Position the camshaft and crankshaft gears so that the valve timing marks align **(see illustration)**. With the marks aligned in this position, the engine is positioned with the number four (4) cylinder at TDC in the firing position.

24 Install the camshaft thrust plate mounting screws and tighten them to the specified torque.

25 Complete the installation by reversing the removal procedure. When installing the distributor, rotate the crankshaft 360-degrees to position the number one (1) cylinder at TDC in the firing position (see Chapter 5).

26 Have the air-conditioning system recharged (if so equipped).

15 Flywheel/driveplate and rear main bearing oil seal - removal and installation

Refer to illustrations 15.4, 15.5 and 15.8

1 The rear main oil seal can be replaced without removal of the oil pan or crankshaft.

2 Refer to Chapter 7, follow all precautionary notes and remove the transmission.

3 If equipped with a manual transmission, remove the clutch pressure plate and disc.

4 Remove the flywheel/driveplate mounting bolts and separate it from the crankshaft **(see illustration)**.

5 Using a screwdriver or pry bar, carefully remove the oil seal from the block **(see illustration)**.

6 Thoroughly clean and dry the block-to-seal mating surface.

7 Apply a light coat of engine oil to the

15.8 Carefully seat the outer edge of the seal squarely in the bore with a hammer and blunt tool

outside surface of the new seal.

8 Using a blunt tool, press the new seal evenly into position in the block **(see illustration)**.

9 Install the flywheel/driveplate and tighten the bolts to the specified torque.

10 If equipped with a manual transmission, reinstall the clutch disc and pressure plate and tighten the pressure plate bolts to the specified torque.

11 Reinstall the transmission as described in Chapter 7.

16 Engine mounts - replacement

1 If the rubber mounts have become hard, split or separated from the metal backing, they must be replaced. This operation may be carried out with the engine/transmission still in the vehicle. See Section 10 for the proper way to raise the engine while it is still in place.

Front mounts

2 Remove the throughbolt and nut.

3 Raise the engine slightly using a hoist or jack with a wood block under the oil pan, then remove the mount and frame bracket assembly from the crossmember.

4 Position the new mount, install the throughbolt and nut, then tighten all the bolts to the specified torque.

Rear mount

5 Remove the crossmember-to-mount bolts, then raise the transmission slightly with a jack.

6 Remove the mount-to-transmission bolts, followed by the mount.

7 Install the new mount, lower the transmission and align the crossmember-to-mount bolts.

8 Tighten all the bolts to the specified torque.

17 Engine - removal and installation

Note: *On some models, the upper bell-housing-to-engine bolts are difficult to get at, in some cases necessitating cutting of the interior floorpan to expose the bolts. If this is the case with your vehicle and you do not wish to cut the floor pan, read through the procedure in this Section to familiarize yourself with the details of the four-cylinder engine removal, then refer to the engine removal section in Chapter 2, Part B, and remove your engine and transmission as a unit, noting any differences between the V6 and four-cylinder applications and adjusting the procedure accordingly.*

Removal

1 Disconnect the cable from the negative battery terminal. **Caution:** *On models equipped with a Delco Loc II audio system,*

disable the anti-theft feature before disconnecting the battery.

2 Remove the clip retaining the underhood light assembly wire harness to the hood.

3 Disconnect the underhood light electrical connector and unbolt the assembly from the hood.

4 Scribe very light lines on the underside of the hood, around the hood mounting bracket, so the hood can be installed in the same position.

5 Apply white paint around the bracket-to-hood bolts so they can be installed in exactly the same positions during installation.

6 Remove the bracket-to-hood bolts.

7 With the help of an assistant, separate the hood from the vehicle.

8 Drain the radiator by turning the petcock handle located at the lower right-hand side of the radiator.

9 Remove the hose from the radiator coolant recovery tank.

10 Remove the upper radiator hose. Take care not to damage the plastic fittings when detaching the hose.

11 If so equipped, remove the automatic transmission cooler lines from the radiator.

12 Remove the bolts securing the top of the radiator fan shroud to the lower half and remove the upper shroud.

13 If equipped with an automatic transmission, remove the radiator from the vehicle. This step is not necessary with manual transmission-equipped vehicles, though it may be desirable to gain more working room for pulling the engine.

14 Mark the drivebelts to simplify installation.

15 Unbolt and remove the fan from the fan pulley.

16 If so equipped, remove the air-conditioning compressor and brackets and lay them to one side. **Caution:** *Do not disconnect any of the air-conditioning lines unless the system has been depressurized by a dealer service department or air-conditioning technician, as personal injury may occur. Disconnection of the lines should not be necessary in this case.*

17 Disconnect the engine electrical connector at the firewall.

18 Disconnect the EFI inlet and return fuel lines at the flexible hoses,
then plug the lines to prevent loss of fuel.

19 Disconnect the vacuum brake hose from the vacuum booster.

20 Remove the ground strap from the rear of the cylinder head, then disconnect the wires at the oil pressure and coolant temperature sending units.

21 From inside the vehicle, remove the right-hand hush panel, then disconnect the ECM harness from the main connector of the ECM unit.

22 Remove the right-hand splash shield from the right fender and feed the ECM harness through the opening from inside the vehicle.

23 Remove the heater hoses from the

heater core.

24 Disconnect the hoses at the air cleaner. Carefully label them as they are removed to simplify installation.

25 Remove the air cleaner assembly.

26 Remove the canister hose from the EFI.

27 Remove the throttle cable from the EFI.

28 Raise the vehicle and place it securely on jackstands, then drain the engine oil.

29 Disconnect the electrical connectors at the transmission.

30 Unbolt and remove the flywheel/driveplate dust cover.

31 On vehicles equipped with an automatic transmission, remove the now exposed converter-to-driveplate bolts. It will be necessary to turn the crankshaft to bring each of the bolts into view (use a wrench on the large bolt at the front of the crankshaft). Mark the relative position of the converter and driveplate with a scribe so it can be reinstalled in the same position. Engage a long screwdriver in the teeth of the driveplate to prevent movement as the bolts are loosened.

32 Unbolt and remove the exhaust pipe from the exhaust manifold.

33 Unbolt and remove the exhaust pipe support from the bellhousing.

34 Disconnect the catalytic converter at the tailpipe joint, then remove the converter and tailpipe from the vehicle.

35 Disconnect and label the wires at the starter motor.

36 Unbolt and remove the starter motor.

37 If equipped with a manual transmission, remove the clutch fork return spring.

38 Remove the bell-housing-to-engine bolts (refer to the note at the beginning of this Section).

39 Remove the motor mount throughbolt on each side of the engine.

40 Make sure that all the wires and hoses have been disconnected from under the vehicle.

41 Remove the jackstands and lower the vehicle.

42 Move back into the engine compartment and make one last check to ensure that all wires and hoses have been disconnected from the engine and that all accessories have enough clearance.

43 Attach the hoist lifting chains to the lifting "eyes' mounted on the engine. There is one bracket at the front of the engine and one at the rear. Make sure the chain is looped properly through the engine brackets and secured with strong nuts and bolts through the chain links. The hook on the hoist should be over the center of the engine with the lengths of chain equal, to lift the engine straight up.

44 Raise the engine hoist until all slack is out of the chains. Do not lift any further at this time.

45 Support the transmission using a jack with wood blocks as cushions.

46 Raise the engine slightly and then pull it forward to clear the input shaft (manual transmission). Where an automatic transmission is involved, keep the torque converter pushed

well to the rear to retain engagement of the converter tangs with the oil pump inside the transmission.

47   Carefully lift the engine straight up and out of the engine compartment, continually checking clearances around it.

48   The transmission should remain supported by the floor jack or wood blocks while the engine is out of place.

49   Attach the engine to an engine stand.

50   Refer to Chapter 2, Part D, for further disassembly and rebuilding procedures.

Installation

51   Lift the engine off the engine stand with a hoist. The chains should be positioned just as they were during removal, with the engine sitting level.

52   Lower the engine into place inside the engine compartment, closely watching clearances. On manual transmission-equipped vehicles, carefully guide the engine onto the transmission input shaft. The two components should be at the same angle, with the shaft sliding easily into the engine.

53   Install the engine mount throughbolts and the bellhousing bolts. Tighten them to the specified torque.

54   Install the remaining engine components in the reverse order of removal.

55   Fill the cooling system with the specified coolant and water mixture (Chapter 1).

56   Fill the engine with the correct grade of engine oil (Chapter 1).

57   Check the transmission fluid level, adding fluid as necessary.

58   Connect the positive battery cable, followed by the negative cable. If sparks or arcing occur as the negative cable is connected to the battery, make sure that all electrical accessories are turned off (check the dome light first). If arcing still occurs, make sure that all electrical wiring is properly connected to the engine and transmission.

59   Refer to Chapter 2, Part D, for the recommended engine startup sequence.

Chapter 2 Part B
V6 engine

Contents

	Section
Air filter - replacement	See Chapter 1
Camshaft - removal and installation	16
Check engine light	See Chapter 6
Crankcase front cover - removal and installation	13
Compression check	See Chapter 2D
Cylinder heads - removal and installation	8
Drivebelt check and adjustment	See Chapter 1
Engine - removal and installation	18
Engine mounts - replacement	17
Engine oil and filter change	See Chapter 1
Engine overhaul - general information	See Chapter 2D
Engine removal - methods and precautions	See Chapter 2D
Engine repairs possible with the engine in the vehicle	See Chapter 2D
Exhaust manifolds - removal and installation	7

	Section
Front cover oil seal - replacement	14
General information	1
Hydraulic lifters - removal, inspection and installation	5
Intake manifold - removal and installation	4
Oil pan - removal and installation	9
Oil pump - removal and installation	10
Rear main bearing oil seal - replacement	11
Rocker arm covers - removal and installation	2
Spark plug replacement	See Chapter 1
Timing chain and sprockets - inspection, removal and installation	15
Valve lash - adjustment	6
Valve train components - replacement	3
Vibration damper - removal and installation	12
Water pump - removal and installation	See Chapter 3

Specifications

Torque specifications

Ft-lbs (unless otherwise indicated)

Camshaft rear cover bolts	84 in-lbs
Camshaft sprocket bolts	18
Clutch cover-to-flywheel bolts	15
Connecting rod cap nuts	37
Crankshaft pulley bolts	25
Crankshaft pulley hub bolt	75
Cylinder head bolts	
2.8L	70
3.1L	
Step 1	40
Step 2	Tighten an additional 90-degrees
Distributor holddown bolt	25
Driveplate-to-torque converter bolts	27
EGR valve mounting bolts	15
Engine mounting bracket bolts	80
Engine strut bracket	35
Exhaust manifold mounting bolts	25
Flywheel mounting bolts	50
Front cover mounting bolts (small)	15
Front cover mounting bolts (large)	25

Firing order
1-2-3-4-5-6

**Cylinder numbering and
direction distributor rotor
rotates (arrow)**

*The blackened terminal shown on the
distributor cap indicates the Number
One spark plug wire position*

Torque specifications

	Ft-lbs (unless otherwise indicated)
Fuel pump mounting bolts..	15
Intake manifold-to-cylinder head bolts	
2.8L	23
3.1L	19
Main bearing cap bolts..	70
Oil filter bolt...	15
Oil filter connector bolt	29
Oil pan mounting bolts (small)	84 in-lbs
Oil pan mounting bolts (large)...	18
Oil pump mounting bolt	30
Oil pump cover bolts	96 in-lbs
Oil pressure switch	60 in-lbs
Oil drain plug	18
Rear lifting bracket bolt	25
Rocker arm cover bolts	96 in-lbs
Rocker arm stud-to-cylinder head..	45
Spark plugs..	12
Starter motor mounting bolts	32
Strut bracket assembly nut and bolt	35
Timing chain tensioner bolts..	15
Water outlet housing bolts...	25
Water pump mounting bolts (small)...	84 in-lbs
Water pump mounting bolts (medium)	15
Water pump mounting bolts (large)	25
Water pump pulley bolt	15

Note: *Refer to Chapter 2, Part D for additional specifications.*

---

1   General information

The forward Sections in this Part of Chapter 2 are devoted to "in-vehicle' repair procedures for the V6 engine. The latter Sections in this Part of Chapter 2 involve the removal and installation procedures for the V6 engine. All information concerning engine block and cylinder head servicing can be found in Part D of this Chapter.

The repair procedures included in this Part are based on the assumption that the engine is still installed in the vehicle. Therefore, if this information is being used during a complete engine overhaul - with the engine already out of the vehicle and on a stand - many of the steps included here will not apply.

The Specifications included in this Part of Chapter 2 apply only to the engine and procedures found here. For Specifications regarding engines other than the V6, see Part A or C, whichever applies. Part D of Chapter 2 contains the Specifications necessary for engine block and cylinder head rebuilding procedures.

---

2   Rocker arm covers - removal and installation

Carbureted models

Refer to illustrations 2.3 and 2.6

Right side

1   Disconnect the cable from the negative battery terminal. **Caution:** *On models equipped with a Delco Loc II audio system, disable the anti-theft feature before disconnecting the battery.*

2   Remove the air cleaner assembly, tagging each hose to be disconnected with a piece of numbered tape to simplify reinstallation.

3   Remove the bolts retaining the air management valve/coil bracket **(see illustration)**.

4   Disconnect the wires and hoses that would interfere with the removal of the rocker arm covers, tagging them as they are disconnected.

5   Disconnect the accelerator controls at the carburetor, then remove them from the support bracket.

6   Remove the rocker arm cover bolts **(see illustration)**.

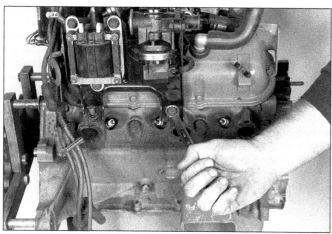

2.3 Remove the AIR management valve/coil bracket

2.6 Remove the rocker arm cover bolts

7 Remove the rocker arm cover. **Note:** *If the cover sticks to the cylinder head, use a block of wood and a rubber hammer to dislodge it. If the cover still will not come loose, pry on it carefully, but do not distort the sealing flange surface.*

8 Before installing the cover, clean all dirt, oil and old gasket material from the sealing surfaces of the cover and cylinder head with a scraper and degreaser.

9 Apply a continuous 3/16inch diameter bead of RTV sealant to the flange of the cover. Be sure to apply the sealant inboard of the bolt holes.

10 Place the rocker arm cover on the cylinder head while the sealant is still wet and install the mounting bolts. Tighten the bolts a little at a time to the specified torque.

11 Complete the installation by reversing the removal procedure.

Left side

12 Disconnect the cable from the negative battery terminal. **Caution:** *On models equipped with a Delco Loc II audio system, disable the anti-theft feature before disconnecting the battery.*

13 Disconnect the hoses at the PCV valve and label them.

14 Remove the air cleaner assembly, tagging each hose to be disconnected with a piece of numbered tape to simplify reinstallation.

15 Disconnect all other wires and hoses that would interfere with the removal of the rocker arm cover, tagging them as they are disconnected.

16 Remove the bolts retaining the solenoid/hose bracket to the front of the cover and set the assembly aside.

17 Remove the rocker arm cover bolts.

18 Disconnect the fuel line at the carburetor, plugging the fitting at the carburetor and the disconnected fuel line to prevent leakage and contamination.

19 Refer to Paragraphs 7 through 11 in this Section.

Fuel injected models

Right side

20 Disconnect the negative battery cable from the battery. **Caution:** *On models equipped with a Delco Loc II audio system, disable the anti-theft feature before disconnecting the battery.*

21 Remove the EGR valve tube from the plenum.

22 Remove the coil and coil mounting bracket from the right cylinder head, marking the connectors with numbered tape to avoid confusion during reinstallation.

23 Remove the plenum and throttle body as outlined in Chapter 4.

24 Remove the rocker arm cover nuts and retainers.

25 Remove the rocker arm cover. If the cover sticks to the head, use a block of wood and a rubber hammer to dislodge it. If the cover still will not come loose, pry on it care-

fully, but do not distort the sealing flange surface.

26 Before installing the cover, clean all dirt, oil and old gasket material from the sealing surfaces of the cover and cylinder head with a scraper and degreaser.

27 Apply a continuous 3/16-inch diameter bead of RTV sealant to the flange of the cover. Be sure to apply the sealant inboard of the bolt holes.

28 Place the rocker arm cover on the cylinder head while the sealant is still wet and install the mounting bolts. Tighten the bolts a little at a time to the specified torque.

29 Complete the installation by reversing the removal procedure.

Left side

30 Disconnect the negative battery cable from the battery. **Caution:** *On models equipped with a Delco Loc II audio system, disable the anti-theft feature before disconnecting the battery.*

32 Disconnect the air management hose (manual transmission only).

33 Remove the plenum and throttle body as outlined in Chapter 4.

34 Remove the air conditioning compressor and bracket and position them out of the way. *Do not, under any circumstances, disconnect the hoses while the system is under pressure!*

35 Remove the rocker arm cover nuts and retainers.

36 Refer to paragraphs 25 through 29 in this Section for the remainder of the procedure.

---

3 Valve train components - replacement

Refer to illustration 3.2

1 Remove the rocker arm cover(s) as described in Section 2.

2 If only the pushrod is to be replaced, loosen the rocker nut enough to allow the

3.2 To remove a pushrod, loosen the rocker arm nut, rotate the rocker arm aside and pull out the pushrod

rocker arm to be rotated away from the pushrod **(see illustration)**. Pull the pushrod out of the hole in the cylinder head.

3 If the rocker arm is to be removed, remove the rocker arm nut and pivot and lift off the rocker arm.

4 If the valve spring is to be removed, remove the spark plug from the affected cylinder.

5 There are two methods that will allow the valve to remain in place while the valve spring is removed. If you have access to compressed air, install an air hose adapter in the spark plug hole. When air pressure is applied to the adapter, the valves will be held in place by the pressure.

6 If you do not have access to compressed air, bring the piston of the affected cylinder to top dead center (TDC) on the compression stroke. Feed a long piece of 1/4inch nylon cord in through the spark plug hole until it fills the combustion chamber. Be sure to leave the end of the cord hanging out of the spark plug hole so it can be removed easily. Rotate the crankshaft with a wrench (in the normal direction of rotation) until slight resistance is felt.

7 Reinstall the rocker arm nut (without the rocker arm).

8 Insert the slotted end of a valve spring compression tool under the nut and compress the spring just enough to remove the spring keepers, then release the pressure on the tool (see Chapter 2A).

9 Remove the retainer, cup shield, O-ring seal, spring, spring damper (if so equipped) and valve stem oil seal (if so equipped).

10 If a rocker arm stud requires replacement, first determine whether it is a screw-in or press-in type. Screw in studs are retained by a nut atop the stud boss, while press-in studs have no retaining nut.

11 Screw-in studs may be replaced by simply removing the damaged one and replacing it with a new one. Be sure to reinstall the pushrod guide (if so equipped) under the stud nut and tighten the nut to the specified torque.

12 Press-in studs that have damaged threads or are loose in the cylinder head should be replaced with new studs that are available in 0.003 inch and 0.013 inch oversize. Press-in stud replacement procedures require the use of special tools and should be performed by an automotive machine shop. If such services are required, remove the cylinder head(s) by referring to Section 8.

13 Inspect the valve train components for wear or damage. Check for a bent pushrod. Replace any defective parts, as necessary.

14 Installation of the valve train components is the reverse of the removal procedure. Install a new valve stem oil seal after compressing the spring and before installing the spring keepers (see Chapter 2A). Before installing the rocker arms, coat the bearing surfaces of the arms and pivots with moly-based grease or engine assembly lube. Be sure to adjust the valve lash as detailed in Section 6.

4   Intake manifold - removal and installation

Refer to illustrations 4.12, 4.13, 4.21, 4.25 and 4.33

1   If the vehicle is equipped with air-conditioning, carefully examine the routing of the hoses and the mounting of the compressor. You may be able to remove the intake manifold without disconnecting the system. If you are in doubt, take the vehicle to a dealer or refrigeration specialist to have the system depressurized. **Warning:** *The air conditioning system is under high pressure. Do not loosen any hose fittings or remove any components until after the system has been discharged. Air conditioning refrigerant must be properly discharged into an EPA-approved recovery/recycling unit at a dealer service department or an automotive air conditioning repair facility. Always wear eye protection when disconnecting air conditioning system fittings.*

2   Disconnect the cable from the negative battery terminal. **Caution:** *On models equipped with a Delco Loc II audio system, disable the anti-theft feature before disconnecting the battery.*

3   Drain the coolant from the radiator (Chapter 1).

Carbureted models

4   Remove the air cleaner assembly, tagging each hose to be disconnected with a piece of numbered tape to simplify reinstallation.

5   Label and disconnect all electrical wires and vacuum hoses at the carburetor.

6   Disconnect the fuel line at the carburetor. Be prepared to catch some fuel, then plug the fuel line to prevent contamination.

7   Disconnect the throttle cable, making careful note of how it was installed.

8   Disconnect the spark plug wires at the spark plugs, referring to the removal technique described in Chapter 1.

9   Disconnect the wires at the coil, again using numbered pieces of tape to label them.

4.12 Remove the power brake tube fitting from the rear of the manifold

10   Remove the distributor cap and the attached spark plug wires (refer to Chapter 5).

11   Remove the distributor (refer to Chapter 5).

12   Remove the power brake fitting from the rear of the manifold **(see illustration)**.

13   Remove the power brake tube mounting bracket from the manifold boss **(see illustration)**.

14   Move the power brake tube/hose aside so it will not interfere with the removal of the manifold.

15   Remove the solenoid/hose bracket from the front of the left rocker arm cover.

16   Remove the mounting bolts from the left rocker arm cover, then remove the cover.

17   Remove the AIR management valve/coil mounting bracket assembly bolts from the right cylinder head, disconnect the AIR management hose and remove the assembly.

18   Remove the bolts from the right rocker arm cover, then remove the cover.

19   Remove the upper radiator hose from the manifold.

20   Disconnect the heater hose at the manifold.

21   Disconnect the vacuum lines from the

4.13 Remove the power brake tube fitting bracket from the manifold

TVS switch at the front of the manifold **(see illustration)**.

22   Remove the electrical connector from the coolant switch at the front of the manifold.

23   Make sure that all wires, vacuum hoses and coolant hoses that would interfere with manifold removal have been disconnected.

24   If the manifold is to be replaced with a new one, the external components remaining on the manifold must be removed for transfer to the new manifold. These components may be removed either before or after the manifold has been separated from the engine. These components include:

a)   *The carburetor choke assembly and carburetor studs or bolts (refer to Chapter 4 for details).*

b)   *The coolant switch.*

c)   *The EGR valve (use a new gasket when installing).*

d)   *The emissions system TVS valve.*

25   Remove the manifold mounting bolts **(see illustration)**.

26   Separate the manifold from the engine. Do not pry between the mating surfaces because it could damage them. Tap the manifold with a hammer and wooden block to

4.21 Disconnect the vacuum lines from the TVS switch

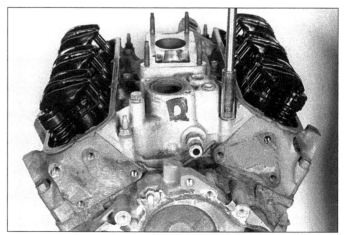

4.25 Remove the manifold mounting bolts

4.33 Intake manifold bolt TIGHTENING sequence

loosen it if necessary.

27 If a new manifold is being installed, transfer the external components from the old manifold to the new one.

28 Before installing the manifold, place clean, lint-free rags in the engine cavity and clean the engine block, cylinder head and manifold gasket surfaces. All gasket material and sealant must be removed prior to installation. Remove all dirt and gasket remnants from the engine cavity.

29 Clean the gasket sealing surfaces with degreaser, then apply a 3/16 inch diameter bead of RTV sealant to the engine block gasket surfaces only.

30 Install the new intake gaskets on the cylinder heads. Notice that the gaskets are marked Right and Left. Be sure to use the correct gasket on each cylinder head.

31 Hold the gaskets in place by extending the bead of RTV up about 1/4inch onto the gasket ends. The new gaskets will have to be cut, as instructed in the gasket set instructions, so they can be installed behind the pushrods.

32 Carefully lower the intake manifold into position, making sure that you do not disturb the gaskets.

33 Install the intake manifold mounting bolts and tighten them following the recommended tightening sequence **(see illustration)**. Tighten the bolts a little at a time until they are all at the specified torque.

34 Install the remaining components in the reverse order of removal.

35 Fill the radiator with coolant, start the engine and check for leaks. Adjust the ignition timing and idle speed as necessary (refer to Chapter 1).

Fuel injected models

36 Remove the port fuel-injection components (plenum, throttle body and fuel rail) (refer to Chapter 4).

37 Disconnect the spark plug wires at the

spark plugs, referring to the procedure described in Chapter 1.

38 Disconnect the wires at the coil, using numbered pieces of tape to label them.

39 Remove the distributor cap and the attached spark plug wires (refer to Chapter 1).

40 Remove the distributor (refer to Chapter 5).

41 Remove the air management hose (manual transmission only).

42 Disconnect the emission canister hoses, marking them with numbered pieces of tape to avoid confusion during installation.

43 Remove the left rocker arm cover.

44 Remove the air management bracket (manual transmission only).

45 Remove the right rocker arm cover.

46 Remove the upper radiator hose from the manifold.

47 Disconnect the heater hose from the manifold.

48 Remove the coolant switches from the front of the manifold.

49 Loosen all of the intake manifold bolts slightly, then remove the bolts.

50 Separate the manifold from the engine. Do not pry between the mating surfaces, because it could damage them. Instead, tap on the manifold using a hammer and a wooden block to break the seal and dislodge the manifold from the engine.

51 Clean the mating surfaces of the manifold and the cylinder heads and remove all traces of old RTV from the engine block ridges.

52 If a new manifold is being installed, transfer the external components from the old manifold to the new one.

53 For the remainder of the procedure, refer to paragraphs 28 through 35.

---

5 Hydraulic lifters - removal, inspection and installation

Refer to illustration 5.8

1 A noisy hydraulic lifter can be isolated when the engine is idling. Place a length of hose or tubing near the position of each valve while listening at the other end of the tube.

2 Assuming that adjustment is correct, the most likely cause of a noisy valve lifter is excessive wear at the lifter foot and camshaft lobe or wear or contamination of the internal components.

3 Remove the rocker arm covers as described in Section 2.

4 Remove the intake manifold as described in Section 4.

5 Loosen the rocker arm nut and rotate the rocker arm away from the pushrod.

6 Remove the pushrod. If more than one pushrod is removed, store them in an organized manner so they can be reinstalled in their original location.

7 To remove the lifter, a special hydraulic lifter removal tool should be used (see Chapter 2A). Do not use pliers or other tools on the

5.8 If the lifters are to be reused, they must be organized to ensure reinstallation in their original positions

outside of the lifter body, as they will damage the finished surface and render the lifter useless.

8 If the lifters are to be reused, they must be stored in an organized manner so they can be reinstalled in their original positions **(see illustration)**.

9 Clean and dry the lifters. Examine them for wear and check carefully for flat spots (see Chapter 2A).

10 If the lifters are worn, they must be replaced with new ones and the camshaft must be replaced as well (see Section 16). If the lifters are in good condition, they can be reinstalled - provided they are installed in their original locations.

11 When installing the lifters, make sure they are replaced in their original bores and coat them with moly-based grease or engine assembly lube.

12 The remaining installation steps are the reverse of removal.

---

6 Valve lash - adjustment

Refer to illustration 6.5

1 Disconnect the cable from the negative battery terminal. **Caution:** *On models equipped with a Delco Loc II audio system, disable the anti-theft feature before disconnecting the battery.*

2 Remove the rocker arm covers (refer to Section 2). Remove the spark plugs.

3 If the valve train components have been serviced just prior to this procedure, make sure that the components are completely reassembled and the rocker arm nuts are loose.

4 Using a breaker bar and socket on the crankshaft pulley center bolt, rotate the crankshaft clockwise until the number one piston is at top dead center (TDC) on the compression stroke. To locate TDC, place your finger over the number one cylinder spark plug hole and rotate the crankshaft. When compression pressure is felt at the spark plug hole, continue rotating the

6.5 Determine the point of zero lash by spinning the pushrod as the nut is slowly tightened - when you just feel a slight amount of resistance at the pushrod, stop tightening the nut - if you tighten the nut to far, engine damage could result when the engine is started

7.8 Remove the right exhaust manifold mounting bolts

8.3 Arrow indicates the location of the left-hand engine drain plug

crankshaft until the timing mark on the vibration damper aligns with the "0" on the timing indicator (see Chapter 1). To make sure that the number one piston is at TDC, remove the distributor cap and check the position of the rotor in the distributor, it should be pointing at the number one cylinder spark plug wire terminal. If it's pointing at the number four terminal, rotate the crankshaft 360-degrees. As an alternate method, watch the number one cylinder rocker arms as you rotate the crankshaft. After the number one cylinder intake valve opens and closes, continue rotating the crankshaft until the timing mark on the vibration damper aligns with the "0" on the timing indicator - the number one piston is at TDC.

5   To adjust a valve, slowly tighten the rocker arm nut until all clearance between the pushrod and rocker arm is removed (zero lash), then tighten the nut 3/4-turn further to preload the lifter. Reaching the point of zero lash can be difficult and requires some expertise, proceed as follows (see illustration):

a) *Move the pushrod up-and-down and tighten the rocker arm nut until all clearance between the pushrod and rocker arm is removed.*

b) *Spin the pushrod between your thumb and index finger and slowly tighten the nut. When you just feel a slight resistance as the pushrod is rotated, stop tightening the nut - you've reached zero lash. Do not exert downward pressure with the tool used to tighten the nut as this will effect the adjustment.*

c) *After the point of zero lash is reached, tighten the nut 3/4-turn to complete the valve adjustment.*

6   With the number one piston at TDC, adjust the following valves: number one, five and six cylinder intake valves and the number one, two and three cylinder exhaust valves,

using the method described.

7   Rotate the crankshaft 360-degrees (number four piston at TDC on the compression stroke) and adjust the following valves: number two, three and four cylinder intake valves and the number four, five and six cylinder exhaust valves, using the method described.

8   Install the rocker arm covers and spark plugs.

---

7   Exhaust manifolds - removal and installation

Right side

Refer to illustration 7.8

1   Disconnect the cable from the negative battery terminal. **Caution:** *On models equipped with a Delco Loc II audio system, disable the anti-theft feature before disconnecting the battery.*

2   Raise the vehicle and support it securely on jackstands.

3   Remove the bolts attaching the exhaust pipe to the exhaust manifold, then separate the pipe from the manifold.

4   Remove the jackstands and lower the vehicle.

5   Disconnect the oxygen sensor pigtail electrical connector.

6   Disconnect the air management hose at the check valve.

7   Disconnect the spark plug wires from the spark plugs, labeling them as they are disconnected to simplify installation.

8   Remove the exhaust manifold mounting bolts and separate the manifold from the engine (see illustration).

9   Installation is the reverse of the removal procedure. Before installing the manifold, be sure to thoroughly clean the mating surfaces on the manifold and cylinder head.

Left side

10   Disconnect the cable from the negative

battery terminal. **Caution:** *On models equipped with a Delco Loc II audio system, disable the anti-theft feature before disconnecting the battery.*

11   Raise the vehicle and place it securely on jackstands.

12   Remove the bolts retaining the exhaust pipe to the manifold, then disconnect the pipe from the manifold.

13   Remove the four bolts and one nut accessible at the rear of the manifold.

14   Remove the jackstands and lower the vehicle.

15   Remove the air cleaner assembly, labeling all hoses.

16   Disconnect the hoses leading to the air management valve.

17   Disconnect and label any wires that will interfere with the removal of the manifold.

18   On some models it may be necessary to remove the power steering pump bracket from the cylinder head. If so, loosen the pump adjusting bracket bolt and remove the pump drivebelt from the pulley first. After removing the bracket from the cylinder head, place the steering pump assembly aside, out of the way.

19   Remove the remaining manifold bolts and separate the manifold and heat shield from the engine.

20   Installation is the reverse of the removal procedure. Be sure to thoroughly clean the cylinder head and manifold surfaces before installing the manifold.

---

8   Cylinder heads - removal and installation

Left side

Refer to illustrations 8.3, 8.9, 8.10, 8.11, 8.12a, 8.12b, 8.14 and 8.19

1   Disconnect the cable from the negative battery terminal. **Caution:** *On models equipped with a Delco Loc II audio system, disable the anti-theft feature before disconnecting the battery.*

8.9 Remove the power steering pump mounting bracket

8.10 Remove the air-conditioner compressor bracket

8.11 Remove the coolant temperature sending unit

8.12a Loosen the rocker arm adjusting nuts

2 Raise the vehicle and place it securely on jackstands.
3 Locate the engine block drain plugs to the rear of the motor mounts (the plug on the left side is just above the oil filter). Remove the plugs and drain the block (see illustration).
4 Disconnect the exhaust pipe from the exhaust manifold.
5 Remove the oil dipstick tube assembly from the side of the engine.
6 Remove the jackstands and lower the vehicle.
7 Remove the intake manifold (refer to Section 4).
8 Remove the exhaust manifold (refer to Section 7).
9 Without disconnecting the hoses, remove the power steering pump and set it aside. Remove the power steering pump bracket from the cylinder head (see illustration).
10 Without disconnecting the refrigerant lines, remove the air conditioning compressor and set it aside. Remove the air conditioner compressor bracket from the front of the

8.12b When removing the pushrods, be sure to store them separately to ensure reinstallation in their original positions

cylinder head (see illustration).
11 Remove the coolant temperature sending unit from the front of the cylinder head (see illustration).
12 Loosen the rocker arm nuts enough to allow removal of the pushrods, then remove

the pushrods (see illustrations).
13 Loosen the head bolts in a sequence opposite the tightening sequence (see illustration 8.19).
14 Remove the cylinder head. To break the gasket seal, insert a bar into one of the

8.14 Use a prybar to break the gasket seal on the right-hand cylinder head

8.19 Cylinder head bolt TIGHTENING sequence

8.26 On models with a serpentine belt, remove the bolts and the belt tensioner

8.27 Remove the alternator bracket from the right-side cylinder head

exhaust ports, then carefully lift on the tool **(see illustration)**.

15   Before installing the cylinder head, the gasket surfaces of both the head and the engine block must be clean and free of nicks and scratches. Also, the threads in the block and on the head bolts must be completely clean, as any dirt remaining in the threads will affect bolt torque.

16   Place the gasket in position over the locating dowels, with the note "This Side Up" visible.

17   Position the cylinder head over the gasket.

18   Coat the cylinder head bolts with an appropriate thread sealer and install the bolts.

19   Tighten the bolts in the proper sequence to the specified torque **(see illustration)**. Work up to the final torque in three steps.

20   Install the pushrods, making sure the lower ends are in the lifter seats, place the rocker arm ends over the pushrods and loosely install the rocker arm nuts.

21   The remaining installation steps are the reverse of those for removal. Before installing the rocker arm covers, adjust the valve lash (refer to Section 6).

Right side

Refer to illustrations 8.26, 8.27 and 8.28

22   Disconnect the cable from the negative battery terminal. **Caution:** *On models equipped with a Delco Loc II audio system, disable the anti-theft feature before disconnecting the battery.* Raise the vehicle and place it securely on jackstands.

23   Locate the engine block drain plugs to the rear of the motor mounts (the plug on the left side is just above the oil filter), remove the plugs and drain the block.

24   Disconnect the exhaust pipe from the exhaust manifold.

25   Remove the jackstands and lower the vehicle.

26   Remove the intake manifold (refer to Section 4). On models with a serpentine belt, remove the belt tensioner **(see illustration)**.

27   Remove the alternator from the alternator bracket, then remove the bracket from the head **(see illustration)**.

28   Remove the lifting "eye" from the rear of the head (necessary only if the head is to be replaced with a new one) **(see illustration)**.

29   Remove the exhaust manifold.

30   Loosen the rocker arm nuts sufficiently

to allow removal of the pushrods, then remove the pushrods.

31   Loosen the head bolts in a sequence opposite to the one used for tightening them **(see illustration 8.19)**.

32   Remove the cylinder head. To break the

8.28 Remove the lifting "eye" from the right-side cylinder head

10.2 Remove the oil pump-to-rear main bearing cap bolt

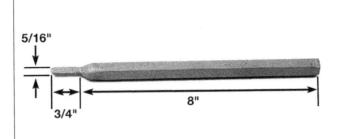

11.3a If the special tool is not available, grind a piece of 1/2-inch diameter brass or aluminum rod to these dimensions as a rear seal driver

gasket seal, insert a bar into one of the exhaust ports, then carefully lift on the tool.

33   To install the head, refer to Steps 15 through 21.

9   Oil pan - removal and installation

1   Disconnect the cable from the negative battery terminal. **Caution:** *On models equipped with a Delco Loc II audio system, disable the anti-theft feature before disconnecting the battery.*

2   On carbureted models, remove the air cleaner assembly, tagging each hose to be disconnected with a piece of numbered tape to simplify reinstallation. On fuel injected models, remove the air intake duct.

3   Remove the distributor cap (refer to Chapter 5) to prevent damage when the engine is raised.

4   Unbolt and remove the top half of the fan shroud.

5   Raise the vehicle and support it on jackstands.

6   Drain the engine oil.

7   Disconnect the exhaust crossover pipe at the exhaust manifold flanges. Lower the exhaust pipes and suspend them from the frame with wire.

8   If equipped with an automatic transmission, remove the converter shroud.

9   If equipped with a manual transmission, remove the flywheel cover.

10   Remove the starter (refer to Chapter 5).

11   Remove the throughbolt at each engine mount.

12   Using an engine hoist, raise the engine to allow the oil pan to clear the crossmember. Check clearances all around the engine as it is raised. Pay particular attention to the distributor and the cooling fan.

14   Place wood blocks between the frame front crossmember and the engine block. The blocks should be approximately three inches thick.

15   Lower the engine onto the wood blocks. Make sure it is firmly supported. If a hoist is being used, keep the lifting chains secured to the engine. **Warning:** *Do not, under any circumstances, work under the engine when supported only by the hoist.*

16   Remove the oil pan bolts. Note the different sizes used and their locations.

17   Remove the oil pan.

18   Before installing the pan, make sure that the sealing surfaces on the pan, block and front cover are clean and free of oil. If the old pan is being reinstalled, make sure that all sealant has been removed from the pan sealing flange and from the blind attaching holes.

19   With all the sealing surfaces clean, place a 1/8-inch bead of RTV sealant on the oil pan sealing flange.

20   Lift the pan into position and install all bolts finger tight. Tighten the end bolts first, then tighten the remaining bolts working from the center out to the ends.

21   Lower the engine onto the mounts and install the throughbolts. Tighten the bolts to the specified torque.

22   Follow the removal steps in reverse order. Fill the crankcase with the correct grade and quantity of oil, start the engine and check for leaks.

10   Oil pump - removal and installation

Refer to illustration 10.2

1   Remove the oil pan (refer to Section 9).

2   Remove the pump-to-rear main bearing cap bolt and separate the pump and extension shaft from the engine **(see illustration)**.

3   To install the pump, move it into position and align the top end of the hexagonal extension shaft with the hexagonal socket in the lower end of the distributor drive gear. The distributor drives the oil pump, so it is essential that this alignment is correct.

4   Install the oil pump-to-rear main bearing cap bolt and tighten it to the specified torque.

5   Reinstall the oil pan.

11   Rear main bearing oil seal - replacement

Note 1: *1985 and later models are equipped with a one-piece rear main oil seal. Refer to*

11.3b Use a packing tool to pack the old rear main bearing oil seal into the groove - 1982 and 1983 models

Chapter 2A for the seal replacement procedure.

Note 2: *The following procedure describes replacing the rear main bearing oil seal with the engine installed in the vehicle. If this method does not repair a rear main seal oil leak, remove the engine and the crankshaft and install a service replacement one-piece rear main seal (available at most auto parts stores) following the instructions provided with the seal kit.*

1982 and 1983 models

Refer to illustrations 11.3a, 11.3b, 11.5 and 11.10

Note: *The special tool kit (designed specially for replacing rope-type seal) used in this procedure should be available at most auto parts stores.*

1   Although the crankshaft must be removed to install a new seal, the upper portion of the seal can be repaired with the crankshaft in place.

2   Remove the oil pan and oil pump (Sections 9 and 10).

3   Using a special tool (available at most auto part stores), drive the old seal gently back into the groove, packing it tight **(see illustrations)**. It will pack in to a depth of 1/4 to 3/4 inch.

4   Repeat the procedure on the other end of the seal.

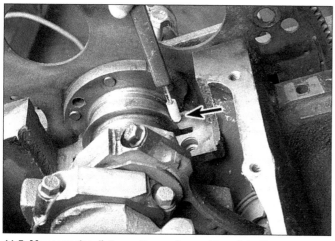

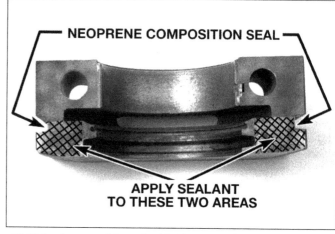

11.5 Measure the distance the seal was driven into the upper seal groove (use a small screwdriver with a sleeve of masking tape as a depth indicator) - 1982 and 1983 models

11.10 Apply anaerobic sealant to the areas shown, but don't get sealant in the grooves or on the seal - 1984 model

5   Measure the amount that the seal was driven up into the groove on one side and add 1/16inch **(see illustration)**. Remove the old seal from the main bearing cap. Use the main bearing cap as a fixture and cut off a piece of the old seal to the predetermined length. Repeat this process for the other side.

6   Install the guide tool on the block.

7   Using the packing tool, work the short pieces of the previously cut seal into the guide tool and pack them into the block groove on each side. The guide and packing tools have been machined to provide a built-in stop. Use of oil on the seal pieces will ease installation.

8   Remove the guide tool.

9   Install a new seal in the main bearing cap.

10   Apply a thin, even coat of anaerobic-type gasket sealant to the areas of the rear main bearing cap as indicated **(see illustra-tion)**. *Caution: Do not get any sealant on the bearing or seal faces.*

11   Apply a light coat of engine oil to the rear main bearing and install the rear main bearing cap. Tighten the rear main bearing cap bolts to the specified torque.

12   Install the oil pump and oil pan.

1984 model

Refer to illustrations 11.17, 11.20 and 11.21

13   Always service both halves of the rear main oil seal. While replacement of this seal is much easier with the engine removed from the vehicle, the job can be done with the engine in place.

14   Remove the oil pan and oil pump as described previously in this Chapter.

15   Remove the rear main bearing cap from the engine.

16   Using a screwdriver, pry the lower half of the oil seal from the bearing cap.

17   To remove the upper half of the seal, use a small hammer and a brass pin punch to roll the seal around the crankshaft journal. Tap one end of the seal with the hammer and

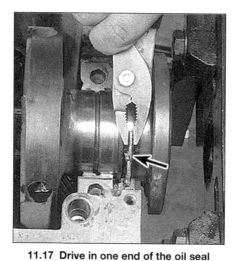

11.17 Drive in one end of the oil seal upper half until the other end protrudes, then pull the seal out with pliers - 1984 model

punch (be careful not to strike the crankshaft) until the other end of the seal protrudes enough to pull the seal out with pliers **(see illustration)**.

18   Remove all sealant and foreign material from the main bearing cap. Do not use an abrasive cleaner for this.

19   Inspect the components for nicks,

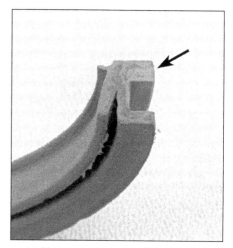

11.20 A very thin coat of RTV gasket sealant should be applied to the area shown (avoid getting sealant on the seal lips) - 1984 model

scratches and burrs at all sealing surfaces. Remove any defects with a fine file or debur-ring tool.

20   Apply a very thin coat of RTV gasket sealant to the outer surface of the upper seal as shown **(see illustration)**. Do not get any sealant on the seal lips.

21   Included in the purchase of the rear

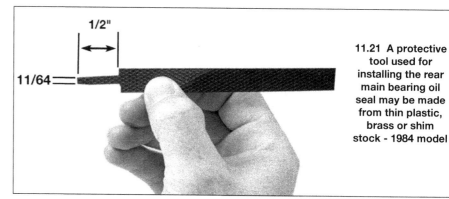

11.21 A protective tool used for installing the rear main bearing oil seal may be made from thin plastic, brass or shim stock - 1984 model

12.6 Use a puller to remove the vibration damper

13.6 Locations of the crankcase front cover mounting bolts

main oil seal should be a small plastic installation tool; if not, a tool may be fashioned **(see illustration)**.

22 With the upper half of the seal positioned so that the seal lip faces toward the front of the engine and the small dust lip faces toward the flywheel, install the seal by rolling it around the crankshaft using the installation tool as a "shoehorn' for protection.

23 Apply sealing compound as described in Step 20 to the other half of the seal and install it in the bearing cap.

24 Apply a thin, even coat of anaerobic-type gasket sealant to the areas of the rear main bearing cap as indicated **(see illustration 11.10)**. **Caution:** *Do not get any sealant on the bearing or seal faces.*

25 Apply a light coat of engine oil to the rear main bearing and install the rear main bearing cap. Tighten the rear main bearing cap bolts to the specified torque.

26 Install the oil pump and oil pan.

12 Vibration damper - removal and installation

Refer to illustration 12.6

1 Remove the hose from the retaining strap atop the radiator shroud by releasing the end of the strap from the underside of the shroud with pliers. Move the hose to the side, out of the way.

2 Remove the bolts and separate the upper radiator shroud from the radiator.

3 Remove the cooling fan from the fan pulley.

4 On models with V-belts, loosen the accessory drivebelt adjusting bolts as necessary, then remove the drivebelts, tagging each one as it is removed to simplify reinstallation. On models with a serpentine belt, rotate the belt tensioner pulley away from the belt and remove the belt (see Chapter 1).

5 Remove the bolt from the center of the

crankshaft pulley and remove the pulley.

6 Attach a puller to the damper. Draw the damper off the crankshaft, being careful not to drop it as it breaks free **(see illustration)**. **Caution:** *A common gear puller should not be used to draw the damper off, as it may separate the outer portion of the damper from the hub. Use only a puller which bolts to the hub.*

7 Before installing the damper, coat the front cover seal area (on the damper) with engine oil or multi-purpose grease.

8 Place the damper in position over the key on the crankshaft. Make sure the damper keyway aligns with the key.

9 Using a damper installation tool, press the damper onto the crankshaft. The special tool distributes the pressure evenly around the hub.

10 Remove the installation tool and install the crankshaft pulley. Tighten the bolts to the specified torque.

11 Follow the removal procedure in the reverse order for the remaining components.

12 Adjust the drivebelts (refer to Chapter 1).

13 Crankcase front cover - removal and installation

Refer to illustration 13.6

1 If equipped with air-conditioning, remove the compressor from the mounting bracket and secure it aside. Do not disconnect any of the air-conditioning system hoses without having the system depressurized by an air conditioning technician. Remove the compressor mounting bracket.

2 Remove the power steering pump, without disconnecting the hoses and secure it aside. Remove the power steering mounting bracket.

3 Remove the vibration damper as described in Section 12.

4 Remove the water pump as described in Chapter 3.

5 Remove the oil pan-to front cover bolts.

On 3.1L models, remove the oil pan as described in Section 9.

6 Remove the front cover mounting bolts and separate the cover from the engine **(see illustration)**.

7 Clean all oil, dirt and old gasket material from the sealing surfaces of the front cover and engine block. Replace the front cover oil seal as described in Section 14.

8 If equipped with a gasket, install a new gasket. If no gasket is used, apply a continuous 3/32-inch bead of anaerobic sealant to both mating surfaces of the front cover (except the cover-to-block mating surface where the cover engages the oil pan lip). Also apply anaerobic sealant to the areas surrounding the coolant passages.

9 Apply a 3/8-inch bead of RTV sealant to the cover-to-oil pan sealing surface.

10 Place the front cover in position on the engine block and install the mounting bolts. Tighten the bolts to the specified torque.

11 The remaining installation procedures are the reverse of removal.

14 Front cover oil seal - replacement

With front cover installed on engine

1 With the vibration damper removed (Section 12), pry the old seal out of the crankcase front cover with a large screwdriver. Be very careful not to damage the surface of the crankshaft.

2 Place the new seal in position with the open end of the seal (seal lip) toward the inside of the cover.

3 Using a seal driver, drive the seal into the cover until it is seated. A section of large-diameter pipe or a large socket could also be used.

4 Be careful not to distort the front cover.

14.7 Drive the seal out of the front cover

14.9 Install the front cover oil seal with a wood block and hammer

With front cover removed from engine

Refer to illustrations 14.7 and 14.9

5   This method is preferred, as the cover can be supported while the old seal is removed and the new one is installed.

6   Remove the crankcase front cover (refer to Section 13).

7   Using a large screwdriver, pry the old seal out of the bore from the front of the cover. Alternatively, support the cover and drive the seal out from the rear **(see illustration)**. Be careful not to damage the cover.

8   With the front of the cover facing up, place the new seal in position with the open end of the seal toward the inside of the cover.

9   Using a wooden block and hammer, drive the new seal into the cover until it is completely seated **(see illustration)**.

10   If the cover was removed, install it by reversing the removal procedure.

---

15 Timing chain and sprockets - inspection, removal and installation

Refer to illustrations 15.9 and 15.13

1   Disconnect the cable from the negative battery terminal. **Caution:** *On models equipped with a Delco Loc II audio system, disable the anti-theft feature before disconnecting the battery.*

2   Remove the vibration damper (refer to Section 12).

3   Remove the crankcase front cover (refer to Section 13).

4   Before removing the chain and sprockets, visually inspect the teeth on the sprockets for signs of wear and the chain for looseness. Also check the condition of the timing chain tensioners.

5   If either or both sprockets show any signs of wear (edges on the teeth of the camshaft sprocket not "square,' bright or blue areas on the teeth of either sprocket,

chipping, pitting, etc.), they should be replaced with new ones. Wear in these areas is very common.

6   If the timing chain has not recently been replaced, or if the engine has over 25000 miles on it, it is almost certainly in need of replacement. Failure to replace a worn timing chain may result in erratic engine performance, loss of power and lowered gas mileage.

7   If any one component requires replacement, all related components, including the tensioners, should be replaced as well.

8   If it is determined that the timing components require replacement, proceed as follows.

9   Rotate the crankshaft until the marks on the camshaft and crankshaft are in exact alignment **(see illustration)**. At this point the number one and four pistons will be at top dead center with the number four piston in the firing position (verify by checking the position of the rotor in the distributor). Do not attempt to remove either sprocket or the timing chain until this is done and do not turn the crankshaft or camshaft after the sprockets/chain are removed.

10   Remove the three camshaft sprocket retaining bolts and lift the camshaft sprocket

15.9 The camshaft and crankshaft sprocket timing marks should be aligned before you remove the timing sprockets and chain

and timing chain off the front of the engine. It may be necessary to tap the sprocket with a soft-faced hammer to dislodge it.

11   If it is necessary to remove the crankshaft sprocket, it can be withdrawn from the crankshaft with a special puller.

12   Attach the crankshaft sprocket to the

15.13 Lubricate the thrust surface of the camshaft sprocket

16.11 Remove the camshaft from the engine block (be very careful not to damage the cam bearings as the shaft is withdrawn)

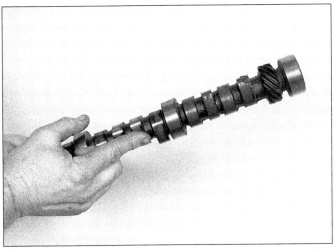

16.13 Be sure to apply camshaft assembly lube to the cam lobes and bearing journals before installing the camshaft

crankshaft using a bolt and washer from the puller set.

13   Lubricate the thrust (rear) surface of the camshaft sprocket with moly-based grease or engine assembly lube **(see illustration)**. Install the timing chain over the camshaft sprocket with slack in the chain hanging down over the crankshaft sprocket.

14   With the timing marks aligned, slip the chain over the crankshaft sprocket and then draw the camshaft sprocket into place with the three retaining bolts. Do not hammer or attempt to drive the camshaft sprocket into place, as it could dislodge the Welch plug at the rear of the engine.

15   With the chain and both sprockets in place, check again to ensure that the timing marks on the two sprockets are properly aligned. If not, remove the camshaft sprocket and move it until the marks align.

16   Lubricate the chain with engine oil and install the remaining components in the reverse order of removal.

16 Camshaft - removal and installation

Refer to illustrations 16.11, 16.13 and 16.14
Note: *Before removing the camshaft, refer to Chapter 2, Part D (Section 15), and measure the lobe lift.*

1   Disconnect the cable from the negative battery terminal. **Caution:** *On models equipped with a Delco Loc II audio system, disable the anti-theft feature before disconnecting the battery.*

2   Drain the oil from the crankcase (refer to Chapter 1).

3   Drain the coolant from the radiator (refer to Chapter 1).

4   Remove the radiator (refer to Chapter 3).

5   If equipped with air-conditioning, remove the condenser (refer to Chapter 3).
Warning: *The air conditioning system is*

under high pressure. Do not loosen any hose fittings or remove any components until after the system has been discharged. Air conditioning refrigerant must be properly discharged into an EPA-approved recovery/recycling unit at a dealer service department or an automotive air conditioning repair facility. Always wear eye protection when disconnecting air conditioning system fittings.

6   Remove the valve lifters (refer to Section 5).

7   Remove the crankcase front cover (refer to Section 13).

8   On carbureted models, remove the fuel pump and pushrod (refer to Chapter 4).

9   Remove the timing chain and sprocket (refer to Section 15).

10   Install a long bolt in one of the camshaft bolt holes to be used as a handle to pull on and support the camshaft.

11   Carefully draw the camshaft out of the engine block. Do this very slowly to avoid damage to the camshaft bearings as the lobes pass over the bearing surfaces. Always support the camshaft with one hand near the engine block **(see illustration)**.

12   Refer to Chapter 2, Part D, for the

camshaft inspection procedures.

13   Prior to installing the camshaft, coat each of the lobes and journals with camshaft assembly lube **(see illustration)**.

14   Slide the camshaft into the engine block, again taking extra care not to damage the bearings **(see illustration)**.

15   Install the camshaft sprocket and timing chain as described in Section 15.

16   Install the remaining components in the reverse order of removal by referring to the appropriate Chapter or Section.

17   Adjust the valve lash (refer to Section 6).

18   Have the air-conditioning system (if so equipped) recharged.

17 Engine mounts - replacement with engine in vehicle

1   If the mounts have become hard, split or separated from the metal backing, they must be replaced. This operation may be carried out with the engine/transmission still in the vehicle. See Section 9 for the proper way to raise the engine while it is still in place.

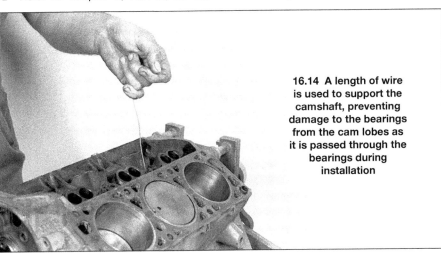

16.14 A length of wire is used to support the camshaft, preventing damage to the bearings from the cam lobes as it is passed through the bearings during installation

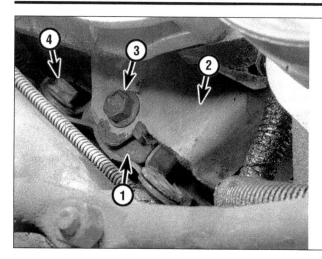

17.2 Front engine mount installation details

1   Engine mount
2   Engine mount-to-engine bracket
3   Through-bolt
4   Engine mount-to-crossmember mounting bolt

18.5 Remove the bracket-to-hood bolts (note white paint around bolt head to ensure correct installation)

Front mount

Refer to illustration 17.2

2   Remove the throughbolt and nut **(see illustration)**.

3   Raise the engine slightly using a hoist or jack with a wood block under the oil pan, then remove the mount and frame bracket assembly from the crossmember.

4   Position the new mount, install the throughbolt and nut, then tighten all the bolts to the specified torque.

Rear mount

5   Remove the crossmember-to-mount bolts, then raise the transmission slightly with a jack.

6   Remove the mount-to-transmission bolts, followed by the mount.

7   Install the new mount, lower the transmission and align the crossmember-to-mount bolts.

8   Tighten all the bolts to the specified torque.

---

18   Engine - removal and installation

Note: *Due to the inaccessibility of the upper*

bellhousing-to-engine bolts in vehicles equipped with a V6 engine, it is recommended that the transmission and engine be removed as a unit, as detailed in this Section. To unbolt the transmission from the engine while both are in the vehicle, in order to remove the engine separately, the floor pan in the interior must be cut away to reveal the upper bellhousing bolts. If this method of removal is pursued, read through the procedure in this Section to familiarize yourself with the V6 engine/transmission removal. Refer to the engine removal section in Chapter 2, Part C, and note the differences between the V6 and V8 and perform the engine removal procedure accordingly - after the floor pan has been cut to reveal the upper bellhousing bolts.

Removal

Refer to illustrations 18.5, 18.11, 18.12, 18.16, 18.23, 18.30, 18.32, 18.34, 18.35, 18.44, 18.51, 18.55 and 18.61

1   Disconnect the cable from the negative battery terminal. **Caution:** *On models equipped with a Delco Loc II audio system, disable the anti-theft feature before disconnecting the battery.*

2   Remove the clip retaining the under-

hood light wire harness to the hood.

3   Disconnect the underhood light electrical connector and unbolt the bulb assembly from the hood.

4   Scribe lines on the underside of the hood, around the hood mounting bracket, so the hood can be installed in the same position.

5   Apply white paint around the bracket-to-hood bolts so they can be aligned quickly and accurately during installation **(see illustration)**.

6   Remove the bracket-to-hood bolts and, along with an assistant, separate the hood from the vehicle.

7   If the vehicle is equipped with a standard transmission, remove the transmission linkage (four-speed) or shift control lever (five-speed). Refer to Chapter 7 for either procedure.

8   Drain the radiator by opening the petcock located at the lower right-hand side of the radiator.

9   Remove the coolant recovery tank hose from the radiator.

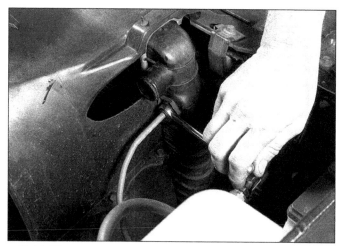

18.11 When removing the automatic transmission cooler lines from the radiator, use a flare nut wrench

18.12 Remove the retaining bolts from the upper radiator shroud

18.16 Unbolt the fan from the fan pulley

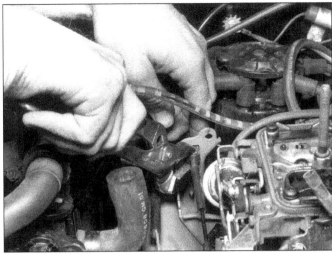

18.23 Remove the carburetor linkage from the linkage bracket

18.30 Remove the solenoid/hose bracket from the left-side rocker arm cover

18.32 Remove the oxygen sensor from the right-side exhaust manifold

10   Remove the upper radiator hose from the radiator. Take care not to damage the plastic fitting when breaking the hose loose.

11   If so equipped, remove the automatic transmission cooler lines at the radiator (see illustration).

12   Remove the bolts securing the top of the radiator fan shroud to the lower half and remove the upper shroud (see illustration).

13   Remove the lower radiator hose from the radiator.

14   Remove the radiator from the vehicle.

15   Mark the drivebelts to simplify installation.

16   Unbolt and remove the fan from the fan pulley (see illustration).

17   If so equipped, remove the air-conditioner compressor without disconnecting the refrigerant lines and secure the compressor aside (refer to Chapter 3).

18   Remove the alternator (refer to Chapter 5).

19   Remove the heater hoses from the water pump and intake manifold outlets.

20   Remove the AIR system air pump (refer to Chapter 6).

21   Disconnect the hoses at the air cleaner. Carefully label them as they are removed to simplify installation.

22   Remove the air cleaner assembly, then disconnect the vacuum brake booster line at the booster.

23   Disconnect the linkage at the carburetor or throttle body (including the cruise control cable, if so equipped). The cables may be removed from the linkage bracket with channel lock pliers (see illustration). Label linkage/cable components as they are disconnected.

24   Disconnect the wiring connectors from the carburetor or throttle body (refer to Chapter 4).

25   Disconnect the wiring connectors and vacuum hoses from the intake manifold.

26   Disconnect the electrical connector for the EFE heater wire at the base of the carburetor (if equipped).

27   Disconnect the wiring harness ground,

located just ahead of the thermostat housing.

28   Disconnect the wire connector at the oil pressure sending unit mounted atop the oil filter.

29   Remove the power steering pump mounting bolts. Lay the pump aside, out of the way, with the hoses still attached.

30   Unbolt the solenoid valve hose bracket from the left-hand rocker arm cover so it will not be damaged when the engine is lifted out of the vehicle (see illustration). Lay the assembly aside.

31   Remove the belts from the water pump pulley, then unbolt and remove the pulley from the front of the water pump.

32   Remove the oxygen sensor from the right side exhaust manifold so it will not be damaged as other components are removed (see illustration). Because it is a very sensitive component, be sure to protect it.

33   Raise the vehicle and place it securely on jackstands.

34   On carbureted models, remove the hoses from the fuel pump, then plug the

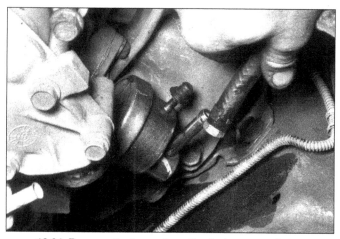

18.34 Remove the hoses from the fuel pump and plug the pump fittings

18.35 Remove the right-side exhaust pipe from the exhaust manifold

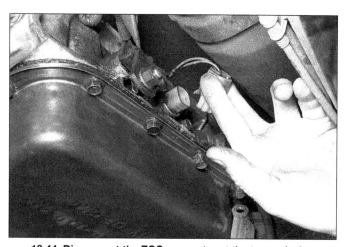

18.44 Disconnect the TCC connector at the transmission (automatic transmission models only)

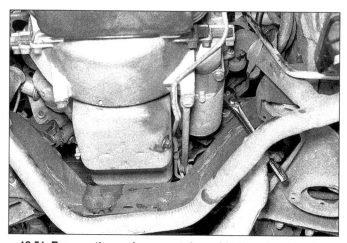

18.51 Remove the engine mount throughbolt on the right side

openings in the pump and hoses to prevent contamination and loss of fuel **(see illustration)**.

35   Remove the bolts securing the exhaust pipes to the exhaust manifolds on both sides of the engine **(see illustration)**.

36   Pull the exhaust pipes off the exhaust manifolds.

37   If equipped with a manual transmission, disconnect the clutch fork return spring.

38   Remove the driveshaft (refer to Chapter 8).

39   Remove the bolts retaining the torque arm to the transmission bracket (refer to Chapter 11 for an illustration if necessary). **Caution:** *Rear spring force will cause the torque arm to move toward the floor pan when the arm is disconnected from the transmission bracket. Carefully place a wooden block between the torque arm and floor pan when disconnecting the arm to avoid possible injury to the hands and fingers and possible damage to the floor pan.*

40   Loosen the rear torque arm bolts and pull the front of the arm free of the bracket, heeding the caution in Step 39.

41   Remove the catalytic converter/exhaust

pipe bracket at the rear of the transmission.

42   Disconnect the speedometer cable from the transmission.

43   If so equipped, remove the automatic transmission throttle valve cable from the bracket on the transmission, then remove the bracket and shift linkage.

44   If so equipped, disconnect the TCC electrical connector at the transmission (automatic transmissions only) **(see illustration)**.

45   Remove the TCC wiring bracket from the transmission.

46   Disconnect the electrical wires at the starter, labeling them to ensure proper installation.

47   Position a movable jack (floor jack or transmission jack) under the transmission oil pan using a block of wood as an insulator. Raise the transmission slightly.

48   Attach the hoist lifting chains to the lifting "eyes' on the engine. There is one bracket at the front of the engine and one at the rear, diagonally opposite each other. Make sure the chain is looped properly through the engine brackets and secured with bolts and nuts through the chain links. The hook on the

lifting hoist should be at the center of the engine. Position the chains so the engine/transmission unit will be at a steep angle with the front higher than the rear.

49   Raise the hoist until all slack is removed from the chains. Do not lift any further at this time.

50   Remove the transmission-to-crossmember bolts and the crossmember-to-frame bolts. Raise the transmission slightly and slide the crossmember to the rear until it can be removed. If equipped with an automatic transmission, insert a plug into the rear to prevent fluid loss when the engine/transmission unit is tilted while being removed. A plastic bag secured with tape will generally work.

51   Remove the engine mount throughbolts **(see illustration)**.

52   Make sure that all wires and hoses, both under the vehicle and in the engine compartment, have been disconnected.

53   Remove the jack from under the transmission.

54   Start pulling the engine out. Pay special attention to the clearance at the distributor and the AIR system assembly. When the dis-

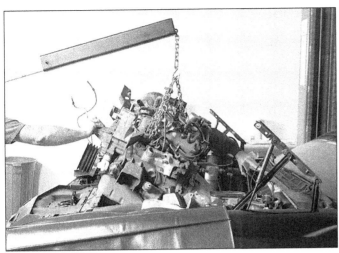

18.55 Lift the engine from the vehicle while checking clearance at the firewall

18.61 Mark the relationship of the driveplate to the torque converter before removing the bolts

tributor is clear of the firewall, if not previously done, detach the wire connector at the rear of the distributor, remove the ground strap at the rear of the left cylinder head and remove the two wiring harness clamps at the bellhousing.

55 Continue to lift the engine at an angle until it is high enough to move forward without hitting the body (see illustration). Work slowly and be on the lookout for hoses and wires that are still connected.

56 Move to an appropriate work area and support the engine and transmission while they are separated.

Separating the engine and transmission

Automatic transmission

57 Remove the transmission cooler lines from the transmission.

58 Remove the starter motor.

59 Remove the strut rods that run between the front of the engine and the transmission dust cover.

60 Remove the dust cover.

61 Using white paint, mark the relationship of the driveplate to the converter (see illustration).

62 Remove the driveplate-to-converter bolts. Use a large screwdriver to turn the driveplate and expose the bolts.

63 When the bolts have all been removed, push the converter toward the rear of the transmission to protect the transmission input seal.

64 Loosen the bellhousing bolts a little at a time.

65 Remove the transmission fluid dipstick and housing by pulling the housing out of the

boss in the transmission, then unbolt the throttle valve cable housing from the transmission.

66 Remove the bellhousing bolts and separate the engine and transmission. Make sure that the converter does not slide forward during the separation procedure.

67 Mount the engine on an engine stand.

68 Refer to the appropriate Sections of this Chapter and Chapter 2, Part D, for further engine disassembly and rebuilding procedures.

Manual transmission

69 Remove the bolts securing the dust cover to the bottom of the bellhousing and remove the dust cover.

70 Loosen the bolts securing the bellhousing to the engine a little at a time.

71 When all bellhousing-to-transmission bolts have been loosened, remove the bolts.

72 Support the weight of the transmission and withdraw it from the engine in a straight line so the clutch disc is not damaged while the input shaft is still engaged in the clutch splines.

73 Refer to Steps 67 and 68.

Installation

74 To rejoin the automatic transmission to the engine, reverse Steps 57 through 66. Be sure to align the mating marks on the torque converter and driveplate and tighten all bolts to the specified torque.

75 To rejoin the manual transmission to the engine, reverse Steps 69 through 72. Be sure to tighten all bolts to the specified torque.

76 Attach the lifting chains to the

engine/transmission assembly just like they were during removal.

77 Tilt and lower the engine/transmission unit into the engine compartment, guiding the engine mounts onto the frame mounts, and at the same time raising the transmission into position.

78 Install the engine mount throughbolts and the rear transmission crossmember. Tighten all bolts to the specified torque.

79 Install the remaining components in the reverse order of removal.

80 If equipped with a manual transmission, adjust the clutch as described in Chapter 1.

81 If equipped with a manual four-speed transmission, adjust the shift linkage as described in Chapter 7A.

82 If equipped with an automatic transmission, adjust the throttle valve and neutral safety and backup light switch as described in Chapter 7B.

83 If so equipped, adjust the cruise control as described in Chapter 10.

84 Fill the cooling system with the specified antifreeze and water mixture (Chapter 1).

85 Fill the engine with the correct grade of engine oil (Chapter 1).

86 Check the fluid level in the transmission and add fluid as necessary.

87 Connect the positive battery cable, followed by the negative cable. If sparks or arcing occur as the negative cable is connected to the battery, make sure that all electrical accessories are turned off (check interior dome lights first). If arcing still occurs, make sure that all electrical wires are properly connected to the engine and transmission.

88 Refer to Chapter 2, Part D, for the engine startup procedure.

Notes

Chapter 2 Part C
V8 engine

Contents

	Section
Air filter - replacement	See Chapter 1
Camshaft - removal and installation	16
Check engine light	See Chapter 6
Compression check	See Chapter 2D
Crankcase front cover - removal and installation	13
Cylinder heads - removal and installation	8
Drivebelt check and adjustment	See Chapter 1
Engine - removal and installation	18
Engine mounts - replacement	17
Engine oil and filter change	See Chapter 1
Engine overhaul - general information	See Chapter 2D
Engine removal - methods and precautions	See Chapter 2D
Exhaust manifolds - removal and installation	7
Front cover oil seal - replacement	14
General information	1

	Section
Hydraulic lifters - removal, inspection and installation	5
Intake manifold - removal and installation	4
Oil pan - removal and installation	9
Oil pump - removal and installation	10
Rear main bearing oil seal - replacement	11
Repair operations possible with the engine in the vehicle	See Chapter 2D
Rocker arm covers - removal and installation	2
Spark plug replacement	See Chapter 1
Timing chain and sprockets - removal and installation	15
Valve lash - adjustment	6
Valve train components - replacement	3
Vibration damper - removal and installation	12
Water pump - removal and installation	See Chapter 3

Specifications

Torque specifications

	Ft-lbs (unless otherwise indicated)
Camshaft sprocket bolts	20
Camshaft rear cover bolts	84 in-lbs
Clutch pressure plate bolts	35
Crankcase front cover bolts	84 in-lbs
Crankshaft pulley bolts	25
Cylinder head bolts	
1988 and earlier	65
1989 and later	68
Distributor clamp bolt	20
Driveplate to torque converter bolts	
1984 and earlier	27
1985 and later	35
Engine mounting bracket bolts	80
Engine strut bracket bolts	35
Exhaust manifold bolts	20
Flywheel-to-crankshaft bolts	
1985 and earlier	60
1986 and later	74

Torque specifications

	Ft-lbs (unless otherwise indicated)
Flywheel housing bolts	
1984 and earlier..	30
1985 and later	55
Flywheel housing cover bolts	84 in-lbs
Fuel pump mounting bolts..	15
Intake manifold bolts	30
Intake manifold cover-to-intake manifold (cross-fire injected models) ...	20
Oil filter bypass valve..	84 in-lbs
Oil pan-to-crankcase bolts	
5/16-inch bolts	14
1/4-inch bolts	84 in-lbs
Rocker arm cover bolts	
1984 and earlier...	48 in-lbs
1985	60 in-lbs
1986 through 1989	75 in-lbs
1990 and later	89 in-lbs
Spark plugs...	22
Vibration damper bolt	60

Note: *Refer to Chapter 2, Part D, for additional specifications.*

2.8a 1986 and earlier rocker arm covers bolts

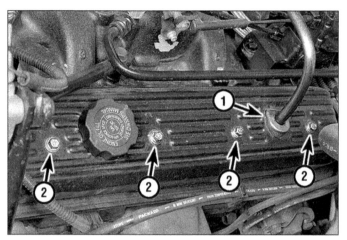

2.8b On 1987 and later models, the rocker arm covers are
retained by bolts in the middle of the cover

1   General information

The forward Sections in this Part of Chapter 2 are devoted to in-vehicle repair procedures for the V8 engine. The latter Sections in this Part of Chapter 2 involve the removal and installation procedures for the V8 engine. All information concerning engine block and cylinder head servicing can be found in Part D of this Chapter.

The repair procedures included in this Part are based on the assumption that the engine is still installed in the vehicle. Therefore, if this information is being used during a complete engine overhaul - with the engine already out of the vehicle and on a stand - many of the steps included here will not apply.

The Specifications included in this Part of Chapter 2 apply only to the engine and procedures found here. For Specifications regarding engines other than the V8, see Part A or B, whichever applies. Part D of Chapter 2 contains the Specifications necessary for engine block and cylinder head rebuilding procedures.

2   Rocker arm covers - removal and installation

Right side

Refer to illustrations 2.8a and 2.8b

1   Disconnect the cable from the negative battery terminal. **Caution:** *On models equipped with a Delco Loc II audio system, disable the anti-theft feature before disconnecting the battery.*

2   Remove the air cleaner assembly, tagging each hose to be disconnected with a piece of numbered tape to simplify reinstallation.

3   Disconnect the AIR management hoses at the manifold.

4   Disconnect the wires and hoses that would interfere with the rocker arm cover removal. Label all hoses and wires to simplify reinstallation.

5   Remove the EGR solenoid.

6   Unbolt the AIR management valve bracket and move it aside.

7   Remove the AIR management tubes.

8   Remove the rocker arm cover bolts **(see illustrations).**

9   Remove the cover. **Note:** *If the cover adheres to the cylinder head, use a block of wood and a rubber hammer to dislodge it. If the cover still will not come loose, pry on it carefully, but do not distort the sealing flange surface.*

10   Prior to installation of the cover, clean all dirt, oil and old gasket material from the sealing surfaces of the cover and cylinder head with a scraper and degreaser.

11   If the rocker cover does not use a gasket, apply a continuous 3/16-inch diameter bead of RTV sealant to the sealing flange of the cover. Be sure to apply the sealant inboard of the bolt holes. Place the rocker arm cover on the cylinder head while the sealant is still wet.

12   If a gasket is used, install a new rocker cover gasket. **Note:** *On models sealed with RTV, rather than reinstalling the covers with RTV, use a new gasket, reinforcement bars and bolts. These parts are available from your dealer or auto supply store.*

13   Install the mounting bolts and tighten them in several stages to the specified torque.

3.5 Install an air hose adapter into the spark plug hole and connect a compressed air source - the air pressure will hold the valves closed when the springs are removed

14   The remaining installation steps are the reverse of removal.

Left side

15   Disconnect the cable from the negative battery terminal. **Caution:** *On models equipped with a Delco Loc II audio system, disable the anti-theft feature before disconnecting the battery.*
16   On TBI models, remove the air cleaner assembly, tagging each hose to be disconnected with a piece of numbered tape to simplify reinstallation.
17   Remove the power brake booster line.
18   Disconnect the AIR hoses.
19   Disconnect the PCV valve hose.
20   Unbolt the wire harness and move it aside.
21   On TBI models, remove the fuel return line from the throttle body.
22   Remove the rocker arm cover bolts.
23   Follow Steps 9 through 14.

---

3   Valve train components - replacement

Refer to illustrations 3.5, 3.8 and 3.14

1   Remove the rocker arm cover(s) as described in Section 2.
2   If only the pushrod is to be replaced, loosen the rocker nut enough to allow the rocker arm to be rotated away from the pushrod. Pull the pushrod out of the hole in the cylinder head.
3   If the rocker arm is to be removed, remove the rocker arm nut and pivot and lift off the rocker arm.
4   If the valve spring is to be removed, remove the spark plug from the affected cylinder.
5   There are two methods that will allow the valve to remain in place while the valve spring is removed. If you have access to compressed air, install an air hose adapter in

3.8 Depress the spring and remove the keepers with a small magnet

the spark plug hole **(see illustration)**. When air pressure is applied to the adapter, the valves will be held in place by the pressure.
6   If you do not have access to compressed air, bring the piston of the affected cylinder to top dead center (TDC) on the compression stroke. Feed a long piece of 1/4inch nylon cord in through the spark plug hole until it fills the combustion chamber. Be sure to leave the end of the cord hanging out of the plug hole so it can be removed easily. Rotate the crankshaft with a wrench (in the normal direction of rotation) until slight resistance is felt.
7   Reinstall he rocker arm nut (without the rocker arm).
8   Insert the slotted end of a valve spring compressing tool under the nut and compress the spring just enough to remove the spring keepers **(see illustration)**.
9   Remove the retainer, cup shield, O-ring seal, spring, spring damper (if so equipped) and valve stem oil seal (if so equipped).
10   If a rocker arm stud requires replacement, first determine whether it is a screw-in or press-in type. Screw-in studs are retained by a nut atop the stud boss, while press-in studs have no retaining nut.
11   Screw-in studs may be replaced by removing the damaged one and replacing it with a new one. Be sure to reinstall the pushrod guide (if so equipped) under the stud nut and tighten the nut to the specified torque.
12   Press-in studs that have damaged threads or are loose in the cylinder head should be replaced with new studs that are available in 0.003-inch and 0.013-inch oversize. Press-in stud replacement procedures require the use of special tools and should be performed by an automotive machine shop. If such services are required, remove the cylinder head(s) by referring to Section 8.
13   Inspect the valve train components for wear or damage. Check for a bent pushrod. Replace any defective parts, as necessary.
14   Installation of the valve train components is the reverse of the removal procedure. Install a new valve stem oil seal after

3.14 Compress the valve spring and install the O-ring seal squarely in the lower groove before installing the keepers

compressing the spring and before installing the spring keepers **(see illustration)**. Before installing the rocker arms, coat the bearing surfaces of the arms and pivots with moly-based grease or engine assembly lube. Be sure to adjust the valve lash as detailed in Section 6.

---

4   Intake manifold - removal and installation

1   If the vehicle is equipped with air-conditioning, carefully examine the routing of the hoses and the mounting of the compressor. You may be able to remove the intake manifold without disconnecting the system. If you are in doubt, take the vehicle to a dealer or refrigeration specialist to have the system depressurized. Do not, under any circumstances, disconnect the hoses while the system is under pressure.
2   Disconnect the cable from the negative battery terminal. **Caution:** *On models equipped with a Delco Loc II audio system, disable the anti-theft feature before disconnecting the battery.*
3   Drain the coolant from the radiator (refer to Chapter 1).
4   Remove the air cleaner assembly, tagging each hose to be disconnected with a piece of numbered tape to simplify reinstallation.

Carbureted models

Refer to illustrations 4.24 and 4.27

5   Label and disconnect all electrical wires and vacuum hoses at the carburetor.
6   Disconnect the fuel line at the carburetor. Be prepared to catch some fuel, then plug the fuel line to prevent contamination.
7   Remove the upper radiator hose from the manifold.
8   Disconnect the heater hose at the manifold.
9   Disconnect the carburetor linkage, making careful note of how it was installed.

4.24 The ends of the intake manifold are sealed with RTV - make sure the bead overlaps the gasket

10   Disconnect the spark plug wires at the spark plugs on the right-hand side, referring to the removal technique described in Chapter 1.

11   Remove the distributor cap and spark plug wires.

12   Remove the distributor (refer to Chapter 5).

13   Remove the cruise control servo and bracket (if so equipped).

14   Remove the alternator upper mounting bracket (refer to Chapter 5).

15   Remove the EGR solenoid and bracket from the rear of the manifold.

16   Disconnect the vacuum line leading to the power brake booster (if so equipped).

17   Remove the manifold mounting bolts.

18   Make sure that all electrical wires, vacuum hoses and coolant hoses that would interfere with manifold removal have been disconnected.

19   Separate the manifold from the engine. Do not pry between the mating surfaces as it could damage them. Tap the manifold with a hammer and wooden block, if necessary.

20   If the intake manifold is to be replaced with another one, transfer all components that have not yet been removed. These include:

a) *Carburetor and carburetor mounting bolts or studs (see Chapter 4 for procedures)*

b) *Thermostat and housing (use a new gasket or RTV, as applicable)*

c) *All necessary switches and fittings*

21   Before installing the manifold, place clean lint-free rags in the engine cavity and clean the engine block, cylinder head and manifold gasket surfaces. All gasket material and sealing compound must be removed prior to installation.

22   Remove all dirt and gasket remnants from the engine cavity.

23   Clean the gasket sealing surfaces with degreaser.

24   Position the gaskets on the cylinder heads and apply a 3/16-inch bead of RTV sealant to the front and rear ridges of the engine block **(see illustration)**.

25   Extend the bead 1/2-inch up each cylinder head to seal and retain the manifold side gaskets. Use sealant at the coolant passages.

26   Carefully lower the intake manifold into position, making sure you do not disturb the gaskets.

27   Install the intake manifold mounting bolts and tighten them following the recommended sequence **(see illustration)**. Tighten the bolts a little at a time until they are all at the specified torque.

28   Install the remaining components in the reverse order of removal.

29   Fill the radiator with coolant, start the engine and check for leaks.

Adjust the ignition timing and idle speed as necessary (refer to Chapter 1).

Throttle body injected models (including cross-fire injection)

Refer to illustration 4.53

30   Disconnect the fuel inlet line at the front TBI unit.

31   Remove the EGR solenoid from the rear of the manifold.

32   Unbolt and remove the alternator adjusting bracket from the front of the manifold.

33   Disconnect the electrical wires from the Idle Air Control (IAC) valve (one on each TBI

unit on cross-fire models).

34   Disconnect the electrical wires from the injectors.

35   Disconnect the electrical wire at the Throttle Position Sensor (TPS).

36   Disconnect the power brake booster lines at the manifold.

37   Disconnect the fuel return line from the throttle body.

38   Remove the air-conditioning strut (if so equipped).

39   Disconnect the throttle cable from the throttle body.

40   Disconnect the automatic transmission throttle valve cable (if so equipped).

41   Disconnect the cruise control cable (if so equipped).

42   Remove the PCV valve.

43   Label and disconnect any vacuum hoses not already disconnected.

44   Make sure that all electrical wires and coolant hoses have been disconnected.

45   On cross-fire models, remove the fuel balance tube.

46   On cross-fire models, remove the nuts and bolts retaining the intake manifold cover to the intake manifold and carefully remove the manifold cover with the TBI units attached.

47   On TBI models, remove the throttle body unit.

48   Remove the distributor (refer to Chapter 5).

49   Remove the radiator hose at the thermostat housing.

50   Disconnect the heater hose at the manifold.

51   Remove the manifold-to-cylinder head bolts and separate the intake manifold from the engine.

52   If the manifold and/or manifold cover is to be replaced with a new one, transfer all components still attached to the old unit to the new one.

53   Follow the procedures outlined in Steps 21 through 29. On single TBI models, use the carbureted model intake manifold bolt tightening sequence **(see illustration 4.27)**. On cross-fire models, use the intake manifold bolt tightening sequence as shown the accompanying illustration **(see illustration)**.

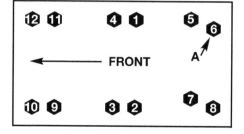

A - Stud bolt

24064-2A-6.15a HAYNES

4.27 Intake manifold bolt tightening sequence - carbureted models

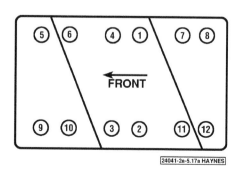

24041-2a-5.17a HAYNES

4.53 Intake manifold bolt tightening sequence - cross-fire injected models

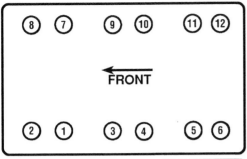

4.70 Intake manifold bolt
tightening sequence -
Tuned Port Injected models

Tuned-port injected models

Refer to illustration 4.70

54   Disconnect the negative battery cable from the battery. **Caution:** *On models equipped with a Delco Loc II audio system, disable the anti-theft feature before disconnecting the battery.*

55   Drain the coolant from the radiator (refer to Chapter 1).

56   Disconnect the accelerator cable, throttle valve cable and cruise control cable from the throttle body by removing the retaining clips and the three bolts from the cable bracket.

57   Remove the air intake duct from the front of the throttle body.

58   Disconnect the coolant hoses at the throttle body.

59   Disconnect the electrical and vacuum connectors from the throttle body, marking them with tape to avoid confusion during reassembly.

60   Remove the intake plenum and runners (refer to Chapter 4).

61   Mark the distributor housing-to-manifold position and rotor-to-distributor housing position, remove the distributor hold-down bolt and lift the distributor from the engine.

62   Disconnect and mark the EGR electrical connector.

63   Mark and disconnect the vacuum hoses from the fuel pressure regulator, EGR valve and canister vacuum valve (near the left front of the intake manifold), then remove the EGR solenoid and the vacuum hoses.

64   Unplug the electrical connectors from the fuel injectors, the ground wire on the thermostat housing and the electrical connectors from the coolant temperature sensors on the front of the intake manifold. Mark the connectors with pieces of numbered tape to avoid confusion during reassembly. Position the wiring harnesses out of the way.

65   Remove the intake manifold bolts, then separate the manifold from the engine. If the manifold sticks, do not pry between the sealing surfaces of the manifold and head. Instead, break it loose using a block of wood and a rubber hammer.

66   Clean the sealing surfaces on the manifold and heads, and also on the ridges between the cylinder banks. Remove all traces of old gasket and sealing material.

67   Install new intake manifold gaskets on the cylinder heads and apply a 3/16-inch bead of RTV sealant to the front and rear ridges of the block **(see illustration 4.24)**. Extend the bead 1 /2-inch up each cylinder head to help hold the gaskets in place.

68   Lower the intake manifold into position and install the bolts.

69   Starting with the center manifold bolts, partially tighten the bolts in a criss-cross pattern to seat the manifold.

70   Final tighten the intake manifold bolts following the recommended sequence to the specified torque **(see illustration)**.

71   The remaining steps are the reverse of removal.

5   Hydraulic lifters - removal, inspection and installation

1   A noisy valve lifter can be isolated when the engine is idling. Place a length of hose or tubing near the position of each valve while listening at the other end of the tube.

2   Assuming that adjustment is correct, the most likely cause of a noisy valve lifter is excessive wear at the lifter foot and camshaft lobe or wear or contamination of the internal components.

1986 and earlier models

Refer to illustrations 5.7, 5.9a and 5.9b

3   Remove the rocker arm covers as described in Section 2.

4   Remove the intake manifold as described in Section 4.

5   Loosen the rocker arm nut and rotate the rocker arm away from the pushrod.

6   Remove the pushrod. If more than one pushrod is removed, store them in an organized manner so they can be reinstalled in their original location.

7   To remove the lifter, a special hydraulic lifter removal tool should be used **(see illustration)**. Do not use pliers or other tools on the outside of the lifter body, as they will damage the finished surface and render the lifter useless.

8   If the lifters are to be reused, they must be stored in an organized manner so they can be reinstalled in their original positions.

9   Clean and dry the lifters. Examine them

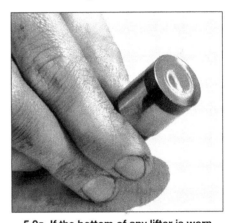

5.7 The lifters on an engine that has accumulated many miles may have to be removed with a special tool (arrow) - store the lifters in an organized manner to make sure they are reinstalled in their original locations

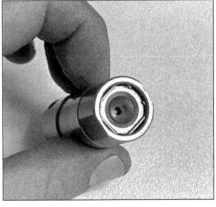

5.9a If the bottom of any lifter is worn concave, scratched or galled, replace the entire set with new lifters (the camshaft must be replaced as well)

5.9b Check the pushrod seat (arrow) in the top of each lifter for wear

for wear or damage and check the foot carefully for wear **(see illustrations)**.

10   If the lifters are worn, they must be replaced with new ones and the camshaft must be replaced as well (see Section 16). If the lifters are in good condition, they can be

5.18 Remove the three bolts and the lifter guide retainer

5.22 The roller on roller lifters must turn freely - check for wear and excessive play as well

6.5 Determine the point of zero lash by spinning the pushrod as the nut is slowly tightened - when you just feel a slight amount of resistance at the pushrod, stop tightening the nut - if you tighten the nut to far, engine damage could result when the engine is started

reinstalled - provided they are installed in their original locations.

11   When installing the lifters, make sure they are replaced in their original bores and coat them with moly-based grease or engine assembly lube.

12   The remaining installation steps are the reverse of removal.

1987 and later models

Refer to illustrations 5.18 and 5.22

13   Beginning in 1987, the V8 engine is equipped with roller type hydraulic valve lifters. Lifter retainers and guides are used to prevent the lifters from rotating on the camshaft lobes.

14   Remove the rocker arm covers as described in Section 2.

15   Remove the intake manifold as described in Section 4.

16   Loosen the rocker arm nut and rotate the rocker arm away from the pushrod.

17   Remove the pushrod. If more than one pushrod is removed, store them in an organized manner so they can be reinstalled in their original location.

18   Remove the valve lifter guide retainer **(see illustration)**.

19   Remove the valve lifter guide.

20   To remove the lifter, a special hydraulic lifter removal tool should be used **(see illustration 5.7)**. Do not use pliers or other tools on the outside of the lifter body, as they will damage the finished surface and render the lifter useless.

21   If the lifters are to be reused, they must be stored in an organized manner so they can be reinstalled in their original positions.

22   Inspection and disassembly are similar to the standard hydraulic lifter. However, the roller must be inspected for freedom of movement, excessive looseness, flat spots or pitting **(see illustration)**. The camshaft must also be inspected if any signs of abnormal wear are indicated.

23   When reinstalling the lifters, make sure they are replaced in their original bores and coat them with moly-based grease or engine

assembly lube.

24   The remaining installation steps are the reverse of removal.

6   Valve lash - adjustment

Refer to illustration 6.5

1   Disconnect the cable from the negative battery terminal. **Caution:** *On models equipped with a Delco Loc II audio system, disable the anti-theft feature before disconnecting the battery.*

2   Remove the rocker arm covers (refer to Section 2). Remove the spark plugs.

3   If the valve train components have been serviced just prior to this procedure, make sure that the components are completely reassembled and the rocker arm nuts are loose.

4   Using a breaker bar and socket on the crankshaft pulley center bolt, rotate the crankshaft clockwise until the number one piston is at top dead center (TDC) on the compression stroke. To locate TDC, place your finger over the number one cylinder spark plug hole and rotate the crankshaft. When compression pressure is felt at the spark plug hole, continue rotating the crankshaft until the timing mark on the vibration damper aligns with the "0" on the timing indicator (see Chapter 1). To make sure that the number one piston is at TDC, remove the distributor cap and check the position of the rotor in the distributor, it should be pointing at the number one cylinder spark plug wire terminal. If it's pointing at the number six terminal, rotate the crankshaft 360-degrees. As an alternate method, watch the number one cylinder rocker arms as you rotate the crankshaft. After the number one cylinder intake valve opens and closes, continue rotating the crankshaft until the timing mark on the vibration damper aligns with the "0" on the timing indicator - the number one piston is at TDC.

5   To adjust a valve, slowly tighten the rocker arm nut until all clearance between the

pushrod and rocker arm is removed (zero lash), then tighten the nut 3/4-turn further to preload the lifter. Reaching the point of zero lash can be difficult and requires some expertise, proceed as follows **(see illustration)**:

  a) *Move the pushrod up-and-down and tighten the rocker arm nut until all clearance between the pushrod and rocker arm is removed.*

  b) *Spin the pushrod between your thumb and index finger and slowly tighten the nut. When you just feel a slight resistance as the pushrod is rotated, stop tightening the nut - you've reached zero lash. Do not exert downward pressure with the tool used to tighten the nut as this will effect the adjustment.*

  c) *After the point of zero lash is reached, tighten the nut 3/4-turn to complete the valve adjustment.*

6   With the number one piston at TDC, adjust the following valves: number one, two, five and seven cylinder intake valves and the number one, three, four and eight cylinder exhaust valves, using the method described.

7   Rotate the crankshaft 360-degrees (number six piston at TDC on the compression stroke) and adjust the following valves: number three, four, six and eight cylinder intake valves and the number two, five, six and seven cylinder exhaust valves, using the method described.

8   Install the rocker arm covers and spark plugs.

7   Exhaust manifolds - removal and installation

1   If the vehicle is equipped with air conditioning, carefully examine the routing of the hoses and the mounting of the compressor. Depending on the system used, you may be able to remove the exhaust manifolds without

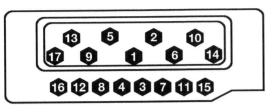

8.17 Cylinder head bolt TIGHTENING sequence

24065-2a-8.16b HAYNES

disconnecting the system. If you are in doubt, take the vehicle to a dealer or air-conditioning technician to have the system depressurized. Do not, under any circumstances, disconnect the hoses while the system is under pressure.

2 Disconnect the negative battery cable from the battery. **Caution:** *On models equipped with a Delco Loc II audio system, disable the anti-theft feature before disconnecting the battery.*

3 On carbureted and TBI models, disconnect the hot air pipe routed from the right exhaust manifold to the air cleaner.

4 Mark each spark plug wire with a strip of numbered tape and then remove the plug wires from the spark plugs. Remember to pull on the rubber boot and not the wire itself.

5 Remove the spark plug heat shields. These four shields are held in place with bolts.

6 Disconnect the AIR hoses at the manifold check valves. The single metal pipe and the AIR manifold can be removed with the exhaust manifold.

7 Without disconnecting the hoses, remove the power steering pump and air conditioning compressor and position them aside. Remove the brackets attached to the left-side manifold.

8 Disconnect the exhaust pipes from the manifold flanges. The exhaust pipes can be hung from the frame with wire.

9 If equipped with bolt locking strips, use a screwdriver to pry the metal locking strips away from the bolt heads. This will enable you to put a standard wrench on the bolts.

10 Remove the exhaust manifold end bolts first, followed by the center bolts. Lift the manifold and AIR injection tubes away from the engine.

11 If you are replacing the exhaust manifolds with new ones, the carburetor hot-air stove and the AIR injection tube system must be transferred to the new manifolds.

12 Before installing the exhaust manifolds, clean the mating surfaces on the cylinder heads and manifolds.

13 Install the manifold and tighten the center bolts first, followed by the end bolts. On some models it is easier to install the left rear spark plug heat shield first due to inaccessibility after the exhaust manifold is in place.

14 Install the remaining components in the reverse order of removal.
Use a new gasket or packing at the exhaust manifold-to-exhaust pipe flange connections.

15 Start the engine and check for exhaust leaks where the manifolds meet the cylinder heads and where the manifolds join the exhaust pipes.

8 Cylinder heads - removal and installation

Refer to illustration 8.17

1 Locate the engine block drain plugs at the center of the block on each side, remove the plugs and drain the coolant.

2 Remove the intake manifold (refer to Section 4).

3 Remove the exhaust manifolds (refer to Section 7).

4 Remove the lower power steering pump adjusting bracket from the left-side cylinder head (if not previously removed). Remove the lower alternator mounting bolt from the right-side cylinder head.

5 Remove the rocker arm covers (refer to Section 2).

6 Loosen the rocker arm nuts enough to allow removal of the pushrods.

7 Remove the pushrods and store them separately so they can be reinstalled in their original positions. A cardboard box with numbered holes can be used effectively.

8 Remove the head bolts in a sequence opposite to the tightening sequence **(see illustration 8.17)**.

9 Remove the cylinder head(s). To break the gasket seal, insert a bar into one of the exhaust ports, then carefully lift on the tool.

10 If a new cylinder head is being installed, transfer the external components from the old head to the new head.

11 Before installing the head(s), the gasket surfaces on both the head and engine block must be clean and free of nicks and scratches. Also, the threads in the block and on the head bolts must be completely clean, as any dirt remaining in the threads will affect the bolt torque.

12 When installing the cylinder heads on engines using a plain steel gasket, coat both sides of the new gasket with a good quality sealer made expressly for head gasket installation. To ensure an even coat of the proper thickness, use a paint roller to spread the sealer. The application of too much sealer may hold the gasket away from the head or block.

13 Do not use sealer on engines requiring a composite-type steel/asbestos gasket.

14 Place the gasket in position over the dowel pins with the bead up.

15 Position the cylinder head over the gasket.

16 Coat the cylinder head bolt threads with an appropriate thread sealer and install the bolts.

17 Tighten the bolts in the proper sequence to the specified torque **(see illustration)**. Work up the final torque in three steps.

18 Install the pushrods, making sure the lower ends are in the lifter seats, place the rocker arm ends over the pushrods and loosely install the rocker arm nuts.

19 The remaining installation steps are the reverse of removal. Before installing the rocker arm covers, adjust the valve lash (refer to Section 6).

9 Oil pan - removal and installation

1 Disconnect the cable from the negative battery terminal. **Caution:** *On models equipped with a Delco Loc II audio system, disable the anti-theft feature before disconnecting the battery.*

2 On carbureted and TBI models, remove the air cleaner assembly. On TPI models, remove the air intake duct.

3 Remove the distributor cap (refer to Chapter 5) to prevent damage when the engine is raised.

4 Unbolt and remove the top half of the fan shroud.

5 Raise the vehicle and support it on jackstands.

6 Drain the engine oil.

7 Disconnect the exhaust crossover pipe at the exhaust manifold flanges. Lower the exhaust pipes and suspend them from the frame with wire.

8 If equipped with an automatic transmission, remove the converter shroud.

9 If equipped with a manual transmission, remove the flywheel cover.

10 Remove the starter (refer to Chapter 5).

11 Turn the bolt at the center of the vibration damper to rotate the crankshaft until the timing mark is facing down, at the 6 o'clock position. This will move the forward crankshaft throw to the top, providing clearance at the front of the oil pan.

12 Remove the throughbolt at each engine mount.

13 At this time the engine must be raised slightly to allow the oil pan to clear the crossmember. The preferred method requires an engine hoist. Hook up the lifting chains as described in Section 18.

14 An alternative method can be used if you are careful. Place a floor jack and a block of wood under the oil pan. The wood block should spread the load across the oil pan to prevent damage or collapse of the oil pan metal. The oil pump pickup and screen are very close to the oil pan bottom, so any collapse of the pan may damage the pickup or prevent the oil pump from drawing oil properly.

10.3 Make sure the nylon sleeve is in place between the oil pump and oil pump driveshaft

15   With either method, raise the engine slowly until wood blocks can be placed between the frame front crossmember and the engine block. The blocks should be approximately three (3) inches thick. Check clearances all around the engine as it is raised. Pay particular attention to the distributor and the cooling fan.
16   Lower the engine onto the wood blocks. Make sure it is firmly supported. If a hoist is being used, keep the lifting chains secured to the engine.
17   Remove the oil pan bolts. Note the different sizes used and their locations.
18   Remove the oil pan.
19   Before installing the pan, make sure that the sealing surfaces on the pan, block and front cover are clean. If the original oil pan is being reinstalled, make sure that all old gasket material has been removed.
20   Always use new gaskets and seals when installing the pan.
21   Lift the pan into position and install all bolts finger tight. Tighten the end bolts first, then tighten the remaining bolts working from the center out to the ends.
22   Lower the engine onto the mounts and install the throughbolts.

23   Follow the removal steps in reverse order. Fill the crankcase with the correct grade and quantity of oil, start the engine and check for leaks.

10   Oil pump - removal and installation

Refer to illustration 10.3
1   Remove the oil pan (refer to Section 9).
2   Remove the pump-to-rear main bearing cap bolt and separate the pump and extension shaft from the engine.
3   If installing a new oil pump, transfer the oil pump driveshaft to the new oil pump. Use a new nylon sleeve to retain the driveshaft to the oil pump **(see illustration)**.
4   To install the pump, move it into position and align the slot on the top end of the extension shaft with the drive tang on the lower end of the distributor driveshaft. The distributor drives the oil pump, so it is essential that this alignment is correct. If necessary rotate the oil pump driveshaft.
5   Install the oil pump-to-main bearing cap bolt and tighten it to the specified torque.
6   Reinstall the oil pan (refer to Section 9).

11   Rear main bearing oil seal - replacement

1985 and earlier models

Refer to illustrations 11.5, 11.9, 11.10, 11.11a, 11.11b, 11.13 and 11.14
1   Early models are equipped with a two-piece neoprene rear main oil seal. Always replace both halves of the rear main oil seal. While replacement of the seal is much easier with the engine removed, the job can be done with the engine in place.
2   Remove the oil pan and oil pump as described previously in this Chapter.
3   Remove the rear main bearing cap from the engine.
4   Using a screwdriver, pry the lower half

11.5 Push one end of the seal into the engine block until you can grab the other end and pull the seal out - use a brass punch or wood dowel so as not to damage the crankshaft

of the oil seal out of the bearing cap.
5   To remove the upper half of the seal, use a small hammer and a brass pin punch to roll the seal around the crankshaft journal. Tap one end of the seal with the hammer and punch (be careful not to strike the crankshaft) until the other end of the seal protrudes enough to pull the seal out with pliers **(see illustration)**.
6   Remove all sealant and foreign material from the block. Do not use abrasives.
7   Inspect the components for nicks, scratches and burrs on all sealing surfaces.
8   Coat the lips of the new seal with clean engine oil. Do not get oil on the seal mating ends.
9   Included with the new rear main oil seal should be a small plastic installation tool. If a tool is not provided, one can be fashioned from thin stiff plastic or brass shim stock **(see illustration)**.
10   Position the narrow end of the installation tool between the crankshaft and the seal seat **(see illustration)**. The idea is to keep the new seal from being damaged by the sharp edge of the seal seat.

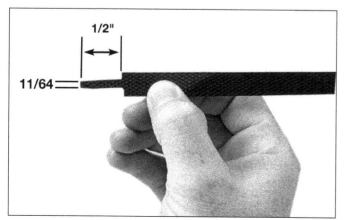

11.9 If the new seal did not include an installation tool, make one from plastic or brass shim stock

11.10 Position the tool to protect the backside of the seal as it passes over the sharp edge of the engine block

11.11a While holding the tool in place, push the new seal around the crankshaft

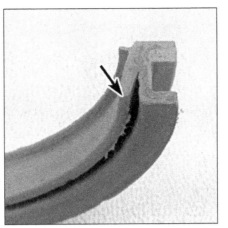

11.11b Make sure the oil seal lip (arrow) is facing toward the front of the engine

11.13 Using the tool as a shoehorn, install the lower half of the seal in the main bearing cap

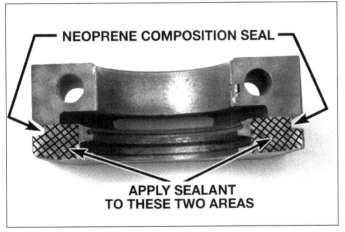

11.14 Apply anaerobic sealant to the areas shown, but don't get sealant in the grooves or on the seal

11.20 Using the notches provided, pry the rear main seal out of the housing with a flat-bladed screwdriver

11 Raise the new upper half of the seal into position with the seal lips facing toward the front of the engine. Push the seal into the seat, using the installation tool as a "shoe-horn' to keep the seal from contacting the sharp edge **(see illustrations)**.

12 Roll the seal around the crankshaft, all the time using the tool for protection. When both ends of the seal are flush with the engine block, remove the installation tool (be careful not to withdraw the seal as well).

13 Install the lower half of the oil seal in the bearing cap, again using the installation tool to protect the seal from the sharp edge **(see illustration)**. Make sure the seal is firmly seated, then withdraw the installation tool.

14 Apply a thin, even film of anaerobic-type sealant to the cap-to-block mating surfaces. Do not get any on the seal lip **(see illustration)**.

15 Install the bearing cap and seal and tighten the bolts to about 10 to12 ft-lbs only. Now tap the end of the crankshaft forward and backward to line up the thrust bearing surfaces. Tighten the bearing cap bolts to the final torque and install the oil pump and oil pan.

1986 and later models

Refer to illustration 11.20

16 Later models are equipped with a one-piece rear main oil seal installed in a bolt-on housing. While replacement of the seal is much easier with the engine removed, the job can be done with the engine in place.

17 Remove the transmission (refer to Chapter 7).

18 On manual transmission models, remove the clutch assembly (refer to Chapter 8).

19 Remove the flywheel or driveplate.

20 Using a flat-bladed screwdriver, carefully pry the oil seal out of the housing **(see illustration)**. Use the notches provided and DO NOT nick or scratch the crankshaft. Note the installed depth and install the new seal to the same depth.

21 Using a one-piece rear main seal installation tool, press the seal squarely into the housing. If the tool is not available, the seal may be driven in using a blunt tool, provided extreme care is taken to drive the seal in squarely and without damage to the seal or crankshaft.

22 The remainder of installation is the reverse of removal.

12 Vibration damper - removal and installation

Refer to illustrations 12.6 and 12.9

1 Remove the hose from the retaining strap atop the radiator shroud by releasing the end of the strap from the underside of the shroud with pliers.

2 Remove the bolts and separate the upper radiator shroud from the radiator.

3 Remove the cooling fan from the fan pulley.

4 Loosen the accessory drivebelt adjusting bolts, as necessary, and remove the drivebelts, tagging each one as it is removed to simplify reinstallation.

5 Remove the bolt from the center of the crankshaft pulley (a screwdriver can be used to lock the starter ring gear on the flywheel), then remove the pulley.

6 Attach a puller to the damper. Draw the damper off the crankshaft, being careful not to drop it as it breaks free. A common gear puller should not be used to draw the damper off, as it may separate the outer portion from the hub. Use only a puller which bolts to the

12.6 Use a puller that bolts to the damper hub to remove the vibration damper - if force is applied the outer edge, the damper may be damaged

12.9 Use a damper installation tool (available at most auto parts stores) to install the vibration damper

hub **(see illustration).**

7   Before installing the damper, coat the front cover seal area (on the damper) with engine oil or multi-purpose grease.

8   Place the damper in position over the key on the crankshaft. Make sure the damper keyway lines up with the key.

9   Using a damper installation tool, push the damper onto the crankshaft. The special tool distributes the pressure evenly around the hub **(see illustration).**

10   Remove the installation tool and install the damper retaining bolt. Tighten the bolt to the specified torque.

11   Follow the removal procedure in the reverse order for the remaining components.

12   Adjust the drivebelts (refer to Chapter 1).

---

13   Crankcase front cover - removal and installation

Refer to illustration 13.4

1   Remove the vibration damper as described in Section 12.

2   Remove the water pump as described in Chapter 3.

3   On 1985 and later models, remove the oil pan as described in Section 9.

4   Remove the crankcase front cover mounting bolts and separate the cover from the engine **(see illustration).**

5   Remove the old gasket and discard it.

6   Clean all oil, dirt and old gasket material from the sealing surfaces of the front cover and engine block. Use a sharp knife to remove any excess oil pan gasket material that may be protruding at the oil pan-to-engine block junction.

7   Replace the front cover seal as described in Section 14.

8   Apply a 1/8-inch bead of RTV sealant to the joint formed by the oil pan and engine block.

9   Coat the new cover gasket with RTV sealant and position it on the engine.

10   Install the cover-to-oil pan gasket, then coat the crankshaft seal with engine assem-

13.4 Remove the front cover bolts and carefully pry the cover off the dowels

bly lube or moly-based grease and position the cover over the crankshaft end.

11   Loosely install the upper cover-to-block retaining bolts.

12   Tighten the bolts in a crisscross pattern while pressing down on the cover so the dowels in the block are aligned with the holes in the cover.

13   Install the remaining cover bolts and tighten them to the specified torque.

14   The remaining steps are the reverse of removal.

---

14   Front cover oil seal - replacement

With front cover installed on engine

1   With the vibration damper removed (Section 12), pry the old seal out of the

14.7 While supporting the cover near the seal bore, drive the seal out from the backside

crankcase front cover with a large screwdriver. Be very careful not to damage the surface of the crankshaft.

2   Place the new seal into position with the open end of the seal and seal lip toward the inside of the cover.

3   Using a seal installation tool, drive the seal into the cover until it is completely seated. Exert even pressure around the entire circumference of the seal as it is hammered into place. A section of large-diameter pipe or a large socket could also be used.

4   Be careful not to distort the front cover.

With front cover removed from engine

Refer to illustrations 14.7 and 14.8

5   This method is preferred because the cover can be supported as the new seal is driven into place, preventing the possibility of cover distortion.

6   Remove the crankcase front cover as described in Section 13.

7   Drive the old seal out of the bore from the backside **(see illustration).**

14.8 Drive the seal squarely into the bore using a seal driver, large socket or section of pipe - be careful not to damage the seal or the cover in the process

8   Support the inside of the cover around the seal area and install the new seal in the same manner as described above **(see illustration)**.

15   Timing chain and sprockets - removal and installation

Refer to illustrations 15.6 and 15.10

1   Disconnect the cable from the negative battery terminal. **Caution:** *On models equipped with a Delco Loc II audio system, disable the anti-theft feature before disconnecting the battery.*
2   Remove the vibration damper (refer to Section 12).
3   Remove the crankcase cover (refer to Section 13).
4   Remove all of the spark plugs from the engine.
5   Temporarily install the crankshaft pulley bolt with several thick washers or a spacer. Rotate the crankshaft in a clockwise direction until the timing marks on the crankshaft and camshaft are aligned **(see illustration 15.10)**.
6   Remove the camshaft sprocket mounting bolts, separate the camshaft sprocket from the camshaft and remove the sprocket and chain **(see illustration)**.
7   If it is necessary to remove the crankshaft sprocket, it can be withdrawn from the crankshaft with a special puller.
8   Attach the crankshaft sprocket to the crankshaft using a bolt and washer from the puller set.
9   Install the timing chain over the camshaft sprocket with slack in the chain hanging down over the crankshaft sprocket.
10   With the timing marks aligned as shown, slip the chain over the crankshaft sprocket and then draw the camshaft sprocket into place with the mounting bolts **(see illustration)**.
11   With the chain and both sprockets in place, check again to ensure that the timing

15.6 Remove the three camshaft sprocket retaining bolts (arrows)

marks on the two sprockets are properly aligned. The camshaft sprocket mark should be at the 6 o'clock position and the crankshaft sprocket mark should be at the 12 o'clock position. A line drawn through the center of the camshaft and crankshaft should pass through each mark. If not, remove the camshaft sprocket and move it until the marks align. **Note:** *Be aware that with the timing marks aligned as shown the engine is positioned with the number six piston at TDC on the compression stroke.*
12   Lubricate the chain with engine oil and install the remaining components in the reverse order of removal.

16   Camshaft - removal and installation

Note: *Before removing the camshaft, refer to Chapter 2, Part D (Section 15), and measure the lobe lift.*

1   If equipped with air conditioning, have

15.10 Install the camshaft sprocket and chain with the timing marks aligned

the system discharged. **Warning:** *The air conditioning system is under high pressure. Do not loosen any hose fittings or remove any components until after the system has been discharged. Air conditioning refrigerant must be properly discharged into an EPA-approved recovery/recycling unit at a dealer service department or an automotive air conditioning repair facility. Always wear eye protection when disconnecting air conditioning system fittings.*
2   Disconnect the cable from the negative battery terminal. **Caution:** *On models equipped with a Delco Loc II audio system, disable the anti-theft feature before disconnecting the battery.*
3   Drain the oil from the crankcase. Drain the coolant from the radiator (refer to Chapter 1).
4   Remove the radiator (refer to Chapter 3).
5   If equipped with air-conditioning, remove the condenser (refer to Chapter 3). Remove the grille.
6   Remove the intake manifold (refer to Section 4).
7   Remove the valve lifters (refer to Section 5).
8   Remove the crankcase front cover (refer to Section 13).
9   On carbureted models, remove the fuel pump and pushrod (refer to Chapter 4).
10   Remove the timing chain and sprocket (refer to Section 15).
11   Install a long bolt in one of the camshaft bolt holes to be used as a handle to pull on and support the camshaft.
12   Carefully withdraw the camshaft from the engine block. Do this very slowly to avoid damage to the camshaft bearings as the lobes pass over the bearing surfaces. Always support the camshaft with one hand near the engine block.
13   Refer to Chapter 2, Part D, for the camshaft inspection procedures.
14   Before installing the camshaft, coat each of the lobes and journals with camshaft assembly lube.
15   Slide the camshaft into the engine block, again taking extra care not to damage the bearings.
16   Install the camshaft sprocket and timing

chain as described in Section 15.

17   Install the remaining components in the reverse order of removal by referring to the appropriate Chapter or Section.

18   Adjust the valve lash (refer to Section 6).

19   Have the air-conditioning system recharged (if so equipped).

17   Engine mounts - replacement

1   If the rubber mounts have become hard, split or separated from the metal backing, they must be replaced. The operation may be carried out with the engine/transmission still in the vehicle. See Section 9 for the proper way to raise the engine while it is still in place.

Front mounts

Refer to illustration 17.2

2   Remove the throughbolt and nut **(see illustration)**.

3   Raise the engine slightly using a hoist or jack with a wood block under the oil pan, then remove the mount and frame bracket assembly from the crossmember.

4   Position the new mount, install the throughbolt and nut, then tighten all the bolts to the specified torque.

Rear mount

5   Remove the crossmember-to-mount bolts, then raise the transmission slightly with a jack.

6   Remove the mount-to-transmission bolts, followed by the mount.

7   Install the mount, lower the transmission and align the crossmember-to-mount bolts.

8   Tighten all the bolts to the specified torque.

18   Engine - removal and installation

Note: *On some models, the upper bellhousing-to-engine bolts are inaccessible, in some cases requiring cutting the floor pan to expose the bolts. If this is the case with your vehicle and you do not want to cut the floor pan, read through the procedure in this Section to familiarize yourself with the V8 engine removal, then refer to the engine removal Section in Chapter 2, Part B, and remove the engine and transmission as a unit, noting any differences between the V6 and V8 engines and adjusting the procedure accordingly (many of the V6 illustrations also apply to the V8).*

Removal

1   If equipped with air conditioning, have the system discharged. **Warning:** *The air conditioning system is under high pressure. Do not loosen any hose fittings or remove any components until after the system has been discharged. Air conditioning refrigerant must be properly discharged into an EPA-approved recovery/recycling unit at a dealer*

17.2 Typical engine mount details

1   Engine mount
2   Engine mount bracket
3   Through-bolt
4   Crossmember

service department or an automotive air conditioning repair facility. Always wear eye protection when disconnecting air conditioning system fittings.

2   Disconnect the cable from the negative battery terminal. **Caution:** *On models equipped with a Delco Loc II audio system, disable the anti-theft feature before disconnecting the battery.*

3   Remove the clip retaining the underhood light wire harness to the hood. Disconnect the underhood light electrical connector and unbolt the bulb assembly from the hood.

4   Scribe lines on the underside of the hood, around the hood mounting bracket, so the hood can be installed in the same position.

5   Apply white paint around the bracket-to-hood bolts so they can be aligned quickly and accurately during installation.

6   Remove the bracket-to-hood bolts.

7   With the help of an assistant, lift the hood and separate it from the vehicle.

8   Drain the radiator by opening the petcock located at the lower right-hand side of the radiator.

9   Remove the coolant recovery tank hose from the radiator.

10   Remove the upper radiator hose from the radiator. Take care not to damage the plastic fitting when breaking the hose loose.

11   If so equipped, remove the automatic transmission cooler lines at the radiator.

12   Remove the bolts securing the top of the radiator fan shroud to the lower half and remove the upper shroud.

13   Remove the lower radiator hose from the radiator.

14   Remove the radiator from the vehicle.

15   Mark the drivebelts to simplify installation.

16   Unbolt and remove the fan from the fan pulley.

17   If so equipped, remove the air conditioning compressor (refer to Chapter 3).

18   Remove the alternator (refer to Chapter 5).

19   Remove the heater hoses from the water pump and intake manifold outlets.

20   Remove the AIR system air pump (refer

to Chapter 6).

21   Disconnect the hoses at the air cleaner. Carefully label them as they are removed to simplify installation.

22   Remove the air cleaner assembly.

23   Disconnect the vacuum brake booster line at the booster.

24   Disconnect the linkage at the carburetor or throttle body, including the cruise control cable (if so equipped). Label the linkage/cable components as they are disconnected.

25   Remove the distributor cap with the wires attached and lay the assembly aside (refer to Chapter 5 if necessary).

26   Disconnect the electrical connectors from the carburetor, throttle body and fuel injectors, as necessary (refer to Chapter 4 for details).

27   Disconnect the electric choke wire connector from the carburetor.

28   Disconnect the electrical connector for the EFE heater wire at the base of the carburetor.

29   On fuel-injected models, remove the fuel hoses at the TBI unit or fuel rail, then plug the openings to prevent fuel loss and contamination.

30   Disconnect the wires at the coolant temperature and oil pressure sending units.

31   Remove the power steering pump mounting bolts and lay the pump aside, with the hoses still attached, out of the way.

32   Unbolt the solenoid valve/hose bracket from the left-hand rocker arm cover so it will not be damaged when the engine is lifted out of the vehicle. Lay the assembly aside.

33   Remove the oxygen sensor from the right-side exhaust manifold so it will not be damaged as other components are removed. Because it is a very sensitive component, be sure to protect it.

34   Raise the vehicle and place it securely on jackstands.

35   Drain the oil from the engine crankcase.

36   Remove the bolts securing the exhaust pipes to the exhaust manifolds on both sides of the engine, then pull the exhaust pipes off the exhaust manifolds.

37   On   manual   transmission-equipped

models, disconnect the clutch fork return spring.

38 Disconnect the electrical wires at the transmission.

39 Unbolt and remove the transmission dust cover.

40 On vehicles equipped with an automatic transmission, remove the converter-to-drive-plate bolts, working through the opening gained by the removal of the cover mentioned in Step 39. Turn the engine over with the bolt at the center of the vibration damper to bring each of the bolts into view. Mark the relative position of the converter with a scribe so it can be reinstalled in the same position. Wedge a screwdriver in the teeth of the flywheel to prevent movement as the bolts are loosened.

41 Disconnect and label the wires at the starter motor.

42 On carburetor-equipped models, disconnect the lines at the fuel pump and plug the openings to prevent loss of fuel and contamination.

43 Remove the AIR system/catalytic converter pipe mounting bracket from the transmission.

44 Remove the bellhousing-to-engine bolts (refer to the note at the beginning of this Section).

45 Remove the motor mount throughbolt on each side of the engine.

46 Make sure that all wires and hoses, both under the vehicle and in the engine compartment, have been disconnected.

47 Remove the jackstands and lower the vehicle.

48 Attach the engine hoist lifting chains to the lifting brackets on the engine.

There is one bracket at the front of the engine and one at the rear, diagonally opposite each other. Make sure the chain is looped properly through the engine brackets and secured with strong nuts and bolts through the chain links. The hook on the hoist should be over the center of the engine with the lengths of chain equal to lift the engine straight up.

49 Raise the engine hoist until all slack is out of the chains. Do not lift any further at this time.

50 Support the transmission using a jack with wood blocks as cushions.

51 Raise the engine slightly and then pull it forward to clear the input shaft (manual transmission). Where an automatic transmission is installed, keep the torque converter pushed well to the rear to ensure engagement of the converter tangs with the oil pump inside the transmission.

52 Carefully lift the engine straight up and out of the engine compartment, continually checking clearances around the engine.

53 The transmission should remain supported by the floor jack or wood blocks while the engine is out of place.

54 Attach the engine to an engine stand.

55 Refer to Chapter 2, Part D, for further disassembly and rebuilding procedures.

Installation

56 Lift the engine off the engine stand with a hoist. The chains should be positioned as they were during removal, with the engine sitting level.

57 Lower the engine into place inside the engine compartment, closely watching clearances. On manual transmissions, carefully guide the engine onto the transmission input shaft. The two components should be at the same angle, with the shaft sliding easily into the engine.

58 Install the engine mount throughbolts and the bellhousing bolts. Tighten the bolts to the specified torque.

59 Install the remaining engine components in the reverse order of removal.

60 Fill the cooling system with the specified antifreeze and water mixture (Chapter 1).

61 Fill the engine with the correct grade of engine oil (Chapter 1).

62 Check the transmission fluid level, adding fluid as necessary.

63 Connect the positive battery cable, followed by the negative cable. If sparks or arcing occur as the negative cable is connected to the battery, make sure that all electrical accessories are turned off (check the dome light first). If arcing still occurs, make sure that all electrical wires are properly connected to the engine and transmission.

64 Refer to Chapter 2, Part D, for the engine startup procedure.

Notes

Chapter 2 Part D
General engine overhaul procedures

Contents

	Section
Camshaft, lifters and bearings - inspection and bearing replacement	15
Compression check	4
Crankshaft - inspection	17
Crankshaft - installation and main bearing oil clearance check	21
Crankshaft - removal	12
Cylinder head - cleaning and inspection	8
Cylinder head - disassembly	7
Cylinder head - reassembly	10
Engine block - cleaning	13
Engine block - inspection	14
Engine overhaul - disassembly sequence	6
Engine overhaul - reassembly sequence	23
Engine rebuilding alternatives	3

	Section
Engine removal - methods and precautions	5
General information	1
Initial start-up and break-in after overhaul	25
Main and connecting rod bearings - inspection	18
Piston/connecting rod assembly - inspection	16
Piston/connecting rod assembly - installation and bearing oil clearance check	22
Piston/connecting rod assembly - removal	11
Piston rings - installation	19
Pre-oiling engine after overhaul	24
Rear main oil seal - installation	20
Repair operations possible with the engine in the vehicle	2
Valves - servicing	9

Specifications

L4 engine

Oil pressure	15 psi minimum at 2000 rpm
Compression pressure	100 to 150 psi
Maximum variation between cylinders	20-percent
Firing order	1-3-2-4

Valves and related components

Valve face angle	45-degrees
Valve seat angle	46-degrees
Valve seat width	
Intake	0.0353 to 0.0747 inch
Exhaust	0.058 to 0.097 inch
Valve seat runout	0.002 inch maximum
Valve stem diameter	
Intake	0.3425 to 0.3418 inch
Exhaust	0.3418 to 0.3425 inch
Stem-to-guide clearance	
Intake and top exhaust	0.001 to 0.0027 inch
Bottom exhaust	0.002 to 0.0037 inch
Valve spring installed height	1.69 inches
Valve spring free length	2.08 inches

Crankshaft and connecting rods

Crankshaft endplay	0.0035 to 0.0085 inch
Connecting rod endplay (side clearance)	0.006 to 0.022 inch
Main bearing journal diameter	2.300 inches
Main bearing oil clearance	0.0005 to 0.0022 inch
Connecting rod bearing journal diameter	2.000 inches
Connecting rod bearing oil clearance	0.0005 to 0.0026 inch
Crankshaft journal taper/out-of-round limit	0.0005 inch

L4 engine (continued)

Engine block

Cylinder bore diameter	4.000 inches
Out-of-round limit	0.0014 inch
Taper limit	0.0005 inch

Pistons and rings

Piston diameter	3.9971 to 3.9975 inches
Piston-to-bore clearance	0.0014 to 0.0022 inch
Compression ring side clearance	0.003 inch
Piston ring end gap	
Top ring	0.015 to 0.025 inch
Second ring	0.010 to 0.020 inch
Oil ring	0.015 to 0.055 inch

Camshaft

Bearing journal diameter	1.869 inches
Bearing oil clearance	0.0007 to 0.0027 inch
Camshaft endplay	0.0015 to 0.005 inch
Lobe lift (intake and exhaust)	0.398 inch

Torque specifications

	Ft-lbs (unless otherwise indicated)
Rocker arm nuts/bolts	20
Main bearing cap bolts	70
Connecting rod cap nuts	32
Cylinder head bolts	
1984 and earlier	85
1985 and later	92
Oil pump driveshaft retainer plate bolts	120 in-lbs

V6 engine

Oil pressure	15 psi minimum at 2000 rpm
Compression pressure	100 to 150 psi
Maximum variation between cylinders	20-percent
Firing order	1-2-3-4-5-6

Valves and related components

Valve face angle	45-degrees
Valve seat angle	46-degrees
Valve seat width	
Intake	0.049 to 0.059 inch
Exhaust	0.063 to 0.075 inch
Valve seat runout	0.002 inch maximum
Valve stem diameter	Not available
Stem-to-guide clearance	0.001 to 0.0028 inch
Valve spring installed height	1.57 inches
Valve spring free length	1.91 inches

Crankshaft and connecting rods

Crankshaft endplay	0.002 to 0.007 inch
Connecting rod endplay (side clearance)	0.006 to 0.017 inch
Main bearing journal diameter	
1982 (all)	2.4937 to 2.4946 inches
1983 and 1984	
Journals 1, 2 and 4	2.4937 to 2.4946 inches
Journal 3	2.4932 to 2.4941 inches
1985 and later	2.647 to 2.648 inches
Main bearing oil clearance	
1982 (all)	0.0017 to 0.0030 inch
1983 and 1984	
Journals 1, 2 and 4	0.0016 to 0.0032 inch
Journal 3	0.0021 to 0.0033 inch
1985 and later	0.0012 to 0.0030 inch
Connecting rod bearing journal diameter	1.9983 to 1.9994 inches
Connecting rod bearing oil clearance	0.0012 to 0.0037 inch
Crankshaft journal taper/out-of-round limit	0.0002 inch

Engine block

Cylinder bore diameter	3.504 to 3.507 inches
Out-of-round limit	0.0005 inch
Taper limit	0.0005 inch

Pistons and rings

Piston diameter	Not available
Piston-to-bore clearance	0.0012 to 0.0026 inch
Compression ring side clearance	
1989 and earlier	
Top ring	0.0012 to 0.0027 inch
Second ring	0.0016 to 0.0037 inch
1990 and later	0.0020 to 0.0035 inch
Oil ring side clearance	0.0080 inch maximum
Piston ring end gap	
Compression rings	0.0010 to 0.0020 inch
Oil ring	0.055 inch maximum

Camshaft

Bearing journal diameter	1.868 to 1.870 inches
Bearing oil clearance	0.001 to 0.004 inch
Camshaft endplay	Not available
Lobe lift	
1984 and earlier	
Intake	0.231 inch
Exhaust	0.262 inch
1985 and later	
Intake	0.262 inch
Exhaust	0.273 inch

Torque specifications

	Ft-lbs
Rocker arm stud-to-cylinder head	45
Main bearing cap bolts	73
Connecting rod cap nuts	39
Cylinder head bolts	
1989 and earlier	70
1990 and later	
Step 1	40
Step 2	Tighten and turn additional 90-degrees

V8 engine

Oil pressure	15 psi minimum at 2000 rpm
Compression pressure	100 to 150 psi
Maximum variation between cylinders	20-percent
Firing order	1-8-4-3-6-5-7-2

Valves and related components

Valve face angle	45-degrees
Valve seat angle	46-degrees
Valve seat width	
Intake	1/32 to 1/16 inch
Exhaust	1/16 to 3/32 inch
Valve seat runout	0.002 inch maximum
Stem-to-guide clearance	
Standard	0.001 to 0.0027 inch
Service limit	
Intake	0.0037 inch maximum
Exhaust	0.0047 inch maximum
Valve spring installed height	1-23/32 inch
Valve spring free length	
1984 and earlier	1.86 inch
1985 and later	2.03 inch

Crankshaft and connecting rods

Main journal diameter	
Journal 1	2.4484 to 2.4493 inches
Journals 2, 3 and 4	2.4481 to 2.4490 inches
Journal 5	2.4479 to 2.4488 inches

V8 engine (continued)

Crankshaft and connecting rods (continued)

Journal taper
   Standard.. 0.0002 inch
   Service limit... 0.0010 inch maximum
Journal out-of-round
   Standard.. 0.0002 inch
   Service limit... 0.0010 inch maximum
Main bearing oil clearance
   Journal 1... 0.0008 to 0.0020 inch
   Journals 2, 3 and 4.. 0.0011 to 0.0023 inch
   Journal 5... 0.0017 to 0.0032 inch
Crankshaft endplay.. 0.002 to 0.006 inch
Connecting rod journal diameter .... 2.0986 to 2.0998 inches
Connecting rod journal taper
   Standard.. 0.0005 inch
   Service limit... 0.001 inch maximum
Connecting rod journal out-of-round
   Standard.. 0.0002 inch
   Service limit... 0.001 inch maximum
Rod bearing oil clearance .... 0.0018 to 0.0039 inch
Connecting rod endplay (side clearance) .... 0.008 to 0.014 inch

Engine block

Bore diameter
   5.0L .... 3.7350 to 3.7385 inches
   5.7L .... 3.9995 to 4.0025 inches
Out-of-round
   Standard.. 0.0010 inch
   Service limit... 0.0020 inch maximum
Taper
   Standard.. 0.0005 inch
   Service limit... 0.0015 inch maximum

Pistons and rings

Piston-to-bore clearance
   Standard.. 0.0007 to 0.0017 inch
   Service limit... 0.0027 inch maximum
Piston ring side clearance
   Top ring
      Standard .... 0.0012 to 0.0032 inch
      Service limit... 0.0042 inch maximum
   Second ring
      Standard .... 0.0012 to 0.0032 inch
      Service limit... 0.0042 inch maximum
Oil ring
   Standard.. 0.002 to 0.007 inch
   Service limit... 0.008 inch maximum
Piston ring end gap
   Top ring
      Standard .... 0.010 to 0.020 inch
      Service limit... 0.030 inch maximum
   Second ring
      Standard .... 0.010 to 0.025 inch
      Service limit... 0.035 inch maximum
Oil ring
   Standard.. 0.015 to 0.055 inch
   Service limit... 0.065 inch maximum

Camshaft

Bearing journal diameter.. 1.8682 to 1.8692 inches
Bearing oil clearance .... 0.001 to 0.004 inch
Camshaft endplay... 0.004 to 0.012 inch

Lobe lift
    1982
        Carbureted
            Intake .... 0.238 inch
            Exhaust .... 0.260 inch
        Throttle Body Injected
            Intake .... 0.260 inch
            Exhaust .... 0.273 inch
    1983 and 1984
        Carbureted
            Standard model
                Intake .... 0.234 inch
                Exhaust .... 0.257 inch
            High Output model
                Intake .... 0.269 inch
                Exhaust .... 0.276 inch
        Throttle Body Injected
            Intake .... 0.257 inch
            Exhaust .... 0.269 inch
    1985 through 1987
        Carbureted
            Standard model
                Intake .... 0.234 inch
                Exhaust .... 0.257 inch
            High Output model
                Intake .... 0.269 inch
                Exhaust .... 0.276 inch
        Tuned Port Injected
            5.0L
                Intake .... 0.269 inch
                Exhaust .... 0.276 inch
            5.7L
                Intake .... 0.273 inch
                Exhaust .... 0.282 inch
    1988 through 1990
        5.0L
            Throttle Body Injected
                Intake .... 0.234 inch
                Exhaust .... 0.257 inch
            Tuned Port Injected
                Intake .... 0.269 inch
                Exhaust .... 0.276 inch
        5.7L
            Intake .... 0.273 inch
            Exhaust .... 0.282 inch
    1991 and 1992
        Throttle Body Injected
            Intake .... 0.234 inch
            Exhaust .... 0.257 inch
        Tuned Port Injected
            Intake .... 0.275 inch
            Exhaust .... 0.285 inch

Torque specifications

Ft-lbs (unless otherwise indicated)

Main bearing cap bolts	70
Connecting rod cap nuts	45
Cylinder head bolts	
1988 and earlier..	65
1989 and later	68
Oil pump-to-rear main bearing cap bolt	65
Rear main bearing oil seal retainer bolts	132 in-lbs

1   General information

Included in this portion of Chapter 2 are the general overhaul procedures for the cylinder head(s) and internal engine components. The information ranges from advice concerning preparation for an overhaul and the purchase of replacement parts to detailed, step-by-step procedures covering removal and installation of internal engine components and the inspection of parts.

The following Sections have been written based on the assumption that the engine has been removed from the vehicle. For information concerning in-vehicle engine repair, as well as removal and installation of the external components necessary for the overhaul, see Part A, B or C of Chapter 2 (depending on engine type) and Section 2 of this Part.

The Specifications included here in Part D are only those necessary for the inspection and overhaul procedures which follow. Refer to Part A, B or C for additional Specifications related to the various engines covered in this manual.

It is not always easy to determine when, or if, an engine should be completely overhauled, as a number of factors must be considered. High mileage is not necessarily an indication that an overhaul is needed while low mileage, on the other hand, does not preclude the need for an overhaul. Frequency of servicing is probably the single most important consideration. An engine that has regular (and frequent) oil and filter changes, as well as other required maintenance, will most likely give many thousands of miles of reliable service. Conversely, a neglected engine may require an overhaul very early in its life.

Excessive oil consumption is an indication that piston rings and/or valve guides are in need of attention (make sure that oil leaks are not responsible before deciding that the rings and guides are bad). Have a cylinder compression or leak-down test performed by an experienced tune-up mechanic to determine for certain the extent of the work required.

If the engine is making obvious knocking or rumbling noises, the connecting rod and/or main bearings are probably at fault. Check the oil pressure with a gauge (installed in place of the oil pressure sending unit) and compare it to the Specifications. If it is extremely low, the bearings and/or oil pump are probably worn out.

Loss of power, rough running, excessive valve train noise and high fuel consumption rates may also point to the need for an overhaul (especially if they are all present at the same time). If a complete tune-up does not remedy the situation, major mechanical work is the only solution.

An engine overhaul generally involves restoring the internal parts to the specifications of a new engine. During an overhaul, the piston rings are replaced and the cylinder walls are reconditioned (rebored and/or honed). If a rebore is done, then new pistons are also required. The main and connecting rod bearings are replaced with new ones and, if necessary, the crankshaft may be reground to restore the journals. Generally, the valves are serviced as well since they are usually in less than-perfect condition at this point. While the engine is being overhauled, other components such as the carburetor, the distributor, the starter and the alternator can be rebuilt also. The end result should be a like-new engine that will give as many trouble-free miles as the original.

Before beginning the engine overhaul, read through the entire procedure to familiarize yourself with the scope and requirements of the job. Overhauling an engine is not that difficult, but it is time consuming. Plan on the vehicle being tied up for a minimum of two weeks, especially if parts must be taken to an automotive machine shop for repair or reconditioning. Check on availability of parts and make sure that any necessary special tools and equipment are obtained in advance. Most work can be done with typical shop hand tools, although a number of precision measuring tools are required for inspecting parts to determine if they must be replaced. Often a reputable automotive machine shop will handle the inspection of parts and offer advice concerning reconditioning and replacement. **Note:** *Always wait until the engine has been completely disassembled and all components, especially the engine block, have been inspected before deciding what service and repair operations must be performed by an automotive machine shop.* Since the block's condition will be the major factor to consider when determining whether to overhaul the original engine or buy a rebuilt one, never purchase parts or have machine work done on other components until the block has been thoroughly inspected. As a general rule, time is the primary cost of an overhaul, so it does not pay to install worn or sub-standard parts.

As a final note, to ensure maximum life and minimum trouble from a rebuilt engine, everything must be assembled with care in a spotlessly clean environment.

2   Repair operations possible with the engine in the vehicle

Many major repair operations can be accomplished without removing the engine from the vehicle.

It is a very good idea to clean the engine compartment and the exterior of the engine with some type of pressure washer before any work is begun. A clean engine will make the job easier and will prevent the possibility of getting dirt into internal areas of the engine.

Remove the hood (Chapter 12) and cover the fenders to provide as much working room as possible and to prevent damage to the painted surfaces.

If oil or coolant leaks develop, indicating a need for gasket or seal replacement, the repairs can generally be made with the engine in the vehicle. The oil pan gasket, the cylinder head gasket(s), intake and exhaust manifold gaskets, timing chain cover gaskets and the front and rear crankshaft oil seals are accessible with the engine in place.

Exterior engine components, such as the water pump, the starter motor, the alternator, the distributor, the fuel pump and the carburetor, as well as the intake and exhaust manifolds, are quite easily removed for repair with the engine in place.

Since the cylinder head(s) can be removed without pulling the engine, valve component servicing can also be accomplished with the engine in the vehicle.

Replacement of, repairs to or inspection of the timing gears or sprockets and chain and the oil pump are all possible with the engine in place.

In extreme cases caused by a lack of necessary equipment, repair or replacement of piston rings, pistons, connecting rods and rod bearings and reconditioning of the cylinder bores is possible with the engine in the vehicle. However, this practice is not recommended because of the cleaning and preparation work that must be done to the components involved.

Detailed removal, inspection, repair and installation procedures for the above mentioned components can be found in the appropriate Part of Chapter 2 or the other Chapters in this manual.

3   Engine rebuilding alternatives

The do-it-yourselfer is faced with a number of options when performing an engine overhaul. The decision to replace the engine block, piston/connecting rod assemblies and crankshaft depends on a number of factors, with the number one consideration being the condition of the block. Other considerations are cost, access to machine shop facilities, parts availability, time required to complete the project and experience.

Some of the rebuilding alternatives include:

Individual parts - If the inspection procedures reveal that the engine block and most engine components are in reusable condition, purchasing individual parts may be the most economical alternative. The block, crankshaft and piston/connecting rod assemblies should all be inspected carefully. Even if the block shows little wear, the cylinder bores should receive a finish hone; a job for an automotive machine shop.

Master kit (crankshaft kit) - This rebuild package usually consists of a reground crankshaft and a matched set of pistons and connecting rods. The pistons will already be installed on the connecting rods. Piston rings and the necessary bearings may or may not be included in the kit. These kits are commonly available for standard cylinder bores, as well as for engine blocks which have been bored to a regular oversize.

4.6 A compression gauge with a threaded fitting for the spark plug hole is preferred over the type that requires hand pressure to maintain the seal

Short block - A short block consists of an engine block with a crankshaft and piston/connecting rod assemblies already installed. All new bearings are incorporated and all clearances will be correct. Depending on where the short block is purchased, a guarantee may be included. The existing camshaft, valve train components, cylinder head(s) and external parts can be bolted to the short block with little or no machine shop work necessary.

Long block - A long block consists of a short block plus an oil pump, oil pan, cylinder head(s), rocker arm cover(s), camshaft and valve train components, timing gears or sprockets and chain and timing gear/chain cover. All components are installed with new bearings, seals and gaskets incorporated throughout. The installation of manifolds and external parts is all that is necessary. Some form of guarantee is usually included with the purchase.

Give careful thought to which alternative is best for you and discuss the situation with local automotive machine shops, auto parts dealers or dealership partsmen before ordering or purchasing replacement parts.

4   Compression check

Refer to illustration 4.6

1   A compression check will tell you what mechanical condition the upper end (pistons, rings, valves, head gaskets) of the engine is in. Specifically, it can tell you if the compression is down due to leakage caused by worn piston rings, defective valves and seats or a blown head gasket. **Note:** *The engine must be at normal operating temperature and the battery must be fully charged for this check.*

2   Begin by cleaning the area around the spark plugs before you remove them. Compressed air should be used, if available, otherwise a small brush or even a bicycle tire pump will work. The idea is to prevent dirt from getting into the cylinders as the compression check is being done.

3   Remove all of the spark plugs from the engine (see Chapter 1).

4   Block the throttle wide open.

5   Disable the ignition system by disconnecting the harness connector from the ignition coil (see Chapter 5). On fuel injected models, disable the fuel pump by disconnecting the fuel pump harness connector at the fuel pump (see Chapter 4B).

6   Install the compression gauge in the number one spark plug hole **(see illustration)**.

7   Crank the engine over at least seven compression strokes and watch the gauge. The compression should build up quickly in a healthy engine. Low compression on the first stroke, followed by gradually increasing pressure on successive strokes, indicates worn piston rings. A low compression reading on the first stroke, which doesn't build up during successive strokes, indicates leaking valves or a blown head gasket (a cracked head could also be the cause). Deposits on the undersides of the valve heads can also cause low compression. Record the highest gauge reading obtained.

8   Repeat the procedure for the remaining cylinders, turning the engine over for the same length of time for each cylinder, and compare the results to this Chapter's Specifications.

9   If the readings are below normal, add some engine oil (about three squirts from a plunger-type oil can) to each cylinder, through the spark plug hole, and repeat the test.

10   If the compression increases after the oil is added, the piston rings are definitely worn. If the compression doesn't increase significantly, the leakage is occurring at the valves or head gasket. Leakage past the valves may be caused by burned valve seats and/or faces or warped, cracked or bent valves.

11   If two adjacent cylinders have equally low compression, there's a strong possibility the head gasket between them is blown. The appearance of coolant in the combustion chambers or the crankcase would verify this condition.

12   If one cylinder is more than 20-percent lower than the others, and the engine has a slightly rough idle, a worn exhaust lobe on the camshaft could be the cause.

13   If the compression is unusually high, the combustion chambers are probably coated with carbon deposits. If that's the case, the cylinder heads should be removed and decarbonized.

14   If compression is way down or varies greatly between cylinders, it would be a good idea to have a leak-down test performed by an automotive repair shop. This test will pinpoint exactly where the leakage is occurring and how severe it is.

5   Engine removal - methods and precautions

If it has been decided that an engine must be removed for overhaul or major repair

work, certain preliminary steps should be taken.

Locating a suitable work area is extremely important. A shop is, of course, the most desirable place to work. Adequate work space along with storage space for the vehicle is very important. If a shop or garage is not available, at the very least a flat, level, clean work surface made of concrete or asphalt is required.

Cleaning the engine compartment and engine prior to removal will help keep tools clean and organized.

An engine hoist or A-frame will also be necessary. Make sure that the equipment is rated in excess of the combined weight of the engine and its accessories. Safety is of primary importance, considering the potential hazards involved in lifting the engine out of the vehicle.

If the engine is being removed by a novice, a helper should be available. Advice and aid from someone more experienced would also be helpful. There are many instances when one person cannot simultaneously perform all of the operations required when lifting the engine out of the vehicle.

Plan the operation ahead of time. Arrange for or obtain all of the tools and equipment you will need prior to beginning the job. Some of the equipment necessary to perform engine removal and installation safely and with relative ease are (in addition to an engine hoist) a heavyduty floor jack, complete sets of wrenches and sockets as described in the front of this manual, wooden blocks and plenty of rags and cleaning solvent for mopping up the inevitable spills. If the hoist is to be rented, make sure that you arrange for it in advance and perform beforehand all of the operations possible without it. This will save you money and time.

Plan for the vehicle to be out of use for a considerable amount of time. A machine shop will be required to perform some of the work which the do-it-yourselfer cannot accomplish due to a lack of special equipment. These shops often have a busy schedule so it would be wise to consult them before removing the engine in order to accurately estimate the amount of time required to rebuild or repair components that may need work.

Always use extreme caution when removing and installing the engine; serious injury can result from careless actions. Plan ahead. Take your time and a job of this nature, although major, can be accomplished successfully.

6   Engine overhaul disassembly sequence

1   It is much easier to disassemble and work on the engine if it is mounted on a portable engine stand. These stands can often be rented for a reasonable fee from an equipment rental yard. Before the engine is mounted on a stand, the flywheel/driveplate

should be removed from the engine (refer to Chapter 8).

2   If a stand is not available, it is possible to disassemble the engine with it blocked up on a sturdy workbench or on the floor. Be extra careful not to tip or drop the engine when working without a stand.

3   If you are going to obtain a rebuilt engine, all external components must come off first in order to be transferred to the replacement engine (just as they will if you are doing a complete engine overhaul yourself). These include:

> Alternator and brackets
> Emissions control components
> Distributor, spark plug wires and spark plugs
> Thermostat and housing cover
> Water pump
> Carburetor/fuel injection components
> Intake/exhaust manifolds
> Oil filter
> Fuel pump
> Engine mounts
> Flywheel/driveplate

Note: *When removing the external components from the engine, pay close attention to details that may be helpful or important during installation. Note the installed position of gaskets, seals, spacers, pins, washers, bolts and other small items.*

4   If you are obtaining a short block (which consists of the engine block, crankshaft, pistons and connecting rods all assembled), then the cylinder heads, oil pan and oil pump will have to be removed also. See Engine rebuilding alternatives for additional information regarding the different possibilities to be considered.

5   If you are planning a complete overhaul, the engine must be disassembled and the internal components removed in the following order:

> Rocker arm cover(s)
> Rocker arms and pushrods
> Cylinder head(s)
> Valve lifters
> Timing chain/gear cover
> Timing chain/sprockets or gears
> Camshaft
> Oil pan
> Oil pump
> Piston/connecting rod assemblies
> Crankshaft

6   Before beginning the disassembly and overhaul procedures, make sure the following items are available:

> Common hand tools
> Small cardboard boxes or plastic bags for storing parts
> Gasket scraper
> Ridge reamer
> Vibration damper puller
> Micrometers
> Small hole gauges
> Telescoping gauges
> Dial indicator set
> Valve spring compressor
> Cylinder surfacing hone

7.2 When checking the valve spring height on V6 engine exhaust valves and all four-cylinder and V8 engine valves, the measurement is made up to the bottom inside surface of the oil shedder

> Piston ring groove cleaning tool
> Electric drill motor
> Tap and die set
> Wire brushes
> Cleaning solvent

7   Cylinder head - disassembly

Refer to illustrations 7.2 and 7.3

Note: *New and rebuilt cylinder heads are commonly available for most engines at dealerships and auto parts stores. Due to the fact that some specialized tools are necessary for the disassembly and inspection procedures, and replacement parts may not be readily available, it may be more practical and economical for the home mechanic to purchase a replacement head (or heads) rather than taking the time to disassemble, inspect and recondition the original head(s).*

1   Cylinder head disassembly involves removal and disassembly of the intake and exhaust valves and their related components. If they are still in place, remove the nuts or bolts and pivot balls, then separate the rocker arms from the cylinder head. Label the parts or store them separately so they can be reinstalled in their original locations.

2   Before the valves are removed, arrange to label and store them, along with their related components, so they can be kept separate and reinstalled in the same valve guides they are removed from. Also, measure the valve spring installed height (for each valve) and compare it to the Specifications **(see illustration)**. If it is greater than specified, the valve seats and valve faces need attention.

3   Compress the valve spring on the first valve with a spring compressor and remove the keepers **(see illustration)**. Carefully release the valve spring compressor and remove the retainer (or rotator), the shield (if so equipped), the springs, the valve guide seal and/or O-ring seal, the spring seat and the valve from the head. If the valve binds in

7.3 Use a valve spring compressor to compress the springs, then remove the keepers from the valve stem

the guide (won't pull through), push it back into the head and deburr the area around the keeper groove with a fine file or whetstone.

4   Repeat the procedure for the remaining valves. Remember to keep together all the parts for each valve so they can be reinstalled in the same locations **(see illustration)**.

5   Once the valves have been removed and safely stored, the head should be thoroughly cleaned and inspected. If a complete engine overhaul is being done, finish the engine disassembly procedures before beginning the cylinder head cleaning and inspection process.

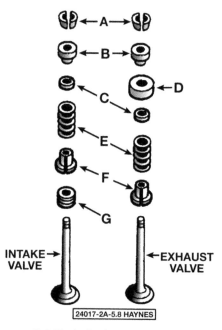

`24017-2A-5.8 HAYNES`

7.4 Typical valve components

A	Keepers
B	Retainers
C	O-ring seals
D	Oil shield
E	Springs
F	Spring dampers
G	Valve stem oil seal

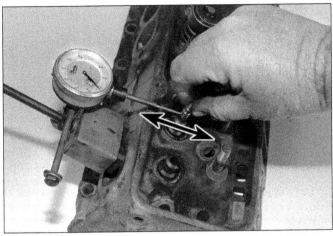

8.12 Check the cylinder head gasket surface for warpage by trying to slip a feeler gauge under the straightedge (see this Chapter's Specifications for the maximum warpage allowed and use a feeler gauge of that thickness)

8.14 A dial indicator can also be used to determine the valve stem-to-guide clearance

8   Cylinder head - cleaning and inspection

Refer to illustrations 8.12, 8.14, 8.19, 8.20, 8.21a and 8.21b

1   Thorough cleaning of the cylinder head and related valve train components, followed by a detailed inspection, will enable you to decide how much valve service work must be done during the engine overhaul.

Cleaning

2   Scrape away all traces of old gasket material and sealing compound from the head gasket, intake manifold and exhaust manifold sealing surfaces.
3   Remove any built-up scale around the coolant passages.
4   Run a stiff wire brush through the oil holes to remove any deposits that may have formed in them.
5   It is a good idea to run an appropriate size tap into each of the threaded holes to remove any corrosion and thread sealant that may be present. If compressed air is available, use it to clear the holes of debris produced by this operation.
6   Clean the exhaust and intake manifold stud threads in a similar manner with an appropriate size die. Clean the rocker arm pivot bolt or stud threads with a wire brush.
7   Next, clean the cylinder head with solvent and dry it thoroughly. Compressed air will speed the drying process and ensure that all holes and recessed areas are clean. **Note:** *Decarbonizing chemicals are available and may prove very useful when cleaning cylinder heads and valve train components. They are very caustic and should be used with caution. Be sure to follow the instructions on the container.*
8   Clean the rocker arms, pivot balls and pushrods with solvent and dry them thoroughly. Compressed air will speed the drying

process and can be used to clean out the oil passages.
9   Clean all the valve springs, keepers, retainers, rotators, shields and spring seats with solvent and dry them thoroughly. Do the components from one valve at a time to avoid mixing up the parts.
10   Scrape off any heavy deposits that may have formed on the valves, then use a motorized wire brush to remove deposits from the valve heads and stems. Again, make sure the valves do not get mixed up.

Inspection
Cylinder head
11   Inspect the head very carefully for cracks, evidence of coolant leakage and other damage. If cracks are found, a new cylinder head should be obtained.
12   Using a straightedge and feeler gauges, check the head gasket mating surface for warpage **(see illustration)**. If the head is warped beyond the limits given in the Specifications, it can be resurfaced at an automotive machine shop.
13   Examine the valve seats in each of the combustion chambers. If they are pitted, cracked or burned, the head will require valve service that is beyond the scope of the home mechanic.
14   Measure the inside diameters of the valve guides (at both ends and the center of each guide) with a small hole gauge and a micrometer. Record these measurements for future reference. These measurements, along with the valve stem diameter measurements, will enable you to compute the valve stem-to-guide clearances. These clearances, when compared to the Specifications, will be one factor that will determine the extent of valve service work required. The guides are measured at the ends and at the center to determine if they are worn in a bell-mouth pattern (more wear at the ends). If they are, guide reconditioning or replacement is necessary. As an alternative, use a dial indicator to mea-

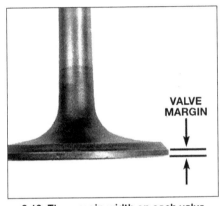

8.19 The margin width on each valve must be as specified (if no margin exists, the valve must be replaced)

sure the lateral movement of each valve stem with the valve in the guide and approximately 1/16-inch off the seat **(see illustration)**.

Rocker arm components
15   Check the rocker arm faces (that contact the pushrod ends and valve stems) for pits, wear and rough spots. Check the pivot contact areas as well.
16   Inspect the pushrod ends for scuffing and excessive wear. Roll the pushrod on a flat surface, such as a piece of glass, to determine if it is bent.
17   Any damaged or excessively worn parts must be replaced with new ones.

Valves
18   Carefully inspect each valve face for cracks, pits and burned spots. Check the valve stem and neck for cracks. Rotate the valve and check for any obvious indication that it is bent. Check the end of the stem for pits and excessive wear. The presence of any of these conditions indicates the need for valve service by a properly equipped professional.
19   Measure the width of the valve margin (on each valve) **(see illustration)**. Any valve with a margin narrower than 1/32-inch will have to be replaced with a new one.

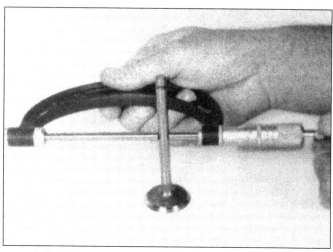

8.20 **Measure the valve stem diameter at three points**

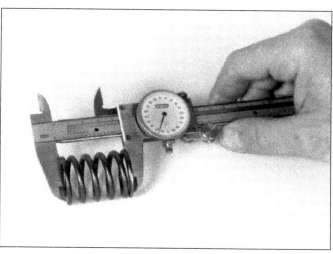

8.21a **Measure the free length of each valve spring with a dial or Vernier caliper**

20   Measure the valve stem diameter **(see illustration)**. **Note:** *The exhaust valves used in the four-cylinder engine have tapered stems and are approximately 0.001-inch larger at the tip end than at the head end.* By subtracting the stem diameter from the corresponding valve guide diameter, the valve stem-to-guide clearance is obtained. Compare the results to the Specifications. If the stem-to-guide clearance is greater than specified, the guides will have to be reconditioned and new valves may have to be installed, depending on the condition of the old ones.

Valve components

21   Check each valve spring for wear (on the ends) and pits. Measure the free length **(see illustration)** and compare it to the Specifications. Any springs that are shorter than specified have sagged and should not be reused. Stand the spring on a flat surface and check it for squareness **(see illustration)**.
22   Check the spring retainers or rotators and keepers for obvious wear and cracks. Any questionable parts should be replaced with new ones, as extensive damage will occur in the event of failure during engine operation.

23   If the inspection process indicates that the valve components are in generally poor condition and worn beyond the limits specified, which is usually the case in an engine that is being overhauled, reassemble the valves in the cylinder head and refer to Section 9 for valve servicing recommendations.
24   If the inspection turns up no excessively worn parts, and if the valve faces and seats are in good condition, the valve train components can be reinstalled in the cylinder head without major servicing. Refer to the appropriate Section for cylinder head reassembly procedures.

9   Valves - servicing

1   Because of the complex nature of the job and the special tools and equipment needed, servicing of the valves, the valve seats and the valve guides (commonly known as a "valve job') is best left to a professional.
2   The home mechanic can remove and disassemble the head, do the initial cleaning and inspection, then reassemble and deliver the head to a dealer service department or a reputable automotive machine shop for the actual valve servicing.

3   The dealer service department, or automotive machine shop, will remove the valves and springs, recondition or replace the valves and valve seats, recondition the valve guides, check and replace the valve springs, spring retainers or rotators and keepers (as necessary), replace the valve seals with new ones, reassemble the valve components and make sure the installed spring height is correct. The cylinder head gasket surface will also be resurfaced if it is warped.
4   After the valve job has been performed by a professional, the head will be in like-new condition. When the head is returned, be sure to clean it again, very thoroughly (before installation on the engine), to remove any metal particles and abrasive grit that may still be present from the valve service or head resurfacing operations. Use compressed air, if available, to blow out all the oil holes and passages.

10   Cylinder head - reassembly

Refer to illustrations 10.4 and 10.6
1   Regardless of whether or not the head was sent to an automotive machine shop for valve servicing, make sure it is clean before beginning reassembly.
2   If the head was sent out for valve servicing, the valves and related components will already be in place. Begin the reassembly procedure with Step 6.
3   Lay all the spring seats in position, then lubricate and install new seals (or deflectors) on each of the valve guides (V6 intake valves and all V8 valves). Using a hammer and an appropriate size deep socket, gently tap each seal into place until it is properly seated on the guide. Do not twist or cock the seals during installation or they will not seal properly on the valve stems.
4   Coat the valve stems with clean moly-based grease or engine assembly lube and install the valves (taking care not to damage the new valve guide seals). Install the springs,

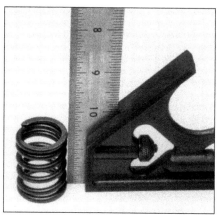

8.21b **Check each valve spring for squareness**

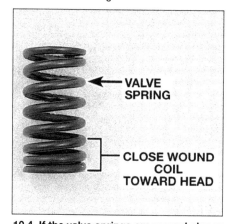

10.4 **If the valve springs are wound closer at one end, install them as shown**

VALVE SPRING

CLOSE WOUND COIL TOWARD HEAD

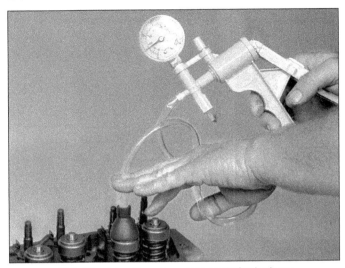

10.6 Checking the valve stem seals for leakage

11.2 A special tool is required to remove the ridge from the top of each cylinder (do it before removing the piston)

the shields (if equipped) and the retainers or rotators **(see illustration)**. Compress the spring and install the O-ring seals (if equipped) in the lower valve stem groove (make sure they are not twisted - a light coat of oil will prevent twisting). When compressing the springs with the valve spring compressor, do not let the retainers contact the valve guide seals. Install the keepers and release the spring compressor. Make certain that the keepers are securely locked in their retaining grooves by tapping on the spring retainer with a soft-faced hammer.

5   Double-check the installed valve spring height. If it was correct before disassembly, it should still be within the specified limits.

6   Check the valve stem O-ring seals with a vacuum pump and adapter **(see illustration)**. A properly installed seal should not leak.

7   When install the rocker arms, be sure to lubricate the ball pivots with moly-based grease or engine assembly lube.

11   Piston/connecting rod assembly - removal

Refer to illustrations 11.2, 11.6 and 11.8

1   Prior to removal of the piston/connecting rod assemblies, the engine should be positioned upright.

2   Using a ridge reamer, completely remove the ridge at the top of each cylinder (follow the manufacturer's instructions provided with the ridge reaming tool) **(see illustration)**. Failure to remove the ridge before attempting to remove the piston/connecting rod assemblies will result in piston breakage.

3   After all of the cylinder wear ridges have been removed, turn the engine upside-down.

4   Before the connecting rods are removed, check the endplay as follows. Mount a dial indicator with its stem in line with the crankshaft and touching the side of the number one cylinder connecting rod cap.

5   Push the connecting rod forward, as far

as possible, and zero the dial indicator. Next, push the connecting rod all the way to the rear and check the reading on the dial indicator. The distance that it moves is the endplay. If the endplay exceeds the service limit, a new connecting rod will be required. Repeat the procedure for the remaining connecting rods.

6   An alternative method is to slip feeler gauges between the connecting rod and the crankshaft throw until the play is removed **(see illustration)**. The endplay is then equal to the thickness of the feeler gauge(s).

7   Check the connecting rods and connecting rod caps for identification marks. If they are not plainly marked, identify each rod and cap using a small punch to make the appropriate number of indentations to indicate the cylinders they are associated with.

8   Loosen each of the connecting rod cap nuts approximately 1/2-turn each. Remove the number one connecting rod cap and bearing insert. Do not drop the bearing insert out of the cap. Slip a short length of plastic or rubber

11.6 Checking connecting rod endplay with a feeler gauge

11.8 To prevent damage to the crankshaft journals and cylinder walls, slip sections of hose over the rod bolts before removing the pistons

12.2 Checking crankshaft endplay with a dial indicator

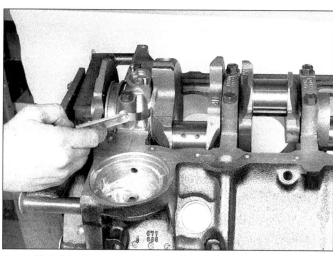

12.3 Checking crankshaft endplay with a feeler gauge

hose over each connecting rod cap bolt (to protect the crankshaft journal and cylinder wall when the piston is removed) **(see illustration)** and push the connecting rod/piston assembly out through the top of the engine. Use a wooden tool to push on the upper bearing insert in the connecting rod. If resistance is felt, doublecheck to make sure that all of the ridge was removed from the cylinder.

9   Repeat the procedure for the remaining cylinders. After removal, reassemble the connecting rod caps and bearing inserts in their respective connecting rods and install the cap nuts finger tight. Leaving the old bearing inserts in place until reassembly will help prevent the connecting rod bearing surfaces from being accidentally nicked or gouged.

12 Crankshaft - removal

Refer to illustrations 12.2, 12.3 and 12.4

1   Before the crankshaft is removed, check the endplay as follows. Mount a dial indicator with the stem in line with the crankshaft and just touching one of the crank throws

(see illustration 12.2).

2   Push the crankshaft all the way to the rear and zero the dial indicator **(see illustration)**. Next, pry the crankshaft to the front as far as possible and check the reading on the dial indicator. The distance that it moves is the endplay. If it is greater than specified, check the crankshaft thrust surfaces for wear. If no wear is apparent, new main bearings should correct the endplay.

3   If a dial indicator is not available, feeler gauges can be used. Gently pry or push the crankshaft all the way to the front of the engine. Slip feeler gauges between the crankshaft and the front face of the thrust main bearing **(see illustration)** to determine the clearance (which is equivalent to crankshaft endplay).

4   Loosen each of the main bearing cap bolts 1/4 of a turn at a time, until they can be removed by hand. Check the main bearing caps to see if they are marked as to their locations. They are usually numbered consecutively (beginning with 1) from the front of the engine to the rear. If they are not, mark them with number stamping dies or a center

punch **(see illustration)**. Most main bearing caps have a cast-in arrow, which points to the front of the engine.

5   Gently tap the caps with a soft-faced hammer, then separate them from the engine block. If necessary, use the main bearing cap bolts as levers to remove the caps. Try not to drop the bearing insert if it comes out with the cap.

6   On models with a one piece seal retainer, remove the retainer and seal.

7   Carefully lift the crankshaft out of the engine. It is a good idea to have an assistant available, since the crankshaft is quite heavy. With the bearing inserts in place in the engine block and in the main bearing caps, return the caps to their respective locations on the engine block and tighten the bolts finger tight.

13 Engine block - cleaning

Refer to illustrations 13.1a, 13.1b and 13.10

1   Remove the soft plugs from the engine block. To do this, knock the plugs sideways into the block (using a hammer and punch),

12.4 Mark the bearing caps with a center punch before removing them

13.1a A hammer and large punch can be used to drive the soft plugs sideways into the block

13.1b Using pliers to remove a soft plug from the block

13.10 A large socket on an extension can be used to drive the new soft plugs into their bores

then grasp them with large pliers and pull them back through the holes **(see illustration)**.

2   Using a gasket scraper, remove all traces of gasket material from the engine block. Be very careful not to nick or gouge the gasket sealing surfaces.

3   Remove the main bearing caps and separate the bearing inserts from the caps and the engine block. Tag the bearings according to which cylinder they removed from (and whether they were in the cap or the block) and set them aside.

4   Using a hex wrench of the appropriate size, remove the threaded oil gallery plugs from the front and back of the block.

5   If the engine is extremely dirty, it should be taken to an automotive machine shop to be steam cleaned or hot tanked. Any bearings left in the block (such as the camshaft bearings) will be damaged by the cleaning process, so plan on having new ones installed while the block is at the machine shop.

6   After the block is returned, clean all oil holes and oil galleries one more time (brushes for cleaning oil holes and galleries are available at most auto parts stores). Flush the passages with warm water until the water runs clear, dry the block thoroughly and wipe all machined surfaces with a light, rust-preventative oil. If you have access to compressed air,

use it to speed the drying process and to blow out all the oil holes and galleries.

7   If the block is not extremely dirty or sludged up, you can do an adequate cleaning job with warm soapy water and a stiff brush. Take plenty of time and do a thorough job. Regardless of the cleaning method used, be very sure to thoroughly clean all oil holes and galleries, dry the block completely and coat all machined surfaces with light oil.

8   The threaded holes in the block must be clean to ensure accurate torque readings during reassembly. Run the proper size tap into each of the holes to remove any rust, corrosion, thread sealant or sludge and to restore any damaged threads. If possible, use compressed air to clear the holes of debris produced by this operation. Now is a good time to thoroughly clean the threads on the head bolts and the main bearing cap bolts as well.

9   Reinstall the main bearing caps and tighten the bolts finger tight.

10   After coating the sealing surfaces of the new soft plugs with a good quality gasket sealer, install them in the engine block **(see illustration)**. Make sure they are driven in straight and seated properly or leakage could result. Special tools are available for this purpose, but equally good results can be

obtained using a large socket (with an outside diameter that will just slip into the soft plug) and a large hammer.

11   If the engine is not going to be reassembled right away, cover it with a large plastic trash bag to keep it clean.

14   Engine block - inspection

Refer to illustrations 14.4a, 14.4b, 14.4c, 14.7a and 14.7b

1   Thoroughly clean the engine block as described in Section 13 and double-check to make sure that the ridge at the top of each cylinder has been completely removed.

2   Visually check the block for cracks, rust and corrosion. Look for stripped threads in the threaded holes. It is also a good idea to have the block checked for hidden cracks by an automotive machine shop that has the special equipment to do this type of work. If defects are found, have the block repaired, if possible, or replaced.

3   Check the cylinder bores for scuffing and scoring.

4   Using the appropriate precision measuring tools, measure each cylinder's diameter at the top (just under the ridge), center and bottom of the cylinder bore, parallel to the

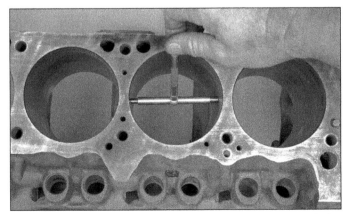

14.4a A telescoping gauge can be used to determine the cylinder bore diameter

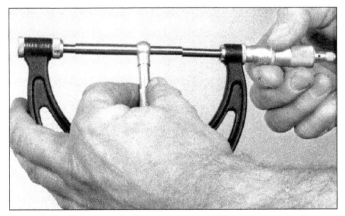

14.4b The gauge is then measured with a micrometer to determine the bore size in inches

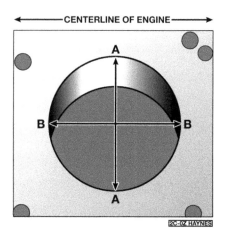

CENTERLINE OF ENGINE

14.4c Measure the diameter of each cylinder at a right angle to engine centerline (A) and parallel to engine centerline (B) - out-of-round is the difference between A and B; taper is the difference between the diameter at the top of the cylinder and the diameter at the bottom of the cylinder

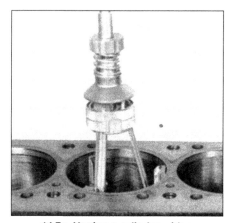

14.7a Honing a cylinder with a surfacing hone

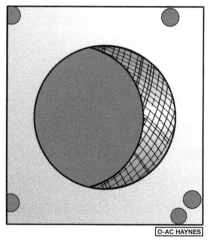

14.7b The cylinder hone should leave a cross-hatch pattern with the lines intersecting at approximately a 60 angle

crankshaft axis **(see illustrations)**. Next, measure each cylinder's diameter at the same three locations across the crankshaft axis. Compare the results to the Specifications. If the cylinder walls are badly scuffed or scored, or if they are out-of-round or tapered beyond the limits given in the Specifications, have the engine block rebored and honed at an automotive machine shop. If a rebore is done, oversize pistons and rings will be required as well.

5   If the cylinders are in reasonably good condition and not worn to the outside of the limits, and if the piston-to-cylinder clearances can be maintained properly, then they do not have to be rebored; honing is all that is necessary.

6   Before honing the cylinders, install the main bearing caps (without the bearings) and tighten the bolts to the specified torque.

7   To perform the honing operation, you will need the proper size flexible hone (with fine stones), plenty of light oil or honing oil,

some rags and an electric drill motor. Mount the hone in the drill motor, compress the stones and slip the hone into the first cylinder **(see illustrations)**. Lubricate the cylinder thoroughly, turn on the drill and move the hone up and down in the cylinder at a pace which will produce a fine crosshatch pattern on the cylinder walls (with the cross-hatch lines intersecting at approximately a 60-degree angle). Be sure to use plenty of lubricant and do not take off any more material than is absolutely necessary to produce the desired finish. Do not withdraw the hone from the cylinder while it is running. Instead, shut off the drill and continue moving the hone up and down in the cylinder until it comes to a complete stop, then compress the stones and withdraw the hone. Wipe the oil out of the cylinder and repeat the procedure on the remaining cylinders. Remember, do not remove too much material from the cylinder wall. If you do not have the tools or do not desire to perform the honing operation, most automotive machine shops will do it for a reasonable fee.

8   After the honing job is complete, chamfer the top edges of the cylinder bores with a small file so the rings will not catch when the pistons are installed.

9   Next, the entire engine block must be thoroughly washed again with warm, soapy

water to remove all traces of the abrasive grit produced during the honing operation. Be sure to run a brush through all oil holes and galleries and flush them with running water. After rinsing, dry the block and apply a coat of light rust preventative oil to all machined surfaces. Wrap the block in a plastic trash bag to keep it clean and set it aside until reassembly.

15 Camshaft, lifters and bearings - inspection and bearing replacement

Refer to illustrations 15.3, 15.9 and 15.12

Camshaft

1   The most critical camshaft inspection procedure is lobe lift measurement, which must be done before the engine is disassembled.

2   Remove the rocker arm cover(s), then remove the nuts/bolts and separate the rocker arms and ball pivots from the cylinder head(s).

3   Beginning with the number one (1) cylinder, mount a dial indicator with the stem resting on the end of, and directly in line with, the exhaust valve pushrod **(see illustration)**.

4   Rotate the crankshaft very slowly in the direction of rotation until the lifter is on the heel of the cam lobe. At this point the pushrod will be at its lowest position.

5   Zero the dial indicator, then very slowly rotate the crankshaft in the direction of rotation until the pushrod is at its highest position. Note and record the reading on the dial indicator, then compare it to the lobe lift specifications.

6   Repeat the procedure for each of the remaining valves. Note that intake and exhaust valves may have different lobe lift specifications.

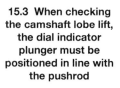

15.3 When checking the camshaft lobe lift, the dial indicator plunger must be positioned in line with the pushrod

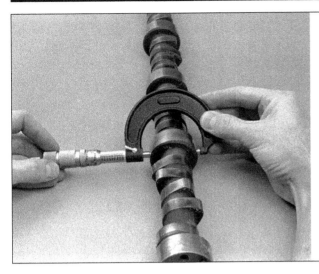

15.9 The camshaft bearing journal diameter is subtracted from the bearing inside diameter to obtain the oil clearance, which must be as specified

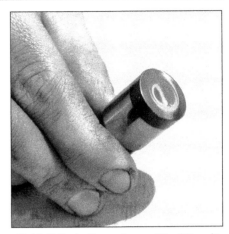

15.12 If the bottom of any of the lifters is worn concave, scratched or galled, they should all be replaced with new ones

7   If the lobe lift measurements are not within 0.002-inch of the specified lobe lift, a new camshaft should be installed.

8   After the camshaft has been removed from the engine, cleaned with solvent and dried, inspect the bearing journals for uneven wear, pitting and evidence of seizure. If the journals are damaged, the bearing inserts in the block are probably damaged as well. Both the camshaft and bearings will have to be replaced with new ones. Measure the inside diameter of each camshaft bearing and record the results (take two measurements, 90° apart, at each bearing).

9   Measure the bearing journals with a micrometer **(see illustration)** to determine if they are excessively worn or out-of-round. If they are more than 0.001-inch out-of-round, the camshaft should be replaced with a new one. Subtract the bearing journal diameter(s) from the corresponding bearing inside diameter measurement to obtain the oil clearance. If it is excessive, new bearings must be installed.

10   Check the camshaft lobes for heat discoloration, score marks, chipped areas, pitting and uneven wear. If the lobes are in good condition and if the lobe lift measurements (Steps 1 through 7) were as specified, the camshaft can be reused.

Lifters

11   Clean the lifters with solvent and dry them thoroughly without mixing them up.

12   Check each lifter wall, pushrod seat and foot for scuffing, score marks and uneven wear. Each lifter foot (the surface that rides on the cam lobe) must be slightly convex - if they are concave **(see illustration)**, the lifters and camshaft must be replaced with new ones. If the lifter walls are damaged or worn (which is not very likely), inspect the lifter bores in the engine block as well. If the pushrod seats are worn, check the pushrod ends. On models with roller lifters, inspect the roller and bearings for wear.

13   If new lifters are being installed, a new camshaft must also be installed. If a new camshaft is installed, then use new lifters as

well. Never install used lifters unless the original camshaft is used and the lifters can be installed in their original locations.

Bearing replacement

14   Camshaft bearing replacement requires special tools and expertise that place it outside the scope of the do-it-yourselfer. Take the block to an automotive machine shop to ensure that the job is done correctly.

---

16   Piston/connecting rod assembly - inspection

Refer to illustrations 16.4, 16.10 and 16.11

1   Before the inspection process can be carried out, the piston/connecting rod assemblies must be cleaned and the original piston rings removed from the pistons. **Note:** *Always use new piston rings when the engine is reassembled.*

2   Using a piston ring installation tool, carefully remove the rings from the pistons. Do not nick or gouge the pistons in the process.

3   Scrape all traces of carbon from the top (or crown) of the piston. A hand-held wire brush or a piece of fine emery cloth can be used once the majority of the deposits have been scraped away. Do not, under any circumstances, use a wire brush mounted in a drill motor to remove deposits from the pistons. The piston material is soft and will be eroded away by the wire brush.

4   Use a piston ring groove cleaning tool to remove any carbon deposits from the ring grooves. If a tool is not available, a piece broken off the old ring will do the job. Be very careful to remove only the carbon deposits. Do not remove any metal and do not nick or scratch the sides of the ring grooves **(see illustration)**.

5   Once the deposits have been removed, clean the piston/rod assemblies with solvent and dry them thoroughly. Make sure that the oil hole in the big end of the connecting rod and the oil return holes in the back sides of

the ring grooves are clear.

6   If the pistons are not damaged or worn excessively, and if the engine block is not rebored, new pistons will not be necessary. Normal piston wear appears as even vertical wear on the piston thrust surfaces and slight looseness of the top ring in its groove. New piston rings, on the other hand, should always be used when an engine is rebuilt.

7   Carefully inspect each piston for cracks around the skirt, at the pin bosses and at the ring lands.

8   Look for scoring and scuffing on the thrust faces of the skirt, holes in the piston crown and burned areas at the edge of the crown. If the skirt is scored or scuffed, the engine may have been suffering from overheating and/or abnormal combustion, which caused excessively high operating temperatures. The cooling and lubrication systems should be checked thoroughly. A hole in the piston crown, an extreme to be sure, is an indication that abnormal combustion (preignition) was occurring. Burned areas at the edge of the piston crown are usually evidence of spark knock (detonation). If any of the above

16.4 Clean the piston ring grooves with a piston ring groove cleaning tool

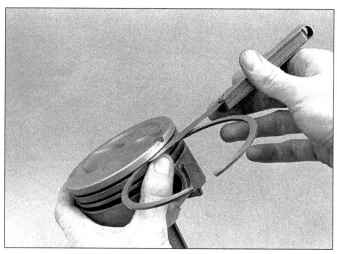

16.10 Check the piston ring side clearance with a feeler gauge

16.11 Measure the piston diameter directly in line with the piston pin hole

problems exist, the causes must be corrected or the damage will occur again.

9   Corrosion of the piston (evidenced by pitting) indicates that coolant is leaking into the combustion chamber and/or the crankcase. Again, the cause must be corrected or the problem may persist in the rebuilt engine.

10   Measure the piston ring side clearance by laying a new piston ring in each ring groove and slipping a feeler gauge in beside it **(see illustration)**. Check the clearance at three or four locations around each groove. Be sure to use the correct ring for each groove; they are different. If the side clearance is greater than specified, new pistons and/or rings will have to be used.

11   Check the piston-to-bore clearance by measuring the bore (see Section 14) and the piston diameter **(see illustration)**. Make sure that the pistons and bores are correctly matched. Measure the piston across the skirt, on the thrust faces (at a 90-degree angle to the piston pin), directly in line with the center of the pin hole. Subtract the piston diameter from the bore diameter to obtain the clearance. If it is greater than specified, the block will have to be rebored and new pistons and rings installed. Check the piston-to-rod clearance by twisting the piston and rod in opposite directions. Any noticeable play indicates that there is excessive wear, which must be corrected. The piston/connecting rod assemblies should be taken to an automotive machine shop to have new piston pins installed and the pistons and connecting rods rebored.

12   If the pistons must be removed from the connecting rods, such as when new pistons must be installed, or if the piston pins have too much play in them, they should be taken to an automotive machine shop. While they are there, it would be convenient to have the connecting rods checked for bend and twist, as automotive machine shops have special equipment for this purpose. Unless new pistons or connecting rods must be installed, do

17.2 Measure the diameter of each crankshaft journal at several points to detect taper and out-of-round conditions

not disassemble the pistons from the connecting rods.

13   Check the connecting rods for cracks and other damage. Temporarily remove the rod caps, lift out the old bearing inserts, wipe the rod and cap bearing surfaces clean and inspect them for nicks, gouges and scratches. After checking the rods, replace the old bearings, slip the caps into place and tighten the nuts finger tight.

17 Crankshaft - inspection

Refer to illustration 17.2

1   Clean the crankshaft with solvent and dry it thoroughly. Be sure to clean the oil holes with a stiff brush and flush them with solvent. Check the main and connecting rod bearing journals for uneven wear, scoring, pitting and cracks. Check the remainder of the crankshaft for cracks and damage.

2   Using an appropriate size micrometer, measure the diameter of the main and connecting rod journals **(see illustration)** and compare the results to the Specifications. By measuring the diameter at a number of points around the journal's circumference, you will

be able to determine whether or not the journal is worn out-of-round. Take the measurement at each end of the journal, near the crank throw, to determine whether the journal is tapered.

3   If the crankshaft journals are damaged, tapered, out-of-round or worn beyond the limits given in the Specifications, have the crankshaft reground by a reputable automotive machine shop. Be sure to use the correct undersize bearing inserts if the crankshaft is reconditioned.

4   Refer to Section 18 and examine the main and rod bearing inserts.

If the bearing inserts and journals are all in good condition, do not decide to reuse the bearings until the oil clearances have been checked.

18 Main and connecting rod bearings - inspection

Refer to illustration 18.1

1   Even though the main and connecting rod bearings should be replaced with new ones during the engine overhaul, the old bearings should be retained for close exami-

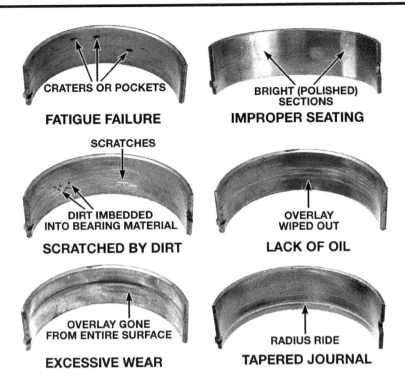

FATIGUE FAILURE
CRATERS OR POCKETS

IMPROPER SEATING
BRIGHT (POLISHED) SECTIONS

SCRATCHED BY DIRT
SCRATCHES
DIRT IMBEDDED INTO BEARING MATERIAL

LACK OF OIL
OVERLAY WIPED OUT

EXCESSIVE WEAR
OVERLAY GONE FROM ENTIRE SURFACE

TAPERED JOURNAL
RADIUS RIDE

18.1 Typical bearing failures

nation, as they may reveal valuable information about the condition of the engine **(see illustration)**.

2   Bearing failure occurs mainly because of lack of lubrication, the presence of dirt or other foreign particles, overloading the engine and corrosion. Regardless of the cause of bearing failure, it must be corrected before the engine is reassembled to prevent it from happening again.

3   When examining the bearings, remove them from the engine block,
the main bearing caps, the connecting rods and the rod caps and lay them out on a clean surface in the same general position as their location in the engine. This will enable you to match any noted bearing problems with the corresponding crankshaft journal.

4   Dirt and other foreign particles get into the engine in a variety of ways. If may be left in the engine during assembly, or it may pass through filters or breathers. It may get into the oil, and from there into the bearings. Metal chips from machining operations and normal engine wear are often present. Abrasives are sometimes left in engine components after reconditioning, especially when parts are not thoroughly cleaned using the proper cleaning methods. Whatever the source, these foreign objects often end up embedded in the soft bearing material and are easily recognized. Large particles will not embed in the bearing and will score or gouge the bearing and shaft. The best prevention for this cause of bearing failure is to clean all parts thoroughly and keep everything spotlessly clean during engine assem-

bly. Frequent and regular engine oil and filter changes are also recommended.

5   Lack of lubrication (or lubrication breakdown) has a number of interrelated causes. Excessive heat (which thins the oil), overloading (which squeezes the oil from the bearing face) and oil leakage or throw off (from excessive bearing clearances, worn oil pump or high engine speeds) all contribute to lubrication breakdown. Blocked oil passages, which usually are the result of misaligned oil holes in a bearing shell, will also oil-starve a bearing and destroy it. When lack of lubrication is the cause of bearing failure, the bearing material is wiped or extruded from the steel backing of the bearing. Temperatures

may increase to the point where the steel backing turns blue from overheating.

6   Driving habits can have a definite effect on bearing life. Full-throttle, low-speed operation (or "lugging' the engine) puts very high loads on bearings, which tends to squeeze out the oil film. These loads cause the bearings to flex, which produces fine cracks in the bearing face (fatigue failure). Eventually the bearing material will loosen in pieces and tear away from the steel backing. Short-trip driving leads to corrosion of bearings because insufficient engine heat is produced to drive off the condensed water and corrosive gases. These products collect in the engine oil, forming acid and sludge. As the oil is carried to the engine bearings, the acid attacks and corrodes the bearing material.

7   Incorrect bearing installation during engine assembly will lead to bearing failure as well. Tight-fitting bearings leave insufficient bearing oil clearance and will result in oil starvation. Dirt or foreign particles trapped behind a bearing insert result in high spots on the bearing which lead to failure.

19   Piston rings - installation

Refer to illustrations 19.3a, 19.3b, 19.9a, 19.9b and 19.12

1   Before installing the new piston rings, the ring end gaps must be checked. It is assumed that the piston ring side clearance has been checked and verified correct (Section 16).

2   Lay out the piston/connecting rod assemblies and the new ring sets so the ring sets will be matched with the same piston and cylinder during the end gap measurement and engine assembly.

3   Insert the top (number one) ring into the first cylinder and square it up with the cylinder walls by pushing it in with the top of the piston **(see illustration)**. The ring should be near the bottom of the cylinder at the lower limit of ring travel. To measure the end gap, slip a feeler gauge between the ends of the ring **(see illustration)**. Compare the mea-

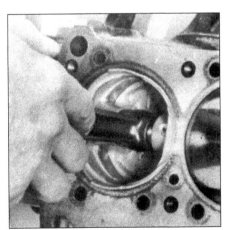

19.3a Use the piston to square up the ring in the cylinder prior to checking the ring end gap

19.3b Measure the ring end gap with a feeler gauge

19.5 If the end gap is too small, clamp a file in a vise and file the ring ends (from the outside in only) to enlarge the gap slightly

19.9a Install the spacer/expander in the oil control ring groove

19.9b Do not use a piston ring tool when installing the oil ring side rails

19.12 Install the compression rings with the special tool - the mark (arrow) must face up

surement to the Specifications.

4   If the gap is larger or smaller than specified, double-check to make sure that you have the correct rings before proceeding.

5   If the gap is too small, it must be enlarged or the ring ends may come in contact with each other during engine operation, which can cause serious damage to the engine. The end gap can be increased by filing the ring ends very carefully with a fine file **(see illustration)**. Mount the file in a vise equipped with soft jaws, slip the ring over the file with the ends contacting the file face and slowly move the ring to remove material from the ends. When performing this operation, file only from the outside in.

6   Excess end gap is not critical unless it is greater than 0.030 to 0.035-inch. Again, double-check to make sure you have the correct rings for your engine.

7   Repeat the procedure for each ring that will be installed in the first cylinder and for each ring in the remaining cylinders. Remember to keep rings, pistons and cylinders matched up.

8   Once the ring end gaps have been checked/corrected, the rings can be installed on the pistons.

9   The oil control ring (lowest one on the piston) is installed first. It is composed of three separate components. Slip the spacer expander into the groove **(see illustration)**, then install the upper side rail. Do not use a piston ring installation tool on the oil ring side rails, as they may be damaged. Instead, place one end of the side rail into the groove between the spacer expander and the ring land, hold it firmly in place and slide a finger around the piston while pushing the rail into the groove **(see illustration)**. Next, install the lower side rail in the same manner.

10   After the three oil ring components have been installed, check to make sure that both the upper and lower side rails can be turned smoothly in the ring groove.

11   The number two (middle) ring is installed next. It should be stamped with a mark so it

can be readily distinguished from the top ring.

Note: *Always follow the instructions printed on the package that the new rings are in - different manufacturers may require different approaches. Do not mix up the top and middle rings, as they have different cross sections.*

12   Use a piston ring installation tool and make sure that the identification mark is facing the top of the piston, then slip the ring into the middle groove on the piston **(see illustration)**. Do not expand the ring any more than is necessary to slide it over the piston.

13   Finally, install the number one (top) ring in the same manner. Make sure the identifying mark is facing up.

14   Repeat the procedure for the remaining pistons and rings. Be careful not to confuse the number one and number two rings.

20   Rear main oil seal - installation

Four-cylinder engine

1   The four-cylinder engine uses a one-piece rear main seal and is installed after final crankshaft installation. Refer to Chapter 2A for the seal installation procedure.

V6 engine

2   1984 and earlier V6 engines were originally equipped with a two-piece rear main seal. The two-piece seal is prone to leakage and should be replaced with a one-piece service replacement rear main seal kit. On 1984 and earlier models, obtain the service replacement kit from an auto parts supplier or dealership parts department and install it following the instructions provided with the kit. The one-piece service replacement seal is installed during crankshaft installation.

3   1985 and later V6 engines are equipped with a one-piece rear main seal similar to the four-cylinder design. On 1985 and later models, refer to Chapter 2A and install the rear main seal after final crankshaft installation.

V8 engine

1985 and earlier models

6   Inspect the bearing cap and engine block mating surfaces and seal grooves for nicks, burrs and scratches. Remove any defects with a fine file or deburring tool.

7   Install the lower seal section in the block with the lip facing the front of the engine. Leave one end protruding from the block approximately 1/4 to 3/8-inch and make sure

21.10 Lay the Plastigage strips on the main bearing journals, parallel to the crankshaft centerline

21.14 Compare the width of the crushed Plastigage to the scale on the envelope to determine the main bearing oil clearance (always take the measurement at the widest point of the Plastigage); be sure to use the correct scale - standard and metric ones are included

it is completely seated.

8   Repeat the procedure to install the remaining seal half in the rear main bearing cap. In this case, leave the opposite end of the seal protruding from the cap approximately the same distance the block seal is protruding from the block.

9   During final installation of the crankshaft (after the main bearing oil clearances have been checked with Plastigage) as described in Section 21, apply a thin, even film of anaerobic-type gasket sealant to the main bearing cap split-line **(see Chapter 2C, illustration 11.14)**. **Caution:** *Do not get any sealant on the bearing face, crankshaft journal or seal lip.* Lubricate the seal lips with clean engine oil and install the cap.

1986 and later models

10   On1986 and later models, the rear main seal is installed after final crankshaft installation.

11   Carefully pry the old oil seal out of the retainer. Clean the retainer of all gasket material and sealant.

12   Install the retainer to the engine block with a new gasket and tighten the bolts to the specified torque.

13   Lubricate the seal lip with clean engine oil. Using a one-piece rear main seal installation tool, press the seal squarely into the housing. If the tool is not available, the seal may be driven in using a blunt tool, provided extreme care is taken to drive the seal in squarely and without damage to the seal or crankshaft.

21   Crankshaft - installation and main bearing oil clearance check

Refer to illustrations 21.10 and 21.14

1   Crankshaft installation is generally one of the first steps in engine reassembly; it is assumed at this point that the engine block and crankshaft have been cleaned, inspected and repaired or reconditioned. **Note:** *On 1984 and earlier V6 and 1985 and earlier V8 engines, refer to Section 20 and prepare the rear main*

oil seal for installation before preceding.

2   Position the engine with the bottom facing up.

3   Remove the main bearing cap bolts and lift out the caps. Lay them out in the proper order to help ensure that they are installed correctly.

4   If they are still in place, remove the old bearing inserts from the block and the main bearing caps. Wipe the main bearing surfaces of the block and caps with a clean, lint-free cloth (they must be kept spotlessly clean).

5   Clean the back sides of the new main bearing inserts and lay one bearing half in each main bearing saddle in the block. Lay the other bearing half from each bearing set in the corresponding main bearing cap. Make sure the tab on the bearing insert fits into the recess in the block or cap. Also, the oil holes in the block and cap must line up with the oil holes in the bearing insert. Do not hammer the bearing into place and do not nick or gouge the bearing faces. No lubrication should be used at this time.

6   The flanged thrust bearing must be installed in the number three (3) cap and saddle on V6 engines and the number five (5-rear) cap and saddle on V8 and four-cylinder engines.

7   Clean the faces of the bearings in the block and the crankshaft main bearing journals with a clean, lint-free cloth. Check or clean the oil holes in the crankshaft, as any dirt here can only go one way - straight through the new bearings.

8   Once you are certain that the crankshaft is clean, carefully lay it in position (an assistant would be very helpful here) in the main bearings with the counterweights lying sideways.

9   Before the crankshaft can be permanently installed, the main bearing oil clearance must be checked.

10   Trim several pieces of the appropriate

type of Plastigage (so they are slightly shorter than the width of the main bearings) and place one piece on each crankshaft main bearing journal, parallel with the journal axis **(see illustration)**. Do not lay them across the oil holes.

11   Clean the faces of the bearings in the caps and install the caps in their respective positions (do not mix them up) with the arrows pointing toward the front of the engine. Do not disturb the Plastigage.

12   Starting with the center main and working out toward the ends,
tighten the main bearing cap bolts, in three steps, to the specified torque. Do not rotate the crankshaft at any time during this operation.

13   Remove the bolts and carefully lift off the main bearing caps. Keep them in order. Do not disturb the Plastigage or rotate the crankshaft. If any of the main bearing caps are difficult to remove, tap them gently from side-to-side with a soft-faced hammer to loosen them.

14   Compare the width of the crushed Plastigage on each journal to the scale printed on the Plastigage container **(see illustration)** to obtain the main bearing oil clearance. Check the Specifications to make sure it is correct.

15   If the clearance is not correct, double-check to make sure you have the right size bearing inserts. Also, make sure that no dirt or oil was between the bearing inserts and the main bearing caps or the block when the clearance was measured.

16   Carefully scrape all traces of the Plastigage material off the main bearing journals and/or the bearing faces. Do not nick or scratch the bearing faces.

17   Carefully lift the crankshaft out of the engine. Clean the bearing faces in the block, then apply a thin, uniform layer of clean, high-quality moly-based grease or engine assembly lube to each of the bearing faces. Be sure to coat the thrust flange faces as well as the journal face of the thrust bearing.

18   On 1985 and earlier V8 models, lubricate the rear main oil seal (where it contacts the crankshaft) with clean engine oil. On 1984 and earlier V6 engines, install the service replacement rear main seal onto the crankshaft (see Section 20). **Note:** *On all four-cylinder engines, 1985 and later V6 engines and 1986 and later V8 engines, the oil seal is installed after the crankshaft is in place.*

19   On 1984 and earlier V6 and 1985 and earlier V8 engines, apply anaerobic-type gasket sealant to the rear main bearing cap **(see Chapter 2C, illustration 11.14)**. Make sure the crankshaft journals are clean, then lay the crankshaft back in place in the block. Clean the faces of the bearings in the caps, then apply a thin, uniform layer of clean, moly-based grease to each of the bearing faces and install the caps in their respective positions with the arrows pointing toward the front of the engine. Install the bolts and tighten them to the specified torque, starting with the center main and working out toward

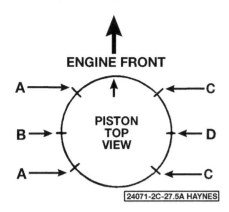

ENGINE FRONT

A →

B →

A →

PISTON
TOP
VIEW

← C

← D

← C

24071-2C-27.5A HAYNES

**22.5 Before installing the pistons,
position the ring end gaps as shown**

A   *Oil ring rail gaps*
B   *Second compression ring gap*
C   *Oil ring spacer gap (position in-between marks)*
D   *Top compression ring gap*

**22.8 The mark on top of the piston must
face the front of the engine**

the ends. Work up to the final torque in three steps.

20   Rotate the crankshaft a number of times by hand and check for any obvious binding.

21   The final step is to check the crankshaft endplay. This can be done with a feeler gauge or a dial indicator set. Refer to Section 12 for the procedure.

22   Piston/connecting rod assembly - installation and bearing oil clearance check

Refer to illustration 22.5, 22.8, 22.10, 22.12 and 22.14

1   Before installing the piston/connecting rod assemblies, the cylinder walls must be perfectly clean, the top edge of each cylinder must be chamfered, and the crankshaft must be in place.

2   Remove the connecting rod cap from the end of the number one connecting rod. Remove the old bearing inserts and wipe the bearing surfaces of the connecting rod and cap with a clean, lint-free cloth (they must be kept spotlessly clean).

3   Clean the back side of the new upper bearing half, then lay it in place in the con-

necting rod. Make sure that the tab on the bearing fits into the recess in the rod. Do not hammer the bearing insert into place and be very careful not to nick or gouge the bearing face. Do not lubricate the bearing at this time.

4   Clean the back side of the other bearing insert and install it in the rod cap. Again, make sure the tab on the bearing fits into the recess in the cap, and do not apply any lubricant. It is critically important that the mating surfaces of the bearing and connecting rod are perfectly clean and oil-free when they are assembled.

5   Position the piston ring gaps as shown **(see illustration)**, then slip a section of plastic or rubber hose over the connecting rod cap bolts.

6   Lubricate the piston and rings with clean engine oil and attach a piston ring compressor to the piston. Leave the skirt protruding about 1/4-inch to guide the piston into the cylinder. The rings must be compressed as far as possible.

7   Rotate the crankshaft until the number one connecting rod journal is as far from the number one cylinder as possible (bottom dead center), and apply a uniform coat of engine oil to the cylinder walls.

8   With the notch on top of the piston facing to the front of the engine **(see illustration)**,

gently place the piston/connecting rod assembly into the number one cylinder bore and rest the bottom edge of the ring compressor on the engine block. Tap the top edge of the ring compressor to make sure it is contacting the block around its entire circumference. **Note:** *On V8 engines, the bearing tab recess in the rod must be opposite the camshaft.*

9   Clean the number one connecting rod journal on the crankshaft and the bearing faces in the rod.

10   Carefully tap on the top of the piston with the end of a wooden hammer handle **(see illustration)** while guiding the end of the connecting rod into place on the crankshaft journal. The piston rings may try to pop out of the ring compressor just before entering the cylinder bore, so keep some downward pressure on the ring compressor. Work slowly, and if any resistance is felt as the piston enters the cylinder, stop immediately. Find out what is hanging up and fix it before proceeding. Do not, for any reason, force the piston into the cylinder, as you will break a ring and/or the piston.

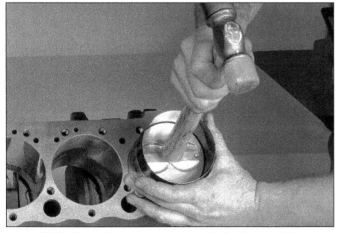

**22.10 If resistance is encountered when tapping the
piston/connecting rod assembly into the block, stop immediately
and make sure the rings are fully compressed**

**22.12 Position the Plastigage strip on the bearing journal, parallel
to the journal axis**

22.14 The crushed Plastigage is compared to the scale printed on the container to obtain the bearing oil clearance

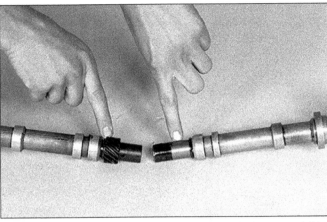

24.3 The pre-oiling tool (right) has the gear and advance weights ground off

11   Once the piston/connecting rod assembly is installed, the connecting rod bearing oil clearance must be checked before the rod cap is permanently bolted in place.

12   Cut a piece of the appropriate type Plastigage slightly shorter than the width of the connecting rod bearing and lay it in place on the number one connecting rod journal, parallel with the journal axis (it must not cross the oil hole in the journal) **(see illustration)**.

13   Clean the connecting rod cap bearing face, remove the protective hoses from the connecting rod bolts and gently install the rod cap in place. Make sure the mating mark on the cap is on the same side as the mark on the connecting rod. Install the nuts and tighten them to the specified torque, working up to it in three steps. Do not rotate the crankshaft at any time during this operation.

14   Remove the rod cap, being very careful not to disturb the Plastigage. Compare the width of the crushed Plastigage to the scale printed on the Plastigage container to obtain the oil clearance **(see illustration)**. Compare it to the Specifications to make sure the clearance is correct. If the clearance is not correct, double-check to make sure that you have the correct size bearing inserts. Also, recheck the crankshaft connecting rod journal diameter and make sure that no dirt or oil was between the bearing inserts and the connecting rod or cap when the clearance was measured.

15   Carefully scrape all traces of the Plastigage material off the rod journal and/or bearing face (be very careful not to scratch the bearing- use your fingernail or a piece of hardwood). Make sure the bearing faces are perfectly clean, then apply a uniform layer of clean, high quality moly-based grease or engine assembly lube to both of them. You will have to push the piston into the cylinder to expose the face of the bearing insert in the connecting rod; be sure to slip the protective hoses over the rod bolts first.

16   Slide the connecting rod back into place on the journal, remove the protective hoses from the rod cap bolts, install the rod cap and tighten the nuts to the specified torque. Again, work up to the torque in three steps.

17   Repeat the entire procedure for the remaining piston/connecting rod assemblies. Keep the back sides of the bearing inserts and the inside of the connecting rod and cap perfectly clean when assembling them. Make sure you have the correct piston for the cylinder and that the notch, arrow or F on the piston faces to the front of the engine when the piston is installed. Remember, use plenty of oil to lubricate the piston before installing the ring compressor and cap. Also, when installing the rod caps for the final time, be sure to lubricate the bearing faces adequately.

18   After all the piston/connecting rod assemblies have been properly installed, rotate the crankshaft a number of times by hand and check for any obvious binding.

19   As a final step, the connecting rod endplay must be checked. Refer to Section 11 for the procedure to follow. Compare the measured endplay to the Specifications to make sure it is correct.

23 Engine overhaul - reassembly sequence

1   Before beginning engine reassembly, make sure you have all the necessary new parts, gaskets and seals as well as the following items on hand:

Common hand tools
A 1/2-inch drive torque wrench
Piston ring installation tool
Piston ring compressor
Short lengths of rubber or plastic hose to fit over connecting rod bolts
Plastigage
Feeler gauges
A fine-tooth file
New, clean engine oil
Engine assembly lube or moly-based grease
RTV-type gasket sealant
Anaerobic-type gasket sealant
Thread locking compound

2   In order to save time and avoid problems, engine reassembly must be done in the following order.

Rear main oil seal (two-piece seal)
Crankshaft and main bearings
Piston rings
Piston/connecting rod assemblies
Oil pump
Oil pan
Camshaft
Timing chain/sprockets or gears
Timing chain/gear cover
Valve lifters
Cylinder head(s)
Pushrods and rockers arms (adjust the valve lash on V6 and V8 engines - see Chapter 2B or 2C)
Intake and exhaust manifolds
Oil filter
Pre-oil the engine (Section 24)
Rocker arm cover(s)
Fuel pump
Water pump
Rear main oil seal (one-piece seal)
Flywheel/driveplate
Carburetor/fuel injection components
Thermostat and housing cover
Distributor, spark plug wires and spark plugs
Emissions control components
Alternator

24 Pre-oiling engine after overhaul

1   After an overhaul it is a good idea to pre-oil the engine before it is installed and initially started. This will reveal any problems with the lubrication system at a time when corrections can be made easily and without major engine damage. Pre-oiling the engine will also allow the parts to be lubricated thoroughly in a normal fashion, but without the heavy loads associated with the combustion process placed upon them.

2   The engine should be assembled completely with the exception of the distributor and the rocker arm covers.

3   On V6 and V8 engines, a modified distributor will be needed for this procedure. This pre-oil tool is a distributor body with the bottom gear ground off and the advance

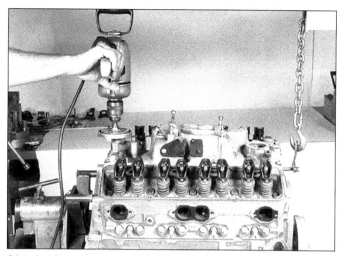

24.4 A drill motor connected to the modified distributor drives the oil pump

24.5 Oil and assembly lube or grease will spurt out of the holes in the rocker arms if the lubrication system is functioning properly

weight assembly removed from the top of the shaft **(see illustration)**.

4   Place the pre-oiler into the distributor shaft access hole at the rear of the intake manifold and make sure the bottom of the shaft mates with the oil pump. Clamp the modified distributor into place just as you would an ordinary distributor. Now attach an electric drill motor to the top of the shaft **(see illustration)**.

5   With the oil filter installed, all oil ways plugged (oil-pressure sending unit at rear of block) and the crankcase full of oil as shown on the dipstick, rotate the pre-oiler with the drill. Make sure the rotation is in a clockwise direction. Soon, oil should start to flow from the rocker arms, signifying that the oil pump and lubrication system are functioning properly. It may take two or three minutes for oil to flow to all of the rocker arms **(see illustration)**. Allow the oil to circulate through the engine for a few minutes, then shut off the drill motor.

6   Check for oil leaks at the filter and all gasket and seal locations.

7   Remove the pre-oil tool, then install the distributor and rocker arm covers.

25   Initial start-up and break-in after overhaul

1   Once the engine has been properly installed in the vehicle, doublecheck the engine oil and coolant levels.

2   With the spark plugs out of the engine and the coil high-tension lead grounded to the engine block, crank the engine over until oil pressure registers on the gauge (if so equipped) or until the oil light goes off.

3   Install the spark plugs, hook up the plug wires and the coil high tension lead.

4   Make sure the carburetor choke plate is closed, then start the engine. It may take a few moments for the gasoline to reach the carburetor, but the engine should start without a great deal of effort.

5   As soon as the engine starts, it should be set at a fast idle (to ensure proper oil circulation) and allowed to warm up to normal operating temperature. While the engine is warming up, make a thorough check for oil and coolant leaks.

6   Shut the engine off and recheck the engine oil and coolant levels.

Also, check the ignition timing and the engine idle speed (refer to Chapter 1) and make any necessary adjustments.

7   Drive the vehicle to an area with minimum traffic, accelerate from 30 to 50 mph, then allow the vehicle to slow to 30 mph with the throttle closed. Repeat the procedure 10 or 12 times. This will load the piston rings and cause them to seat properly against the cylinder walls. Check again for oil and coolant leaks.

8   Drive the vehicle gently for the first 500 miles (no sustained high speeds) and keep a constant check on the oil level. It is not unusual for an engine to use oil during the break-in period.

9   At approximately 500 to 600 miles, change the oil and filter, retorque the cylinder head bolts and recheck the valve clearances (if applicable).

10   For the next few hundred miles, drive the vehicle normally. Do not pamper it or abuse it.

11   After 2000 miles, change the oil and filter again and consider the engine fully broken in.

Chapter 3
Cooling, heating and air conditioning systems

Contents

	Section
Air conditioning system - servicing	11
Air conditioning compressor - removal and installation	13
Air conditioning condenser - removal and installation	12
Antifreeze - general information	2
Coolant level check	See Chapter 1
Coolant temperature sending unit - check and replacement	8
Cooling system check	See Chapter 1
Cooling system servicing (draining, flushing and refilling)	Chapter 1

	Section
Electric cooling fan - removal and installation	14
General information	1
Heater blower motor - removal and installation	9
Heater core - removal and installation	10
Radiator - removal and installation	5
Thermostat - check	4
Thermostat - replacement	3
Water pump - check	6
Water pump - removal and installation	7

Specifications

Cooling system

Coolant capacity	See Chapter 1
Radiator cap rating	14 to 17 psi
Thermostat rating	195-degrees

Torque specifications

Ft-lbs (unless otherwise indicated)

Fan-to-water pump bolts	20
Water pump bolts	
L4	15
2.8L V6	
Long bolts and nut	15
Short bolts	84 in-lbs
3.1L V6	
Long bolts	33
Short bolts	192 in-lbs
Small bolt	89 in-lbs
V8	30
Water outlet (thermostat housing cover) bolts	
L4	20
V6	18
V8	25
Coolant temperature sending unit	20

3.20 Remove the thermostat housing (L4 engine shown)

3.21 Remove the thermostat from the engine (L4 engine shown)

1   General information

The cooling system is very conventional in design, utilizing a cross flow radiator, an engine driven water pump and thermostat controlled coolant flow. On early models, the engine driven fan is equipped with a clutch, which allows it to draw air through the radiator at lower speeds. At higher speeds the fan is not needed for cooling, so the clutch automatically lowers the fan speed and reduces the engine power required for fan operation. Later models are equipped with an electric cooling fan. The electric cooling fan is controlled by either a coolant switch or the ECM, depending on model. The fan is energized when the coolant reaches a preset temperature and shuts off when the coolant has cooled sufficiently.

The water pump is mounted on the front of the engine and is driven by a belt from the pulley mounted on the front of the crankshaft. The belt is also used to drive other components.

The heater utilizes the heat produced by the engine and absorbed by the coolant to warm the interior of the vehicle. It is manually controlled from inside by the driver or passenger.

Air conditioning is optional equipment. All components of the system are mounted in the engine compartment and the system is driven by a belt from the pulley mounted on the front of the crankshaft. Output of the system is controlled from inside the vehicle.

2   Antifreeze - general information

Caution: *Do not allow antifreeze to come in contact with your skin or painted surfaces of the vehicle. Flush contacted areas immediately with plenty of water. Antifreeze can be fatal to children and pets. They like because it is sweet. Just a few licks can cause death. Wipe up garage floor and drip pan coolant spills immediately. Keep antifreeze containers covered and repair leaks in your cooling*

system immediately.

The cooling system should be filled with a water/ethylene glycol-based antifreeze solution, which will prevent freezing down to at least -20 degrees F at all times. It also provides protection against corrosion and increases the coolant boiling point.

The cooling system should be drained, flushed and refilled at least every other year (see Chapter 1). The use of antifreeze solutions for periods of longer than two years is likely to cause damage and encourage the formation of rust and scale in the system.

Before adding antifreeze to the system, check all hose connections and gaskets for leaks, antifreeze tends to search out and leak through very minute openings.

The exact mixture of antifreeze-to-water which you should use depends on the relative weather conditions. The mixture should contain at least 50 percent antifreeze, but should never contain more than 70 percent antifreeze.

3   Thermostat - replacement

Caution: *The engine must be completely cool before beginning this procedure. Also, when working in the vicinity of the electric fan, disconnect the negative battery cable from the battery to prevent the fan from coming on accidentally.*

Four-cylinder engine

1   Refer to the Caution in Section 2.
2   Disconnect the cable from the negative battery terminal.
3   Drain the cooling system until the level is below the thermostat by opening the petcock at the bottom right side of the radiator. Close the petcock when enough coolant has drained.
4   Remove the upper radiator hose from the water outlet housing.
5   Remove the two housing bolts and separate the housing from the engine.
6   Remove the thermostat from the thermostat housing.

7   Before installing the thermostat, clean the gasket sealing surfaces on the water outlet and thermostat housing.
8   To install the thermostat, place it in the housing, install a new gasket (not RTV), install the water outlet and tighten the water outlet bolts to the specified torque.
9   Install the upper radiator hose.
10   Fill the cooling system with the proper antifreeze/water mixture (refer to Chapter 1).
11   Reconnect the battery cable and start the engine.
12   Run the engine with the radiator cap removed until the upper radiator hose is hot (thermostat open).
13   With the engine idling, add coolant to the radiator until the level reaches the bottom of the filler neck.
14   Install the radiator cap, making sure the arrows on the cap line up with the radiator overflow tube.

V6 and V8 engines

Refer to illustrations 3.20 3.21 and 3.22

15   Refer to the Caution in Section 2.
16   Disconnect the cable from the negative battery terminal.
17   On carbureted and TBI models, remove the air cleaner, tagging all hoses as they are removed to simplify installation. On TPI models, remove the air intake duct.
18   Drain the cooling system until the level is below the thermostat (refer to Step 3).
19   Disconnect the upper radiator hose from the thermostat housing.
20   Remove the thermostat housing bolts and separate the housing from the engine **(see illustration)**. On some models, the alternator mounting bracket will have to be disconnected first, as it is attached to the housing mounting stud. Also, late model vehicles may have a TVS switch installed in the thermostat housing. If so, disconnect each of the vacuum hoses from the switch (noting their installed positions) and then unscrew the switch from the housing.
21   After separating the thermostat housing from the engine, the thermostat will be visible and can be lifted out **(see illustration)**. Note

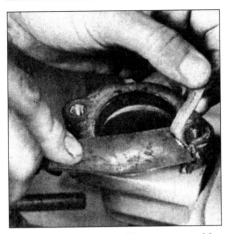

3.22 Using a putty knife to remove old gasket material from the thermostat gasket sealing surface

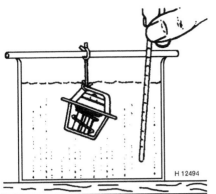

4.5 Testing the thermostat opening temperature

how the thermostat is positioned in the recess, as it must be replaced in the same position.

22   Before installation, use a gasket scraper or putty knife to carefully remove all traces of the old gasket from the thermostat housing and the engine sealing surface **(see illustration)**. Do not allow the gasket pieces to drop down into the intake manifold.

23   Apply a 1/8-inch bead of RTV sealant to the sealing surface on the engine and place the thermostat into the recess.

24   Immediately place the thermostat housing with sealant and a new gasket into position and tighten the bolts to the specified torque.

25   Where applicable, install the alternator brace and/or the TVS switch and vacuum hoses.

26   Connect the upper radiator hose and tighten the hose clamp securely.

27   Reinstall the air cleaner.

28   To complete the installation, refer to Steps 10 through 14.

4   Thermostat - check

Refer to illustration 4.5

1   Before assuming the thermostat is to blame for a cooling system problem, check the coolant level, drivebelt tension (see Chapter 1) and temperature gauge operation.

2   If the engine seems to be taking a long time to warm up (based on heater output or temperature gauge operation), the thermostat is probably stuck open. Replace the thermostat with a new one.

3   If the engine runs hot, use your hand to check the temperature of the upper radiator hose. If the hose isn't hot, but the engine is, the thermostat is probably stuck closed, preventing the coolant inside the engine from escaping to the radiator. Replace the thermostat. **Caution:** *Don't drive the vehicle without a thermostat. The computer may stay in open loop and emissions and fuel economy will suffer.*

4   If the upper radiator hose is hot, it means that the coolant is flowing and the thermostat is open. Consult the Troubleshooting section at the front of this manual for cooling system diagnosis.

5   To accurately test the thermostat, remove it from the engine and suspend the (closed) thermostat on a length of string in a container of cold water, with a cooking thermometer beside it **(see illustration)**. Ensure that neither touches the sides or bottom of the container.

6   Heat the water and check the temperature at which the thermostat opens. Compare this value with the value listed in this Chapter's Specifications. Make sure the thermostat is fully open as the water boils. Remove the container from the heat, allow it to cool and confirm the thermostat closes fully.

7   If the thermostat does not open at the specified temperature or does not fully open and close as described, replace the thermostat.

5   Radiator - removal and installation

Refer to illustrations 5.8, 5.9 and 5.16
Caution: *The engine must be completely cool before beginning this procedure. Also, when working in the vicinity of the electric fan, disconnect the negative battery cable from the battery to prevent the fan from coming on accidentally.*

1   Refer to the Caution in Section 2.

2   Disconnect the cable from the negative battery terminal.

3   Drain the radiator (refer to Chapter 1, if necessary).

4   On models with an engine driven fan, unbolt the fan from the fan pulley hub and remove the fan assembly. On vehicles equipped with a fan clutch, store the clutch in an upright position to prevent seal leakage. On models with an electric fan, remove the fan assembly (refer to Section 14).

5   Disconnect the upper and lower radiator hoses from the radiator.

6   On vehicles equipped with an automatic transmission, disconnect the fluid cooler lines at the radiator and immediately plug the lines.

7   Remove the clamp securing the coolant reservoir hose to the radiator outlet. On V6 models, remove the hose from the upper radiator shroud retainers.

8   Remove the bolts securing the top of the upper radiator shroud to the metal support **(see illustration)**.

9   Remove the bolts securing the bottom of the radiator shroud to the chassis, if so equipped **(see illustration)**.

5.8 Typical radiator fan upper shroud mounting bolt locations (arrows)

5.9 Remove the lower radiator shroud bolts (V6 engine shown)

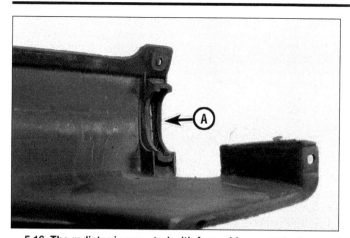

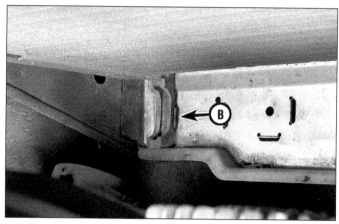

5.16 The radiator is mounted with four rubber mounts, two are located at each end of the upper radiator shroud (view A) - the lower mounts are located at the chassis core support (view B)

10   Pull straight up on the radiator and shroud assembly and remove it from the vehicle.

11   Carefully examine the radiator for evidence of leaks and damage. It is recommended that any necessary repairs be performed by a reputable radiator repair shop.

12   With the radiator removed, brush accumulations of insects and leaves from the fins and examine and replace, if necessary, any hoses or clamps which have deteriorated.

13   The radiator can be flushed as described in Chapter 1.

14   Check the pressure rating of the radiator cap and have it tested.

15   If you are installing a new radiator, transfer the fittings from the old unit to the new one.

16   Installation is the reverse of the removal procedure. When setting the radiator in the chassis, make sure the bottom of the radiator is seated correctly in the two bottom cradles secured to the radiator support **(see illustration)**.

17   After installing the radiator, refill it with the proper coolant mixture (refer to Chapter 1), then start the engine and check for leaks.

18   If equipped with a new automatic transmission, check the transmission fluid level (refer to Chapter 1).

6   Water pump - check

Refer to illustration 6.4

1   A failure in the water pump can cause overheating and serious engine damage (the pump will not circulate coolant through the engine).

2   There are three ways to check the operation of the water pump while it is still installed on the engine. If the pump is defective, it should be replaced with a new or rebuilt unit.

3   With the engine at normal operating temperature, squeeze the upper radiator hose. If the water pump is working properly, a pressure surge will be felt as the hose is released.

4   Water pumps are equipped with "weep' or vent holes **(see illustration)**. If a pump seal failure occurs, small amounts of coolant will leak from the weep holes. In most cases it will be necessary to use a flashlight from under the vehicle to see evidence of leakage

from this point on the pump body.

5   If the water pump shaft bearings fail, there may be a squealing sound emitted from the front of the engine while it is running. Shaft wear can be felt if the water pump pulley is forced up and down. Do not mistake drivebelt slippage (which also causes a squealing sound) for water pump failure.

7   Water pump - removal and installation

Refer to illustrations 7.4, 7.7, 7.10a, 7.10b, 7.10c and 7.10d

Caution: *The engine must be completely cool before beginning this procedure. Also, when working in the vicinity of the electric fan, disconnect the negative battery cable from the battery to prevent the fan from coming on accidentally.*

1   Refer to the Caution in Section 2.

2   Disconnect the cable from the negative battery terminal.

3   Drain the cooling system (refer to Chapter 1, if necessary).

4   On models with multiple V-belts, mark

6.4 The water pump weep hole (arrow) will drip coolant when the pump shaft seal fails

7.4 Mark the drivebelts before removing them to ensure reinstallation on the correct pulleys

7.7 Various brackets and/or accessories must be removed before the water pump can be detached

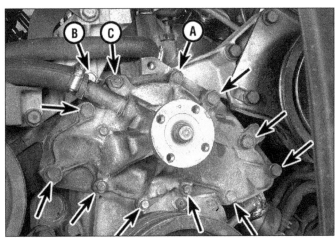

7.10a On 2.8L V6 engines in which stud C goes through both the timing cover and the water pump, remove the water pump mounting bolts (arrows) after using bolt A to attach the tool mentioned in the text to the right cylinder head at hole B - this will ensure the front cover is not dislodged during water pump removal

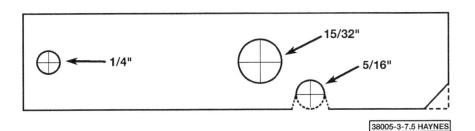

7.10b Template for 2.8L V6 front cover retaining bracket (actual size) - it should be fabricated from 3/8-inch flat stock and drilled and cut as shown

the accessory drivebelts with white paint to simplify installation (see illustration).

5   On models with an engine driven fan, remove the upper fan shroud (refer to Section 5, if necessary).

6   On models with multiple V-belts, remove the accessory drivebelts (alternator, air con- ditioning compressor, AIR system pump, power steering pump, as applicable), by loosening the pivot and adjusting bolts and pushing the accessory toward the engine. On models with a single serpentine belt, rotate the belt tensioner pulley away from the belt to release the tension on the belt, remove the belt and release the tensioner.

7   On some engines, it may be necessary to remove one or more accessories to gain access to the water pump (see illustration). Refer to the appropriate Chapter(s) to remove them.

8   On models with an engine driven fan, remove the bolts retaining the fan to the fan pulley hub, then separate the fan and pulley hub from the water pump.

9   Disconnect the heater and lower radiator hoses from the water pump.

10   Remove the water pump mounting bolts and separate the water pump from the engine (see illustrations). Caution: On some early 2.8L V6 engines, the water pump bolts hold the front cover to the block. Removal of the water pump may break the chemical seal and allow coolant into the oil. To prevent this, before removing the water pump, secure the front cover to the block with the clamping device shown in the accompanying illustrations.

7.10c The bracket is bolted in place on the right-hand cylinder head - note the 1/4 x 1-inch bolt inserted at the left side of the bracket and secured with a nut, it serves as a spacer

7.10d Water pump mounting bolt locations - V8 engine

8.6 Disconnect the wire connector from the coolant temperature sending unit prior to removing it from the cylinder head (V6 engine shown)

11   Prior to installing the water pump, remove all old gasket material and sealant from the gasket sealing surfaces. Clean the threaded holes in the block as well.
12   Installation is the reverse of the removal procedure.
13   If a new water pump is being installed, transfer the heater hose fitting from the old pump to the new one.
14   On L4 engines, use a new gasket or RTV type sealant when installing the pump.
15   On V6 and V8 engines, use a new gasket when installing the pump.
16   Tighten the bolts to the specified torque after coating the threads with RTV sealant to prevent leaks. Follow a crisscross pattern and work up to the final torque in three (3) steps.
17   Adjust all drivebelts (see Chapter 1).
18   Connect the negative battery cable and fill the radiator with a mixture of antifreeze and water. Start the engine and allow it to idle until the upper radiator hose gets hot. Check for leaks. With the engine hot, fill the radiator with more coolant mixture until the level is at the bottom of the filler neck. Install the radiator

cap and check the coolant level periodically during the first few miles of driving.

8   Coolant temperature sending unit - check and replacement

Refer to illustration 8.6
1   The coolant temperature indicator system is composed of a light mounted in the instrument panel and a coolant temperature sending unit located at the left rear corner of the cylinder head on L4 engines and at the front of the left cylinder head on V6 and V8 engines. If a temperature gauge is included in the instrument cluster, the temperature sending unit is replaced by a transducer.
2   **Caution:** *Since the ignition key will be in the On position for some of the diagnostic steps, be especially careful to stay clear of the electric fan blades.*
3   If overheating occurs, check the coolant level in the system and then make sure that the wiring between the light or gauge and the sending unit is secure.
4   When the ignition switch is turned on and the starter motor is turning, the indicator light should be on (overheated engine indication). If the light is not on, the bulb may be burned out, the ignition switch may be faulty or the circuit may be open.
5   As soon as the engine starts, the light should go out and remain out unless the engine overheats. Failure of the light to go out may be due to grounded wiring between the light and the sending unit, a defective sending unit or a faulty ignition switch.
6   If the sending unit is to be replaced, disconnect the wiring connector, unscrewed the sending unit from the cylinder head and install the replacement **(see illustration)**. Make sure that the engine is cool before removing the defective sending unit. There will be some coolant loss, so check the level after the replacement has been installed.

9   Heater blower motor - removal and installation

Refer to illustrations 9.2 and 9.6
1   Disconnect the cable from the negative battery terminal.
2   Working in the engine compartment, disconnect the wires at the lower motor and resistor **(see illustration)**.
3   Disconnect the blower motor cooling tube.
4   On some models, it may be necessary to disconnect the radio capacitor to allow removal of the blower motor.
5   Remove the blower motor mounting bolts and separate the motor/cage assembly from the case.
6   While holding the cage, remove the cage retaining nut and slide the cage off the motor shaft **(see illustration)**.
7   Installation is the reverse of the removal procedure.

10   Heater core - removal and installation

Refer to illustrations 10.3 and 10.4a through 10.14b
Warning: *If equipped with a Supplemental Inflatable Restraint system (SIR), more commonly known as airbags, disable the airbag system before working in the vicinity of airbag system components to avoid the possibility of accidental deployment of the airbag, which could result in personal injury (see Chapter 10).*
1   Disconnect the cable from the negative battery terminal.
2   Drain the cooling system (see Chapter 1).
3   Remove the heater hoses from the heater core inlet and outlet pipes **(see illustration)**. Use care in removing the hoses so as not to damage the heater pipes. If necessary, cut the

9.2 To remove the heater blower motor, remove the mounting bolts (A), the motor and resistor wires (B) and the cooling tube (C)

9.6 Remove the nut (arrow) to separate the cage from the blower motor shaft

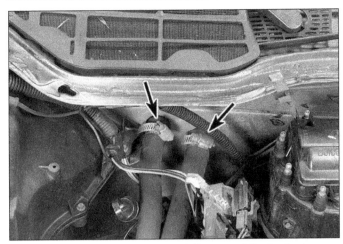

10.3 Loosen the hose clamps and disconnect the heater hoses from the heater core tubes at the firewall

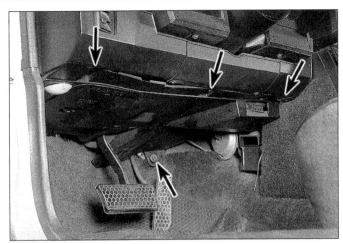

10.4a Remove the insulating panel from under the left side . . .

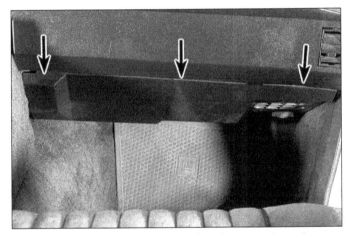

10.4b . . . and the right side of the instrument panel

10.5a To remove the instrument panel pad, remove the screws in the defroster vents . . .

hoses off the pipes and replace the hoses with new ones.

4   Remove the lower insulating panels from under the left and right side of the instrument panel **(see illustrations)**.

5   Remove the instrument panel pad. Retaining screws are located in the defroster vents and along the underside of the rear edge of the pad **(see illustrations)**.

6   Remove the retaining nuts to both front

speakers and remove the speakers. Also remove the right speaker bracket **(see illustration)**.

7   Remove the screws retaining the side window defrost ducts, the front carrier braces

10.5b . . . and the screws along the underside of the front edge

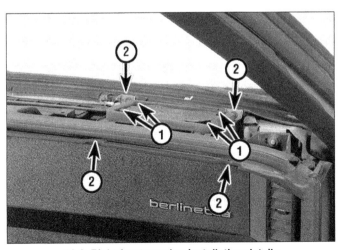

10.6 Right front speaker installation details

1   *Speaker mounting screws*
2   *Speaker bracket mounting screws*

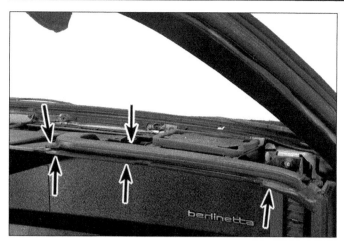

10.7 Remove the screws (arrows) and the side window defroster ducts

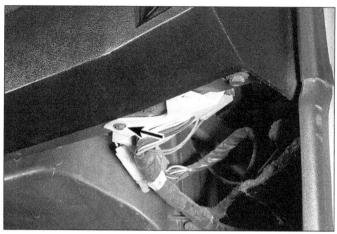

10.8 Remove the screw and carefully disconnect the electrical connectors to the ECM

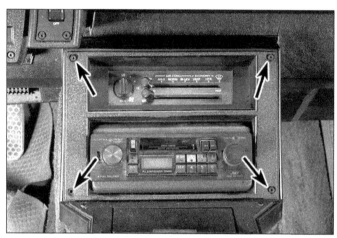

10.9 Remove the screws (arrows) and the center trim panel

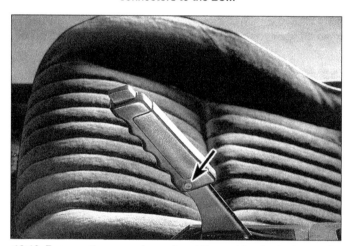

10.10 Remove the screw (arrow) and the parking brake lever grip

and carrier shelf and remove the side defrost ducts **(see illustration)**.

8 Carefully disconnect the Electronic Control Module and set it safely aside **(see illustration)**.

9 Remove the radio trim plate, the upper console trim, and the glove box assembly **(see illustration)**.

10 Remove the parking brake handle grip, the screws retaining the console body, and position the console back out of the way **(see illustration)**.

11 Remove the trim plate under the steering column. Remove the two nuts retaining the steering column and carefully lower and support the steering column out of the way **(see illustration)**.

12 Remove the bolts/nuts retaining the instrument panel to the cowl **(see illustrations)**.

13 Move the instrument panel back far

10.11 Remove the two nuts (arrows) retaining the steering column to the instrument panel brace and lower the steering column

10.12a Remove the instrument panel mounting bolts/nuts at the top (arrows) . . .

10.12b ... and at each end (arrow)

10.13a Pull the instrument panel back to access the heater module

1 Instrument panel assembly
2 Heater core module assembly

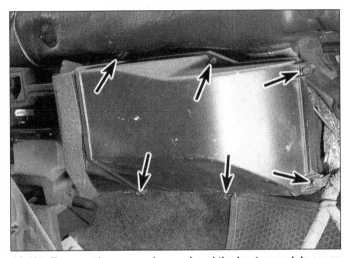

10.13b Remove the screws (arrows) and the heater module cover

10.14a Remove the screws (arrows) retaining the heater core shroud to the module and withdraw the heater core

enough to access the heater module and remove the heater module cover **(see illustrations)**.

14 Remove the screws retaining the heater core shroud to the module and withdraw the heater core from the module **(see illustrations)**.

15 Installation is the reverse of removal. Replace the seals with new ones. Refill the cooling system and test for leaks.

11 Air conditioning system - servicing

1 Regularly inspect the condenser fins (located ahead of the radiator) and brush away leaves and bugs.

2 Clean the evaporator drain tubes.

3 Check the condition of the system hoses. If there is any sign of deterioration or hardening, have them replaced by your dealer.

4 At the recommended intervals, check and adjust the compressor drivebelt as described in Chapter 1.

5 Because of the special tools, equipment

and skills required to service air conditioning systems, and the differences between the various systems that may be installed on vehicles, major air conditioner servicing procedures cannot be covered in this manual.

6 We will cover component removal, as the home mechanic may realize a substantial

savings in repair costs if he removes components himself, takes them to a professional for repair, and/or replaces them with new ones.

7 Problems in the air conditioning system should be diagnosed, and the system refrigerant evacuated, by an air conditioning tech-

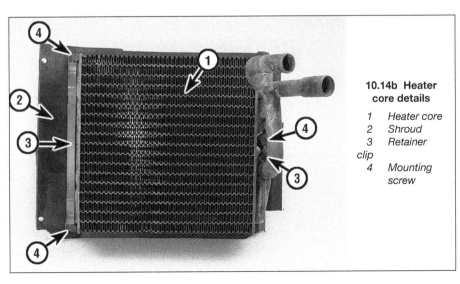

10.14b Heater core details

1 Heater core
2 Shroud
3 Retainer clip
4 Mounting screw

12.4 Typical air conditioning condenser installation details

1	Condenser	3	Radiator support
2	Upper mounts	4	Lower mounts

13.1a Loosen the compressor pivot bolt

13.1b Remove the compressor adjusting bolt

13.2 Disconnecting the compressor wiring connector

nician before component removal/replacement is attempted.

8   Once the new or reconditioned component has been installed, the system should then be recharged and checked by an air conditioning technician.

9   Before indiscriminately removing the air conditioning system components, get more than one estimate of repair costs from reputable air conditioning service centers. You may find it to be cheaper and less trouble to let the entire operation be performed by someone else.

12   Air conditioning condenser - removal and installation

Refer to illustration 12.4

Warning: *The air conditioning system is under high pressure. Do not loosen any hose fittings or remove any components until after the system has been discharged. Air conditioning refrigerant must be properly dis-*

charged *into an EPA-approved recovery/recycling unit at a dealer service department or an automotive air conditioning repair facility. Always wear eye protection when disconnecting air conditioning system fittings.*

1   After having the system evacuated, disconnect the coupled hose and liquid line fittings.

2   Remove the screws retaining the top radiator shroud.

3   Remove the bolts holding the condenser to the top radiator support.

4   Carefully tilt the radiator to the rear and lift the condenser out of the bottom cradle supports **(see illustration)**.

5   If the condenser is being replaced with a new one, transfer the brackets and mounts from the old unit to the new one.

6   The installation procedures are the reverse of those for removal. When installing the hose and fittings, use new O-rings and lubricate them with refrigerant oil.

7   After the condenser is installed, have the system recharged.

13   Air conditioning compressor - removal and installation

Refer to illustrations 13.1a, 13.1b, 13.2 and 13.5

Warning: *The air conditioning system is under high pressure. Do not loosen any hose fittings or remove any components until after the system has been discharged. Air conditioning refrigerant must be properly discharged into an EPA-approved recovery/recycling unit at a dealer service department or an automotive air conditioning repair facility. Always wear eye protection when disconnecting air conditioning system fittings.*

1   After having the system evacuated, remove the compressor pivot and adjusting bolts **(see illustrations)**.

2   Disconnect the electrical connector from the compressor **(see illustration)**.

3   Remove the drivebelt, routing the lower loop behind the vibration damper to gain additional slack, if necessary.

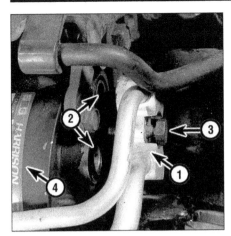

13.5 Air conditioning compressor refrigerant line attachments

1 Fitting and hose assembly
2 Gasket O-rings
3 Hose fitting assembly attaching bolt
4 Air conditioning compressor

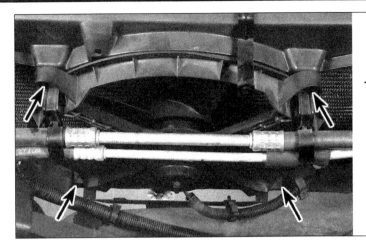

14.3 Remove the electric cooling fan mounting bolts (arrows)

4   Remove the compressor from the mount.

5   Remove the fitting block (coupled hose assembly) bolt at the rear of the compressor **(see illustration)**.

6   If the compressor is being replaced with a new one, transfer the usable switches from the old unit to the new one.

7   Installation procedures are the reverse of those for removal. When installing the fitting block, use new O-rings and lubricate them with refrigerant oil.

8   After the compressor is installed, have the system recharged.

14   Electric cooling fan - removal and installation

Refer to illustration 14.3

1   Disconnect the negative battery cable from the battery.

2   Unplug the electrical connector from the fan motor and position the wiring harness out of the way.

3   Remove the fan frame-to-radiator support bolts and lift the fan assembly out of the engine compartment, taking care not to damage the radiator fins **(see illustration)**.

4   If the fan blade is to be replaced, remove the fan retaining nut and remove the fan from the motor shaft.

5   If the motor is to be replaced, remove the fan retaining nut and the fan, then remove the four motor-to-frame mounting bolts and detach the motor from the frame.

6   Installation is the reverse of removal.

Notes

Chapter 4 Part A
Fuel and exhaust systems - carbureted models

Contents

	Section
Carburetor - removal and installation...	7
Carburetor (E2SE) - overhaul...	8
Carburetor (E4ME) - overhaul...	9
Electronic feedback carburetor system	10
Engine idle speed check and adjustment... .... See Chapter 1
Exhaust system check... See Chapter 1
Exhaust system components - removal and installation...	11
Fuel filter replacement... .... See Chapter 1
Fuel line - repair and replacement...	4

	Section
Fuel pump - check...	2
Fuel pump - removal and installation...	3
Fuel system check... See Chapter 1
Fuel tank - removal and installation...	5
Fuel tank - repair...	6
General information...	1
Thermo-controlled Air Cleaner (THERMAC) check... ...... See Chapter 1
Throttle linkage check... .... See Chapter 1

Specifications

Torque specifications

Carburetor mounting nuts/bolts	
Long	84 in-lbs
Short...	132 in-lbs
Fuel inlet nut-to-carburetor..	18 Ft-lbs
Fuel tank strap-to-chassis bolts..	25 Ft-lbs

1 General information

The fuel system consists of a rear mounted fuel tank, a mechanically operated fuel pump, a carburetor and an air cleaner.

The carburetor may be a two or four-barrel type, depending on the engine. The V6 engine uses the E2SE two-barrel carburetor. The V8 engine uses the E4ME 4-barrel carburetor. Both carburetors are electronically controlled feedback carburetors.

The exhaust system includes a catalytic converter, muffler, related emissions equipment and associated pipes and hardware.

2 Fuel pump - check

Warning: *Gasoline is extremely flammable, so extra precautions must be taken when working on any part of the fuel system. Do not smoke or allow open flames or bare light bulbs near the work area. Also, do not work in a garage if a natural gas-type appliance with a pilot light is present.*

1   On V6 models, the fuel pump is located on the left-front side of the engine.

2   On V8 models, the fuel pump is located on the right-front side of the engine.

3   Both types of fuel pumps are sealed and no repairs are possible. However, the fuel pump can be inspected and tested on the vehicle as follows.

4   Make sure that there is fuel in the fuel tank.

5   Check for leaks at all gasoline line connections between the fuel tank and the carburetor. Tighten any loose connections. Inspect all hoses for flat spots and kinks which would restrict the fuel flow. Air leaks or restrictions on the suction side of the fuel pump will greatly affect the pump's output.

6   Check for leaks at the fuel pump diaphragm flange.

7   Disconnect the ignition primary connec-

3.2 When removing the fuel lines from the pump, hold the fitting on the pump with an open-end wrench while turning the fuel line fitting with a flare-nut wrench (V6 engine shown)

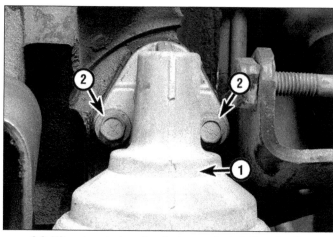

3.3 V8 carbureted engine fuel pump installation details

1 Fuel pump	*2 Mounting bolts*

tor at the distributor or ignition coil. Disconnect the fuel inlet line from the carburetor, attach a length of rubber fuel hose to the line and place the other end of the hose in an approved gasoline container.

8   Crank the engine a few revolutions and make sure that well-defined spurts of fuel are ejected from the open end of the line. If not, the fuel line is clogged or the fuel pump is defective.

9   Connect a fuel pressure gauge to the end of the fuel line. Crank the engine several revolutions and note the pressure of the gauge. Normal fuel pressure should be 5 to 7 psi. If the fuel pressure is low and the fuel line is not clogged, replace the fuel pump with a new one.

3   Fuel pump - removal and installation

Refer to illustrations 3.2 and 3.3

Warning: *Gasoline is extremely flammable, so extra precautions must be taken when working on any part of the fuel system. Do not smoke or allow open flames or bare light bulbs near the work area. Also, do not work in a garage if a natural gas-type appliance with a pilot light is present.*

1   Disconnect the cable from the negative battery terminal. **Caution:** *On models equipped with a Delco Loc II audio system, disable the anti-theft feature before disconnecting the battery.*

2   Remove the fuel inlet and outlet lines. Use two wrenches, an open-end wrench on the fuel pump and a flare-nut wrench on the fuel line fitting to prevent damage to the pump and connections **(see illustration)**.

3   Remove the fuel pump mounting bolts, the pump and the gasket **(see illustration)**.

4   On V6 models, remove the pushrod. On V8 models, remove the two bolts and the fuel pump adapter and gasket, then remove the pushrod.

5   Clean the gasket material and sealant

from the fuel pump, adapter and engine block. Coat the pushrod with heavy grease (to hold it in position) and install it in the engine block.

6   On V8 models, install the adapter plate with a new gasket.

7   Install the pump using a new gasket. Use gasket sealant on the screw threads.

8   Connect the fuel lines, start the engine and check for leaks.

4   Fuel line - repair and replacement

Warning: *Gasoline is extremely flammable, so extra precautions must be taken when working on any part of the fuel system. Do not smoke or allow open flames or bare light bulbs near the work area. Also, do not work in a garage if a natural gas-type appliance with a pilot light is present.*

1   If a section of metal fuel line must be replaced, only brazed, seamless steel tubing should be used, since copper or aluminum does not have enough durability to withstand normal operating vibrations.

2   If only one section of a metal fuel line is damaged, it can be cut out and replaced with a piece of rubber hose. Be sure to use only reinforced fuel resistant hose, identified by the word "Fluroelastomer' on the hose. The inside diameter of the hose should match the outside diameter of the metal line. The rubber hose should be cut four (4) inches longer than the section it's replacing, so there are two (2) inches of overlap between the rubber and metal line at either end of the section. Hose clamps should be used to secure both ends of the repaired section.

3   If a section of metal line longer than six (6) inches is being removed, use a combination of metal tubing and rubber hose so the hose lengths will be no longer than 10 inches.

4   Never use rubber hose within four (4) inches of any part of the exhaust system or within 10 inches of the catalytic converter.

5   When replacing clamps, make sure the replacement clamp is identical to the one being replaced, as different clamps are used depending on location.

5   Fuel tank - removal and installation

Refer to illustrations 5.12 and 5.18

Warning: *Gasoline is extremely flammable, so extra precautions must be taken when working on any part of the fuel system. Do not smoke or allow open flames or bare light bulbs near the work area. Also, do not work in a garage if a natural gas-type appliance with a pilot light is present.*

1   Remove the cable from the negative battery terminal. **Caution:** *On models equipped with a Delco Loc II audio system, disable the anti-theft feature before disconnecting the battery.*

2   Remove the filler cap from the fuel tank.

3   Drain all fuel from the tank into a clean container. Since there are no drain plugs on the tank, siphon the fuel through the filler neck or drain the fuel through the fuel line running to the carburetor or fuel injection unit. **Warning:** *Use an approved siphoning kit, DO NOT start the siphoning process with your mouth because serious personal injury could result.*

4   Raise the vehicle and support it securely on jackstands.

5   Disconnect the exhaust pipe from the rear of the catalytic converter.

6   Disconnect the rear exhaust pipe hanger and allow the exhaust system to hang over the rear axle assembly.

7   Remove the exhaust pipe and muffler heat shields.

8   Remove the fuel filler neck shield.

9   Remove the rear suspension track bar (refer to Chapter 11).

10   Remove the rear suspension track bar brace.

11   Disconnect the fuel gauge wiring har-

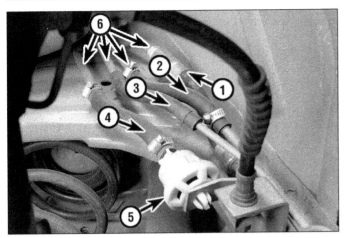

5.12 Typical fuel tank-to-fuel line connections

1   Fuel feed hose
2   Fuel return hose
3   Fuel vapor hose
4   Ventilator hose

5   Ventilator valve assembly
6   Fuel lines (part of fuel level
    sending unit assembly)

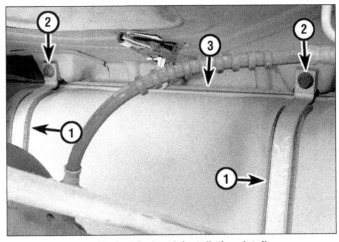

5.18 Typical fuel tank installation details

1   Fuel tank support straps
2   Support strap mounting bolts
3   Fuel tank

ness connector.
12   Disconnect the tank ventilator hose, the fuel vapor hose and the fuel feed and return hoses from the sender assembly by loosening the hose clamps and pulling the hoses off the fittings **(see illustration)**.
13   Remove the fuel line retaining bracket on the left side, then remove the brake line clip from the bracket.
14   Position a floor jack under the rear axle assembly and raise it just enough to take the weight off the axle.
15   Disconnect the lower ends of the rear shock absorbers.
16   If so equipped, remove the rear stabilizer bar from the frame.
17   Lower the jack under the axle assembly enough to allow removal of the rear coil springs.
18   Remove the fuel tank supporting strap bolts **(see illustration)**.
19   Remove the fuel tank by rotating the front of the tank down and sliding the tank to the right. It may be necessary to further lower the jack under the rear axle to remove the tank, but care should be taken to ensure that the brake lines and cables are not stretched or damaged.
20   Installation is the reverse of the removal procedure.
21   When the installation is complete, carefully check all lines, hoses and fittings for leaks.

6   Fuel tank - repair

1   Any repairs to the fuel tank or filler neck should be carried out by a professional who has experience in this critical and potentially dangerous work. Even after cleaning and flushing of the fuel system, explosive fumes can remain and ignite during repair of the tank.

2   If the fuel tank is removed from the vehicle, it should not be placed in an area where sparks or open flames could ignite the fumes coming out of the tank. Be especially careful inside garages where a natural gas-type appliance is located, because the pilot light could cause an explosion.

7   Carburetor - removal and installation

Refer to illustration 7.4
Warning: *Gasoline is extremely flammable, so extra precautions must be taken when working on any part of the fuel system. Do not smoke or allow open flames or bare light bulbs near the work area. Also, do not work in a garage if a natural gas-type appliance with a pilot light is present.*
1   Remove the cable from the negative battery terminal. **Caution:** *On models equipped with a Delco Loc II audio system, disable the anti-theft feature before disconnecting the battery.*

7.4 Typical accelerator cable and cruise control cable installation details

1   Cruise control throttle cable
2   Carburetor throttle lever
3   Accelerator cable
4   Cable bracket
5   Retainer clip
6   Pin

2   Remove the air cleaner.
3   Disconnect the fuel and vacuum lines from the carburetor, noting their locations.
4   Disconnect the accelerator cable and cruise control cable (if equipped) **(see illustration)**.
5   Disconnect the throttle valve cable (automatic transmission).
6   Remove all hoses and wires, making very careful note of how they are attached. Tags or coded pieces of tape will help.
7   Remove the carburetor mounting nuts and/or bolts and separate the carburetor from the manifold.
8   Remove the gasket and/or EFE insulator.
9   Installation is the reverse of the removal procedure, but the following points should be noted:

a) *By filling the carburetor bowl with fuel, the initial startup will be easier and less drain on the battery will occur.*
b) *New gaskets should be used.*
c) *Idle speed and mixture settings should be checked and adjusted, if necessary.*

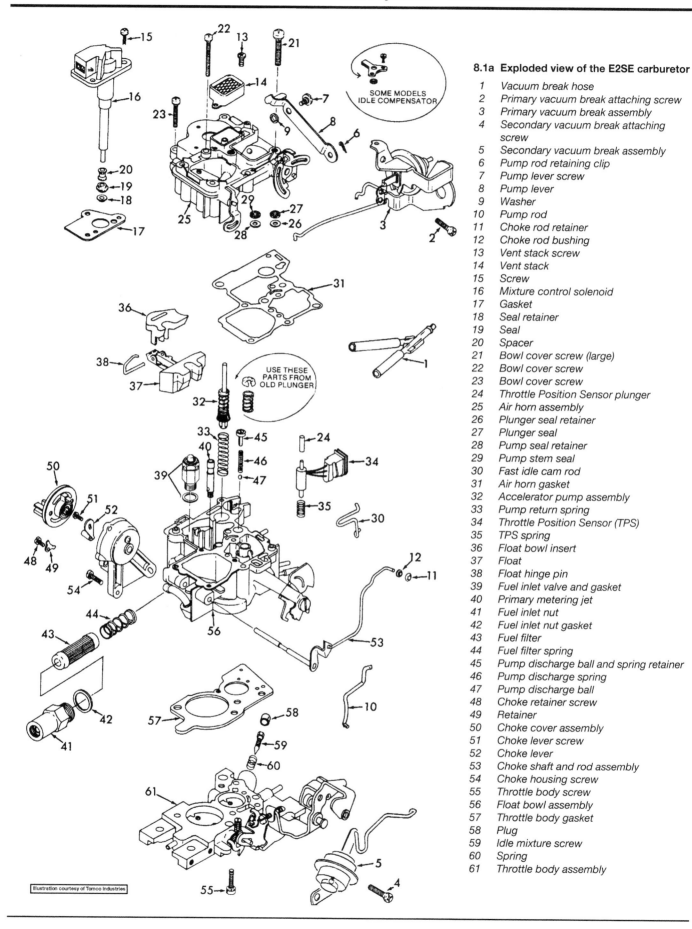

8.1a Exploded view of the E2SE carburetor

1   Vacuum break hose
2   Primary vacuum break attaching screw
3   Primary vacuum break assembly
4   Secondary vacuum break attaching screw
5   Secondary vacuum break assembly
6   Pump rod retaining clip
7   Pump lever screw
8   Pump lever
9   Washer
10   Pump rod
11   Choke rod retainer
12   Choke rod bushing
13   Vent stack screw
14   Vent stack
15   Screw
16   Mixture control solenoid
17   Gasket
18   Seal retainer
19   Seal
20   Spacer
21   Bowl cover screw (large)
22   Bowl cover screw
23   Bowl cover screw
24   Throttle Position Sensor plunger
25   Air horn assembly
26   Plunger seal retainer
27   Plunger seal
28   Pump seal retainer
29   Pump stem seal
30   Fast idle cam rod
31   Air horn gasket
32   Accelerator pump assembly
33   Pump return spring
34   Throttle Position Sensor (TPS)
35   TPS spring
36   Float bowl insert
37   Float
38   Float hinge pin
39   Fuel inlet valve and gasket
40   Primary metering jet
41   Fuel inlet nut
42   Fuel inlet nut gasket
43   Fuel filter
44   Fuel filter spring
45   Pump discharge ball and spring retainer
46   Pump discharge spring
47   Pump discharge ball
48   Choke retainer screw
49   Retainer
50   Choke cover assembly
51   Choke lever screw
52   Choke lever
53   Choke shaft and rod assembly
54   Choke housing screw
55   Throttle body screw
56   Float bowl assembly
57   Throttle body gasket
58   Plug
59   Idle mixture screw
60   Spring
61   Throttle body assembly

SOME MODELS
IDLE COMPENSATOR

USE THESE
PARTS FROM
OLD PLUNGER

Illustration courtesy of Tomco Industries

8.1b Location of carburetor
identification number

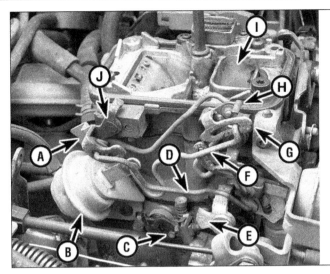

8.1c **E2SE carburetor
choke system**

A   Intermediate
    choke lever
B   Secondary
    vacuum break unit
C   Fast idle screw
D   Air valve rod
E   Fast idle cam
F   Intermediate
    choke link
G   Vacuum break
    lever
H   Choke lever
I   Choke valve
J   Air valve lever

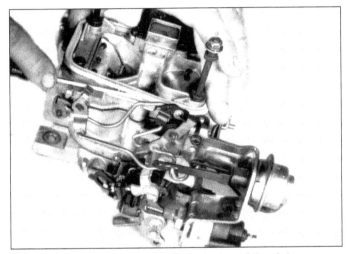

8.3/1 Remove the gasket from the top of the air horn

8.3/2 Remove the fuel inlet nut, fuel filter and spring

8   Carburetor (E2SE) - overhaul

Note: *Carburetor overhaul is an involved procedure that requires some experience. The
home mechanic without much experience
should have the overhaul done by a dealer
service department or repair shop. Because
of running production changes, some details
of the unit overhauled here may not exactly
match those of your carburetor, although the
home mechanic with previous experience
should be able to detect the differences and
modify the procedure.*

Disassembly

*Refer to illustrations 8.1a, 8.1b, 8.1c, 8.3/1
through 8.3/29 and 8.6*

1   Before disassembling the carburetor,
purchase a carburetor rebuild kit for your particular model (the model number will be
found on a metal tag at the side of the carburetor) **(see illustrations)**. This kit will have all
the necessary replacement parts for the overhaul procedure as well as all the necessary
adjustment specifications listed in the
instruction sheet.

8.3/3 Remove the pump lever
attaching screw

2   It will be necessary to have a relatively
large, clean workbench to lay out all of the
parts as they are removed. Many of the parts
are very small and can be lost easily if the
work space is cluttered.

3   Carburetor disassembly is illustrated in

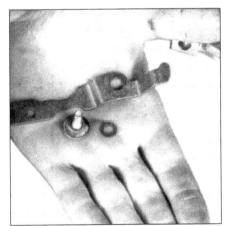

8.3/4 Disconnect the pump rod from the
pump lever and remove the pump lever

the following step-by-step photo sequence
to make the operation as easy as possible
(see illustrations). Work slowly through the
procedure and if at any point you feel the
reassembly of a certain component may
prove confusing, stop and make a rough

8.3/5 Disconnect the primary vacuum break diaphragm hose from the throttle body

8.3/6 Remove the screws that retain the idle speed solenoid/vacuum break diaphragm bracket

8.3/7 Lift off the idle speed solenoid/vacuum break diaphragm assembly and disconnect the air valve link from the vacuum break plunger (repeat this step for the secondary vacuum break assembly, disconnecting the link from the slot in the choke lever)

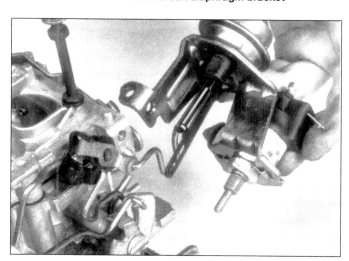

8.3/8 Disconnect the vacuum break and air valve links from the levers (Note: It is not necessary to disconnect the links from the vacuum break plungers unless either the rods or vacuum break units are being replaced)

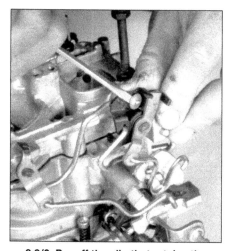

8.3/9 Pry off the clip that retains the intermediate choke link to the choke lever and separate the link from the lever

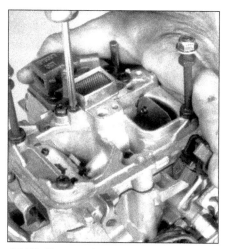

8.3/10 Remove the screws that retain the vent/screen assembly to the air horn and lift off the assembly

8.3/11 Remove the screws that retain the mixture control solenoid and, using a slight twisting motion, lift the solenoid out of the air horn

8.3/12 Remove the screws securing the air horn to the float bowl, noting their various lengths and positions to simplify installation

8.3/13 Rotate the fast idle cam up, lift off the air horn and disconnect the fast idle cam link from the fast idle cam

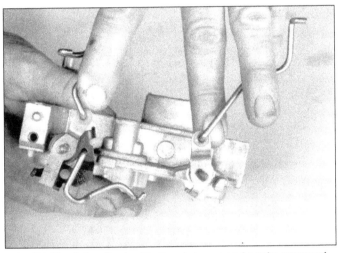

8.3/14 The links attached to the air horn need not be removed unless their replacement or removal is required to service other components

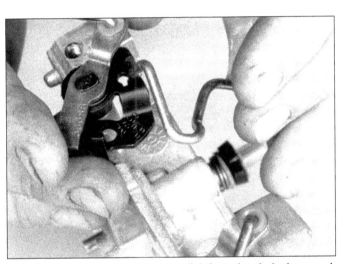

8.3/15 Disengage the fast idle cam link from the choke lever and save the bushing for reassembly

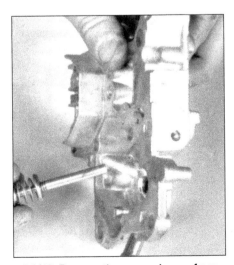

8.3/16 Remove the pump plunger from the air horn or the pump well in the float bowl

8.3/17 Compress the pump plunger spring and separate the spring retainer clip and spring from the piston

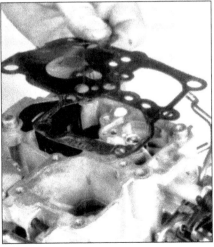

8.3/18 Remove the air horn gasket from the float bowl

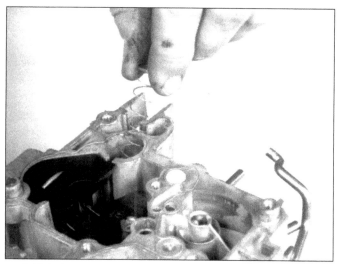

8.3/19 Remove the pump return spring from the pump well

8.3/20 Remove the plastic filler block that covers the float valve

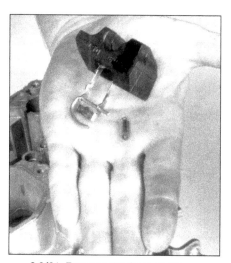

8.3/21 Remove the float and lever assembly, float valve and stabilizing spring (if used) by pulling up on the hinge pin

8.3/22 Remove the float valve seat and gasket (left) and the extended metering jet (right) from the float bowl

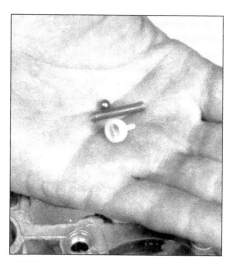

8.3/23 Using the needlenose pliers, pull out the white plastic retainer and remove the pump discharge spring and check ball (do not pry on the retainer to remove it)

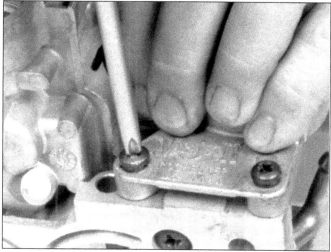

8.3/24 Remove the screws that retain the choke housing to the throttle body

8.3/25 Remove the screws that retain the float bowl to the throttle body

8.3/26 Separate the float bowl from the throttle body

8.3/27 Carefully file the heads off the pop rivets that retain the choke cover to the choke housing, remove the cover and tap out the remainder of the rivets

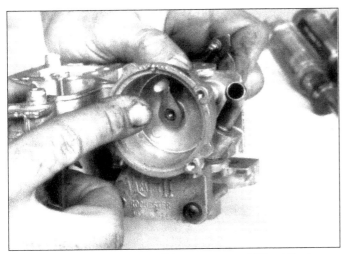

8.3/28 Remove the choke coil lever screw and lift out the lever

8.3/29 Remove the intermediate shaft and lever assembly by sliding it out the lever side of the float bowl (refer to Step 4 in the text)

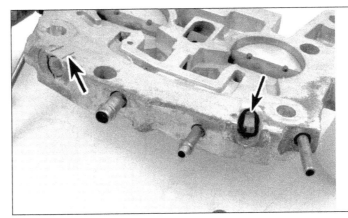

8.6 The idle mixture screws on the E2SE carburetor are concealed behind two plugs - carefully cut where indicated (arrow), then knock the plug out

sketch or apply identification marks. The time to think about reassembling the carburetor is when it is being taken apart. The disassembly photo sequence begins with photo 8.3/1.

4   The final step in disassembly involves the idle mixture needle. It is recessed in the throttle body and sealed with a hardened steel plug. The plug should not be removed unless the needle requires replacement or normal cleaning procedures fail to clean the idle mixture passages. If the idle mixture needle must be removed proceed as follows.

5   Secure the throttle body in a vise so it is inverted with the manifold side up. Use blocks of wood to cushion the throttle body.

6   Locate the idle mixture needle and plug. It should be marked by an indented locator point on the underside of the throttle body. Using a hacksaw, make two parallel cuts in the throttle body on either side of the locator mark **(see illustration)**. The cuts should be deep enough to touch the steel plug, but should not extend more than 1/8inch beyond

the locator point.

7   Position a flat punch at a point near the ends of the saw marks. Holding it at a 45-degree angle, drive it into the throttle body until the casting breaks away, exposing the steel plug.

8   Use a center punch to make an indenta-

tion in the steel plug. Holding it at a 45-degree angle, drive the plug from the throttle body casting. If the plug breaks apart, be sure to remove all of the pieces.

9   Use a 3/16inch deep socket to remove the idle mixture needle and spring from the throttle body.

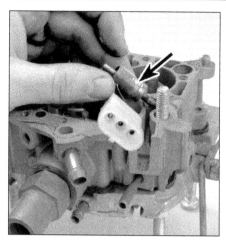

8.11 TPS removal on the E2SE carburetors - push up from the bottom of the electrical connector and remove the TPS and connector assembly from the float bowl; also, remove the spring from the bottom of the float bowl

Cleaning and inspection

Refer to illustrations 8.11 and 8.22

10   Clean the air horn, float bowl, throttle body and related components with clean solvent and blow them out with compressed air. A can of compressed air can be used if an air compressor is not available. Do not use a piece of wire for cleaning the jets and passages.

11   The idle speed solenoid, mixture control solenoid, Throttle Position Sensor, electric choke, pump plunger, diaphragm, plastic filler block and other electrical, rubber and plastic parts should not be immersed in carburetor cleaner because they will harden, swell or distort **(see illustration)**.

12   Make sure all fuel passages, jets and other metering components are free of burrs and dirt.

13   Inspect the upper and lower surfaces of the air horn, float bowl and throttle body for damage. Be sure all material has been removed.

14   Inspect all lever holes and plastic bushings for excessive wear and an out-of-round condition and replace them if necessary.

15   Inspect the float valve and seat for dirt, deep wear grooves and scoring and replace it if necessary.

16   Inspect the float valve pull clip for proper installation and adjust it if necessary.

17   Inspect the float, float arms and hinge pin for distortion and binding and correct or replace as necessary.

18   Inspect the rubber cup on the pump plunger for excessive wear and cracks.

19   Check the choke valve and linkage for excessive wear, binding and distortion and correct or replace as necessary.

20   Inspect the choke vacuum diaphragm for leaks and replace if necessary.

21   Check the choke valve for freedom of movement.

22   Check the mixture control solenoid in the following manner.

   a)   *Connect one end of a jumper wire to either end of the solenoid connector and the other end to the positive terminal of the battery.*

   b)   *Connect another jumper wire between the other terminal of the solenoid connector and the negative terminal of the battery.*

   c)   *Remove the rubber seal and retainer from the end of the solenoid stem and attach a hand vacuum pump to it* **(see illustration)**.

   d)   *With the solenoid fully energized (lean position), apply at least 25 in-Hg of vacuum and time the leakdown rate from 20 to 15 in-Hg. The leakdown rate should not exceed 5 in-Hg in five (5) seconds. If leakage exceeds that amount, replace the solenoid.*

   e)   *To check if the solenoid is sticking in the down position, again apply about 25 in-Hg of vacuum to it, then disconnect the jumper lead to the battery and watch the pump gauge reading. It should fall to zero in less than one (1) second.*

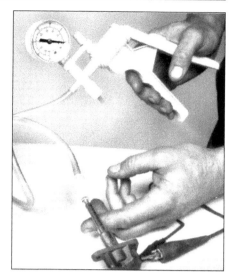

8.22 Attach a hand vacuum pump to the end of the mixture control solenoid

Reassembly

Refer to illustrations 8.25 through 8.64j

23   Before reassembling the carburetor, compare all old and new gaskets back-to-back to make sure they match perfectly. Check especially that all the necessary holes are present and in the proper positions in the new gaskets.

24   If the idle mixture needle and spring have been removed, reinstall them by lightly seating the needle, then back it off three (3) turns. This will provide a preliminary idle mixture adjustment. Final idle mixture adjustment must be made on the vehicle. Proper adjustment must be done using special emission sensing equipment, making it impractical for the home mechanic. To have the mixture settings checked or readjusted, take your vehicle to a dealer or other qualified mechanic with the proper equipment.

25   Install a new gasket on the bottom of the float bowl **(see illustration)**.

26   Mount the throttle body on the float bowl

8.25 Install the gasket on the bottom of the float bowl

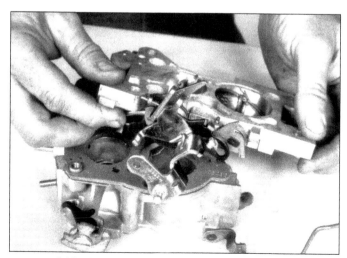

8.26a Mount the throttle body on the float bowl

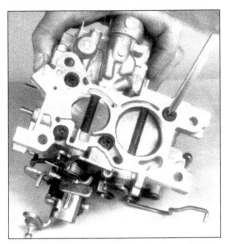

8.26b Install the throttle body-to-float bowl attaching screws

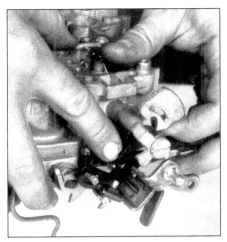

8.27 Check the engagement of the lockout tang in the secondary lockout lever

8.28a Attach the choke housing to the throttle body

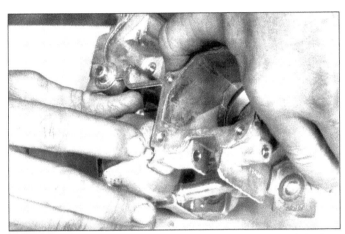

8.28b The lug at the rear of the choke housing should sit in the bowl recess

8.30 Install the thermostatic coil lever so it is in the 12 o'clock position when the intermediate choke lever is facing up

so it is properly installed over the locating dowels on the bowl (see illustration), reinstall the screws and tighten them evenly and securely (see illustration). Be sure that the steps on the fast idle cam face toward the fast idle screw on the throttle lever when installed.

27   Inspect the linkage to make sure that the lockout tang properly engages in the slot of the secondary lockout lever and that the linkage moves freely without binding (see illustration).

28   Attach the choke housing to the throttle body, making sure the locating lug on the rear of the housing sits in the recess in the float bowl (see illustrations).

29   Install the intermediate choke shaft and lever assembly in the float bowl by pushing it through from the throttle lever side.

30   Position the intermediate choke lever in the up position and install the thermostatic coil lever on the end sticking into the choke housing. The coil lever is properly aligned when the coil pickup tang is in the 12 o'clock position (see illustration). Install the screw in the end of the intermediate shaft to secure the coil lever.

31   Three self-tapping screws supplied in

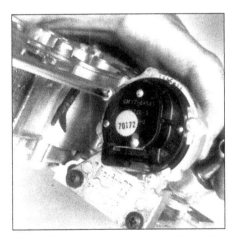

8.31 The choke cover is reinstalled with the self-tapping screws supplied in the overhaul kit

the overhaul kit are used in place of the original pop rivets to secure the choke cover and coil assembly to the choke housing. Thread the screws into the housing, making sure they start easily and are properly aligned (see

8.32 Be sure the notch in the choke cover is aligned with the raised casting projection on the housing cover flange

illustration), then remove them.

32   Place the fast idle screw on the highest step of the fast idle cam, then install the choke cover on the housing, aligning the notch in the cover with the raised casting

8.35 Install the main metering jet

8.36 Install the float valve seat assembly

projection on the housing cover flange **(see illustration)**. When installing the cover, be sure the coil pickup tang engages the inside choke lever. **Note:** *The thermostatic coil tang is formed so that it will completely encircle the coil pickup lever. Make sure the lever is inside of the tang when installing the cover.*

33   With the choke cover in place, install the self-tapping screws and tighten them securely.

34   Install the pump discharge check ball and spring in the passage next to the float chamber, then place a new plastic retainer in the hole so that its end engages the spring and tap it lightly into place until the retainer top is flush with the bowl surface.

35   Install the main metering jet in the bottom of the float chamber **(see illustration)**.

36   Install the float valve seat assembly and gasket **(see illustration)**.

37   To make float level adjustments easier, bend the float arm up slightly at the notch before installing the float **(see illustration)**.

38   Install the float valve onto the float arm by sliding the lever under the pull clip. Install the float pin in the float lever **(see illustration)**. Install the float assembly by aligning the valve and seat and the float retaining pin

8.37 Prior to installation, bend the float arm up slightly at the point shown

and locating channels in the float bowl.

39   To adjust the float level, hold the float pin firmly in place, push down on the float arm at the outer end, against the top of the float valve, and see if the top of the float is the specified distance from the float bowl surface **(see illustration)**. Bend the float arm as necessary to achieve the proper measure-

8.38 Install the float retaining pin in the float lever

ment by pushing down on the pontoon. Refer to the carburetor overhaul kit instruction sheet for the proper float measurement for your vehicle. Check the float level visually following adjustment.

40   Install the plastic filler block over the float valve so that it is flush with the float bowl surface **(see illustration)**.

8.39 Measure the float level

8.40 Install the plastic float block

8.42 Install a new air horn gasket on the float bowl

8.43 Install the pump return spring in the pump well

8.44 Install the pump plunger in the pump well

8.49 Engage the fast idle cam link in the fast idle cam prior to installation of the air horn

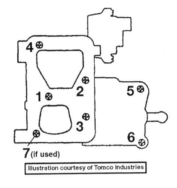

8.50 Recommended air horn screw tightening sequence

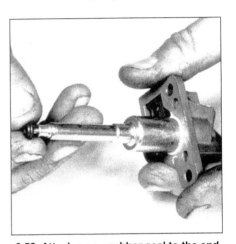

8.52 Attach a new rubber seal to the end of the mixture control solenoid

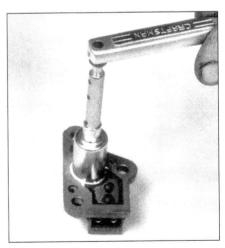

8.53 Use a hammer and hollow tool to tap the seal retainer onto the mixture control solenoid stem

41 If the carburetor is equipped with a Throttle Position Sensor, install the TPS return spring in the bottom of the well in the float bowl. Then install the TPS and connector assembly by aligning the groove in the electrical connector with the slot in the float bowl. When properly installed, the assembly should sit below the float bowl surface.

42 Install a new air horn gasket on the float bowl **(see illustration)**.

43 Install the pump return spring in the pump well **(see illustration)**.

44 Reassemble the pump plunger assembly, lubricate the plunger cap with a thin coat of engine oil and install the pump plunger in the pump well **(see illustration)**.

45 If used, remove the old pump plunger seal and retainer and the old TPS plunger seal and retainer from the air horn. Install new seals and retainers in both locations and lightly stake both seal retainers in three (3) places other than the original staking locations.

46 Install the fast idle cam rod in the lower hole of the choke lever.

47 If so equipped, apply a light coat of sili-

cone grease or engine oil to the TPS plunger and push it through the seal in the air horn so that about one-half of the plunger extends above the seal.

48 Before installing the air horn, apply a light coat of silicone grease or engine oil to the pump plunger stem to aid in slipping it through the seal in the air horn.

49 Rotate the fast idle cam to the Up position so it can be engaged with the lower end of the fast idle cam rod **(see illustration)**. While holding down on the pump plunger assembly, carefully lower the air horn onto the float bowl and guide the pump plunger stem through the seal.

50 Install the air horn retaining screws and washers, making sure the different length screws are inserted into their respective holes, then tighten them in the sequence illustrated **(see illustration)**.

51 If so equipped, install a new seal in the recess of the float bowl and attach the hot idle compensator valve.

52 Install a new rubber seal on the end of the mixture control solenoid stem until it is up against the boss on the stem **(see illustration)**.

53 Using a 3/16inch socket and a hammer **(see illustration)**, drive the retainer over the

mixture control solenoid stem just far enough to retain the rubber seal, while leaving a slight clearance between them for seal expansion.

54 Apply a light coat of engine oil to the rubber seal and, using a new gasket, install the mixture control solenoid in the air horn.

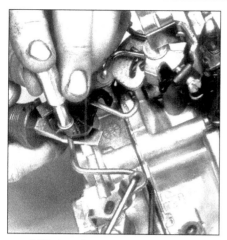

8.56 Attach a retaining clip to the intermediate choke rod to secure it to the choke lever

8.57 Install the idle speed solenoid/vacuum break diaphragm assembly

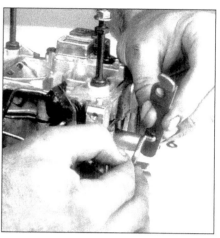

8.58a Engage the pump rod with the pump rod lever

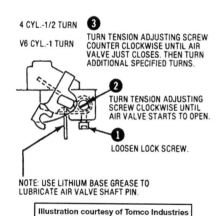

8.58b Insert the pump rod mounting screw through the pump rod prior to installation

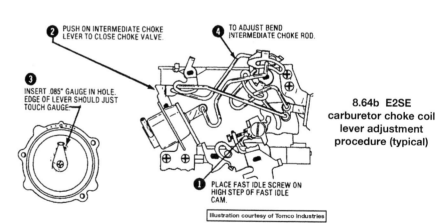

8.62 Install a new gasket on top of the air horn

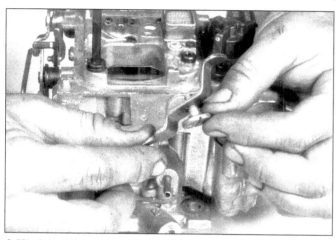

4 CYL.-1/2 TURN **③**

V6 CYL.-1 TURN

TURN TENSION ADJUSTING SCREW COUNTER CLOCKWISE UNTIL AIR VALVE JUST CLOSES. THEN TURN ADDITIONAL SPECIFIED TURNS.

②

TURN TENSION ADJUSTING SCREW CLOCKWISE UNTIL AIR VALVE STARTS TO OPEN.

①

LOOSEN LOCK SCREW.

NOTE: USE LITHIUM BASE GREASE TO LUBRICATE AIR VALVE SHAFT PIN.

Illustration courtesy of Tomco Industries

8.64a E2SE carburetor air valve spring adjustment procedure

② PUSH ON INTERMEDIATE CHOKE LEVER TO CLOSE CHOKE VALVE.

④ TO ADJUST BEND INTERMEDIATE CHOKE ROD.

③ INSERT .085" GAUGE IN HOLE. EDGE OF LEVER SHOULD JUST TOUCH GAUGE.

① PLACE FAST IDLE SCREW ON HIGH STEP OF FAST IDLE CAM.

Illustration courtesy of Tomco Industries

8.64b E2SE carburetor choke coil lever adjustment procedure (typical)

Use a slight twisting motion while installing the solenoid to help the rubber seal slip into the recess.

55   Install the vent/screen assembly on the air horn.

56   Install a plastic bushing in the hole in the choke lever, with the small end facing out, then, with the intermediate choke lever at the 12 o'clock position, install the intermediate choke rod in the bushing. Install a new retaining clip on the end of the rod. Use a broad flat-blade screwdriver and a 3/16inch socket as shown **(see illustration)**. Make sure the clip is not seated tightly against the bushing

and that the linkage moves freely.

57   Reattach the primary and secondary vacuum break links and install the vacuum break and idle speed solenoid assemblies **(see illustration)**.

58   Engage the pump rod with the pump rod lever **(see illustration)**, install a new retaining clip on the pump rod and install the pump lever on the air horn with the washer between

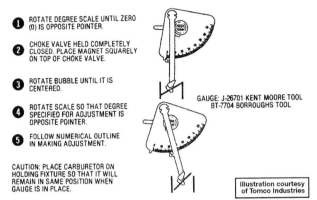

❶ ROTATE DEGREE SCALE UNTIL ZERO (0) IS OPPOSITE POINTER.

❷ CHOKE VALVE HELD COMPLETELY CLOSED. PLACE MAGNET SQUARELY ON TOP OF CHOKE VALVE.

❸ ROTATE BUBBLE UNTIL IT IS CENTERED.

❹ ROTATE SCALE SO THAT DEGREE SPECIFIED FOR ADJUSTMENT IS OPPOSITE POINTER.

❺ FOLLOW NUMERICAL OUTLINE IN MAKING ADJUSTMENT.

GAUGE: J-26701 KENT MOORE TOOL
BT-7704 BORROUGHS TOOL

CAUTION: PLACE CARBURETOR ON HOLDING FIXTURE SO THAT IT WILL REMAIN IN SAME POSITION WHEN GAUGE IS IN PLACE.

Illustration courtesy of Tomco Industries

ANGLE GAUGE: BASIC ADJUSTMENT
CONTINUE NUMERICAL OUTLINE IN EACH ADJUSTMENT USING DEGREE SETTING.

8.64c Measuring the E2SE carburetor choke valve angle with an angle gauge

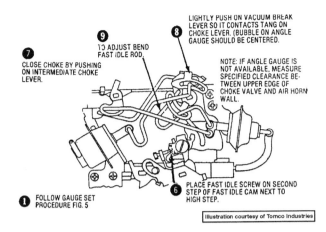

❾ TO ADJUST BEND FAST IDLE ROD.

❽ LIGHTLY PUSH ON VACUUM BREAK LEVER SO IT CONTACTS TANG ON CHOKE LEVER. (BUBBLE ON ANGLE GAUGE SHOULD BE CENTERED.)

❼ CLOSE CHOKE BY PUSHING ON INTERMEDIATE CHOKE LEVER.

NOTE: IF ANGLE GAUGE IS NOT AVAILABLE, MEASURE SPECIFIED CLEARANCE BETWEEN UPPER EDGE OF CHOKE VALVE AND AIR HORN WALL.

❶ FOLLOW GAUGE SET PROCEDURE FIG. 5.

❻ PLACE FAST IDLE SCREW ON SECOND STEP OF FAST IDLE CAM NEXT TO HIGH STEP.

Illustration courtesy of Tomco Industries

8.64d E2SE carburetor choke rod fast idle cam adjustment procedure

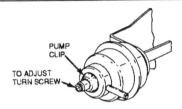

PUMP CLIP

TO ADJUST TURN SCREW

1. REMOVE VACUUM BREAK ASSY. AND CAREFULLY GRIND OFF ADJUSTMENT SCREW CAP.
2. FOLLOW PRI. OR SEC. VAC. BREAK ADJUSTMENT PROCEDURE.
3. ON PRIMARY SIDE MAKE SURE VAC. DIAPHRAGM IS FULLY SEATED. BEND AIR VALVE ROD TO OBTAIN A SLIGHT CLEARANCE BETWEEN ROD AND END OF SLOT. (FOLLOW PRI. VAC. BREAK ADJ. WITH AIR VALVE ROD. ADJ.)
4. ON DELAY MODELS, PLUG AIR BLEED HOLE WITH A VERAJET TYPE PUMP PLUNGER CUP.
5. TO ADJUST USE A 1/8" HEX WRENCH TO TURN SCREW IN REAR COVER UNTIL BUBBL EIS CENTERED. (REMOVE CUP AFTER ADJUSTMENT.)
6. SEAL ADJ. SCREW WITH A SEALER. (SUCH AS A SILICONE SEALANT RTV RUBBER OR EQUIVALENT.)

Illustration courtesy of Tomco Industries

8.64e E2SE carburetor vacuum break adjustment details

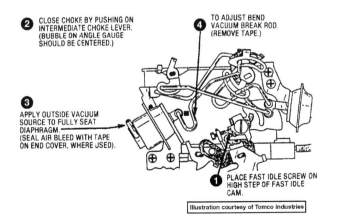

❽ HOLD CHOKE VALVE TOWARDS CLOSED POSITION BY PUSHING ON INTERMEDIATE CHOKE LEVER. (BE SURE PLUNGER BUCKING SPRING IS COMPRESSED AND SEATED.) BUBBLE ON ANGLE GAUGE SHOULD BE CENTERED.

❾ TO ADJUST BEND VACUUM BREAK ROD. (REMOVE TAPE.)

NOTE: IF ANGLE GAUGE IS NOT AVAILABLE, MEASURE SPECIFIED CLEARANCE BETWEEN UPPER EDGE OF CHOKE VALVE AND AIR HORN WALL.

❶ FOLLOW GAUGE SET PROCEDURE. FIG. 6.

❻ PLACE FAST IDLE SCREW ON HIGH STEP OF FAST IDLE CAM.

❼ APPLY OUTSIDE VACUUM SOURCE TO FULLY SEAT DIAPHRAGM. NOTE: PLUG END COVER BLEED HOLE WITH TAPE.

Illustration courtesy of Tomco Industries

8.64f E2SE carburetor primary vacuum break adjustment procedure for models with dual vacuum break units

the lever and the air horn **(see illustration)**.

59   Reconnect the vacuum break hoses.

60   Install the fuel filter with the hole facing toward the inlet nut.

61   Place a new gasket on the inlet nut and install and tighten it securely. Take care not to over tighten the nut, as it could damage the gasket, leading to a fuel leak.

62   Install a new gasket on the top of the air horn **(see illustration)**.

63   Check that all linkage hookups have been made and that they do not bind.

64   For external linkage adjustment procedures, refer to the accompanying illustrations **(see illustrations)**.

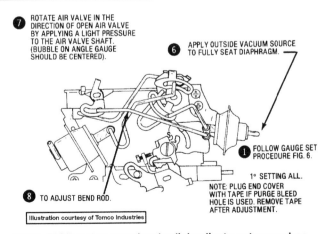

❼ ROTATE AIR VALVE IN THE DIRECTION OF OPEN AIR VALVE BY APPLYING A LIGHT PRESSURE TO THE AIR VALVE SHAFT. (BUBBLE ON ANGLE GAUGE SHOULD BE CENTERED).

❻ APPLY OUTSIDE VACUUM SOURCE TO FULLY SEAT DIAPHRAGM.

❽ TO ADJUST BEND ROD.

❶ FOLLOW GAUGE SET PROCEDURE FIG. 6.

1° SETTING ALL.
NOTE: PLUG END COVER WITH TAPE IF PURGE BLEED HOLE IS USED. REMOVE TAPE AFTER ADJUSTMENT.

Illustration courtesy of Tomco Industries

8.64g E2SE carburetor air valve link adjustment procedure

❷ CLOSE CHOKE BY PUSHING ON INTERMEDIATE CHOKE LEVER. (BUBBLE ON ANGLE GAUGE SHOULD BE CENTERED.)

❹ TO ADJUST BEND VACUUM BREAK ROD. (REMOVE TAPE.)

❸ APPLY OUTSIDE VACUUM SOURCE TO FULLY SEAT DIAPHRAGM. (SEAL AIR BLEED WITH TAPE ON END COVER, WHERE USED).

❶ PLACE FAST IDLE SCREW ON HIGH STEP OF FAST IDLE CAM.

Illustration courtesy of Tomco Industries

8.64h E2SE carburetor secondary vacuum break adjustment procedure

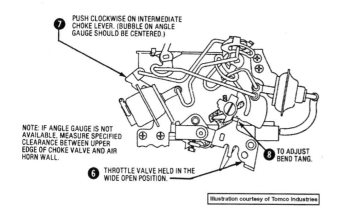

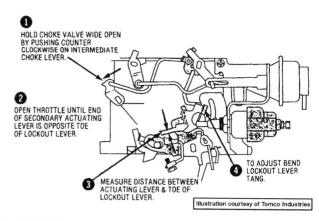

8.64i E2SE carburetor unloader adjustment procedure

8.64j E2SE carburetor secondary lockout adjustment procedure

9   Carburetor (E4ME) - overhaul

Because of the complexity of these carburetors and the special tools required to overhaul and adjust them, it is recommended that major service work be left to a dealer service department or a carburetor repair specialist.

10   Electronic feedback carburetor system

General information

1   The electronic feedback carburetor system relies on an electronic signal, which is generated by an exhaust gas oxygen sensor, to control a variety of devices and keep emissions within limits. The system works in conjunction with a three-way catalyst to control the levels of carbon monoxide, hydrocarbons and oxides of nitrogen. The feedback carburetor system also works in conjunction with the computer. The two systems share certain sensors and output actuators; therefore, diagnosing this system will require a thorough check of all the feedback carburetor components.

2   The system operates in two modes: open loop and closed loop. When the engine is cold, the air/fuel mixture is controlled by the computer in accordance with a program designed in at the time of production. The air/fuel mixture during this time will be richer to allow for proper engine warm-up. When the engine is at operating temperature, the system operates in closed loop and the air/fuel mixture is varied depending on the information supplied by the exhaust gas oxygen sensor.

3   Here is a list of the various sensors and output actuators involved with these feedback carburetor systems:

Coolant temperature sensor
Exhaust Gas Oxygen sensor
High energy Electronic Ignition system
Early Fuel Evaporation system
Mixture Control Solenoid
Electronic Control Module
Throttle Position Sensor
Vehicle Speed Sensor
Barometric Pressure Sensor or Manifold
Absolute Pressure sensor
Idle Speed Control switch
Park/Neutral switch

Mixture control (M/C) solenoid

Refer to illustrations 10.12, 10.16a and 10.16b

4   The mixture control (M/C) solenoid is a device that controls fuel flow from the bowl to the main well and at the same time controls the idle circuit air bleed.

5   The mixture control (M/C) solenoid is located in the float bowl where the power piston used to be. It is equipped with a spring loaded plunger that moves up and down like a power piston but more rapidly. Certain areas on the plunger head contact the metering rods and an idle air bleed valve. Plunger movement controls both the metering rods and the idle air bleed valve simultaneously.

6   When the mixture control solenoid is energized it moves down causing the metering rods to move into the jets and restrict the flow of fuel into the main well. The idle air bleed plunger opens the air bleed and allows air into the idle circuit. Both these movements reduce fuel flow and thereby LEANS out the system.

7   When the plunger is de-energized, the M/C solenoid moves up, causing the metering rods to move out of the jets and allow more fuel to the main well, less idle air and increased fuel flow. Here the solenoid is in the RICH position.

8   The mixture control solenoid varies the air/fuel ratio based on the electrical input from the ECM. When the solenoid is ON, the fuel is restricted and the air is admitted. This gives a lean air/fuel ratio (approximately 18:1). When the solenoid is OFF, fuel is admitted and the air/fuel ratio is approximately 13:1. During closed loop operation, the ECM controls the M/C solenoid to approximately 14.7:1 by controlling the ON

and OFF time of the solenoid.

9   As the solenoid "on time" changes, the up time and down time of the metering rods also changes. When a lean mixture is desired, the M/C solenoid will restrict fuel flow through the metering jet 90-percent of the time, or, in other words, a lean mixture will be provided to the engine.

10   This lean command will read as 54-degrees on the dwell meter (54-degrees is 90-percent of 60-degrees), and means the M/C solenoid has restricted fuel flow 90-percent of the time. A rich mixture is provided when the M/C solenoid restricts it only 10-percent of the time and allows a rich mixture to flow to the engine. A rich command will have a dwellmeter reading of 6-degrees (10-percent of 60-degrees); the M/C solenoid has restricted fuel flow 10-percent of the time. On some engines dwellmeter readings can vary between 5-degrees and 55-degrees, rather than between 6-degrees and 54-degrees. The ideal mixture would be shown on the dwellmeter with the needle varying or swinging back and forth, anywhere between 10-degrees and 50-degrees. "Varying" means the needle continually moves up and down the scale. The amount it moves does not matter, only the fact that it does move. The dwell is being varied by the signal sent to the ECM by the oxygen sensor in the exhaust manifold.

11   The following checks assume the engine has been tuned and the ignition system is in order.

12   The dwellmeter is used to diagnose the M/C solenoid system. This is done by connecting a dwellmeter to the dwell test lead (a green single wire connector in the M/C solenoid wiring harness) **(see illustration)**.

13   In the older style contact-points ignition systems, the dwellmeter reads the period of time that the points were closed (or "dwelled" together). That period of time was when voltage flowed to the ignition coil. In the feedback carburetor system, the dwellmeter is used to read the time that the ECM closes the M/C solenoid circuit to ground, allowing voltage to operate the M/C solenoid. Dwell, as used in feedback system performance

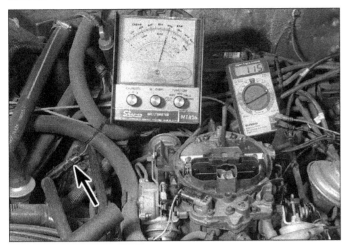

10.12 Connect the dwellmeter to the terminal in the green connector (arrow) near the carburetor

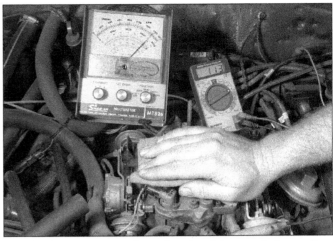

10.16a Place a rag over the carburetor air horn and observe the dwell indicate LEAN to compensate for a rich condition

diagnosis, is the time that the M/C solenoid circuit is closed (energized). The dwellmeter will translate this time into degrees. The "6-cylinder" (0-degrees to 60-degrees) scale on the dwellmeter is used for this reading. The ability of the dwellmeter to make this kind of conversion makes it an ideal tool to check the amount of time the ECM internal switch is closed, thus energizing the M/C solenoid. The only difference is that the degree scale on the meter is more like percent of solenoid "on time" rather than "degrees of dwell".

14 First set the dwellmeter on the 6-cylinder position, then connect it to the M/C solenoid dwell test lead to measure the output of the ECM. Do not allow the terminal to touch ground, including hoses. You must use the 6-cylinder position when diagnosing all engines, whether the engine you're working on is a 4, 6, or 8 cylinder engine. **Note:** *Some older dwellmeters may not work properly on the feedback carburetor systems. Don't use any dwellmeter which causes a change in engine operation when it is connected to the solenoid lead.*

The 6-cylinder scale on the dwellmeter provides evenly divided points, for example:

> 15-degrees = 1/4 scale
> 30-degrees = midscale
> 45-degrees = 3/4 scale

Connect the positive clip lead of the dwellmeter to the M/C solenoid dwell test connector. Attach the other dwellmeter clip lead to ground. Do not allow the clip leads to contact other conductive cables or hoses which could interfere with accurate readings.

15 After connecting the dwellmeter to a warm, operating engine the dwell at idle and part throttle will read between 5-degrees and 55-degrees and will be varying. That is, the needle will move continuously up and down the scale. What matters is that the needle does move, not how much it moves. Typical needle movement will occur between the 30 and 35-degree range in a normal operating feedback system. Needle movement indicates that the engine is in "closed-loop," and that the dwell is being varied by signals from

10.16b Disconnect a vacuum line and observe the dwell indicate RICH to compensate for a lean condition

the ECM. However, if the engine is cold, has just been restarted, or the throttle is wide open, the dwell will be fixed and the needle will be steady. Those are signs that the engine is in "open-loop."

16 Diagnostic checks to find a condition without a trouble code are usually made on a warm engine (in "closed-loop") easily checked by making sure the upper radiator hose is hot. There are three ways of distinguishing "open" from "closed-loop" operation.

a) *A variation in dwell will occur only in "closed-loop."*

b) *Test for closed loop operation by restricting airflow to the carburetor to choke the engine* **(see illustration).** **Warning:** *Use a thick rag (not your bare hand) and keep your face away from the carburetor. If the dwellmeter moves up scale, that indicates "closed-loop."*

c) *If you create a large vacuum leak and the dwell drops down, that also indicates "closed-loop"* **(see illustration).**

17 Basically, "closed loop" indicates that the ECM is using information from the exhaust oxygen sensor to influence operation of the Mixture Control (M/C) solenoid. The ECM still considers other information such as

engine temperature, RPM, barometric and manifold pressure and throttle position along with the exhaust oxygen sensor information.

18 During "open-loop" all information except the exhaust oxygen sensor input is considered by the ECM to control the M/C solenoid. Accurate readings from the oxygen sensor are not attained until the sensor is completely hot (600-degrees F).

Note: *The exhaust oxygen sensor may cool below its operational temperature during prolonged idling. This will cause an open loop condition, and make the diagnostic information not usable during diagnosis. Engine RPM must be increased to warm the exhaust oxygen sensor, and again re-establish "closed loop". Diagnosis should begin over at the first step after "closed-loop" is resumed.*

System performance test

Refer to illustrations 10.19a and 10.19b

19 Start the engine and then ground the diagnostic terminal (see Chapter 6). This will allow the feedback system to run on the preset default values, thereby not allowing the computer to make any running adjustments. **Caution:** *Do not ground the diagnostic connector before starting the engine. Disconnect the purge hose from the charcoal canister*

10.19a Clamp off the charcoal canister purge hose to prevent entrance of fuel vapors into the carburetor

10.19b With the dwell test lead grounded and a tachometer properly installed, disconnect the M/C solenoid connector and observe the engine speed

and plug it (or simply clamp it shut) (see illustration). Disconnect the bowl vent hose at the carburetor and plug the hose on the canister side. These two steps will not allow any recirculated crankcase vapors to enter the carburetor during testing. Connect a tachometer according to the manufacturers' instructions. Disconnect the M/C solenoid and ground the M/C solenoid dwell test lead (see illustration). Run the engine at 3,000 rpm and while holding the throttle steady, reconnect the M/C solenoid and observe the rpm. Unground the dwell lead before returning the engine to idle.

20   The test results will be either:

   a) *The engine speed will decrease at least 100 rpm*
   b) *The engine speed does not decrease (at least 100 rpm) and may increase rpm*

21   If the engine speed does decrease (at least 100 rpm), check the wiring on the M/C solenoid for any damaged connectors and if they are OK, check the carburetor adjustments. Also check for fuel in the charcoal canister or crankcase or any other condition that would cause richness.

22   If the engine decrease's at least 100 rpm, connect a dwellmeter to the M/C solenoid dwell lead (see illustration 10.12). Be sure to read the information on "closed loop" operation and dwellmeters. Set the carburetor on the high step of the fast idle cam and run the engine for one minute or until the dwell starts to vary (whichever happens first). Return the engine to idle and observe the dwell. In most cases the dwell should vary between 10 and 50-degrees but there are several types of problems that may occur. **Note:** *The following tests must be performed with the diagnostic connector still grounded unless otherwise indicated.*

Fixed dwell under 10-degrees

23   This condition indicates that the feedback system is responding RICH to offset a very lean condition in the engine. One way to separate the problem is to choke the carburetor with the engine at part throttle. This will either increase or decrease the dwell.

24   If the dwell increases, check for a vacuum leak in the hoses, gaskets, AIR system etc. Also, check for an exhaust leak near the oxygen sensor, any hoses that are misrouted and an EGR valve that is not operating or that is leaking.

25   If the dwell does not increase, check the oxygen sensor, the wiring harness from the ECM to the oxygen sensor, TPS, TPS voltage and/or the ECM. If necessary, have the system diagnosed by a dealer service department or other repair shop.

Fixed dwell between 10 and 50-degrees

26   This condition indicates that the feedback system is stuck in one mode (open loop) because of a faulty coolant temperature sensor, oxygen sensor or TPS. Start the diagnosis by running the engine at part throttle for one minute. Then with the engine idling, observe the dwell, remove the connector from the coolant sensor and jump the terminals on the connector. This will ground the signal and indicate to the ECM that the sensors are functioning (grounded). The dwell reading should not be fixed with the sensor grounded. Check each sensor for the correct resistance and voltage signal from its respective electrical connector.

Fixed dwell over 50-degrees

27   This condition indicates that the feedback system is responding LEAN to offset a very rich condition in the engine. Start the diagnosis by running the engine at fast idle for about two minutes and then let it return to idle. This procedure makes sure that the feedback system is in closed-loop (warmed-up). Next, disconnect the large vacuum hose to the PCV valve and cause a major vacuum leak. Do not allow the engine to stall. The dwell should drop by approximately 20-degrees. If it does, check the carburetor adjustments. Also, check the evaporative canister for fuel overload or leaks in the purge control system that would cause an over-rich condition in the carburetor.

28   If the dwell does not drop, check the

oxygen sensor, oxygen sensor signal voltage and their respective wiring circuits for any problems.

Feedback carburetor adjustments

Mixture control solenoid plunger adjustment

Refer to illustrations 10.29, 10.32, 10.36a, 10.36b, 10.41a, 10.41b and 10.43

Measuring plunger travel

Note: *If you're reassembling the carburetor after overhaul, just set the air horn in place when measuring the plunger travel (there's no need to install the gasket or screws, and you may need to remove the air horn again to adjust the travel).*

29   Insert a special carburetor float gauge tool (available from auto parts stores and tool manufacturers) into the vertical "D"-shaped vent hole in the air horn casting next to the idle air bleed valve cover (see illustration). It may be necessary to file material off the side of the gauge to make clearance for the tool to enter without binding.

10.29 To measure plunger travel, first measure the plunger height with no pressure applied to the gauging tool, then . . .

10.32 . . . press lightly on the gauge until it bottoms the plunger and measure again. The difference between the two measurements is the travel

10.36a To adjust the lean-stop, place the gauging tool (being held at right) on the metering jet and under the solenoid plunger, as shown, then depress the plunger and turn the adjusting screw until the plunger bottoms against the lean stop and touches the top of the gauging tool simultaneously

30 First, press down on the gauge and make sure the gauge moves up and down freely without any binding.

31 With the gauge resting against the plunger (no pressure applied), sight across the air horn and record the mark on the gauge that lines up with the top of the air horn casting (upper edge).

32 Lightly press DOWN on gauge until the plunger bottoms. Record this measurement **(see illustration)**.

33 Now, subtract the first measurement from the second measurement and record the difference. This reading is the total plunger travel.

34 If the difference in travel is between 3/32 and 5/32-inch, no adjustment is necessary. If the difference is less than 3/32-inch (not enough travel) or greater than 5/32-inch (too much travel), adjust the plunger travel, as follows.

Adjusting plunger travel

Note 1: *This adjustment usually is not required after overhaul if you kept the adjustment at its original setting. It is only necessary if the travel is not correct or the duty cycle (percent "on time") is incorrect as determined by the System performance test.*

Note 2: *The air horn must be removed from the carburetor during this procedure; however, the procedure can be performed with the carburetor on or off the vehicle.*

Note 3: *These carburetors are equipped with either a* **four-point** *mixture control solenoid adjustment system (earlier feedback systems) or a* **two-point** *mixture control solenoid adjustment system (mainly on 1985 and later models). The four-point system is equipped with separate lean-stop and rich-stop adjustments that correctly adjust the travel of the solenoid plunger from rich to lean. The* **two-point** *system uses an integral rich limit stop bracket that essentially limits the travel of the plunger going up (rich) as well as the plunger going down (lean). Therefore, only one adjustment is required.*

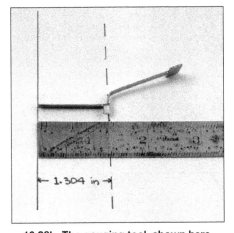

10.36b The gauging tool, shown here, allows you to precisely set the lean-stop at 1.304 inches

35 Remove the air horn from the carburetor. Also remove the solenoid plunger and the plastic float-bowl insert, then reinstall the plunger (removing the insert will allow access to the float-bowl area). **Note:** *On two-point systems, it will be necessary to remove the solenoid adjusting screw and rich-stop before the solenoid plunger and insert can be removed. Prior to removing the screw, turn it clockwise until the plunger bottoms, counting the number of turns. After the insert is removed, reassemble the plunger and stop, setting the adjusting screw the same number of turns from the lightly-bottomed position.*

36 Now install a special gauging tool (available from auto parts stores and automotive tool suppliers) onto the base of the metering jet and directly under the solenoid plunger. This tool is made specifically to measure the solenoid plunger travel **(see illustrations)**, which should be 1.304 inches (approximately 1-5/16 inches).

37 Apply very slight pressure to the solenoid plunger with the tip of your finger to bottom it. Slowly turn the lean-stop screw with the special adjustment tool until it (the

10.41a On four-point systems, if the travel is not correct after setting the lean-stop, unscrew . . .

solenoid plunger) just touches the gauging tool when it's bottomed against the lean stop. This is the factory-specified distance for the lean-stop adjustment.

38 If you have a four-point system, proceed to Step 11. If you have a two-point system, the adjustment is complete. Turn the adjusting screw clockwise to the lightly-bottomed position, counting the number of turns, then unscrew and remove the screw, rich stop and plunger. Install the insert and reassemble the parts, turning the adjusting screw in the same number of turns. Install the air horn and proceed to Step 16.

Models with four-point adjustment systems only

39 Reinstall the air horn, along with a new gasket, onto the carburetor main body, but attach it temporarily with two screws only.

40 Recheck the solenoid plunger travel as described earlier in this procedure. If the travel is correct, completely assemble the air horn and proceed to Step 44 to finish the job.

41 If the plunger travel is still not correct, remove the air horn, invert it and remove the

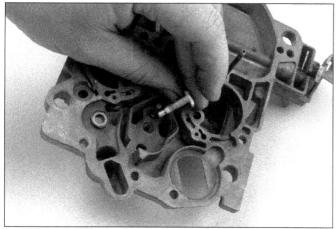

10.41b . . . and remove the rich-stop screw from the carburetor air horn

10.43 Use the special tool to adjust the rich-stop screw on a four-point system

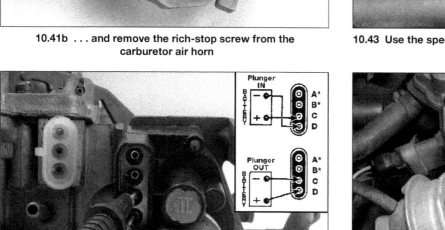

10.49 Using jumper wires, apply battery voltage to terminals C and D to extend or retract the ISC solenoid plunger

10.51 Location of the idle-stop screw (arrow) on feedback carburetors (on non-feedback carburetors, this is the curb-idle speed screw)

rich-stop screw **(see illustrations)**.

42 With the rich-stop screw removed, drive the plug out of the air horn using a punch and a hammer. Reinstall the rich-stop screw.

43 Reinstall the carburetor air horn. Insert the special float gauge into the "D"-shaped vent hole (see *Measuring plunger travel* earlier in this procedure) and, using the special carburetor tool, turn the rich-stop screw **(see illustration)** until the total solenoid plunger travel is 1/8-inch.

All feedback models

44 With the plunger travel correctly set and the air horn reinstalled, install the replacement plugs into the carburetor body. Be sure to install the hollow end down.

Idle speed control (ISC) system

Refer to illustrations 10.49, 10.51, 10.53a and 10.53b

45 If the carburetor has just been overhauled, the ISC solenoid has just been replaced with a new one or the vehicle is having a fluctuating idle problem, it is necessary to check the ISC system adjustments.

46 First, place the transmission in Neutral,

set the parking brake and block the drive wheels. Connect a tachometer according to the manufacturer's instructions.

47 With the A/C off (if equipped), start the engine and warm it up to normal operating temperature (closed loop).

48 Turn the ignition OFF and unplug the connector from the ISC motor.

49 Fully retract the ISC plunger by applying 12 volts to terminal C (run a fused jumper wire from the battery's positive terminal) of the ISC motor connector and grounding terminal D **(see illustration)**. **Note 1:** *Do not allow the battery voltage (12V) to contact terminal C any longer than necessary to retract the ISC plunger. Prolonged contact will cause damage to the motor. Also, it is very important to never connect a voltage source across terminals A and B, as it may damage the internal throttle contact switch.* **Note 2:** *If the ISC solenoid does not respond when battery voltage is applied to the C and D terminals, it is faulty and should be replaced.*

50 Start the engine and allow it to return to normal operation (closed loop). Have an assistant place his foot firmly on the brake and place the transmission in DRIVE.

51 With the plunger completely retracted (from Step 5), turn the idle stop screw **(see illustration)** to attain the minimum idle speed. The minimum base idle speed is usually about 450 rpm in DRIVE. Consult your VECI label under the hood of the engine compartment to verify the idle specifications.

52 Now, with the transmission In DRIVE and an assistant holding his foot firmly on the brake pedal, fully extend the plunger by applying battery positive voltage (12V) to terminal D of the ISC connector and grounding terminal C **(see illustration 10.49)**. **Note:** *Do not allow the battery voltage (12V) to contact terminal D any longer than necessary to retract the ISC plunger. Prolonged contact will cause damage to the motor. Also, it is very important to never connect a voltage source across terminals A and B as it may damage the internal throttle contact switch.*

53 With the ISC plunger fully extended, check the idle speed, which should be approximately 900 rpm in DRIVE. If necessary, adjust the plunger length **(see illustrations)**. Consult your VECI label under the hood of the engine compartment to verify the idle specifications. **Note:** *When the ISC*

10.53a Adjust the ISC plunger by turning the plunger end with pliers (carburetor removed from engine for clarity), but . . .

10.53b . . . do not unscrew the plunger end more than 5/16-inch from its all-the-way-in setting

10.58 Be sure to protect the carburetor when drilling out the air-bleed rivets (otherwise, metal chips could fall into the carburetor!)

10.61 The special gauging tool shown is necessary to check the idle air bleed valve adjustment. Use a large screwdriver, as shown here, to adjust the valve

plunger is in the fully extended position, the plunger must also be in contact with the throttle lever (on carburetor) to prevent possible internal damage to the ISC solenoid.

54   If the ISC motor was removed and the plunger length changed, double-check the maximum allowable speed (see Step 33). If necessary, readjust the plunger until the correct speed is obtained.

55   Reconnect the harness connector to the ISC motor.

Idle air bleed valve (E4ME)

Refer to illustrations 10.58 and 10.61

56   To gain access to the idle air bleed valve, it is necessary to first remove the idle air bleed valve cover. Often, if the carburetor has been overhauled before, the cover has already been removed. If it has not been removed before, the air bleed is probably still at its factory setting and should not be tampered with unless it's known to be incorrect.

57   With the engine off, cover all the bowl vents, air inlets and air intakes with masking tape to prevent any metal chips from falling into the carburetor.

58   Carefully align a Number 35 drill bit (0.110 inches) on one end of the steel rivet heads and drill only enough to remove the rivet head from the carburetor. Use a drift and hammer to remove the remaining rivet from the assembly **(see illustration)**.

59   Lift off the cover and remove any pieces of metal, rivets or debris from the carburetor body. Discard the idle air bleed valve cover.

60   Using compressed air (if available) blow out any remaining pieces of metal from the carburetor area. **Warning:** *Be sure to wear safety goggles whenever using compressed air.*

61   Install a special gauging tool (available from auto parts stores and automotive tool suppliers) into the "D"-shaped vent hole **(see illustration)**. The upper end of the tool should be positioned over the open cavity next to the idle air bleed valve.

62   While holding the gauging tool down lightly, engage the solenoid plunger against the solenoid stop. Adjust the idle air bleed valve so that the gauging tool will pivot over and just contact the top of the valve.

63   Remove the gauging tool. Next, it will be

necessary to check the duty cycle (on-time) of the mixture control solenoid to verify that the air bleed is adjusted properly or if the idle mixture must be adjusted (see Section 15).

64   Disconnect the canister purge hose and plug the end (canister side). Start the engine and allow it to reach normal operating temperature (closed loop). Follow the procedure for checking the M/C solenoid duty cycle in Section 15. If the dwell average is not within 25 to 35 degrees, then it will be necessary to adjust the idle mixture (refer to the procedure in this Chapter).

Idle load compensator (ILC) (some later models)

65   The idle load compensator (ILC) is used instead of an ISC solenoid on some later engines to control curb idle speed. The ILC uses manifold vacuum to sense changes in engine load and compensates by changing the idle speed. The idle load compensator is adjusted at the factory. It is not necessary to make any adjustments unless the curb idle speed is out of adjustment or the ILC solenoid was faulty and had to be replaced.

10.78 To adjust the TPS, drill out and remove the access plug (carburetor shown removed from vehicle for clarity) . . .

10.82 . . . then use a small screwdriver to turn the adjusting screw until the voltmeter readings are correct

Before adjusting the ILC, make sure the vacuum lines to the anti-diesel solenoid, vacuum regulator, ILC solenoid and all the components that require vacuum are in order and not leaking. Check the vacuum schematic on the VECI label under the hood for the correct vacuum routing.

66   Connect a tachometer according to the manufacturer's instructions. Remove the air cleaner and plug the hose to the thermal vacuum valve (TVV). Disconnect and plug the vacuum hose to the EGR valve. Disconnect and plug the vacuum hose to the canister purge port. Disconnect and plug the vacuum hose to the ILC.

67   Back out the idle-stop screw on the carburetor three turns **(see illustration 10.51)**.

68   Turn the A/C system OFF (if equipped). Place the transmission in PARK (automatic) or NEUTRAL (manual), apply the emergency brake and place a block under each of the drive wheels.

69   Start the engine and allow it to reach normal operating temperature. On automatic transmission models, have an assistant place his foot firmly on the brake pedal and place the transmission in DRIVE. The ILC plunger should be fully extended with no vacuum applied. Using back-up wrenches, adjust the ILC plunger to obtain 750 rpm, plus or minus 50 rpm.

70   Next, measure the distance from the jam nut to the tip of the plunger. It should not exceed 1-inch. If the plunger measures more than 1-inch, check for other carburetor problems.

71   Remove the plug from the vacuum hose and reconnect it to the ILC and observe the idle speed. Idle speed should be 450 rpm (vehicle in NEUTRAL (manual) or DRIVE with the brake applied). If the idle speed is correct, proceed to Step 74).

72   If the idle speed is not correct, it will be necessary to adjust the ILC diaphragm. Stop the engine and remove the ILC. Remove the center cap from the center outlet tube. Using a 3/32-inch Allen wrench, insert it through the open center tube to engage the idle speed

adjusting screw inside. If the idle speed was low, turn the adjusting screw counterclockwise ONE turn to increase the idle speed approximately 75 rpm. Conversely, turn the screw clockwise ONE turn to lower the idle speed about 75 rpm. **Note:** *Turn the adjusting screw TWO full turns to raise or lower the idle speed approximately 150 rpm and consequently use the same ratio to calculate the necessary rpm change desired for the situation.*

73   Re-install the ILC onto the carburetor and attach the springs and other related parts. Recheck the idle speed. Make sure the engine is completely warmed up (closed loop). If it is not correct, repeat the ILC adjustment procedure.

74   The last adjustment must be performed on the engine after the TPS value has been reset by the ECM. This can be accomplished by turning off the ignition for 10 seconds or more. Using a hand-held vacuum pump, apply vacuum to the ILC vacuum tube inlet to fully retract the plunger.

75   Adjust the idle stop screw on the carburetor to obtain 450 rpm with the vehicle in NEUTRAL (manual) or DRIVE with the brake applied.

76   Place the vehicle in PARK and stop the engine. Remove the plug from the ILC vacuum hose and install the hose onto the ILC. Reconnect all the vacuum hoses, install the air cleaner and gaskets.

Throttle position sensor (TPS)

Refer to illustrations 10.78 and 10.82

77   The throttle position sensor is equipped with an adjustment screw that is covered over with a factory-installed plug. Do not remove the plug unless it has been determined with careful testing that the TPS is out of adjustment. Refer to Section 15 in this Chapter for complete diagnostic procedures for checking the TPS. This is a critical adjustment that must be performed correctly.

78   Using a 0.076-inch (5/64-inch) drill bit, carefully drill a hole in the aluminum plug covering the TPS adjustment screw **(see illus-**

tration). Drill only enough metal to start a self-tapping screw.

79   Start a number 8 X 1/2-inch long self-tapping screw into the drilled hole in the plug, turning the screw just enough to ensure good thread engagement in the hole.

80   Place a wide blade section of screwdriver between the screw head and the air horn casting. Carefully pry against the screw head to remove the plug. It is also possible to use a slide hammer.

81   Install jumper wires by backprobing the electrical connector onto the correct terminals of the TPS electrical connector, then connect a digital voltmeter between the wires connected to the B (center) and C (bottom) terminals of the TPS connector on the carburetor.

82   With the engine off and ignition on and the throttle closed, turn the screw until the voltage reading is approximately 0.25 volts (E2SE) or 0.50 volts (E4ME) **(see illustration)**.

83   Make sure the voltage is correct. Then install a new plug into the air horn. Drive the plug into place until it is flush with the raised boss on the casting.

11   Exhaust system components - removal and installation

Caution: *The vehicle's exhaust system generates very high temperatures and should be allowed to cool down completely before any of the components are touched. Be especially careful around the catalytic converter, where the highest temperatures are generated. Due to the high temperatures and exposed locations of the exhaust system components, rust and corrosion can "freeze' parts together. Liquid penetrating oils are available to help loosen frozen fasteners; however, in some cases it may be necessary to cut the pieces apart with a hacksaw or cutting torch. The latter method should be employed only by persons experienced in this work.*

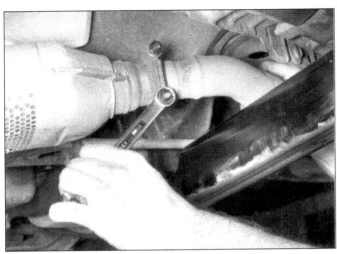

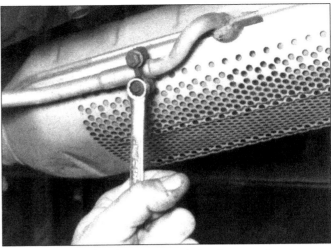

11.6 Remove the catalytic converter-to-front exhaust pipe clamp

11.7 Remove the air management pipe-to-converter clamp

1   Raise the vehicle and place it securely on jackstands.

Front exhaust pipe

2   Remove the bolts securing the exhaust pipe to the exhaust manifold(s).
3   Remove the clamp securing the exhaust pipe to the catalytic converter.
4   Separate the exhaust pipe from the exhaust manifold(s) and from the catalytic converter.
5   Installation is the reverse of the removal procedure. Be sure to install new bolts, gaskets and nuts.

Catalytic converter

Refer to illustrations 11.6, 11.7. and 11.8

Caution: *Make sure the catalytic converter has been allowed sufficient time to cool before attempting removal.*
6   Remove the converter-to-front exhaust pipe clamp **(see illustration)**.
7   Remove the air management pipe-to-converter clamp **(see illustration)**.
8   Remove the bolts securing the rear of the converter to the converter/intermediate pipe bracket and remove the converter **(see illustration)**.
9   Installation is the reverse of the removal procedure. Be sure to use new gaskets, bolts and nuts.

Intermediate exhaust pipe

10   Remove the clamp securing the intermediate pipe to the muffler/tailpipe assembly.
11   Remove the bolts securing the intermediate pipe to the intermediate pipe/converter bracket and remove the intermediate pipe.
12   Installation is the reverse of the removal procedure. Be sure to use new gaskets, bolts and nuts.

Muffler and tailpipe

13   The tailpipe is actually part of the muffler assembly.
14   Remove the bolts and nuts which connect the muffler to the intermediate exhaust pipe.
15   Remove the muffler/tailpipe assembly

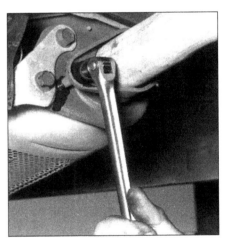

11.8 Remove the bolts from the converter/intermediate pipe bracket

rear hanger bolts and separate the muffler/tailpipe assembly from the vehicle.
16   Installation is the reverse of the removal procedure. Be sure to use new nuts and bolts.

Notes

Chapter 4 Part B
Fuel and exhaust systems - fuel injected models

Contents

	Section
Air filter replacement	See Chapter 1
Exhaust manifolds - removal and installation	See Chapter 2
Exhaust system check	See Chapter 1
Exhaust system - removal and installation	See Chapter 4A
Fuel filter replacement	See Chapter 1
Fuel injection system - check	5
Fuel pressure relief	2
Fuel pump - check	3
Fuel pump - removal and installation	4
Fuel system check	See Chapter 1
Fuel tank - cleaning and repair	See Chapter 4A

	Section
Fuel tank - removal and installation	See Chapter 4A
General information	1
Minimum idle speed - adjustment	9
TBI mounting torque check	See Chapter 1
Throttle Body Injection system - component replacement	6
Thermostatic air cleaner check	See Chapter 1
Throttle cable - removal and installation	8
Throttle linkage inspection	See Chapter 1
Multi Port Injection system - component replacement	7
Underhood hose check and replacement	See Chapter 1

Specifications

Throttle Body Injection (includes Crossfire fuel injection)

Fuel pressure	9 to 13 psi
Fuel injector resistance	1.0 to 2.0 ohms
Maximum shoulder-to-ball clearance (Crossfire)	0.200 inch
Minimum throttle rod-to-casting clearance (Crossfire)	0.040 inch

Torque specifications

	Ft-lbs (unless otherwise noted)
Idle Air Control (IAC) assembly	156 in-lbs
Fuel meter body screws	35 in-lbs
Fuel inlet and return fittings	21
Fuel connecting tube nuts	24
Fuel meter cover screws	28 in-lbs
TBI mounting bolts	96 to 168 in-lbs
Intake cover plate bolts and stud (Crossfire)	15 to 25

Multi Port Fuel Injection (includes Tuned Port Injection)

Fuel pressure	34 to 47 psi
Fuel injector resistance	11.0 to 14.0 ohms
Minimum idle speed idle stop screw adjustment	
V6 engine	
automatic transmission...	550 rpm (in Drive)
manual transmission...	650 rpm (in Neutral)
V8 engine	
automatic transmission...	400 rpm (in Drive)
manual transmission...	450 rpm (in Neutral)
Throttle position sensor (TPS) adjustment	
V6 engine	0.35 to 0.67 volt
V8 engine	0.46 to 0.62 volt

Torque specifications

	Ft-lbs (unless otherwise noted)
Intake plenum-to-intake manifold bolts (V6)...	18
Intake runner-to-plenum bolts (V8)...	18
Intake runner-to-manifold bolts (V8)...	18
Fuel filter fittings	22
Fuel feed and return line-to-fuel rail fittings......................................	22
Fuel rail-to-intake manifold bolts..	18
Fuel pressure connection assembly..	88 in-lbs
Throttle body-to-plenum bolts..	18
Idle air control (IAC) valve (except 3.1L V6)..	13
Idle air control (IAC) valve screws (3.1L V6)	30 in-lbs
EGR valve bolts	14
EGR control solenoid nut..	17

1   General information

Fuel system

The fuel system consists of a rear mounted fuel tank, an electrically operated in-tank fuel pump, a fuel injection system, and the fuel feed and return lines between the fuel injection system and the tank. Two types of fuel injection systems are used depending on model and engine type; throttle body fuel injection and multi port fuel injection.

Fuel is delivered to the fuel injector(s) at a constant pressure. To maintain the fuel pressure at a constant level, excess fuel is returned to the fuel tank through a fuel pressure regulator.

The injectors are similar to an electronic solenoid and are controlled by the ECM. The amount of fuel injected is varied by the length of time the injector plunger is held open. To determine how much fuel is required at a give moment, the ECM processes signals from sensors recording engine coolant temperature, exhaust oxygen content, throttle position, intake air mass, engine rpm, vehicle speed and accessory load.

Throttle Body Injection

Models equipped with Throttle Body Injection utilize a centrally located injector in a throttle body housing. All four cylinder models are equipped with a single-point Throttle Body Injection system The Crossfire Fuel Injection system on 1982 through 1984 V8 models consists of two Throttle Body

Injection (TBI) units mounted diagonally on a common intake manifold. The Throttle Body Injection system used on 1988 and later V8 models uses two injectors instead of one mounted in the throttle body unit.

Multi Port Fuel Injection

Models equipped with Multi Port Fuel Injection utilize one injector per cylinder. The injectors are mounted on the intake manifold, above each intake port. The throttle body serves only to control the amount of air entering the engine. Because each cylinder is equipped with an injector mounted immediately adjacent to the intake valve, much better control of the fuel/air mixture is possible. 1985 and later V6 models are equipped with a multi port fuel injection system The multi port fuel injection system used on 1985 and later V8 models is referred to as Tuned Port Injection (TPI).

Exhaust system

The exhaust system consists of exhaust manifold(s), exhaust pipes, catalytic converter and muffler(s). The catalytic converter is an emissions control device added to the exhaust system to reduce pollutants.

2   Fuel pressure relief

Warning: *Gasoline is extremely flammable, so take extra precautions when you work on any part of the fuel system. Don't smoke or allow open flames or bare light bulbs near the work area, and don't work in a garage where*

a natural gas-type appliance (such as a water heater or a clothes dryer) with a pilot light is present. Since gasoline is carcinogenic, wear latex gloves when there's a possibility of being exposed to fuel, and, if you spill any fuel on your skin, rinse it off immediately with soap and water. Mop up any spills immediately and do not store fuel-soaked rags where they could ignite. The fuel system is under constant pressure, so, if any fuel lines are to be disconnected, the fuel pressure in the system must be relieved first. When you perform any kind of work on the fuel system, wear safety glasses and have a Class B type fire extinguisher on hand.

Note: *After the fuel pressure has been relieved. it's a good idea to lay a shop towel over any fuel connection to be disassembled, to absorb the residual fuel that may leak out when servicing the fuel system.*

1   The fuel system remains under pressure even after the engine has been shut off for an extended time. Therefore it is necessary to relieve the pressure in the fuel system before any work is done on fuel injection components or lines.

2   Remove the cover from the fuse block located in the right end of the dashboard.

3   Pull the fuel pump fuse from the fuse block to disable the pump. The fuel pump fuse is identified by the letters FP.

4   Start the engine and let it run until it dies from lack of fuel, then crank the engine over for several seconds with the starter to insure that all pressure has been eliminated from the system.

5   Be sure to replace the fuel pump fuse when work is completed on the fuel system.

3.2 The Assembly Line Diagnostic Link (ALDL) diagnostic connector is located underneath the instrument panel - to energize the fuel pump, apply battery voltage to terminal G (arrow)

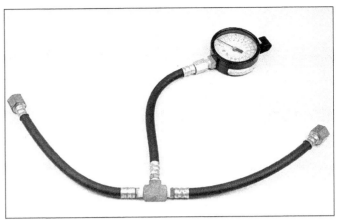

3.6 To check the fuel pressure of a TBI-equipped vehicle, use a gauge capable of reading between 9 and 13 psi and short lengths of hose with a plastic T-fitting and hose clamps

3   Fuel pump - check

Refer to illustrations 3.2, 3.6, 3.12 and 3.15

1   Before testing the fuel pump, check the fuel pump fuse. If the fuse is blown, replace it before proceeding.

2   Remove the plastic cover from the Assembly Line Diagnostic Link (ALDL) located underneath the instrument panel, just to the left of the console. Energize the fuel pump by applying battery voltage to terminal G **(see illustration)**. You should hear the pump come on inside the fuel tank. If the pump does not come on, either there is a loose connection or an open in the fuel pump circuit, or the fuel pump is faulty. Check the entire fuel pump electrical circuit (see Wiring Diagrams) before diagnosing the fuel pump as faulty. If the circuit checks out, replace the pump.

3   If the pump runs, check the fuel pressure.

TBI fuel pressure check

4   Remove the air cleaner and plug the Thermac intake vacuum port on the TBI unit.

5   Relieve fuel pressure (Section 2). See the **Warning** in Section 2. Remove the fuel line between the front and rear throttle bodies. Use a back-up wrench to prevent the fuel nut from turning when loosening the fuel line fittings.

6   Connect a fuel injection pressure gauge to two short sections of rubber hose with a T-fitting **(see illustration)** and install it between the two TBI balance line fittings. The gauge must be capable of reading fuel pressure between 9 and 13 psi.

7   Start the engine and observe the fuel pressure reading. It should be between 9 and 13 psi.

8   If there is no pressure (even though the fuse is okay and the pump is running), the inline fuel filter is probably clogged. Replace it and check the pressure again. If there is still zero pressure, the fuel feed line itself may be plugged. If the fuel feed line is clear, check

3.12 Measure fuel flow at the fuel feed line - on TPI models it's necessary to remove the intake ducting attach the hose to the fuel feed line

the fuel pump inlet filter.

9   If the pressure is below 9 psi, the inline fuel filter may be restricted. Replace it and check the pressure again. If it is still low, check for a loose fuel line fitting or a punctured fuel line. Also inspect both TBI injectors and the fuel meter cover for a leak. If nothing is leaking, disconnect the electrical connectors to both injectors, block the fuel return line by pinching the flexible hose portion and apply 12 volts to the ALDL pump test terminal G and note the indicated pressure. If the pressure is now above 13 psi, check the line between the fuel pump and the test gauge. If the line is okay, replace the rear TBI fuel meter cover. If the pressure is still below 9 psi, replace the fuel pump.

10   If the pressure is above 13 psi, the problem is either a plugged fuel return line or a faulty fuel meter cover.

11   Remove the fuel pressure gauge and install the steel fuel line between the throttle bodies. Remove the plug covering the Thermac vacuum port on the TBI unit and install the air cleaner.

3.15 To determine the fuel pressure, attach a pressure gauge capable of reading between 34 and 47 psi to the Schrader valve on the fuel rail

MPFI and TPI fuel flow check

12   Relieve fuel pressure (Section 2). See the **Warning** in Section 2. Disconnect the fuel feed line from the fuel rail and connect a hose to the feed line. Run the other end of the hose into a suitable unbreakable container **(see illustration)**.

13   Have an assistant to monitor the hose and container as you activate the fuel pump at the Assembly Line Diagnostic Link (ALDL) underneath the dashboard. Remove the plastic cover from the ALDL diagnostic connector and apply battery voltage to terminal G of the fuel pump test connector **(see illustration 3.2)**.

14   The fuel pump should supply 1/2 pint or more in 15 seconds. If the fuel system flow is below the minimum, inspect the fuel system for a restriction between the pump and the fuel rail. If there is no restriction, check the fuel pump pressure.

MPFI and TPI fuel pressure check

15   Relieve fuel pressure (Section 2). See

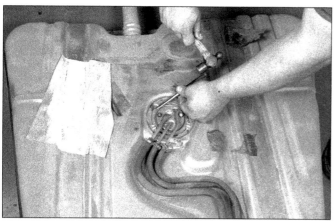

4.4 Using a hammer and brass punch, tap the lock-ring counterclockwise to loosen it

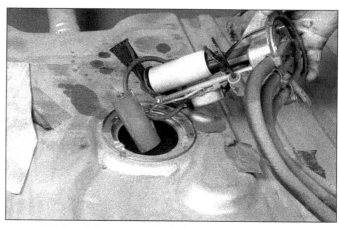

4.5 Carefully remove the fuel sending unit assembly through the opening

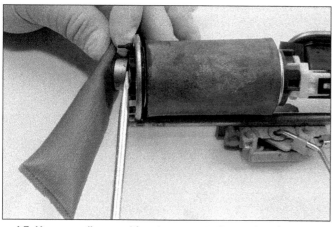

4.7 Use a small screwdriver to separate the strainer from the bottom of the fuel pump inlet

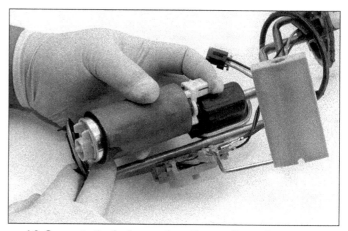

4.9 Separate the fuel pump from the sending unit assembly

the **Warning** in Section 2. With the ignition off attach a fuel pressure gauge to the fuel pressure fitting on the fuel rail **(see illustration)**.

16   While observing the fuel gauge, turn the ignition on. The fuel pump will run approximately 2 seconds.

17   As the pump runs the pressure should be 34 to 47 psi. When the pump stops running the pressure should drop slightly and hold steady.

18   If the pressure is within specification, start the engine and observe the fuel pressure. It should be 3 to 10 psi lower than in Step 22.

19   If the fuel pressure is less than specified check the fuel quantity, the fuel filter and fuel lines for restrictions and leaks, if none are found replace the fuel pump.

20   If the fuel pressure is in excess of specified, check the fuel return line for restrictions, if none are found replace the pressure regulator.

---

4   Fuel pump - removal and installation

Refer to illustrations 4.4, 4.5, 4.7 and 4.9
Warning: *Gasoline is extremely flammable, so take extra precautions when you work on*

any part of the fuel system. Don't smoke or allow open flames or bare light bulbs near the work area, and don't work in a garage where a natural gas-type appliance (such as a water heater or a clothes dryer) with a pilot light is present. Since gasoline is carcinogenic, wear latex gloves when there's a possibility of being exposed to fuel, and, if you spill any fuel on your skin, rinse it off immediately with soap and water. Mop up any spills immediately and do not store fuel-soaked rags where they could ignite. The fuel system is under constant pressure, so, if any fuel lines are to be disconnected, the fuel pressure in the system must be relieved first. When you perform any kind of work on the fuel system, wear safety glasses and have a Class B type fire extinguisher on hand.

1   Disconnect the cable from the negative terminal of the battery **Caution:** *On models equipped with a Delco Loc II audio system, disable the anti-theft feature before disconnecting the battery.*

2   Remove the fuel tank (see Chapter 4A).

3   The fuel pump/sending unit assembly is located inside the fuel tank. It is held in place by a cam lock ring mechanism consisting of an inner ring with three locking cams and an outer ring with three tangs. The outer ring is

welded to the tank and can't be turned.

4   To unlock the fuel pump/sending unit assembly, turn the inner ring counterclockwise until the locking cams are free of the tangs **(see illustration)**. A special tool is available to loosen the lock ring, but if the tool is not available a hammer and brass punch can be used. **Warning:** *If using a punch to loosen the lock ring, use a brass punch. Do not use a steel punch - a spark could cause an explosion!*

5   Lift the fuel pump/sending unit assembly from the fuel tank **(see illustration)**. **Caution:** *The fuel level float and sending unit are delicate. Do not bump them against the tank during removal or the accuracy of the sending unit may be affected.*

6   Inspect the condition of the rubber gasket around the opening of the tank. If it is dried, cracked or deteriorated, replace it.

7   Remove and inspect the strainer on the lower end of the fuel pump **(see illustration)**. If it's dirty, replace it.

8   Disconnect the electrical connector from the fuel pump.

9   Remove the pump from the sending unit by pushing the fuel pump up into the rubber coupler and sliding the bottom of the pump away from the bottom support **(see illustration)**. After the

pump is clear of the bottom support, pull it out of the rubber coupler. Care should be taken to prevent damage to the rubber insulator during removal.

10   Inspect the rubber coupler and sound insulator and replace them if necessary. Reassemble the fuel pump onto the sending unit assembly.

11   Insert the fuel pump/sending unit assembly into the fuel tank.

12   Turn the inner lock ring clockwise until the locking cams are fully engaged by the retaining tangs. **Note:** *If you have installed a new O-ring type rubber gasket, it may be necessary to push down on the inner lock ring until the locking cams slide under the retaining tangs.*

13   Install the fuel tank.

5   Fuel injection system - check

Warning: *Gasoline is extremely flammable, so take extra precautions when you work on any part of the fuel system. Don't smoke or allow open flames or bare light bulbs near the work area, and don't work in a garage where a natural gas-type appliance (such as a water heater or a clothes dryer) with a pilot light is present. Since gasoline is carcinogenic, wear latex gloves when there's a possibility of being exposed to fuel, and, if you spill any fuel on your skin, rinse it off immediately with soap and water. Mop up any spills immediately and do not store fuel-soaked rags where they could ignite. The fuel system is under constant pressure, so, if any fuel lines are to be disconnected, the fuel pressure in the system must be relieved first (see Section 2). When you perform any kind of work on the fuel system, wear safety glasses and have a Class B type fire extinguisher on hand.*

Note: *The following procedure is based on the assumption that the fuel pump is working and the fuel pressure is adequate (see Section 3).*

1   Check to see that the battery is fully charged, as the control unit and sensors depend on an accurate supply voltage in order to properly meter the fuel.

2   Check all electrical connectors that are related to the system. Loose electrical connectors and poor grounds can cause many problems that resemble more serious malfunctions. Check the ground wire connections on the intake manifold for tightness.

3   Check the system fuses. If a blown fuse is found, replace it and see if it blows again. If it does, search for a grounded wire in the harness to the fuel pump.

4   Check the air filter element - a dirty or partially blocked filter will severely impede performance and economy (see Chapter 1).

5   Check the air intake duct to the intake manifold for leaks, which will result in an excessively lean mixture. Also check the condition of all vacuum hoses connected to the intake manifold.

6   Remove the air cleaner assembly or air intake duct from the throttle body and check for dirt, carbon or other residue build-up. If it's dirty, clean it with carburetor cleaner and a toothbrush.

7   On TBI models, start the engine and check the fuel spray pattern at the tip of the injector. A well defined cone of fuel should be evident. If the fuel spay is intermittent or broken, the injector may be faulty.

7   On MPFI models, start the engine and place an automotive stethoscope against each injector, one at a time, and listen for a clicking sound, indicating operation. If you don't have a stethoscope, place the tip of a screwdriver against the injector and listen through the handle.

8   Unplug the injector electrical connector(s) and test the resistance of each injector. Compare the values to the Specifications listed in this Chapter.

9   Install an injector test light ("noid" light) into each injector electrical connector, one at a time. Crank the engine over. Confirm that the light flashes evenly on each connector. This will test for voltage to the injectors. If no voltage is reaching the injectors the ECM or circuitry is defective.

6   Throttle Body Injection system - component replacement

Warning: *Gasoline is extremely flammable, so take extra precautions when you work on any part of the fuel system. Don't smoke or allow open flames or bare light bulbs near the work area, and don't work in a garage where a natural gas-type appliance (such as a water heater or a clothes dryer) with a pilot light is present. Since gasoline is carcinogenic, wear latex gloves when there's a possibility of being exposed to fuel, and, if you spill any fuel on your skin, rinse it off immediately with soap and water. Mop up any spills immediately and do not store fuel-soaked rags where they could ignite. The fuel system is under constant pressure, so, if any fuel lines are to be disconnected, the fuel pressure in the system must be relieved first (see Section 2). When you perform any kind of work on the fuel system, wear safety glasses and have a Class B type fire extinguisher on hand.*

1   Relieve system fuel pressure (see Section 2).

2   Disconnect the cable from the negative terminal of the battery. **Caution:** *On models equipped with a Delco Loc II audio system, disable the anti-theft feature before disconnecting the battery.*

3   Remove the air cleaner housing assembly, adapter and gaskets.

Fuel meter cover/fuel pressure regulator assembly

Refer to illustrations 6.5a, 6.5b and 6.7

Note: *The fuel pressure regulator is housed in the fuel meter cover. Whether you are replacing the meter cover or the regulator itself, the entire assembly must be replaced. The regulator must not be removed from the cover.*

4   Unplug the electrical connectors to the fuel injectors.

5   Remove the long and short fuel meter cover screws and remove the fuel meter cover **(see illustrations)**.

6   Remove the fuel meter outlet passage gasket, cover gasket and pressure regulator seal. Carefully remove any old gasket material that is stuck with a razor blade. **Caution:** *Do not attempt to re-use either of these gaskets.*

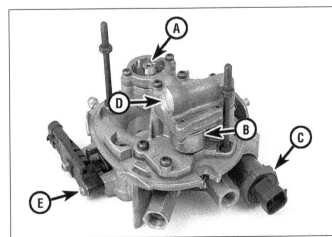

6.5a Typical TBI assembly

A   Fuel injector
B   Fuel pressure regulator
C   Idle Air Control (IAC) valve
D   Fuel meter cover
E   Throttle Position Sensor (TPS)

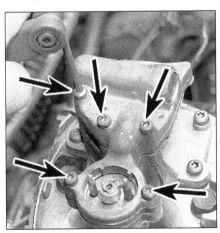

6.5b Remove the fuel meter cover screws (arrows)

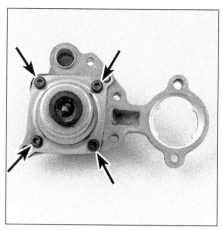

6.7 The fuel pressure regulator and fuel meter cover an integral assembly that cannot be serviced separately - DO NOT remove the four screws (arrows) under any circumstances

6.16 Place a screwdriver shank on top of the fuel meter cover gasket surface to serve as a fulcrum and, using another screwdriver, carefully pry the fuel injector from the fuel meter body

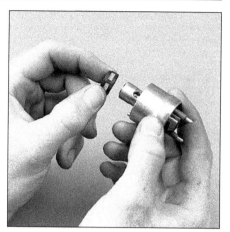

6.21 Slide the new filter onto the nozzle of the fuel injector

7   Inspect the cover for dirt, foreign material and casting warpage. If it is dirty, clean it with a clean shop rag soaked in solvent. Do not immerse the fuel meter cover in cleaning solvent - it could damage the pressure regulator diaphragm and gasket. **Warning:** *Do not remove the four screws* **(see illustration)** *securing the pressure regulator to the fuel meter cover. The regulator contains a large spring under compression which, if accidentally released, could cause injury. Disassembly might also result in a fuel leak between the diaphragm and the regulator housing. The new fuel meter cover assembly will include a new pressure regulator.*

8   Install the new pressure regulator seal, fuel meter outlet passage gasket and cover gasket.

9   Install the fuel meter cover using Loctite 262 or equivalent on the screws. **Note:** *The short screws go next to the injectors.*

10   Attach the electrical connectors to both injectors.

11   Attach the cable to the negative terminal of the battery.

12   With the engine off and the ignition on, check for leaks around the gasket and fuel line couplings.

13   Install the air cleaner, adapter and gaskets.

Fuel injector(s)

Refer to illustrations 6.16, 6.21, 6.22, 6.23, 6.24 and 6.25

Note: *When replacing a fuel injector, be sure to use one having the identical part number. Injectors from other models are calibrated with different flow rates but can be interchanged.*

14   To unplug the electrical connectors from the fuel injectors, squeeze the plastic tabs and pull straight up.

15   Remove the fuel meter cover/pressure regulator assembly. **Note:** *Do not remove the fuel meter cover assembly gasket - leave it in place to protect the casting from damage during injector removal.*

16   Use two screwdrivers **(see illustration)**

to pry out the injector(s).

17   Remove the upper (larger) and lower (smaller) O-rings and filter from the injector(s).

18   Remove the steel backup washer from the top of each injector cavity.

19   Inspect the fuel injector filters for evidence of dirt and contamination. If present, check for the presence of dirt in the fuel lines and fuel tank.

20   Be sure to replace the fuel injector with an identical part. Injectors from other models can fit in the TBI assembly but are calibrated for different flow rates.

21   Slide the new filter into place on the nozzle of the injector **(see illustration)**.

22   Lubricate the new lower (smaller) O-ring with automatic transmission fluid and place it on the small shoulder at the bottom of the fuel injector cavity in the fuel meter body **(see illustration)**.

23   Install the steel back-up washer in the injector cavity **(see illustration)**.

24   Lubricate the new upper (larger) O-ring with automatic transmission fluid and install it on top of the steel back-up washer **(see illustration)**. **Note:** *The backup washer and the large O-ring must be installed before the injector. If they aren't, improper seating of the*

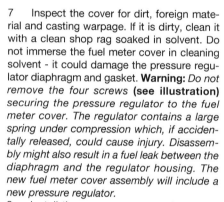

6.22 Lubricate the lower O-ring with transmission fluid then place it on the shoulder in the bottom of the injector cavity

6.23 Place the steel back-up washer on the shoulder near the top of the injector cavity

6.24 Lubricate the upper O-ring with transmission fluid then install it on top of the steel washer

6.25 Make sure that the lug is aligned with the groove in the bottom of the fuel injector cavity

6.31 The TPS is mounted to the throttle body with two screws

large O-ring could cause fuel leakage.

25 To install an injector, align the raised lug on the injector base with the notch in the fuel meter body cavity (see illustration). Push down on the injector until it is fully seated in the fuel meter body. Note: *The electrical terminals should be parallel with the throttle shaft.*

26 Install the fuel meter cover assembly and gasket.

27 Attach the cable to the negative terminal of the battery.

28 With the engine off and the ignition on, check for fuel leaks.

29 Attach the electrical connectors to the fuel injectors.

30 Install the air cleaner housing assembly, adapter and gaskets.

Throttle Position Sensor (TPS)

Refer to illustration 6.31

31 Remove the two TPS attaching screws and retainers and remove the TPS from the throttle body (see illustration).

32 If you intend to re-use the same TPS, do not attempt to clean it by soaking it in any liquid cleaner or solvent. The TPS is a delicate electrical component and can be damaged by solvents.

33 Install the TPS on the throttle body while lining up the TPS lever with the TPS drive lever.

34 Install the two TPS attaching screws and retainers.

35 Install the air cleaner housing assembly, adapter and gaskets.

36 Attach the cable to the negative terminal of the battery.

Idle Air Control (IAC) valve

Refer to illustrations 6.37, 6.40 and 6.41

37 Unplug the electrical connector from the IAC valve and remove the IAC valve (see illustration).

38 Remove and discard the old IAC valve gasket. Clean any old gasket material from the surface of the throttle body assembly to insure proper sealing of the new gasket.

39 Most IAC valve pintles have the same dual taper. However, the pintles on some units have a 12 mm diameter and the pintles on others have a 10 mm diameter. A replacement IAC valve must have the appropriate pintle taper and diameter for proper seating of the valve in the throttle body.

40 Measure the distance between the tip of the pintle and the housing mounting surface (see illustration). If dimension "A" is greater than 1-1/8 inches, it must be reduced to prevent damage to the valve.

41 To adjust the pintle of an IAC valve, first determine if it is a Type 1 or a Type 2 IAC valve. Type 1 has a collar at the end where the electrical connector connects. Type 2

6.37 The IAC valve can be removed with an adjustable wrench (shown) or a 1-1/4 inch wrench

does not have a collar. To adjust a Type 1 valve, grasp the valve and exert firm pressure on the pintle with the thumb (see illustration). Use a slight side-to-side movement on the pintle as you press it in with your thumb. To adjust a Type 2 valve, push the pintle in and turn it clockwise until it is properly set.

6.40 Distance (A) should be less than 1-1/8 inch before installing the IAC valve

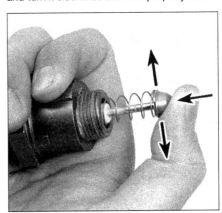

6.41 To adjust the Type 1 IAC valve pintle distance, gently press the pintle in while rocking it back-and-forth

6.50 Remove the fuel inlet and outlet nuts from the fuel meter body

6.52 Once the fuel inlet and outlet nuts are off, pull the fuel meter body straight up to separate it from the throttle body

6.65 When disconnecting the fuel feed and return lines from the fuel inlet and outlet nuts, be sure to use a backup wrench to prevent damage to the lines

42   Install the IAC valve and tighten it to the specified torque. Attach the electrical connector.
43   Install the air cleaner housing assembly, adapter and gaskets.
44   Attach the cable to the negative terminal of the battery.
45   Start the engine and allow it to reach operating temperature, then turn it off. No adjustment of the IAC valve is required after installation.

Fuel meter body assembly

Refer to illustrations 6.50 and 6.52
46   Unplug the electrical connectors from the fuel injectors.
47   Remove the fuel meter cover/pressure regulator assembly, fuel meter cover gasket, fuel meter outlet gasket and pressure regulator seal.
48   Remove the fuel injectors.
49   Unscrew the fuel inlet and return line threaded fittings, detach the lines and remove the O-rings.
50   Remove the fuel inlet and outlet nuts and gaskets from the fuel meter body assembly **(see illustration)**. Note the locations of the nuts to ensure proper reassembly. The inlet nut has a larger passage than the outlet nut.
51   Remove the gasket from the inner end of each fuel nut.
52   Remove the fuel meter body-to-throttle body attaching screws and remove the fuel meter body from the throttle body **(see illustration)**.
53   Install the new throttle body-to-fuel meter body gasket. Match the cut-out portions in the gasket with the openings in the throttle body.
54   Install the fuel meter body on the throttle body. Coat the fuel meter body-to-throttle body attaching screws with thread locking compound before installing them.
55   Install the fuel inlet and outlet nuts, with new gaskets, in the fuel meter body and tighten the nuts to the specified torque. Install the fuel inlet and return line threaded

fittings with new O-rings. Use a backup wrench to prevent the nuts from turning.
56   Install the fuel injectors.
57   Install the fuel meter cover/pressure regulator assembly.
58   Attach the cable to the negative terminal of the battery.
59   Attach the electrical connectors to the fuel injectors.
60   With the engine off and the ignition on, check for leaks around the fuel meter body, the gasket and around the fuel line nuts and threaded fittings.
61   Install the air cleaner housing assembly, adapters and gaskets.

Throttle body assembly

Refer to illustrations 6.65 and 6.66
62   Unplug all electrical connectors - the IAC valve, TPS and fuel injectors. Detach the grommet with the wires from the throttle body.
63   Detach the throttle linkage, return spring(s), transmission control cable (automatic) and, if equipped, cruise control.
64   Clearly label, then detach, all vacuum hoses.
65   Using a backup wrench, detach the inlet and outlet fuel line nuts **(see illustration)**.

6.66 To remove the throttle body from the intake manifold, remove the three nuts/bolts (arrows)

Remove the fuel line O-rings from the nuts and discard them.
66   Remove the TBI mounting bolts **(see illustration)** and lift the TBI unit from the intake manifold. Remove and discard the TBI manifold gasket.
67   Place the TBI unit on a holding fixture. **Note:** *If you don't have a holding fixture, and decide to place the TBI directly on a work bench surface, be extremely careful when servicing it. The throttle valve can be easily damaged.*
68   Remove the fuel meter body-to-throttle body attaching screws and separate the fuel meter body from the throttle body.
69   Remove the throttle body-to-fuel meter body gasket and discard it.
70   Remove the TPS.
71   Invert the throttle body on a flat surface for greater stability and remove the IAC valve.
72   Clean the throttle body assembly in a cold immersion cleaner. Clean the metal parts thoroughly and blow dry with compressed air. Be sure that all fuel and air passages are free of dirt or burrs. **Caution:** *Do not place the TPS, IAC valve, pressure regulator diaphragm, fuel injectors or other components containing rubber in the solvent or cleaning bath. If the throttle body requires*

cleaning, soaking time in the cleaner should be kept to a minimum. Some models have throttle shaft dust seals that could lose their effectiveness by extended soaking.

73   Inspect the mating surfaces for damage that could affect gasket sealing. Inspect the throttle lever and valve for dirt, binds, nicks and other damage.

74   Invert the throttle body on a flat surface for stability and install the IAC valve and the TPS.

75   Install a new throttle body-to-fuel meter body gasket and place the fuel meter body assembly on the throttle body assembly. Coat the fuel meter body-to-throttle body attaching screws with thread locking compound and tighten them securely.

76   Install the TBI unit and tighten the mounting bolts to the specified torque. Use a new TBI-to-manifold gasket.

77   Install new O-rings on the fuel line nuts. Install the fuel line and outlet nuts by hand to prevent stripping the threads. Using a backup wrench, tighten the nuts to the specified torque once they have been correctly threaded into the TBI unit.

78   Attach the vacuum hoses, throttle linkage, return spring(s), transmission control cable (automatics) and, if equipped, cruise control cable. Attach the grommet, with wire harness, to the throttle body.

79   Plug in all electrical connectors, making sure that the connectors are fully seated and latched.

80   Check to see if the accelerator pedal is free by depressing the pedal to the floor and releasing it with the engine off.

81   Connect the negative battery cable, and, with the engine off and the ignition on, check for leaks around the fuel line nuts.

82   Check the TPS output (see Chapter 6).

83   Install the air cleaner housing assembly, adapter and gaskets.

Crossfire Fuel Injection

Refer to illustration 6.109

Note: *This section must be followed when replacing one or both throttle bodies with new ones. Replacement of one or more of these assemblies requires throttle valve synchronization and checking of throttle rod alignment.*

84   If tamper resistant plugs over the throttle stop screws have been installed, remove them in accordance with the following procedure. **Note:** *Replacement throttle bodies do not have throttle stop screw plugs installed.*

 a) *Use a punch to establish the location, over the centerline of the throttle stop screw, for the hole to be drilled.*

 b) *Drill a 5/32-inch (4 mm) diameter hole through the throttle body casting to the hardened plug.*

 c) *Use a 1/16-inch (2 mm) diameter punch to drive through the drilled hole and knock out the plug.*

85   If the front TBI unit throttle synchronizing screw has a welded retaining collar, grind off the weld. (A replacement throttle body

may have a throttle synchronizing screw with a non-welded collar).

86   Block movement of the throttle lever by relieving the force of the heavy spring against the throttle synchronizing screw to prevent the levers from coming into contact. An Allen wrench can be inserted under the stop screw plate to do this. **Caution:** *If the lever is not blocked before the throttle synchronizing screw is removed, the screw may be damaged and reinstallation will be accomplished only with great difficulty.*

87   Remove the synchronizing screw and collar, discard the collar and reinstall the screw. Remove the block from the throttle lever. **Caution:** *The collar must be removed to prevent possible interference with the air cleaner.*

88   Install the throttle body mounting gaskets.

89   With either throttle body in place, install the throttle rod end bearing on the front unit throttle lever stud.

90   Install the other throttle body on the manifold cover locator pins.

91   Tighten all mounting bolts to the specified torque. Do not install the throttle rod and bearing assembly retaining clip yet.

Throttle valve synchronization - preliminary adjustment

Note: *The following adjustment procedure may be performed with the entire assembly on the vehicle or as a bench procedure.*

92   Turn both front and rear unit throttle stop screws counterclockwise enough to break contact with the throttle lever tangs.

93   Adjust the throttle synchronizing screw to allow both throttle valves to close. The throttle rod end bearing will move freely on the front unit throttle lever stud when both valves are closed.

94   Turn the front throttle stop screw clockwise slowly until it makes contact with the throttle lever tang. Turn it clockwise an additional 1/4-turn.

95   Turn the rear throttle stop screw clockwise slowly until it makes contact with the throttle lever tang. Turn it clockwise an additional 1/2-turn.

96   Final adjustments are made with the assembly installed on the vehicle and with the engine running.

Throttle rod alignment check

97   Actuate the rear TBI unit throttle lever to bring both units to the wide open throttle position by rotating the throttle valves and load them in the direction shown. Move the throttle rod toward the front unit casting boss.

98   Check the clearance between the shoulder of the stud and the side of the ball surface. Maximum clearance must not exceed 0.200-inch (5.0 mm). If the clearance is greater, the replacement assembly (manifold cover or throttle body) must be exchanged and the preliminary throttle valve synchronizing adjustment must be repeated.

99   Use needle-nose pliers to carefully

install a new throttle rod and bearing assembly retaining clip (supplied in the service package).

100 Move both the front and rear throttle levers through the total throttle travel while simultaneously loading the throttle valves in the forward direction.

101 Check the clearance between the throttle rod and front throttle body casting boss. If the minimum clearance is less than 0.040-inch (1.02 mm), the replacement assembly must be exchanged. If the minimum clearance is at least 0.040-inch (1.02 mm), proceed to the Final installation steps.

102 If the assemblies were replaced in the previous Step, repeat the Throttle valve synchronization - preliminary adjustment and the Throttle rod alignment check Steps to ensure that both clearances are still correct.

Final installation steps

103 Reinstall the connecting fuel tube between the throttle body assemblies and tighten it to the specified torque. Use a back-up wrench to prevent the fuel nuts on the throttle bodies from turning.

104 Reconnect the fuel inlet and return line fittings to the front TBI assembly, using a back-up wrench on the fuel nuts to prevent them from turning. Make sure that the fuel line O-rings are not nicked, cut or damaged before connecting the lines.

105 Reattach the throttle cable, transmission detent and, if equipped, the cruise control linkage.

106 Reinstall the related vacuum hoses.

107 Connect the electrical connectors to the IAC valves, the fuel injectors and the TPS switch. Reinstall the wiring harness on the throttle body.

108 With the engine off, depress the accelerator pedal to the floor and release it, checking for free return of the pedal.

109 With the throttle plate in the closed (idle) position, install the TPS switch onto the throttle body with the retainers and the two new

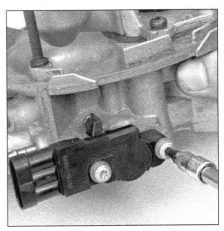

6.109 Make sure that you use the new TPS switch attaching screws (Torx), lockwashers and retainers that are included in the service kit and that the TPS pickup lever is located above the tang on the throttle actuator lever

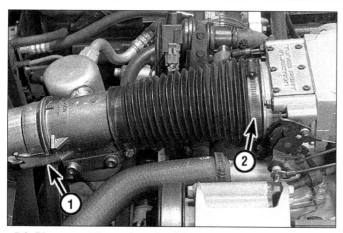

7.3 Disconnect the spring clips (1) attaching the MAF sensor to the air intake duct and loosen the hose clamp (2) attaching the flexible ducting to the throttle body

7.4 Two retainer clips must be removed - one locks the cruise control cable (the one being disconnected in the photo) and the other locks the throttle cable to the throttle shaft lever arm

screws and lockwashers which came in the service kit **(see illustration)**. **Caution:** *Do not reuse the old screws. Be sure to use the thread-locking agent supplied in the service kit on the screw threads.*

110 Make sure that the TPS is lined up in accordance with the alignment mark you made prior to removing it and that the throttle position sensor pickup lever is located above the tang on the throttle actuator lever. **Note:** *The following TPS adjustment procedure cannot be performed until the vehicle is running, and it requires the use of a digital volt-meter capable of reading 0.525 ± 0.075 volts. If you don't have this tool, the following procedure is impossible to perform.*

111 Disconnect the TPS harness connector from the TPS switch. Use three jumper wires to reconnect the TPS connector to the TPS switch.

112 Hook up the digital voltmeter to terminals A and B and, with the ignition On but the engine not running, rotate the TPS switch to obtain a reading of 0.525 ± 0.075 volts at the closed throttle position. If you cannot obtain this reading, replace the TPS.

113 Tighten the screws, then recheck your reading to insure that the adjustment has not changed.

114 Turn the ignition Off, remove the jumper wires and reconnect the TPS harness connector to the TPS switch.

115 If the manifold cover or throttle bodies were replaced, adjust the minimum idle speed and synchronize the throttle valves.

116 Install the air cleaner. With the engine off, depress the accelerator pedal to the floor and release, checking for free return of the pedal.

117 Start the engine and check for fuel leaks.

118 Allow the engine to reach normal operating temperature.

119 On manual transmission equipped vehicles the idle speed will automatically be controlled when normal operating temperature is reached.

120 On vehicles with an automatic transmission, the idle air control unit will begin con-

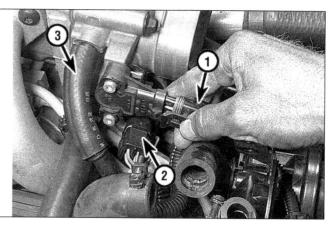

7.5 Disconnect the electrical connectors from the TPS (1) and the IAC valve (2), then detach the PCV hose (3) from the throttle body

trolling idle speed when the engine is at normal operating temperature and the transmission is shifted into Drive.

121 If the idle speed is too high and does not regulate back to normal after a few moments, operate the vehicle at a speed of 45 mph. At that speed the electronic control module will command the idle air control pintle to extend fully to the mating seat in the throttle body, which will allow the electronic control module to establish an accurate reference point with respect to pintle position. Proper idle regulation will result.

---

7   **Multi Port Fuel Injection system - component replacement**

Warning: *Gasoline is extremely flammable, so take extra precautions when you work on any part of the fuel system. Don't smoke or allow open flames or bare light bulbs near the work area, and don't work in a garage where a natural gas-type appliance (such as a water heater or a clothes dryer) with a pilot light is present. Since gasoline is carcinogenic, wear latex gloves when there's a possibility of being exposed to fuel, and, if you spill any fuel on your skin, rinse it off immediately with soap and water. Mop up any spills immediately and do not store fuel-soaked rags where*

they could ignite. The fuel system is under constant pressure, so, if any fuel lines are to be disconnected, the fuel pressure in the system must be relieved first (see Section 2). When you perform any kind of work on the fuel system, wear safety glasses and have a Class B type fire extinguisher on hand.

Throttle body

Refer to illustrations 7.3, 7.4, 7.5, 7.6 and 7.7
Warning: *Wait until the engine is completely cool before beginning this procedure.*

1   Disconnect the cable from the negative terminal of the battery. **Caution:** *On models equipped with a Delco Loc II audio system, disable the anti-theft feature before disconnecting the battery.*

2   Drain the cooling system.

3   Detach the air intake duct **(see illustration)**.

4   Disconnect the accelerator cable from the throttle lever, then detach the cable housing from its bracket **(see illustration)**.

5   Unplug the Idle Air Control (IAC) valve and the Throttle Position Sensor (TPS) electrical connectors **(see illustration)**.

6   Loosen the clamps and disconnect the coolant hoses from the underside of the throttle body **(see illustration)**. Be prepared for some coolant spillage and plug the ends of the hoses.

7.6 Loosen the hose clamps and detach the coolant hoses (arrows) from the throttle body coolant cover

7.7 The vacuum hose (arrow) between the throttle body and the vacuum damper valve (the small black plastic cylinder) must be disconnected

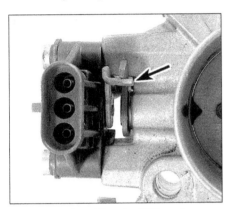

7.15 When installing the TPS, make sure that the TPS pickup lever (arrow) lines up with the tang on the throttle actuator lever

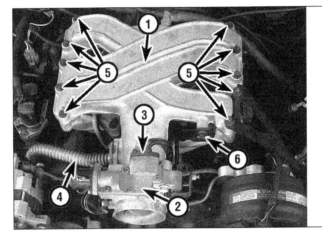

7.28 Intake manifold plenum installation details - V6 models

1 Intake manifold plenum
2 Throttle body assembly
3 Vacuum harness
4 EGR tube
5 Plenum mounting bolts
6 Throttle cable bracket

7   Mark and disconnect any vacuum hoses connected to the throttle body. Also detach the breather hose, if equipped (see illustration).
8   Remove the throttle body bolts and detach the throttle body.
9   Clean off all traces of old gasket material from the throttle body and the plenum.
10   Install the throttle body and a new gasket and tighten the bolts to the torque listed in this Chapter's Specifications.
11   The remainder of the procedure is the reverse of removal.

Idle Air Control (IAC) valve

12   Replacement of the IAC valve is identical to the procedure described in Section 6 with the exception of the 3.1L V6 models. On 3.1L V6 models, the valve is retained by two screws - remove the screws and remove the valve.
13   Refer to Section 6 for adjustment of the valve pintle prior to installation. Adjustment of the 3.1L V6 valve is identical to the Type1 IAC valve.

Throttle Position Sensor (TPS)

Refer to illustration 7.15
14   Replacement of the TPS is identical to the procedure described in Section 6 with the

exception of; on all except 3.1L V6 models, adjust the TPS as follows.

TPS adjustment

15   With the throttle valve in the closed idle position, install the TPS on the throttle body assembly, making sure that the TPS lever lines up with the tang on the throttle actuator lever (see illustration).
16   Install the retainers and loosely install the TPS mounting screws.
17   Install three jumper wires between the TPS and the harness connector.
18   With the ignition turned to On, and a digital voltmeter connected to the terminals marked A and B on the connector (the top and middle terminals), adjust the TPS to obtain a reading of 0.54 to 0.56 volt.
19   Tighten the TPS mounting screws and recheck your reading to insure that the adjustment has not changed.
20   Turn the ignition Off, remove the jumper wires and connect the wiring harness electrical connector to the TPS.
21   Connect the wiring harness electrical connector to the terminal on the underside of the MAF sensor and install the MAF sensor/flexible ducting assembly between the air intake assembly and the throttle body. Snap the spring clips into place to lock the MAF sensor to the air intake and tighten the hose clamp that attaches the flexible ducting

to the throttle body.
22   Start the engine and then turn it off. The IAC valve will automatically reset itself.

Fuel rail and injectors

Warning: *Before any work is performed on the fuel lines, fuel rail or injectors, the fuel system pressure must be relieved (see Section 2).*
23   Detach the cable from the negative terminal of the battery. **Caution:** *On models equipped with a Delco Loc II audio system, disable the anti-theft feature before disconnecting the battery.*

V6 models

Refer to illustrations 7.28, 7.30, 7.35 and 7.40
Note: *An eight digit identification numbers stamped on the side of the fuel rail assembly. Refer to this number if servicing or part replacement is required.*
24   Detach the air intake duct from the throttle body. Remove the vacuum lines and throttle cable bracket bolts.
25   Remove the EGR pipe-to-EGR valve base bolts.
26   Remove the throttle body.
27   Remove the air conditioning compressor-to-plenum bracket, if equipped.
28   Remove the bolts retaining the plenum to the intake manifold and remove the

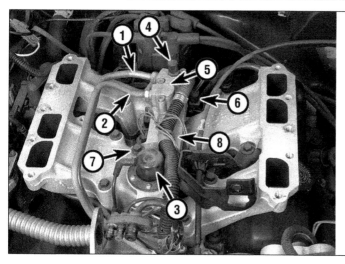

7.30 Fuel rail installation details - V6 models

1   Fuel inlet line
2   Fuel outlet lie
3   Fuel pressure regulator
4   Fuel pressure test connector
5   Fuel block
6   Fuel injector
7   Fuel rail mounting bolt
8   Fuel rail

V8 models

Refer to illustrations 7.50, 7.51, 7.54, 7.61, 7.63, 7.70, 7.72a, 7.72b, 7.54 and 7.76

Note: *Whether you are replacing a malfunctioning fuel injector, a leaking injector O-ring seal or the fuel pressure regulator assembly, it is essential that all 0-ring seals at both ends of all injectors be inspected. In view of the relative inaccessibility of these items, it is a good idea to replace all of them, since you have to remove the injectors to inspect the 0 rings anyway.*

50   Label and disconnect the two vacuum lines from the right rear corner of the plenum **(see illustration)**.

51   Unscrew the power brake booster vacuum line threaded fitting at the left rear corner of the plenum **(see illustration)**.

52   Unplug the wiring harness electrical connector from the Manifold Air Temperature (MAT) sensor protruding from the underside of the plenum, just forward of the pressure regulator.

53   Disconnect the PCV fresh air intake assembly from the right rocker arm cover, disconnect it from the PCV assembly and remove it. Remove the two flanged screws from the right fuel injector wire harness shroud and remove the shroud.

54   Remove the four Torx bolts which attach the upper end of the right intake runner

---

plenum and gaskets **(see illustration)**.

29   On 2.8L models, disconnect the cold start valve line fitting at the fuel rail.

30   Remove the fuel lines at the fuel rail. Remove the vacuum line at the fuel pressure regulator **(see illustration)**.

31   Label each injector connector with its corresponding cylinder number. Unplug the injector electrical connectors.

32   Remove the fuel rail retaining bolts and carefully remove the fuel rail and injectors. **Caution**: *Use care when handling the fuel rail assembly to avoid damaging the injectors.*

33   To remove the cold start valve on 2.8L models, disconnect the electrical connector and remove the cold start valve retaining bolt.

34   Disconnect the valve from the tube and body assembly by bending the tab back to permit unscrewing of the valve.

35   Install a new cold start valve 0-ring seal and body 0-ring seal on the cold start valve **(see illustration)**.

36   Install a new 0-ring seal on the tube and body assembly.

37   Turn the valve completely into the body assembly. Turn the valve back one full turn, until the electrical connector is at the top position.

38   Bend the tang of the body forward to limit rotation of the valve to less than one full turn.

39   Before reinstalling the valve assembly, coat the 0-ring seals with engine oil.

40   To remove the fuel injectors, rotate the injector retaining clip(s) to the released position **(see illustration)**. **Caution**: *To prevent dirt from entering the engine, the area around the injectors should be cleaned before servicing.* Remove the injector(s).

41   Inspect the injector 0-ring seal(s). Replace them if they are damaged. Install the new 0-ring seal(s), as required, on the injector(s) and lubricate them with engine oil.

42   Install the injectors on the fuel rail. Rotate the injector retainer clips to the locked position.

43   The pressure regulator is factory adjusted and is not serviceable. Do not attempt to remove the regulator from the fuel rail.

44   Before installing the fuel rail, lubricate all injector 0-ring seals with engine oil.

45   Install the fuel rail and injector assembly. Install the fuel rail retaining bolts and tighten them to the specified torque.

46   Plug in the injector electrical connectors. Install the vacuum line at the regulator.

47   Install the fuel lines at the fuel rail. Install the cold start valve line at the fuel rail.

48   Install the plenum.

49   Attach the negative battery cable to the battery.

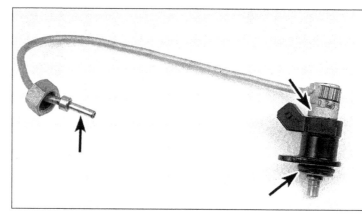

7.35 Cold start valve assembly - 2.8L V6 models (arrows indicate where O-rings should be replaced)

7.40 Rotate the fuel injector retaining clip (arrow) to release the injector from the fuel rail

7.50 Label and disconnected the vacuum lines at the right rear corner of the plenum

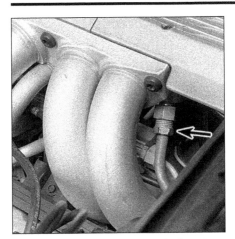

7.51 Unscrew the power brake booster vacuum line threaded fitting (arrow) and pull it back slightly from the plenum

7.54 There are four Torx bolts attaching the upper end of each intake runner assembly to the plenum (only three are visible in this photo of the right runner)

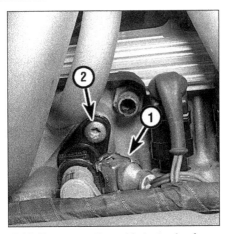

7.61 To remove the cold start valve from the intake manifold, disconnect the fuel injector wiring harness electrical connector (1) from the cold start valve electrical terminal and remove the cold start valve flange bolt (2)

assembly to the plenum and the five Torx bolts which attach the lower end of the runners to the intake manifold. One of the bolts holding the runners to the manifold can only be removed by reaching under the plenum from the left side of the engine **(see illustration)**.

55   Remove the four Torx bolts which attach the upper end of the left intake runner assembly to the plenum.

56   Remove the plenum.

57   Remove the right rocker arm cover upper nuts to free the PCV line brackets. Remove the PCV line to provide clearance for the intake runner flanges. Remove the right intake runner assembly.

58   Label and disconnect the vacuum hoses at the EGR valve and fuel pressure regulator.

59   The left runner has one Torx bolt that is accessible only from the opposite side of the engine. It's located just below the pressure regulator. Remove it.

60   Disconnect the PCV valve assembly from the left rocker arm cover and the PCV line assembly and remove it. Remove the two flanged screws from the left fuel injector wire harness shroud and remove the shroud.

61   Disconnect the wire harness electrical connector from the cold start valve **(see illustration)**. Disconnect the connectors from the left bank of fuel injectors.

62   Remove the left rocker arm cover upper nuts to free the PCV line brackets. Remove the PCV line assembly to provide clearance for removal of the left intake runner assembly.

63   Using a backup wrench on the fuel rail fitting, disconnect the cold start valve fuel line fitting from the rear of the left side fuel rail **(see illustration)**. Remove the cold start valve mounting bolt. Clean the area around the cold start valve connection, then remove the cold start valve assembly.

64   To replace the cold start valve or check the 0-ring between the cold start valve and the elbow fitting of the tube and body assembly, bend back the tab on the elbow fitting of the tube and body assembly and unscrew the valve from the tube. Before disassembling the valve from the tube and body assembly, note the relationship of the electrical connector to the tube and body assembly. This is important because the connector must be in the same position once the new valve is

adjusted in order for the cold start valve assembly to fit properly.

65   Inspect the 0-ring between the cold start valve and the elbow fitting of the tube and body assembly. Replace it if it's damaged.

66   To adjust the new cold start valve, screw it into the elbow fitting of the tube and body assembly until it bottoms. Be sure to lubricate the O-ring with engine oil. Turn the valve back one full turn, until the electrical connector is at the top position you noted before separating the valve from the tube and body assembly.

67   Bend the tab forward to limit rotation of the valve to less than a full turn.

68   Remove the four remaining Torx bolts which attach the lower end of the left intake runner assembly to the intake manifold.

69   Remove the left intake runner assembly.

70   Disconnect the fuel line fittings at the bracket on the right front of the intake manifold, then remove both bracket bolts from the manifold **(see illustration)**.

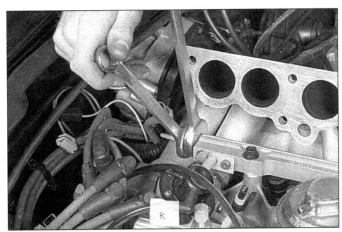

7.63 Using a backup wrench on the fuel rail nut, unscrew the cold start valve line threaded fitting from the left fuel rail

7.70 To disconnect the fuel feed and return lines from the intake manifold, unscrew both threaded fittings (1 and 2), then remove both bracket bolts (arrows)

7.72a To remove the fuel rail assembly from the intake manifold, remove the four fuel rail mounting bolts (arrows)

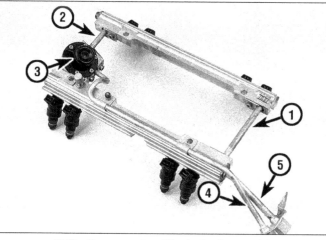

7.72b Components of the fuel rail assembly

1	Front crossover tube	4	Fuel outlet pipe
2	Rear crossover tube	5	Fuel inlet pipe
3	Fuel pressure regulator		

71   Disconnect the wiring harness electrical connectors from the right bank of fuel injectors.

72   Remove the four mounting bolts from the fuel rail and remove the fuel rail assembly **(see illustrations)**.

73   Before disassembly or removal of any component(s), the fuel rail assembly should be thoroughly cleaned with a spray type cleaner. **Caution:** *Do not immerse the fuel rail assembly or its components in solvent.*

74   Each port injector is located and held in position by a retainer clip that must be rotated to release and/or lock the injector **(see illustration 7.40)**. Rotate the injector retaining clips to the unlocked position **(see illustration)**.

75   Pull the injectors from the fuel rail with a slight twisting motion.

76   Inspect all injector O-ring seals for wear. Remove and discard damaged O-rings by prying them off either end of the injector with a small screwdriver **(see illustration)**.

77   To install the fuel injectors, lubricate all injector 0-ring seals with engine oil, push the injectors back into the fuel rail and rotate each injector retainer clip to the locked position.

78   The remainder of installation is the reverse of removal.

Fuel pressure regulator (V8 models)

Note 1: *On V6 models, the fuel pressure regulator is part of the fuel rail and not serviced separately.*

Note 2: *The fuel pressure regulator is factory adjusted. Do not attempt to remove the regulator cover. If the regulator is malfunctioning, it must be replaced.*

79   Remove either front crossover tube retainer screw and crossover tube retainer. Remove the rear retainer-to-regulator base screw and crossover tube retainer.

80   Separate the left and right fuel rail

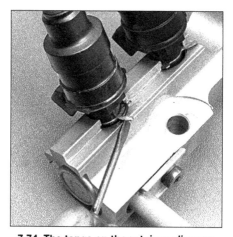

7.74 The tangs on the retainer clips are difficult to push with your finger - use a small screwdriver instead

assemblies.

81   Remove the regulator bracket-to-right fuel rail screw (just below the fuel pressure connection), the bracket-to-pressure regulator base screws and the pressure regulator mounting bracket.

82   Remove the fuel outlet tube rear retaining bracket stud.

83   Remove the pressure regulator base-to-right fuel rail mounting screw (on the side of the regulator base, just in front of the rear crossover tube).

84   Remove the pressure regulator and base assembly from the right fuel rail assembly.

85   Carefully rotate and pull on the regulator to remove it from the fuel outlet tube.

86   Remove the regulator base-to-fuel rail connector.

87   Remove the remaining front and rear crossover tube retainers and separate the crossover tubes from the fuel rails.

88   Inspect all 0-rings on both ends of the connector, the fuel outlet tube and the

7.76 The O-rings on the ends of the injectors should be inspected and, if damaged, should be replaced

crossover tubes for any damage that could prevent proper sealing. Replace any damaged 0-ring seals. **Caution:** *The fuel rail assembly O-rings are not all the same size. When removing O-rings, be sure to note their location and size to ensure proper replacement.*

89   Lubricate all 0-rings with engine oil and install them on the connector, the fuel outlet tube and the front and rear crossover tubes.

90   Insert the front and rear crossover tubes into their respective holes in the left fuel rail assembly. Install the tube retainers and screws and tighten them securely.

91   Install the base-to-right fuel rail connector in the pressure regulator base.

92   Push the regulator and base assembly onto the fuel outlet tube.

93   Rotate the regulator and base assembly to position the base properly against the right fuel rail assembly.

94   Install and tighten securely the pressure regulator base-to-right fuel rail mounting screw.

8.1 Remove the clip to disengage the throttle cable from the throttle body lever

8.5 To detach the throttle cable from the accelerator pedal, pull the small plastic retainer loose from the upper end of the pedal lever, the lift the cable up and out of the groove in the lever

95   Install the pressure regulator and base assembly mounting bracket, the bracket-to-base mounting screws and the bracket-to-fuel rail mounting screw.
96   Install and tighten securely the fuel outlet tube rear retaining bracket stud.
97   Reconnect the left fuel rail assembly and previously installed crossover tubes to the right fuel rail assembly by gently pushing the crossover tube ends into the right fuel rail.
98   Install the rear crossover tube-to-pressure regulator base retainer and mounting screw. Tighten the screw securely.
99   Install the front crossover tube-to-right fuel rail assembly retainer and mounting screw. Tighten the screw securely.

8   Throttle cable - removal and installation

Refer to illustrations 8.1 and 8.5
1   Remove the throttle cable retaining clip from the throttle body throttle shaft lever arm **(see illustration)**.
2   Push the rear throttle body throttle shaft lever forward slightly and disconnect the cable from the throttle linkage.
3   Disconnect the plastic cable retainer from the bracket on the intake manifold.
4   Peel away the body sealing compound from the retainer at the firewall plate. Pinch the tabs on the top and bottom of the retainer and disconnect the retainer from the plate.
5   Disconnect the throttle cable from the accelerator pedal **(see illustration)**.
6   Remove the cable.
7   Installation is the reverse of the removal procedure.
8   There are no linkage adjustments because the replacement cable is a specific length. Therefore, it is imperative that the correct cable be installed. Only the specific replacement cable will work properly on this vehicle.
9   After assembly, the accelerator linkage must operate freely without binding between

the closed and wide open positions. Check for correct opening and closing by operating the accelerator pedal and, if any binding is present, check the routing of the cable and make sure there are no kinks in the cable.

9   Minimum idle speed - adjustment

Crossfire (TBI) system

Note: *The following procedure cannot be performed without two special tools. One is a balance manometer and the other is an idle air valve plug set.*
1   The throttle position of each throttle body must be balanced so that the throttle plates are synchronized and open simultaneously.
2   Adjustment should be performed only when a manifold cover or throttle body has been replaced, or if there is any indication of tampering with the throttle valve synchronizing screw on the front unit.
3   Remove the air cleaner and air cleaner-to-throttle body gaskets. Plug the Thermac vacuum port on the rear throttle body.
4   If in place, remove the tamper resistant plugs covering both throttle stop screws.
5   Block the drive wheels and apply the parking brake.
6   Connect a tachometer.
7   Disconnect the Idle Air Control (IAC) valve electrical connector.
8   Plug the idle air passages of each throttle body with a special plug available at most auto supply stores. Be certain that the plugs are fully seated and that no air leaks exist.
9   Start the engine and allow engine rpm to stabilize at normal operating temperature.
10   Place the transmission in Drive. Engine rpm should decrease below curb idle speed. If engine rpm does not decrease, check for a vacuum leak.
11   Remove the cap from the ported tube on the rear throttle body and connect the water manometer.

12   Adjust the rear unit throttle stop screw to obtain approximately six inches of water on the manometer. If unable to adjust to this level, be sure that the front unit stop screw is not limiting throttle travel.
13   Remove the manometer from the rear throttle body and install the cap on the ported tube.
14   Remove the cap from the ported tube on the front throttle body and connect the manometer. The reading should also be approximately six inches of water. If the adjustment is required, proceed as follows:

a) *Locate the throttle synchronizing screw and collar on the front throttle body. The screw retaining collar is welded to the throttle lever to discourage tampering with this adjustment.*
b) *If the collar is in place, grind off the weld from the screw collar and throttle lever.*
c) *Block movement of the throttle lever by relieving the force of the heavy spring against the throttle synchronizing screw to prevent the levers from coming into contact.* **Caution:** *If the lever is not blocked before the throttle synchronizing screw is removed, the screw may be damaged and reinstallation will be done only with great difficulty.*
d) *Remove the screw and collar. Discard the collar.*
e) *Reinstall the throttle synchronizing screw, using thread locking compound (supplied in the service repair package) in accordance with the directions. If the compound is not available, use Loctite 262 or an equivalent.*
f) *Immediately adjust the screw to obtain approximately six inches of water on the manometer.*

15   Remove the manometer from the front throttle body and install the cap on the ported tube.
16   Adjust the rear throttle stop screw on the rear unit to obtain 475 rpm.
17   Turn the ignition to Off and place the transmission in Neutral.

18   Adjust the front throttle stop screw to obtain 0.005-inch clearance between the front throttle stop screw and the throttle lever tang.
19   Remove the idle air passage plugs and reconnect the IAC valves.
20   Start the engine. It may run at high rpm at first, but the rpm will decrease when the IAC valves close the air passages. Stop the engine when the idle rpm has decreased.
21   Check the TPS voltage and adjust if necessary.
22   Install the air cleaner gasket, connect the vacuum line to the throttle body and install the air cleaner.

TPI system

Refer to illustration 9.30
Note: *Before attempting to adjust the minimum air rate the throttle body and throttle plate must be clean and free of debris.*
23   With the engine off, remove the MAF sensor (if equipped) and the air inlet duct. Hold the throttle wide open and spray a generous quantity of aerosol carburetor cleaner on the back of the throttle plate and around the inside of the throttle body. Allow this to soak for several minutes.
24   Soak a corner of a soft rag with carburetor cleaner and scrub the throttle plate and the inside of the throttle body.
25   Using aerosol carburetor cleaner, spray the remaining carbon deposits from the throt-

tle plate and throttle body.
26   Wipe any remaining cleaner from the throttle body.
27   Reinstall the MAF sensor and air inlet duct. Start the engine and allow it to idle.
28   Road test the vehicle and re-evaluate the need to adjust the minimum air rate. Often, cleaning the throttle body and throttle plate will restore minimum air flow and further adjustments are not necessary. If the minimum air must be adjusted proceed as follows:
29   The idle stop screw, which regulates the minimum idle speed of the engine, is adjusted at the factory and covered with a plug to discourage unnecessary adjustment. However, if it is necessary to gain access to the idle stop screw assembly, proceed as follows.
30   Pierce the idle stop screw plug with an awl and apply leverage to remove it **(see illustration)**.
31   With the IAC valve connected, ground the diagnostic terminal of the ALDL connector under the left side of the dash.
32   Turn the ignition to On but do not start the engine. Wait at least 30 seconds.
33   With the ignition On, disconnect the IAC electrical connector.
34   Remove the ground from the diagnostic lead and start the engine. Run the engine until it reaches normal operating temperature.
35   Adjust the idle screw to obtain either

9.30 To remove the idle stop screw plug from the throttle body of the TPI system, pierce it with an awl and apply leverage to remove it

400 rpm in Drive with an automatic transmission or 450 rpm in Neutral with a manual transmission.
36   Turn the ignition to Off and reconnect the IAC valve connector.
37   Following the procedure outlined in Section 7, adjust the TPS. Secure the TPS and recheck the voltage.
38   Start the engine and check for proper idle operation.

Chapter 5
Engine electrical systems

Contents

	Section		Section
Alternator - removal and installation	12	Ignition system - check	5
Alternator brushes and voltage regulator - replacement	13	Ignition system - general information and precautions	1
Battery - emergency jump starting	3	Ignition timing check and adjustment	See Chapter 1
Battery - removal and installation	2	Spark plug replacement	See Chapter 1
Battery cables - check and replacement	4	Spark plug wires, distributor cap and rotor - check	
Charging system - check	11	and replacement	See Chapter 1
Charging system - general information and precautions	10	Starter motor - removal and installation	16
CHECK ENGINE light	See Chapter 6	Starter motor - testing in vehicle	15
Distributor - removal and installation	6	Starter motor brushes - replacement.	18
Ignition coil - check and replacement	9	Starter solenoid - replacement	17
Ignition module - replacement	8	Starting system - general information	14
Ignition pickup coil - check and replacement	7		

1 Ignition system - general information and precautions

The ignition system is composed of the battery, distributor, ignition switch, spark plugs and the primary (low tension) and secondary (high tension) wiring circuits.

A high energy ignition (HEI) distributor is used on all vehicles. Some vehicles use a coil mounted integral with the distributor, while others employ a separately mounted coil.

The HEI works in conjunction with the electronic spark timing (EST) system and uses a magnetic pickup assembly (located inside the distributor) which contains a permanent magnet, a pole piece with internal teeth and a pickup coil in place of the traditional ignition point assembly.

All ignition timing changes in the HEI/EST distributor are accomplished by an electronic control module (ECM), which monitors data from various engine sensors, computes the desired spark timing and signals the distributor to change the timing accordingly. The electronic spark control (ESC) used on some engines retards the spark advance when detonation occurs. This retard mode is held for approximately 20 seconds, after which the spark control will revert to the EST.

There are three basic components involved in the ESC system: the sensor, the controller and the distributor. The ESC sensor detects the presence (or absence) and intensity of detonation through vibrations in the engine and sends the data to the controller. The ESC controller changes the sensor signal into a command signal to the distributor, which then adjusts the spark timing accordingly.

The distributor in the ECS system is modified to respond to the ESC controller signal. This command is delayed when detonation occurs, thus providing the level of retard required. The degree of retard is based on the detonation.

The secondary (spark plug) wire used with the HEI system is a carbon-impregnated cord conductor encased in an 8 mm (5/16inch) diameter rubber jacket. This type of wire will withstand very high temperatures and still provide insulation for the HEI system high voltage. For more information on spark plug wiring, refer to Chapter 1.

Resistor type, tapered seat spark plugs without gaskets are used in all engines. Refer to the underhood Vehicle Emissions Control Information label for the proper plug for your particular engine. **Caution:** *Because of the very high voltage generated by the HEI sys-tem, extreme care should be taken whenever an operation involving ignition components is performed. This not only includes the distributor, coil, control module and spark plug wires, but related items that are connected to the system as well (such as the plug connections, tachometer and testing equipment). Consequently, before any work is performed, the ignition should be turned off or the battery ground cable disconnected. On models equipped with a Delco Loc II audio system, disable the anti-theft feature before disconnecting the battery.*

2 Battery - removal and installation

1 The battery is located at the front of the engine compartment. It is held in place by a holddown clamp near the bottom of the battery case.

2 Hydrogen gas is produced by the battery, so keep open flames and lighted cigarettes away from it at all times.

3 Always keep the battery in an upright position. Spilled electrolyte should be rinsed off immediately with large quantities of water. Always wear eye protection when working around a battery.

4 Always disconnect the negative (-) bat-

tery cable first, followed by the positive (+) cable. **Caution:** *On models equipped with a Delco Loc II audio system, disable the anti-theft feature before disconnecting the battery.*

5   After the cables are disconnected from the battery, remove the holddown clamp.

6   Carefully lift the battery out of the engine compartment.

7   Installation is the reverse of removal. The cable clamps should be tight, but do not overtighten them as damage to the battery case could occur. The battery posts and cable ends should be cleaned prior to connection (see Chapter 1).

3   Battery - emergency jump starting

Refer to the Booster battery (jump) starting procedure at the front of this manual.

4   Battery cables - check and replacement

1   Periodically inspect the entire length of each battery cable for damage, cracked or burned insulation and corrosion. Poor battery cable connections can cause starting problems and decreased engine performance.

2   Check the cable-to-terminal connections at the ends of the cables for cracks, loose wire strands and corrosion. The presence of white, fluffy deposits under the insulation at the cable terminal connection is a sign the cable is corroded and should be replaced. Check the terminals for distortion, missing mounting bolts or nuts and corrosion.

3   If only the positive cable is to be replaced, be sure to disconnect the negative cable from the battery first. **Caution:** *On models equipped with a Delco Loc II audio system, disable the anti-theft feature before disconnecting the battery.*

4   Disconnect and remove the cable(s) from the vehicle. Make sure the replacement cable(s) is the same length and diameter.

5   Clean the threads of the starter or ground connection with a wire brush to remove rust and corrosion. Apply a light coat of petroleum jelly to the threads to ease installation and prevent future corrosion. Inspect the connections frequently to make sure they are clean and tight.

6   Attach the cable(s) to the starter or ground connection and tighten the mounting nut(s) securely.

7   Before connecting the new cable(s) to the battery, make sure they reach the terminals without having to be stretched.

8   Connect the positive cable first, followed by the negative cable. Tighten the nuts and apply a thin coat of petroleum jelly to the terminal and cable connection.

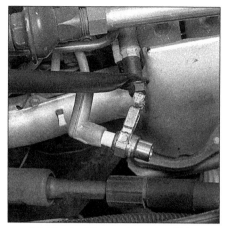

5.1 The ignition system should be checked with a spark tester - if the ignition system produces blue sparks jumping the tester gap, the system is functioning properly

5   Ignition system - check

Refer to illustration 5.1

Caution: *Because of the very high voltage generated by the HEI system, extreme care should be taken whenever an operation is performed involving ignition components. This not only includes the distributor, coil, control module and spark plug wires, but related items that are connected to the system as well, such as the plug connections, tachometer and any test equipment. Consequently, before any work is performed, the ignition should be turned off or the battery ground cable disconnected. On models equipped with a Delco Loc II audio system, disable the anti-theft feature before disconnecting the battery.*

1   If the engine turns over but will not start, remove the spark plug wire from a spark plug, install a calibrated ignition system tester to the wire and clip the tester to a good engine ground point **(see illustration)**. Have an assistant crank the engine and watch the end of the tester for well-defined, bright blue sparks.

2   If there is no spark, check another wire in the same manner. A few sparks, then no spark, should be considered as no spark.

3   If there is good spark, check the spark plugs (refer to Chapter 1) and/or the fuel system (refer to Chapter 4).

4   If there is a weak spark or no spark in a coil-in-cap system:

  a) *Check for battery power at the BAT terminal of the distributor connector with the ignition key ON.*

  b) *Remove the distributor cap and check the carbon button (connection between the distributor rotor and ignition coil) for damage.*

  c) *Check the resistance of the ignition coil.*

  d) *Check the pick-up coil for a short or open circuit.*

  e) *Have the ignition module checked by a dealer or repair shop.*

6.6 Removing the distributor holddown clamp bolt (V6 engine shown out of vehicle)

5   If there is a weak spark or no spark in a system with a remote coil:

  a) *Check for battery power at the BAT terminal of the ignition coil connector with the ignition key ON.*

  b) *Disconnect the coil lead from the distributor, connect the ignition tester to the coil wire and check for spark as described above.*

  c) *If there is a spark, check the distributor cap and/or rotor (refer to Chapter 1).*

  d) *If there is no spark, check the ignition coil resistance and the pick-up coil for a short or open circuit. Have the ignition module checked by a dealer or repair shop.*

6   Further checks of the HEI ignition system must be done by a dealer or repair shop.

6   Distributor - removal and installation

Removal

Refer to illustration 6.6

1   On HEI coil-in-cap models, disconnect the ignition switch/battery feed wire (BAT) and the tachometer lead (TACH) from the distributor cap.

2   Release the coil connectors from the cap. Depress the locking tabs by hand; do not use a screwdriver or other tool.

3   Turn the distributor cap locking latches counterclockwise, remove the cap and position it out of the way.

4   Disconnect the four-terminal ECM wiring harness connector from the distributor.

5   On HEI coil-in-cap models, release the wiring harness latches and remove the wiring harness retainer if necessary for clearance. Note that the spark plug wire numbers are indicated on the retainer.

6   Remove the distributor holddown clamp bolt and the holddown clamp **(see illustration)**.

7   Mark the position of the rotor on the outside of the distributor body, then slowly lift

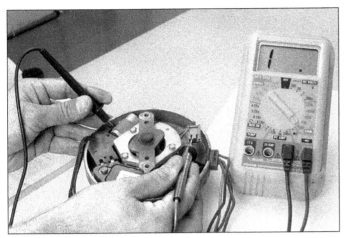

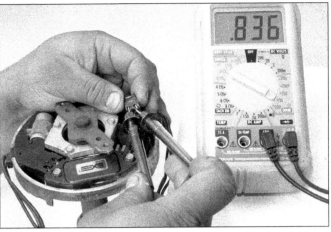

7.3 Check for continuity between the pick-up coil terminals and the distributor body - if continuity is indicated, there is a short-to-ground in the pick-up coil

7.4 Measure the resistance of the pick-up coil - it should be between 500 and 1,500 ohms

the distributor from the engine and note again the position of the rotor when it stops turning. Mark this position also. To ensure correct timing of the distributor upon reinstallation, the rotor should be in the second mark position prior to reinserting the distributor in the engine.

8   Avoid turning the crankshaft with the distributor removed, as the ignition timing will be changed.

Installation if the crankshaft was not turned after distributor removal

9   Position the rotor in the exact location it was in (second mark, Step 7) when the distributor was removed.

10   Lower the distributor into the engine. To mesh the gears at the bottom of the distributor, it may be necessary to turn the rotor slightly.

11   With the base of the distributor all the way down against the engine block, the rotor should point to the first alignment mark (Step 7). If not, refer to Step 17.

12   Place the holddown clamp in position and loosely install the holddown bolt.

13   Connect the ignition feed wire and tachometer wire, if so equipped, to the distributor.

14   Install the distributor cap. If the secondary wiring harness was removed from the cap, install it.

15   Reconnect the coil connector.

16   With the distributor in its original position, tighten the holddown bolt and check the ignition timing (Chapter 1).

Installation if the crankshaft was turned after distributor removal

17   Remove the number 1 spark plug.

18   Place your finger over the spark plug hole while turning the crankshaft with a wrench on the pulley bolt at the front of the engine.

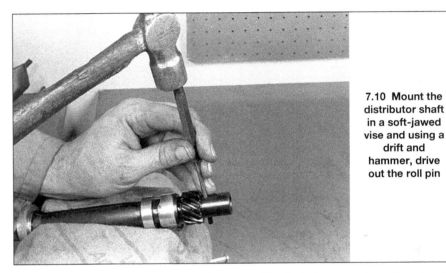

7.10 Mount the distributor shaft in a soft-jawed vise and using a drift and hammer, drive out the roll pin

19   When you feel compression, continue turning the crankshaft slowly until the timing mark on the crankshaft pulley is aligned with the 0 on the engine timing indicator.

20   Position the rotor to point between the number 1 and 8 spark plug terminals in the distributor cap on V8 engines; between the number 1 and 6 terminals on the V6 engine; or between the number 1 and 3 terminals on L4 engines.

21   Complete the installation by referring to Steps 10 through 16.

---

7   Ignition pickup coil - check and replacement

Check

Refer to illustrations 7.3 and 7.4

1   Remove the distributor cap and rotor.

2   Carefully disconnect the pick-up coil wiring connector from the ignition module.

3   Using an ohmmeter, check for continuity between each of the pick-up coil connector terminals and the distributor body (see illustration). If continuity is indicated, the pick-up

coil is shorted to ground - replace the pick-up coil.

4   Measure the resistance across the two terminals of the pick-up coil connector (see illustration). Pick-up coil resistance should be between 500 and 1,500 ohms. Flex the pick-up coil leads and see if the resistance changes. If the resistance is not as specified, replace the pick-up coil.

Replacement

Refer to illustrations 7.10, 7.11, 7.13 and 7.14

5   Remove the distributor from the engine as described previously.

6   Remove the distributor cap and rotor.

7   Disconnect the pickup coil leads from the module.

8   On models with a separately mounted coil and a Hall-effect switch, remove the switch retaining screws and the switch.

9   Mark the distributor gear and shaft so they can be reassembled in the same position.

10   Carefully mount the distributor in a soft-jawed vise and, using a hammer and punch, remove the roll pin from the distributor shaft and gear (see illustration).

7.11 Remove the driven gear and spacers from the end of the shaft; note the assembled order of the parts

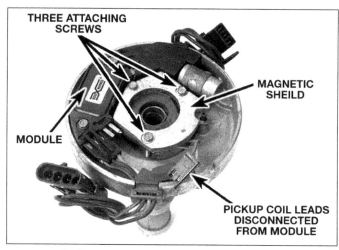

7.13 Ignition pick-up coil installation details

7.14 The pick-up coil and pole piece assembly can be removed after carefully prying out the C-clip

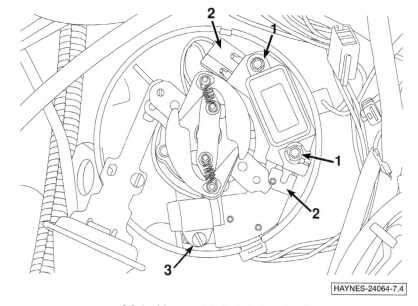

HAYNES-24064-7.4

8.3 Ignition module installation details

1   Module mounting screws
2   Wiring harness connectors
3   Capacitor mounting screws

8.5 Before installing a module, apply a layer silicone grease (included with the module) to the metal base of the module

11   Remove the gear and washers from the shaft **(see illustration)**.
12   Carefully pull the shaft out through the top of the distributor.
13   On distributors with coil-in-cap con-struction, remove the three mounting screws and the magnetic shield **(see illustration)**.
14   Remove the C-washer retaining ring at the center of the distributor and remove the pickup coil **(see illustration)**.
15   Installation is the reverse of the removal procedure.

8   Ignition module - replacement

Refer to illustrations 8.3 and 8.5
Note: *Special equipment is required to prop-erly test an ignition module. Remove the module and have it tested at a properly equipped repair facility. Many auto parts stores will test an ignition module, free of charge.*
1   Remove the distributor.

2   Remove the distributor cap and rotor.
3   On distributors with an internal capacitor, remove the module mounting screws and the capacitor mounting screw, then separate the module, capacitor and harness assembly from the distributor base and disconnect the wiring harness from the module **(see illustration)**.
4   On distributors without an internal capacitor, disconnect the wiring harness from the module, remove the two mounting screws and remove the module.
5   Installation is the reverse of the removal procedure. Before installing the module, apply a layer of silicone lubricant to the metal base the module **(see illustration)**. The grease acts as insulation between the distrib-utor and the module, the module will quickly fail if installed without the grease.

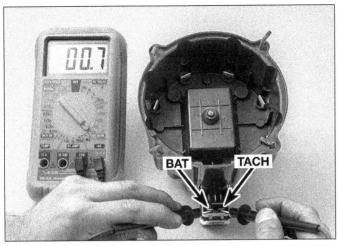

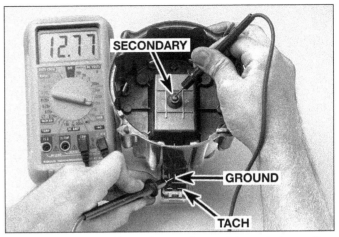

9.2 Measure the ignition coil primary resistance across the BAT and TACH terminals at the distributor cap - the resistance should be very low (zero or near zero)

9.3 Measure the ignition coil secondary resistance between the TACH terminal and the SECONDARY terminal, then the GROUND terminal and the SECONDARY terminal - replace the coil only if <u>both</u> readings are infinite (use the high scale on the meter)

9.6a Remove the screws and detach the coil cover

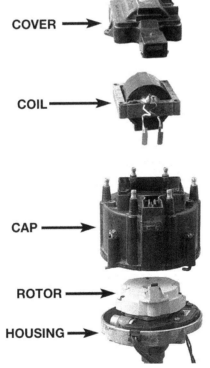

9.6b Ignition coil installation details

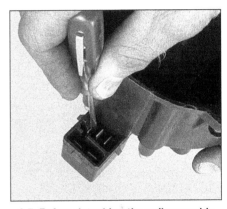

9.7 Before detaching the coil assembly from the cap, push the male spade-type connectors out of the cap

9.8 Replace the carbon button and arc seal

9 Ignition coil - check and replacement

Coil-in-cap models

Refer to illustrations 9.2, 9.3, 9.6a, 9.6b, 9.7 and 9.8

Check

1 Remove the distributor cap (see Steps 4 and 5).

2 Using an ohmmeter set on the low scale, measure the resistance across the TACH and BAT terminals at the distributor cap **(see illustration)**. The resistance should be very low (zero or near zero). If not as specified, replace the coil.

3 Switch the meter to the high scale and measure the resistance between the TACH terminal and the SECONDARY terminal, then the GROUND terminal and the SECONDARY terminal **(see illustration)**. Replace the coil only if <u>both</u> readings are infinite.

Replacement

4 Disconnect the wiring connectors from the distributor cap. Release the spark plug wiring harness retainer clips and remove the wiring retainer with the spark plug wires attached from the distributor cap. Position the retainer and spark plug wire assembly aside.

5 Rotate the four distributor cap locking latches counterclockwise and remove the cap from the distributor.

6 Remove the coil cover mounting screws and lift the cover off **(see illustrations)**.

7 Remove the coil mounting screws and separate the coil and leads from the cap **(see illustration)**.

8 Remove the carbon button and arc seal and replace them **(see illustration)** (new components are usually provided with a replacement coil).

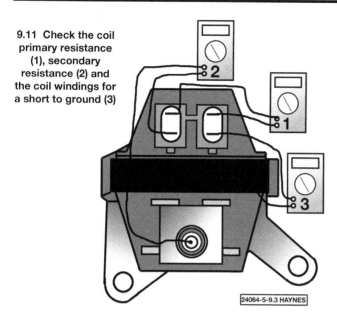

9.11 Check the coil primary resistance (1), secondary resistance (2) and the coil windings for a short to ground (3)

24064-5-9.3 HAYNES

12.3 Arrow (A) indicates the bolt retaining the heater hose clamp to the alternator adjusting bracket; arrow (B) indicates the bolt retaining the negative battery cable (ground) (V6 engine shown)

9   Installation is the reverse of the removal procedure noting the following points:
   a) *Make sure the coil leads are installed in the distributor cap exactly as removed and the ground terminal is properly connected.*
   b) *Be sure the spark plug wires are properly installed and press the retainer down until it latches.*

Models with separately mounted coil

Refer to illustration 9.11

Check

10   Disconnect the wiring connectors and the secondary ignition wire from the coil.
11   Using an ohmmeter, measure the primary and secondary coil resistance and check for an open circuit or short to ground as shown **(see illustration)**. The test results should be as follows:
   a) **Test 1** - *Very low (zero or near zero) - use low scale*
   b) **Test 2** - *Very high (but should not read infinite) - use high scale*
   c) **Test 3** - *Very high (infinite) -use high scale*
12   Replace the coil if it fails any of the tests.

Replacement

13   Disconnect the wiring connectors and the secondary ignition wire from the coil.
14   Loosen the coil bracket mounting nuts and remove the coil.
15   Installation is the reverse of the removal procedure.

10   Charging system - general information and precautions

The charging system is made up of the

alternator, voltage regulator and battery. These components work together to supply electrical power for the engine ignition, lights, radio, etc.

The alternator is turned by a drivebelt at the front of the engine. When the engine is operating, voltage is generated by the internal components of the alternator to be sent to the battery for storage.

The purpose of the voltage regulator is to limit the alternator voltage to a preset value. This prevents power surges, circuit overloads, etc., during peak voltage output. On all models with which this manual is concerned, the voltage regulator is contained within the alternator housing.

The charging system does not ordinarily require periodic maintenance. The drivebelts, electrical wiring and connections should, however, be inspected at the intervals suggested in Chapter 1.

Take extreme care when making circuit connections to a vehicle equipped with an alternator and note the following. When making connections to the alternator from a battery, always match correct polarity. Before using electric-arc welding equipment to repair any part of the vehicle, disconnect the wires from the alternator and the battery terminal. Never start the engine with a battery charger connected. Always disconnect both battery leads before using a battery charger.

11   Charging system - check

1   If a malfunction occurs in the charging circuit, do not immediately assume that the alternator is causing the problem. First check the following items:
   The battery cables where they connect to the battery (make sure the connections are clean and tight)
   The battery electrolyte specific gravity (if it is low, charge the battery)

   Check the external alternator wiring and connections (they must be in good condition)
   Check the drivebelt condition and tension (see Chapter 1)
   Check the alternator mount bolts for tightness
   Run the engine and check the alternator for abnormal noise
2   Using a voltmeter, check the battery voltage with the engine off. It should be approximately 12 volts.
3   Start the engine and check the battery voltage again. It should now be approximately 14 to 15 volts. If it does not rise above the figure recorded in Step 2 when the engine is started or if it exceeds 15 volts and the battery and wiring are in good condition, the alternator is probably defective.
4   Many auto parts stores will bench-test an alternator free of charge. If in doubt about the condition of the alternator, remove the alternator and have it checked further.

12   Alternator - removal and installation

Refer to illustrations 12.3, 12,4a, 12.4b, 12.5 and 12.6

1   Disconnect the cable from the negative battery terminal. **Caution:** *On models equipped with a Delco Loc II audio system, disable the anti-theft feature before disconnecting the battery.*
2   Remove the alternator drivebelt (see Chapter 1).
3   If necessary, remove the bolt retaining the heater hose clamp and the negative battery cable from the alternator adjusting bracket **(see illustration)**.
4   Disconnect the electrical connector and the alternator output wire from the rear of the alternator **(see illustrations)**.

12.4a Disconnect the wiring connector from the alternator

12.4b Remove the nut and disconnect the output wire from the rear of the alternator

12.5 Remove the bolt retaining the alternator to the adjusting bracket

12.6 Remove the alternator pivot bolt (arrow) and remove the alternator from the vehicle

13.3a Separating the alternator front and rear end frames

5   Remove the bolt retaining the alternator to the adjusting bracket **(see illustration)**.
6   Remove the alternator pivot bolt and remove the alternator from the vehicle **(see illustration)**.
7   Installation is the reverse of the removal procedure. Adjust the alternator drivebelt (refer to Chapter 1).

---

13   Alternator brushes and voltage regulator - replacement

Refer to illustration 13.3a, 13.3b. 13.4, 13.5, 13.6 and 13.9
Note: *This procedure applies to 1986 and earlier models equipped with a SI-type alternator. 1987 and later models are equipped with a CS-type alternator. The CS-type alternator is not rebuildable.*

1   Remove the alternator from the vehicle (refer to Section 12).
2   Scribe a match mark on the front and

rear end frame housings of the alternator to facilitate reassembly.
3   Remove the four (4) throughbolts holding the front and rear end frames together,

then separate the end frames **(see illustrations)**.
4   Remove the nuts holding the stator to the rectifier bridge and separate the stator

13.3b Details of the alternator rear end frame

A   *Brush holder*
B   *Paper clip retaining brushes*
C   *Regulator*
D   *Resistor (not all models)*
E   *Diode trio*
F   *Rectifier bridge*

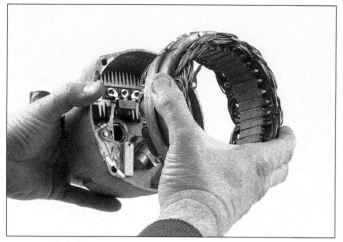

13.4 Separate the stator from the rear end frame

13.5 Remove the diode trio

13.6 Remove the brush holder from the rear end frame

13.9 Insert a paper clip from rear of the alternator to hold the brushes in place during disassembly and reassembly

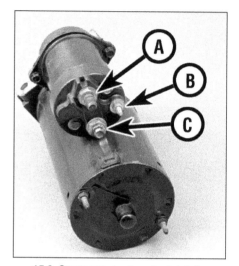

15.6 Starter motor terminal details

A   Battery terminal
B   Switch terminal (S)
C   Motor terminal (M)

from the end frame **(see illustration)**.
5   Remove the screw attaching the diode trio to the end frame and remove the trio **(see illustration)**.
6   Remove the screws retaining the brush holder and resistor (if equipped) to the end frame and remove the brush holder **(see illustration)**.
7   Remove the voltage regulator.
8   Installation is the reverse of the removal procedure, noting the following.
9   When installing the brush holder assembly, press the brushes into the holder and insert a straight section of wire, such as a straightened paper clip, through the rear of the end frame and through the holes in the brush holder to hold the brushes in while reassembly is completed **(see illustration)**. The paper clip should not be removed until the front and rear end frames have been bolted together.

14   Starting system - general information

The function of the starting system is to crank the engine. This system is composed

of a starting motor, solenoid and battery. The battery supplies the electrical energy to the solenoid, which then completes the circuit to the starting motor which does the actual work of cranking the engine.

The solenoid and starting motor are mounted together at the lower right side of the engine. No periodic lubrication or maintenance is required.

The electrical circuitry of the vehicle is arranged so that the starter motor can only be operated when the clutch pedal is depressed (manual transmission) or the transmission selector lever is in Park or Neutral (automatic transmission).

Never operate the starter motor for more than 30 seconds at a time without pausing to allow it to cool for at least two minutes. Excessive cranking can cause overheating, which can seriously damage the starter.

15   Starter motor - testing in vehicle

Refer to illustration 15.6
1   If the starter motor does not turn at all when the switch is operated, make sure that

the shift lever is in Neutral or Park (automatic transmission) or that the clutch pedal is depressed (manual transmission).
2   Make sure that the battery is charged and that all cables, both at the battery and starter solenoid terminals, are secure.
3   If the motor spins but the engine is not being cranked, then the overrunning clutch in the starter motor is slipping and the motor must be removed from the engine and disassembled.
4   If, when the switch is actuated, the starter motor does not operate at all but the solenoid clicks, then the problem is in the main solenoid contacts or the starter motor itself.
5   If the solenoid plunger cannot be heard when the switch is actuated, the solenoid itself if defective or the solenoid circuit is open.
6   To check out the solenoid, connect a jumper lead between the battery (+) and the S terminal on the solenoid **(see illustration)**.

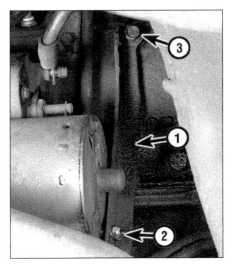

16.3 Typical starter motor support bracket installation details (V8 engine shown)

1   Starter motor support bracket
2   Starter motor-to-bracket nut
3   Support bracket-to-engine block bolt

If the starter motor now operates, the solenoid is OK and the problem is in the ignition or neutral start switches or in the wiring.

7   If the starter motor still does not operate, remove the starter/solenoid assembly for disassembly, testing and repair.

8   If the starter motor cranks the engine at an abnormally slow speed, first make sure that the battery is charged and that all terminal connections are tight. If the engine is partially seized, or has the wrong viscosity oil in it, it will crank slowly also.

9   Run the engine until normal operating temperature is reached, then disconnect the coil wire from the distributor cap and ground it on the engine (remote coil only). On models with the coil in the distributor cap, disconnect the pink wire from the distributor.

10   Connect a voltmeter positive lead to the starter motor terminal of the solenoid and

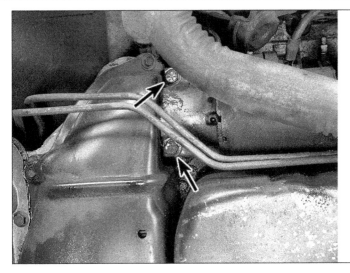

16.4a Starter motor mounting bolts (arrows) - typical V8 engine

then connect the negative lead to ground.

11   Actuate the ignition switch and take the voltmeter readings as soon as a steady figure is indicated. Do not allow the starter motor to turn for more than 30 seconds at a time. A reading of 9 volts or more, with the starter motor turning at normal cranking speed, is normal. If the reading is 9 volts or more but the cranking speed is slow, the motor is faulty. If the reading is less than 9 volts and the cranking speed is slow, the solenoid contacts are probably burned (refer to Section 17).

16   Starter motor - removal and installation

Refer to illustrations 16.3, 16.4a, 16.4b and 16.4c

1   Disconnect the cable from the negative battery terminal. **Caution:** *On models equipped with a Delco Loc II audio system, disable the anti-theft feature before disconnecting the battery.*

2   Raise the vehicle and place it securely on jackstands.

3   Remove the brackets supporting the starter **(see illustration)**.

4   Remove the two starter motor-to-engine bolts and let the starter drop down far enough to allow removal of the nuts attaching the wires to the starter solenoid and battery cable **(see illustrations)**. Support the starter while the wires are removed.

5   Separate the starter from the engine.

6   Installation is the reverse of the removal procedure. Make sure that any removed shims are reinstalled.

17   Starter solenoid - replacement

Refer to illustration 17.3

1   After removing the starter and solenoid as described in Section 16, disconnect the strap from the solenoid MOTOR terminal.

2   Remove the two screws which secure the solenoid housing to the end frame assembly.

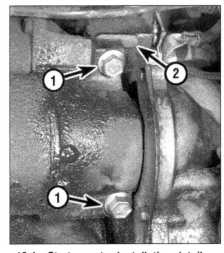

16.4c Starter motor installation details - typical L4 engine

1   Starter motor mounting bolts
2   Starter motor shim (not used on all models)

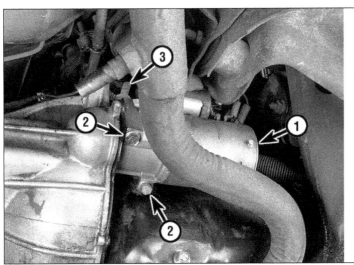

16.4b Starter motor installation details - typical V6 engine

1   Starter motor
2   Mounting bolts
3   Starter motor shim (not used on all models)

17.3 Withdrawing the solenoid assembly from the starter housing

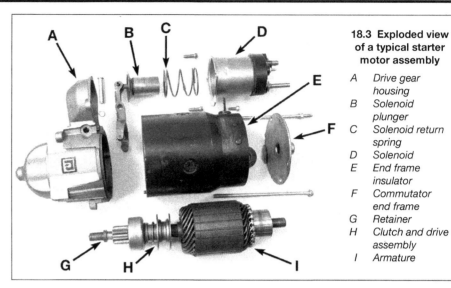

18.3 Exploded view of a typical starter motor assembly

A   Drive gear
    housing
B   Solenoid
    plunger
C   Solenoid return
    spring
D   Solenoid
E   End frame
    insulator
F   Commutator
    end frame
G   Retainer
H   Clutch and drive
    assembly
I   Armature

18.4 Remove the end frame from the field frame housing

18.6 Unbolt the brush and brush support from the brush holder

3   Twist the solenoid in a clockwise direction to disengage the flange key and then withdraw the solenoid **(see illustration)**.
4   Attach the solenoid to the starter motor. First make sure that the return spring is in position on the plunger and then insert the solenoid body into the drive housing and turn the body counterclockwise to engage the flange key.
5   Install the two solenoid screws and connect the MOTOR connector strap.

18   Starter motor brushes - replacement

Refer to illustrations 18.3, 18.4, 18.6 and 18.8
1   Remove the starter and solenoid assembly from the vehicle (refer to Section 16).
2   Remove the solenoid from the starter housing (refer to Section 17).
3   Remove the starter motor throughbolts after marking the relationship of the commutator end frame to the field frame housing to

simplify reassembly **(see illustration)**.
4   Remove the end frame from the field frame housing **(see illustration)**.
5   Mark the relationship of the field frame housing to the drive end housing and pull the field frame housing away from the drive end housing and over the armature.
6   Unbolt the brushes and brush supports from the brush holders in the field frame housing **(see illustration)**.
7   To install new brushes, attach the brushes to the brush supports, making sure they are flush with the bottom of the supports, and bolt the brushes/supports to the brush holders.
8   Install the field frame over the armature, with the brushes resting on the first step of the armature collar at this point **(see illustration)**.
9   Make sure the field frame is properly aligned with the drive end housing, then push the brushes off the collar into place on the armature.
10   The remaining installation steps are the reverse of those for removal.

18.8 Install the field frame over the armature (note the position of the brushes on the armature collar)

Chapter 6
Emissions control systems

Contents

	Section
Air Injection Reaction (AIR) system	8
Computer Command Control System (CCCS)	2
Early Fuel Evaporation (EFE) system	11
Electronic Control Module/PROM - removal and installation	3
Electronic Spark Control (ESC) system	6
Electronic Spark Timing (EST) system	5
Evaporative Emissions Control System (EECS)	9

	Section
Exhaust Gas Recirculation (EGR) system	7
General information	1
Information sensors	4
Positive Crankcase Ventilation (PCV) system	10
Thermostatic Air Cleaner (THERMAC)	12
Transmission Converter Clutch (TCC)	13

---

1 General information

Refer to illustrations 1.1a, 1.1b, 1.1c, 1.6a and 1.6b

To prevent pollution of the atmosphere from burned and evaporating gases, a number of emissions control systems are incorporated on the vehicles covered by this manual. The combination of systems used depends on the year in which the vehicle was manufactured, the locality to which it was originally delivered and the engine type. The major systems incorporated on the vehicles with which this manual is concerned include the **(see illustrations)**:

Fuel Control System
Electronic Spark Timing (EST)
Electronic Spark Control (ESC)
Air Injection Reaction (AIR)
Early Fuel Evaporation (EFE)

1.1a Locations of emission system and related components - carbureted V8 models

1	EGR valve	6	Baro sensor	11	EVAP canister purge valve
2	EGR control solenoid	7	Electronic spark control module	12	Engine coolant temperature sensor
3	Mixture control solenoid	8	AIR check valve	13	AIR pump
4	EST distributor	9	Evaporative emissions canister	14	AIR switching valve
5	Throttle position sensor	10	PCV valve		

1.1b Locations of emissions system and related components - MPFI V6 models

1	MAP sensor	6	Engine coolant temperature sensor	11	AIR pump switching valve
2	Digital EGR valve	7	Throttle position sensor	12	Fuel pump relay
3	EST distributor	8	Evaporative emissions canister	13	Cooling fan relay
4	IAC valve	9	EVAP canister purge valve	14	Air conditioning compressor relay
5	PCV valve	10	AIR pump		

1.1c Locations of emissions system and related components - Tuned Port Injected V8 models

1 Electronic Control Module (ECM)
2 Exhaust oxygen sensor
3 Throttle position sensor
4 Coolant temperature sensor
5 Vehicle speed sensor
6 Knock sensor (ESC)
7 Mass Air Flow sensor (MAF)
8 Manifold air temperature sensor
9 Idle air control motor
10 Fuel pump relay
11 Air control (divert) solenoid
12 Air switching (catalytic converter) solenoid
13 Exhaust Gas Recirculation vacuum solenoid
14 Fuel vapor canister solenoid
15 Electronic Spark Control (ESC) module
16 Engine coolant fan relay

24041-6-2.1b HAYNES

Exhaust Gas Recirculation (EGR)
Evaporative Emissions Control (EECS)
Transmission Converter Clutch (TCC)
Positive Crankcase Ventilation (PCV)
Thermostatic Air Cleaner (THERMAC)

All of these systems are linked, directly or indirectly, to the Computer Command Control System (CCCS).

The Sections in this Chapter include general descriptions, checking procedures (where possible) and component replacement procedures (where applicable) for each of the systems listed above.

Before assuming that an emissions control system is malfunctioning, check the fuel and ignitions systems carefully. In some cases, special tools and equipment, as well as specialized training, are required to accurately diagnose the causes of a rough running or difficult-to start engine. If checking and servicing become too difficult or if a procedure is beyond the scope of the home mechanic, consult your dealer service department. This does not necessarily mean, however, that the emissions control systems are particularly difficult to maintain and

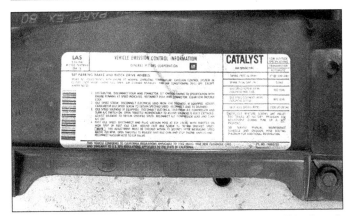

1.6a A Vehicle Emissions Control Information label will be found in the engine compartment of all vehicles

1.6b Typical vacuum hose schematic label

repair. You can quickly and easily perform many checks and do most (if not all) of the regular maintenance at home with common tune-up and hand tools. **Note:** *The most frequent cause of emissions system problems is simply a loose or broken vacuum hose or wiring connection. Therefore, always check the hose and wiring connections first.*

Pay close attention to any special precautions outlined in this Chapter. It should be noted that the illustrations of the various systems may not exactly match the system installed on your particular vehicle, due to changes made by the manufacturer during production or from year to year.

A Vehicle Emissions Control Information label is located in the engine compartment of all vehicles with which this manual is concerned **(see illustrations)**. This label contains important emissions specifications and setting procedures, as well as a vacuum hose schematic with emissions components identified. When servicing the engine or emissions systems, the VECI label in your particular vehicle should always be checked for up-to-date information.

2   Computer Command Control System (CCCS)

Refer to illustrations 2.7, 2.8 and 2.16

General description

This electronically controlled emissions system is linked with as many as nine other related emissions systems. It consists mainly of sensors (as many as 15) and an Electronic Control Module (ECM). Completing the system are various engine components which respond to commands from the ECM.

In many ways, this system can be compared to the central nervous system in the human body. The sensors (nerves) constantly gather information and send this data to the ECM (brain) which processes the data and, if necessary, sends out a command for some type of vehicle (body) change.

Here's a specific example of how one

2.7 The Assembly Line Diagnostic Link (arrow) is located under the instrument panel

portion of this system operates. An oxygen sensor, mounted in the exhaust manifold and protruding into the exhaust gas stream, constantly monitors the oxygen content of the exhaust gas as it travels through the exhaust pipe. If the percentage of oxygen in the exhaust gas is incorrect, an electrical signal is sent to the ECM. The ECM takes this information, processes it and then sends an electrical command to the carburetor Mixture Control (M/C) solenoid telling it to change the fuel/air mixture. To be effective, all this happens in a fraction of a second, and it goes on continuously while the engine is running. The end result is a fuel/air ratio which is constantly kept at a predetermined "exact" proportion, regardless of driving conditions.

Testing

One might think that a system which uses exotic electrical sensors and is controlled by an on-board computer would be difficult to diagnose. This is not necessarily the case.

The Computer Command Control System has a built-in diagnostic system which indicates a problem by flashing a CHECK ENGINE light on the instrument panel. When this light comes on during normal vehicle operation, a fault has been detected.

2.8 Typical Assembly Line Diagnostic Link - the two terminals you will be concerned with are the A (ground) and B (diagnostic) terminals

Perhaps more importantly, the ECM will recognize this fault in a particular system monitored by one of the various information sensors and store it in its memory in the form of a trouble code. Although the trouble code cannot reveal the exact cause of the malfunction, it greatly facilitates diagnosis as you or a dealer mechanic can tap into the ECM memory and be directed to the problem area.

To extract this information from the ECM memory, you must use a short jumper wire to ground a diagnostic terminal. This terminal is part of a wiring connector located under the dashboard **(see illustration)**. A small, rectangular plate is used to cover the connector and must be pried out of place to provide access to the terminals.

With the connector exposed to view, push one end of the jumper wire into the diagnostic terminal (B) and the other end into the ground terminal (A) **(see illustration)**. **Note:** *Do not start the engine with the diagnostic terminal grounded.*

Turn the ignition to the On position - *not* the Start position. The CHECK ENGINE light should flash Trouble Code 12, indicating that the diagnostic system is working. Code 12 will consist of one flash, followed by a short pause, and then two flashes in quick succession. After a longer pause, the code will repeat itself two more times.

If no other codes have been stored, Code 12 will continue to repeat itself until the jumper wire is disconnected. If additional

Trouble Codes have been stored, they will follow Code 12. Again, each Trouble Code will flash three times before moving on.

The ECM can also be checked for stored codes on carbureted models with the engine running. Completely remove the jumper wire from the diagnostic and ground terminals, start the engine and then plug the jumper wire back in. With the engine running, all stored Trouble Codes will flash. However, Code 12 will flash only if there is a fault in the distributor reference circuit.

Once the code(s) have been noted, use the Trouble Code Identification information which follows to locate the source of the fault. **Note:** *Whenever the positive battery cable is disconnected, all stored Trouble Codes in the EMC are erased. Be aware of this before you disconnect the battery for servicing or replacement of electrical components, engine removal, etc.*

It should be noted that the self-diagnosis feature built into this system does not detect all possible faults. If you suspect a problem with the Computer Command Control System, but a CHECK ENGINE light has not come on, have your local dealer perform a "system performance check.'

Furthermore, when diagnosing an engine performance, fuel economy or exhaust emissions problem (which is not accompanied by a CHECK ENGINE light) do not automatically assume the fault lies in this system. Perform all standard troubleshooting procedures, as indicated elsewhere in this manual before turning to the Computer Command Control System.

Finally, since this is an electronic system, you should have a basic knowledge of automotive electronics before attempting any diagnosis. Damage to the ECM, PROM or related components can easily occur is care is not exercised.

Trouble Code Identification

Following is a list of the Trouble Codes which may be encountered while diagnosing the Computer Command Control System. Also included are the most probable causes. If the problem persists, the vehicle must be diagnosed by a professional mechanic who can use specialized diagnostic tools and advanced troubleshooting methods to check the system. Procedures marked with an asterisk (*) indicate component replacements which may not cure the problem in all cases. For this reason, you may want to

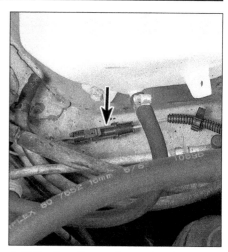

2.16 To clear the trouble codes from ECM memory, disconnect the ECM power harness connector (arrow) near the battery

seek professional advice before purchasing replacement parts.

To clear the Trouble Code(s) from the ECM memory, unplug the ECM electrical pigtail at the positive (+) battery cable **(see illustration)**.

Trouble code	Circuit or system	Probable cause *
Code 12	No distributor reference pulse	This code will flash whenever the diagnostic terminal is grounded with the ignition On and the engine not running. If additional trouble codes are stored, they will appear after this code has flashed three times.
Code 13	Oxygen sensor circuit	Sticking Throttle Position Sensor; poor electrical connection, open or short in circuit; defective oxygen sensor; defective ECM
Code 14	Coolant temperature sensor circuit	Engine overheating; poor electrical connection, open or short in circuit; defective coolant temperature sensor; defective ECM
Code 15	Coolant temperature sensor circuit	Poor electrical connection, open or short in circuit; defective coolant temperature sensor; defective ECM
Code 21	Throttle Position Sensor	Sticking or misadjusted TPS; poor electrical connection, open or short in circuit; defective TPS; defective ECM
Code 22	Throttle Position Sensor	Sticking or misadjusted TPS; poor electrical connection, open or short in circuit; defective TPS; defective ECM
Code 23 (carbureted models)	Mixture control solenoid	Open circuit or short to ground in the mixture solenoid circuit; defective mixture control solenoid; defective ECM - If the mixture control solenoid is shorted to ground, the ECM may have been damaged.
Code 23 (fuel injected models)	Manifold Air Temperature Sensor	Poor electrical connection, open or short in circuit; defective MAT sensor; defective ECM
Code 24	Vehicle Speed Sensor	Poor electrical connection, open or short in circuit; defective VSS; defective ECM - A fault in this circuit should be indicated only when the vehicle is in motion. Disregard Code 24 if it is set when the drive wheels are not turning.
Code 25 (fuel injected models)	Manifold Air Temperature Sensor	Poor electrical connection, open or short in circuit; defective MAT sensor; defective ECM
Code 32 (carbureted models)	Differential pressure sensor	Defective differential pressure sensor; poor electrical connection, open or short in circuit; defective ECM

Trouble code	Circuit or system	Probable cause *
Code 32 (fuel injected models - except 3.1L V6)	EGR system	Restricted vacuum hose to EGR solenoid or valve; poor electrical connection, open or short in circuit; defective EGR solenoid; defective EGR valve; defective ECM
Code 32 (3.1L V6)	Digital EGR system	Poor electrical connection, open or short in circuit; defective EGR valve; defective ECM
Code 33 (All except 1989 and earlier MPFI V6 and TPI V8)	MAP sensor	Leaking or restricted vacuum hoses; poor electrical connection, open or short in circuit; defective MAF sensor; defective ECM
Code 33 (1989 and earlier MPFI V6 and TPI V8)	MAF sensor	Poor electrical connection, open or short in circuit; defective MAF sensor; defective ECM
Code 34 (carbureted models)	Differential pressure sensor	Leaking or restricted vacuum hoses; poor electrical connection, open or short in circuit; defective differential pressure sensor; defective ECM
Code 34 (fuel injected models - except1989 and earlier MPFI V6 and TPI V8)	MAP sensor	Leaking or restricted vacuum hoses; poor electrical connection, open or short in circuit; defective Map sensor; defective ECM
Code 34 (1989 and earlier MPFI V6 and TPI V8)	MAF sensor	Poor electrical connection, open or short in circuit; defective MAF sensor; defective ECM
Code 35	Idle speed control	TPS sticking or misadjusted; poor electrical connection, open or short in circuit; defective IAC valve; defective ECM
Code 36 (1989 and earlier MPFI V6 and TPI V8)	MAF sensor burn-off circuit	Poor electrical connection, open or short in circuit; defective MAF sensor burn-off relay; defective MAF sensor; defective ECM
Code 41 (carbureted models	No distributor reference pulses to ECM with engine running	Poor electrical connection, open or short in circuit; defective distributor pick-up coil; fault in the MAP or differential pressure sensor circuit
Code 41 (fuel injected models)	Cylinder select error	Incorrect or defective MEMCAL or ECM
Code 42	Electronic Spark Timing	Poor electrical connection, open or short in circuit; defective ignition module; defective ECM
Code 43	Electronic Spark Control	Poor electrical connection, open or short in circuit; defective ESC module; defective knock sensor; defective ECM
Code 44	Lean exhaust	Lean condition caused by malfunctioning carburetor/fuel injector, vacuum leak, low fuel pressure, etc.; poor electrical connection, open or short in circuit; defective MAP sensor; defective oxygen sensor; defective ECM
Code 45	Rich exhaust	Rich condition caused by restricted air filter, fuel in evaporative charcoal canister or crankcase, malfunctioning carburetor/fuel injector, high fuel pressure, etc.; defective TPS; defective MAP; defective oxygen sensor; defective ECM
Code 46	Vehicle Anti-Theft System (VATS)	Poor electrical connection, open or short in circuit; incorrect ignition key or starting procedure; defective VATS module; defective ECM
Code 51	PROM	Defective PROM; PROM not properly installed; incorrect PROM installed; defective ECM
Code 52	Fuel CALPAK	Defective CALPAK; CALPAK not properly installed; incorrect CALPAK installed
Code 53 (carbureted models)	EGR control	Leaking, restricted or improperly routed vacuum line; EGR valve stuck or leaking; poor electrical connection, open or short in circuit; defective EGR solenoid; defective ECM

Trouble code	Circuit or system	Probable cause *
Code 53 (fuel injected models)	System over voltage	Charging system problem
Code 54 (carbureted models)	Mixture control solenoid	Open circuit or short to ground in the mixture solenoid circuit; defective mixture control solenoid; defective ECM - If the mixture control solenoid is shorted to ground, the ECM may have been damaged.
Code 54 (fuel injected models)	Fuel pump circuit	Defective fuel pump; defective fuel pump relay; defective oil pressure switch; poor electrical connection, open or short in circuit; defective ECM
Code 55	ECM	Poor electrical connection, open or short in circuit; defective ECM
Code 61	Oxygen sensor	Contaminated oxygen sensor

Component replacement may not cure the problem in all cases. For this reason, you may want to seek professional advice before purchasing replacement parts.

3 Electronic Control Module/PROM - removal and installation

Refer to illustrations 3.1, 3.7, 3.10, 3.11 and 3.12

1   The Electronic Control Module (ECM) is located under the instrument panel **(see illustration)**.

2   Disconnect the negative battery cable from the battery. **Caution:** *On models equipped with a Delco Loc II audio system, disable the anti-theft feature before disconnecting the battery.*

3   Remove the three right hand hush panel screws and detach the panel (under the right side of the dashboard).

4   Disconnect the wire harnesses from the ECM. **Caution:** *The ignition switch must be turned to Off when removing or installing the ECM connectors.*

5   Remove the retaining bolts and carefully detach the ECM.

6   Turn the ECM so that the bottom cover is facing up and carefully place it on a clean work surface.

7   Remove the PROM access cover by removing the screws retaining it **(see illustration)**.

8   If you are replacing the ECM itself, the new ECM will not contain a PROM. It will be necessary to remove the old PROM from the old ECM and install it in the new one.

9   Even if you are simply installing a new PROM in the old ECM, it will still be necessary to remove the old PROM from the old ECM and install a new PROM in its place.

10   Using a PROM removal tool (if available), grasp the PROM carrier at the narrow ends **(see illustration)**. Gently rock the carrier from end to end while carefully pulling up. **Note:** *PROM removal tools usually come supplied with new PROMs and ECMs.* The PROM carrier and PROM should lift out of the PROM socket easily.

11   Note the reference end of the PROM carrier before setting it aside **(see illustration)**.

12   If you are replacing the ECM, remove the new ECM from its container and check the service number to make sure that it is the

3.1 The ECM, on most models, is located under the right-side of the instrument panel

same as the number on the old ECM **(see illustration)**. If the numbers are different, you have the wrong ECM or the wrong PROM.

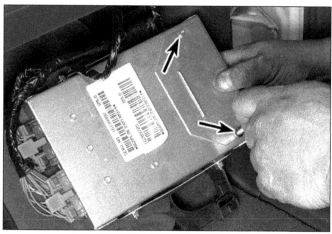

3.7 Remove the screws and the PROM cover

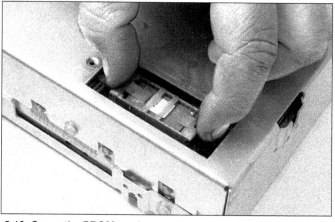

3.10 Grasp the PROM carrier at the narrow ends and gently rock it until the PROM is disconnected from its socket

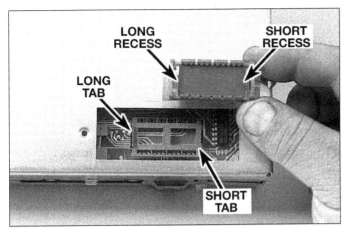

3.11 Note the reference end of the PROM carrier before setting it aside

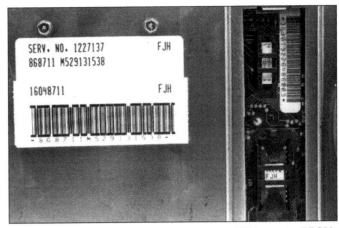

3.12 Make sure the service numbers on the ECM and the PROM are the same

13   If you are replacing the PROM, remove the new PROM from its container and check the service number to make sure that it is the same as the number of the old PROM.

14   Position the PROM/carrier assembly squarely over the PROM socket with the small notched end of the carrier aligned with the small notch in the socket at the pin 1 end. Press on the PROM carrier until it seats firmly in the socket. If the PROM is new, make sure that the notch in the PROM is matched to the small notch in the carrier. **Caution:** *If the PROM is installed backwards and the ignition switch is turned on, the PROM will be destroyed. Using the tool, install the new PROM carrier in the PROM socket of the ECM. The small notch of the carrier should be aligned with the small notch in the socket. Press on the PROM carrier until it is firmly seated in the socket.* **Caution:** *Do not press on the PROM - press only on the carrier.*

15   Attach the access cover to the ECM and tighten the two screws.

16   Install the ECM in the support bracket, plug in the electrical connectors to the ECM and install the hush panel.

17   Start the engine.

18   Enter the diagnostic mode by grounding the diagnostic terminal of the ALDL (see Section 2). If no trouble codes occur, the PROM is correctly installed.

19   If Trouble Code 51 occurs, or if the CHECK ENGINE light comes on and remains constantly lit, the PROM is not fully seated, is installed backwards, has bent pins or is defective.

20   If the PROM is not fully seated, pressing firmly on both ends of the carrier should correct the problem.

21   It is possible to install the PROM backwards. If this occurs, and the ignition key is turned to On, the PROM circuitry will be destroyed and the PROM will have to be replaced.

22   If the pins have been bent, remove the PROM in accordance with the above procedure, straighten the pins and reinstall the PROM. If the bent pins break or crack when

4.1a Engine coolant temperature sensor location (arrow) - carbureted V8 models

you attempt to straighten them, discard the PROM and replace it with a new one.

23   If careful inspection indicates that the PROM is fully seated, has not been installed backwards and has no bent pins, but the CHECK ENGINE light remains lit, the PROM is probably faulty and must be replaced.

4   Information sensors

Refer to illustrations 4.1a, 4.1b, 4.4a, 4.4b, 4.6, 4.8, 4.10 and 4.17

Note: *Note all the sensors listed in this Section are installed on all models.*

Engine coolant temperature sensor

1   A coolant temperature sensor is used on all models. The coolant sensor is located on the thermostat housing or on the front of the intake manifold, depending on model **(see illustrations)**. Do not confuse the engine coolant sensor with the sending unit used for the temperature gauge. The engine coolant sensor uses a two-wire connector (usually a

4.1b On TPI V8 models the engine coolant temperature sensor is located on the front of the intake manifold (arrow) - it may be necessary to remove a bracket to access the sensor

black and yellow wire). The sensor is a thermistor (a resistor which varies the value of its voltage output in accordance with temperature changes), as the coolant temperature increases the sensor resistance decreases. A failure in the coolant sensor circuit should set either a Code 14 or a Code 15. These codes indicate a failure in the coolant temperature circuit, so the appropriate solution to the problem will be either repair of a wire or replacement of the sensor.

2   To remove the sensor, remove the bracket bolts and slide the air control valve assembly forward until you have room to unscrew the sensor. Disengage the locking tab on the connector with a small screwdriver and unplug it from the sensor. Carefully unscrew the sensor itself. **Caution:** *Handle the coolant sensor with care. Damage to this sensor will affect the operation of the entire fuel injection system.*

3   Before installing the new sensor, wrap the threads with Teflon sealing tape to prevent leakage and thread corrosion.

4.4a The MAP sensor is mounted on a bracket at the left side of the engine compartment . . .

4.4b . . . or attached to the cowl on the right side of the engine compartment

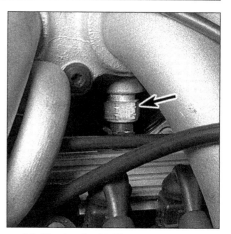

4.6 The Intake Air Temperature (IAT) sensor is located in the underside of the plenum on TPI V8 models (arrow) - to remove it, you must first remove the plenum

Manifold Absolute Pressure (MAP) sensor

4   A Manifold Absolute Pressure (MAP) sensor is used on all models except 1989 and earlier MPFI V6 and TPI V8 models. On carbureted models the sensor is also referred to as a vacuum sensor or a differential pressure sensor. The MAP sensor is typically mounted to a bracket on the firewall or cowl, on either side, depending on model **(see illustrations)**. The MAP sensor monitors the intake manifold pressure changes resulting from changes in engine load and speed and converts the information into a voltage output. The ECM uses the MAP sensor to control fuel delivery and ignition timing.
5   A failure in the MAP sensor circuit should set a Code 33 or a Code 34.

Manifold Air Temperature (MAT) sensor

6   A Manifold Air Temperature (MAT) sensor is used on all MPFI V6, TPI V8 and TBI V8 models. The MAT sensor is located in various locations, depending on model: on MPFI V6 models it is located in the air intake duct or air cleaner housing; on TPI V8 models it is located in the underside of the plenum **(see illustration)**; on TBI V8 models it is located in the air cleaner housing. The MAT sensor is a thermistor (a resistor which changes the value of its voltage output as the temperature changes); as the temperature increases, the resistance of the sensor decreases. The ECM supplies a five-volt signal to the sensor through a resistor in the ECM and measures the actual voltage, which varies with the temperature, at the sensor. Thus the ECM "knows" the manifold air temperature. The ECM uses the MAT sensor signal to delay EGR until the manifold air temperature reaches 40-degrees F.
7   A failure in the MAT sensor circuit should set either a Code 23 or a Code 25.

Mass Air Flow (MAF) sensor

8   The Mass Air Flow (MAF) sensor is used

on 1989 and earlier MPFI V6 and TPI V8 models. The sensor is located between the fresh air intake and the flexible intake duct **(see illustration)**, measures the amount of air entering the engine. The MAF sensor is a hot wire type and is controlled by a MAF sensor relay. Current is supplied to the sensing wire to maintain a calibrated temperature but as air flow increases or decreases, the current will vary. This causes a voltage change within the circuitry of the MAF sensor directly proportional to the air mass. The voltage change, which varies from 0.5-volts at idle to about 4.7-volts at wide open throttle, is processed by the ECM for calculating fuel delivery.
9   A failure in the MAF sensor should set either a Code 33 or a Code 34.

Oxygen sensor

10   The oxygen sensor, which is located in the exhaust pipe **(see illustration)**, monitors the oxygen content of the exhaust gas stream. The oxygen content in the exhaust reacts with the oxygen sensor to produce a voltage output, which varies from 0.1 to 0.9-volts. The ECM monitors this voltage output to determine the ratio of oxygen to fuel in the mixture. The ECM alters the air/fuel mixture

4.8 The Mass Air Flow sensor is located in the air intake duct (arrow)

ratio by controlling the pulse width (open time) of the fuel injectors. A mixture ratio of 14.7 parts air to 1 part fuel is the ideal mixture ratio for minimizing exhaust emissions, thus allowing the catalytic converter to operate at maximum efficiency. It is this ratio of 14.7 to 1 which the ECM and the oxygen sensor attempt to maintain at all times.
11   The oxygen sensor is like an open circuit and produces no voltage when it is below its normal operating temperature of about 600-degrees F (315-degrees C). During this initial period before warm-up, the ECM operates in open loop mode.
12   However, if the engine reaches normal operating temperature and/or has been running for two or more minutes, and if the oxygen sensor is producing a steady signal voltage between 0.35 and 0.55-volts, even though the TPS indicates that the engine is not at idle, the ECM will set a Code 13.
13   An open in the oxygen sensor circuit should set a Code 13. A low voltage in the circuit should set a Code 44. A high voltage in the circuit should set a Code 45. Codes 44 and 45 may also be set as a result of fuel sys-

4.10 The oxygen sensor is located in the exhaust pipe (arrow)

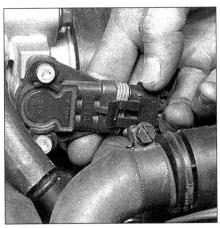

4.17 The Throttle Position Sensor (TPS) is located on the throttle body at the end of the throttle shaft

tem problems. When any of the above codes occur, the ECM operates in the open loop mode. It controls fuel delivery in accordance with a programmed default value instead of feedback information from the oxygen sensor.

14   The proper operation of the oxygen sensor depends on four conditions:

    a)   **Electrical** - *The low voltages and low currents generated by the sensor depend upon good, clean connections, which should be checked whenever a malfunction of the sensor is suspected or indicated.*

    b)   **Outside air supply** - *The sensor is designed to allow air circulation to the internal portion of the sensor. Whenever the sensor is removed and installed or replaced, make sure the air passages are not restricted.*

    c)   **Proper operating temperature** - *The ECM will not react to the sensor signal until the sensor reaches approximately 600-degrees F (315-degrees C). This factor must be taken into consideration when evaluating the performance of the sensor.*

    d)   **Unleaded fuel** - *The use of unleaded fuel is essential for proper operation of the sensor. Make sure the fuel you are using is of this type.*

15   In addition to observing the above conditions, special care must be taken whenever the sensor is serviced.

    a)   *The oxygen sensor has a permanently attached pigtail and connector that should not be removed from the sensor. Damage or removal of the pigtail or connector can adversely affect operation of the sensor.*

    b)   *Grease, dirt and other contaminants should be kept away from the electrical connector and the louvered end of the sensor.*

    c)   *Do not use cleaning solvents of any kind on the oxygen sensor.*

    d)   *Do not drop or roughly handle the sensor.*

    e)   *The silicone boot must be installed in the correct position to prevent the boot from being melted and to allow the sensor to operate properly.*

16   Refer to Chapter 1 for the oxygen sensor replacement procedure.

Throttle position sensor (TPS)

Note: *The following information applies to fuel injected models, refer to Chapter 4A for TPS information on carbureted models.*

17   The throttle position sensor (TPS) is located on the end of the throttle shaft **(see illustration)**. By monitoring the output voltage from the TPS, the ECM can determine fuel delivery based on throttle valve angle (driver demand). A broken or loose TPS can cause intermittent bursts of fuel from the injector and an unstable idle because the ECM thinks the throttle is moving.

18   A problem in any of the TPS circuits will set either a Code 21 or 22. Once a trouble code is set, the ECM will use an artificial default value for TPS and some vehicle performance will return.

19   Should the TPS require replacement, the procedure is described in Chapter 4B.

Park/Neutral switch

21   A Park/Neutral switch is used on all automatic transmission models. The Park/Neutral (P/N) switch indicates to the ECM when the transmission is in Park or Neutral. This information is used for the transmission Converter Clutch (TCC) and the Idle Air Control (IAC) valve operation. **Caution:** *The vehicle should not be driven with the Park/Neutral switch disconnected because idle quality will be adversely affected and a false Code 24 (failure in the Vehicle Speed Sensor circuit) may be set.*

Crank signal

22   The ECM receives a 12-volt signal from the starter solenoid during cranking to allow enrichment and cancel the diagnostic mode until the engine is running or 12-volts is no longer in the circuit. If the signal from the starting solenoid is not available, the vehicle can be difficult to start.

Air conditioning On signal

23   This signal tells the ECM that the A/C selector switch is turned to the On position and that the pressure cycling switch is closed. The ECM uses this information to adjust the idle speed when the air conditioning system is working. If this signal is not available to the ECM, idle may be rough, especially when the A/C compressor cycles.

Vehicle speed sensor

24   The vehicle speed sensor (VSS) sends a pulsing voltage signal to the ECM, which the ECM converts to miles per hour. This sensor controls the operation of the TCC. On 1984 and earlier models the VSS is incorporated into the speedometer. On 1985

and later models, the VSS is mounted on the transmission.

Distributor reference signal

25   The distributor sends a signal to the ECM to tell it both engine rpm and crankshaft position. See Electronic Spark Timing (EST), Section 5, for further information.

---

5   Electronic Spark Timing (EST) system

Refer to illustration 5.4

General description

1   To provide improved engine performance, fuel economy and control of exhaust emissions, the Electronic Control Module (ECM) controls distributor spark advance (ignition timing) with the Electronic Spark Timing (EST) system.

2   The ECM receives a reference pulse from the distributor, which indicates both engine rpm and crankshaft position. The ECM then determines the proper spark advance for the engine operating conditions and sends an EST pulse to the distributor.

Check

3   The ECM will set EST at a specified value when the diagnostic terminal in the ALDL connector is grounded. To check for EST operation, the timing should be checked at 2000 rpm with the terminal ungrounded. Then ground the diagnostic terminal. If the timing changes at 2000 rpm, the EST is operating. A fault in the EST system will usually set Trouble Code 42.

Setting timing

4   To set the initial base timing, disconnect the distributor four-wire connector or the set timing connector (refer to the VECI label for specific instructions) **(see illustration)**.

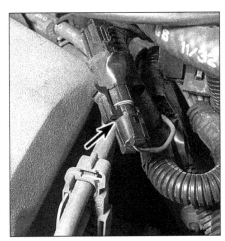

5.4 To set the initial base timing, disconnect the EST by-pass connector (arrow) (usually located at the firewall, near the power brake booster)

6.3a The ESC (knock) sensor is either located at the lower portion of the engine block, just above the oil pan rail (arrow) . . .

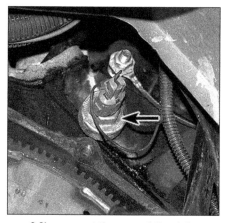

6.3b . . . or at the rear of the right cylinder head (arrow)

6.13 The ESC module is located at the firewall, usually near the power brake booster (arrows)

5　Set the timing as specified on the VECI label. This will cause a Code 42 to be stored in the ECM memory. Be sure to clear the memory after setting the timing (see Section 2).

6　For further information regarding the testing and component replacement procedures for the HEI/EST distributor, refer to Chapter 5.

6　Electronic Spark Control (ESC) system

Refer to illustrations 6.3a, 6.3b and 6.13

General description

1　Irregular octane levels in modern gasoline can cause detonation in a high performance engine. Detonation is sometimes referred to as "spark knock." This condition causes the pistons and rings to vibrate and rattle, producing a characteristic knocking or pinging sound.

2　The Electronic Spark Control (ESC) system is designed to retard spark timing up to 20-degrees to reduce spark knock in the engine. This allows the engine to use maximum spark advance to improve driveability and fuel economy.

3　The ESC knock sensor, which is located either on the lower right side of the engine block (some models use two sensors, one on the right side of the block and one on the left side of the block, service procedures are the same for both sensors) or on the rear of the right cylinder head **(see illustrations)**, sends a voltage signal of 8 to 10-volts to the ECM when no spark knock is occurring and the ECM provides normal advance. When the knock sensor detects abnormal vibration (spark knock), the ESC module turns off the circuit to the ECM. The ECM then retards the EST distributor until spark knock is eliminated.

4　Failure of the ESC knock sensor signal or loss of ground at the ESC module will

cause the signal to the ECM to remain high. This condition will result in the ECM controlling the EST as if no spark knock is occurring. Therefore, no retard will occur and spark knock may become severe under heavy engine load conditions. At this point, the ECM will set a Code 43.

5　Loss of the ESC signal to the ECM will cause the ECM to constantly retard EST. This will result in sluggish performance and cause the ECM to set a Code 43.

Component replacement

ESC sensor

6　Disconnect the cable from the negative terminal of the battery.

7　Raise the vehicle and support it on jackstands. Refer to Chapter 1 and drain the cooling system.

8　Disconnect the wiring harness connector from the ESC sensor.

9　Remove the ESC sensor from the block. Coolant will flow out of the hole, so be careful not to get any in your eyes.

10　Apply thread sealant to the ESC sensor threads.

11　Installation is the reverse of the removal procedure. Be sure to refill the cooling system.

ESC module

12　Disconnect the cable from the negative terminal of the battery.

13　Disconnect the electrical connector from the module **(see illustration)**.

14　Remove the mounting screws and remove the module.

15　Installation is the reverse of removal.

7　Exhaust Gas Recirculation (EGR) system

Refer to illustrations 7.5, 7.14 and 7.16

Note 1: *The following procedure applies to fuel injected models (except 3.1L V6 models), refer to Chapter 1 for information on carbureted models.*

Note 2: *All 1990 and later models with a 3.1 liter V6 engine are equipped with a Digital EGR Valve, which controls EGR flow electronically (through the ECM), rather than directly by intake manifold vacuum. This procedure does not apply to the Digital EGR Valve. The Digital EGR system should be diagnosed and repaired by a dealer or other properly equipped repair shop.*

General description

1　The Exhaust Gas Recirculation (EGR) system is used to lower NOx (oxides of nitrogen) emission levels by decreasing combustion temperature. The main element of the system is the EGR valve, mounted on the intake manifold, which feeds small amounts of exhaust gas back into the combustion chamber.

2　The EGR valve is opened by manifold vacuum to allow exhaust gases to flow into the intake manifold. The EGR valve is usually open during warm engine operation and anytime the engine is running above idle speed. The amount of gas recirculated is controlled by variations in vacuum and exhaust backpressure.

3　The valve used on this engine is called a negative backpressure valve. It varies the amount of exhaust gas flow into the manifold depending on manifold vacuum and variations in exhaust backpressure.

4　The diaphragm on this valve has an internal vacuum bleed hole that is held closed by a small spring when there is no exhaust backpressure. Engine vacuum opens the EGR valve against the pressure of a large spring. When manifold vacuum combines with negative exhaust backpressure, the vacuum bleed hole opens and the EGR valve closes.

5　An ECM controlled solenoid **(see illustration)** is used in the vacuum line in order to maintain finer control of EGR flow. The ECM uses information from the coolant temperature, throttle position and manifold pressure sensors to regulate the vacuum solenoid.

6　During cold operation and at idle, the solenoid circuit is grounded by the ECM to

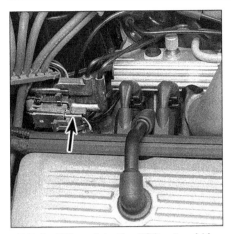

7.5 On TPI models, the EGR solenoid is located at the right rear corner of the intake manifold next to the distributor (arrow)

7.14 The engine side of the EGR solenoid should have at least ten inches of vacuum

7.16 Connect a test light across the EGR solenoid wire harness terminals after checking the solenoid for vacuum - the light should come on

block vacuum to the EGR. When the solenoid circuit is not grounded by the ECM, vacuum is allowed to the EGR.

7   With the engine stopped, turning the ignition key to On will turn on the solenoid, allowing vacuum to the EGR valve. Grounding the ALDL diagnostic terminal **(see illustration 2.8)** will remove power to the solenoid and allow vacuum to the EGR valve.

8   Too much EGR flow weakens combustion, causing the engine to run roughly or stall. During cold operation it may cause the engine to stop after a cold start. At idle, excessive EGR flow may cause the engine to run roughly or stop after decelerating. Too much EGR flow while cruising may cause the engine to surge.

9   Too little or no EGR flow allows combustion temperatures to get too high during acceleration and load conditions. The result can be spark knock (detonation), engine overheating and emission test failure.

Check

Throttle Body Injection models

10   Make a physical inspection of the hoses and electrical connections to ensure that nothing is loose.

11   With the ignition off, check the EGR valve to make sure it is closed by pushing up on the underside of the diaphragm.

12   If the valve is open, disconnect the vacuum hose at the EGR valve and recheck the diaphragm. If the diaphragm is still open, clean the EGR valve and passages as described in Steps 36 on. Replace the EGR valve if necessary. If the diaphragm is closed, connect a vacuum gauge in place of the EGR valve. Proceed to Step 16.

13   If the valve is closed, install a tachometer and bring the engine to normal operating temperature. Do not ground the diagnostic terminal. With the engine idling, the rpm should drop as the EGR valve is opened by pushing up on the underside of the diaphragm. If there is no change in rpm, clean the EGR valve and passages. Replace

the EGR valve if necessary.

14   If the rpm drops, disconnect the EGR control solenoid, then check for movement of the diaphragm as the rpm is increased to 1200 rpm and returned to idle. If there is no movement, check vacuum at the engine side of the EGR solenoid **(see illustration)**. It should have at least ten inches of vacuum. If it doesn't, repair the cause of poor vacuum to the solenoid. If it has ten inches of vacuum, check to make sure that it has at least ten inches of vacuum on the EGR valve side of the solenoid. If it does, check for a restricted hose between the EGR valve and the solenoid. If the hose is okay, the EGR valve is faulty. If it does not have at least ten inches of vacuum on the EGR valve side of the solenoid, the EGR solenoid is faulty.

15   If there is movement, connect a vacuum gauge in place of the EGR valve. There should be at least ten inches of vacuum. If not, perform the sequence of tests outlined in Step 14.

16   If there are at least ten inches of vacuum, leave the vacuum gauge connected, turn the ignition on and, with the engine stopped, connect a test light across the solenoid harness connector terminals **(see illustration)**. The light should go on.

17   If the light does not come on, probe both harness connector terminals with a test light connected to ground. If there is no voltage at either terminal, repair the open ignition circuit to the solenoid. If the light goes on at one terminal, check for an open in the circuit from the EGR solenoid to the ECM (see the wiring diagrams at the end of Chapter 12). If there is no open in the EGR solenoid circuit, either the ECM connector terminal is faulty or the ECM itself is faulty. **Note:** *Before replacing the ECM, use an ohmmeter and check the resistance between the solenoid terminals of the TCC and the EGR. Refer to the ECM wiring diagrams for coil terminal identification for both the solenoid(s) and relay(s) to be checked. Replace any solenoid where the resistance measures less than 20 ohms.*

18   If the light does come on, ground the diagnostic terminal and note whether the light

comes on. If the light comes on, check for a short to ground in the EGR solenoid circuit. If it's okay, the ECM is faulty. **Note:** *Before replacing the ECM, use an ohmmeter to check the resistance between the solenoid terminals of the TCC and the EGR. Refer to the ECM wiring diagrams for coil terminal identification for both the solenoid(s) and relay(s) to be checked. Replace any solenoid where the resistance measures less than 20 ohms. If the light does not come on, reconnect the solenoid, start the engine and note the vacuum gauge reading while the engine is idling. If there is vacuum, replace the EGR control solenoid. If there is no vacuum, the EGR control solenoid is okay. Check the valve for binding, leakage and loose mounting bolts.*

Multi Port Fuel Injection models

19   Before performing the following sequence of test procedures, check for ported vacuum to the EGR solenoid and check the hoses for leaks and restrictions. There should be at least seven inches of vacuum at 2000 rpm. **Note:** *The following test sequence assumes there is no Code 32.*

20   With the ignition on and the engine stopped, ground the diagnostic terminal, disconnect the EGR solenoid vacuum harness and apply ten inches of vacuum to the manifold side of the solenoid. It should be able to hold this vacuum. If it can, proceed to Step 24.

21   If it cannot hold vacuum, disconnect the EGR solenoid electrical connector and connect a test light between the harness terminals.

22   If the light comes on, replace the solenoid.

23   If the light does not come on, probe each harness connector terminal with a test light connected to ground. If there is no light, repair the open in the circuit from the fuse box to the EGR solenoid (see the wiring diagrams). If the light comes on at one terminal, check for an open in the circuit from the EGR solenoid to the ECM. If the circuit has no

open, check the resistance of the solenoid - it should have more than 20 ohms. If it doesn't, replace the solenoid and the ECM. If it does, either the ECM connection or the ECM itself is faulty. **Note:** *Before replacing the ECM, check the resistance of each ECM controlled relay and solenoid coil with an ohmmeter. Refer to the wiring diagrams for coil terminal identification for both the solenoid(s) and relay(s) to be checked. Replace any relay or solenoid if the coil resistance measures less than 20 ohms.*

24   If the solenoid can hold vacuum, unground the diagnostic terminal - the vacuum should drop.

25   If there is no drop in vacuum, disconnect the solenoid electrical connector and note the vacuum reading. If the vacuum reading drops, repair the short to ground in the EGR solenoid circuit. If there is no short to ground in the EGR solenoid circuit, the ECM is faulty. **Note:** *Before replacing the ECM, check the resistance of each ECM controlled relay and solenoid coil with an ohmmeter. Refer to the ECM wiring diagrams (at the end of Chapter 12) for coil terminal identification for both the solenoid(s) and relay(s) to be checked. Replace any relay or solenoid if the coil resistance measures less than 20 ohms. If there is no drop in vacuum reading, replace the EGR solenoid.*

26   If there is a drop in vacuum, turn the ignition off, connect a vacuum pump to the EGR valve and use a mirror to watch the valve diaphragm while applying vacuum. The diaphragm should move freely and hold vacuum for at least 20 seconds. If it doesn't, replace the EGR valve.

27   If the diaphragm does hold vacuum for at least 20 seconds, apply vacuum to the EGR valve, start the engine and immediately note the vacuum reading. The valve is good if it moves to its seated position (valve closed) and if the vacuum reading drops while starting the engine. If the valve does not move to its seated position and/or vacuum doesn't drop while starting the engine, remove the EGR valve and check for plugged passages. If the passages are not plugged, replace the EGR valve. If the valve moves to its seated position and the vacuum reading drops while starting the engine, check the Park/Neutral switch on the automatic transmission (if equipped) by connecting a test light between the EGR solenoid harness terminals, and, with the engine at normal operating temperature, accelerate the engine to about 1500 rpm in Park (observe the light, it should stay on). Repeat this test in Drive - the test light should dim or go out. If both conditions are met, then the switch is okay.

Component replacement

EGR valve

28   If your vehicle is a TBI model, remove the air cleaner. If your vehicle is a MPFI model, remove the plenum (Chapter 4).

29   Disconnect the EGR valve vacuum line from the valve.

30   Remove the EGR valve mounting bolts.

31   Remove the EGR valve from the manifold.

32   If the EGR passages in the manifold have deposits built-up in them they should be cleaned. Care should be taken to ensure that all loose particles are completely removed to prevent them from clogging the EGR valve or from being ingested into the engine. **Caution:** *Do not wash the EGR valve in solvents or degreaser - permanent damage to the valve diaphragm may result. Sand blasting of the valve is also not recommended since it can affect the operation of the valve.*

33   Buff the exhaust deposits from the mounting surface and around the valve with a wire brush.

34   Look for exhaust deposits in the valve outlet. Remove deposit buildup with a screwdriver.

35   Clean the mounting surfaces of the intake manifold and the valve assembly.

36   If the valve itself is operating correctly, it can be reused after it has been cleaned and checked for deposits. Hold the valve in your hand. Tap on the end of the round pintle using a light snapping action with a soft-face hammer. This will remove the exhaust deposits from the valve seat. Remove all loose particles.

37   Clean the mounting surface of the valve and the pintle with a wire brush.

38   Depress the valve diaphragm and check the seating area for cleanliness and signs of rubbing by looking through the valve outlet. If the pintle or the seat are not completely clean, repeat the procedure in Step 36.

39   Hold the bottom of the valve securely and try to rotate the top of the valve back-and-forth. Replace the valve if any looseness is felt.

40   Inspect the valve outlet for deposits. Remove any deposit buildup with a screwdriver or other suitable sharp tool.

41   Clean the manifold mounting surface.

42   Using a new gasket, install the old (cleaned) or new EGR valve on the intake manifold.

43   Install the EGR valve mounting bolts and tighten them securely.

44   Attach the vacuum hose to the valve.

45   If your vehicle is a TBI model, install the air cleaner. If your vehicle is a MPFI model, install the plenum (Chapter 4).

EGR solenoid

46   Disconnect the cable from the negative terminal of the battery.

47   If your vehicle is a TBI model, remove the air cleaner.

48   Disconnect the electrical connector at the solenoid.

49   Disconnect the vacuum hoses from the solenoid.

50   Remove the mounting nut and detach the solenoid.

51   Install the new solenoid and tighten the nut securely. The remainder of the installation procedure is the reverse of removal.

EGR valve cleaning

52   With the EGR valve removed, inspect the passages for excessive deposits.

53   It is a good idea to place a rag securely in the passage opening to keep debris from entering. Clean the passages by hand, using a drill bit.

8   Air Injection Reaction (AIR) system

Refer to illustrations 8.2, 8.22, 8.23, 8.32, 8.41, 8.47 and 8.52

General description

1   The Air Injection Reaction (AIR) system reduces hydrocarbon (HC), carbon monoxide (CO) and nitrous oxide (NOx) emissions in the exhaust and enables the catalytic converter to heat up quickly after engine start-up so that conversion of the exhaust gases can occur as soon as possible.

2   The AIR system includes an air pump, a control valve (which is really two - switching and divert - valves), a check valve, the catalytic converter and the plumbing which connects these components **(see illustration)**. A positive displacement, vane-type pump, driven by a belt on the front of the engine, supplies air to the system. The pump is permanently lubricated and requires no maintenance.

3   Intake air passes through a centrifugal filter fan at the front of the pump where foreign materials are separated from the air by centrifugal force. The fan should not be cleaned nor should it ever be removed from the pump unless it is damaged, as removal will destroy the fan.

4   On carbureted and TBI models, air flows from the pump through an ECM controlled valve called an AIR control valve and through check valves to either the exhaust manifolds or the catalytic converter. There are really two valves inside the AIR control valve housing. The diverter valve sends air to the air cleaner, when necessary, to protect the catalytic converter, or to the other valve, which is called a switching valve. The switching valve directs

8.2 Typical Air Injection Reaction (AIR) pump location (arrow)

air either to the exhaust ports or to the catalytic converter, depending on operating conditions. The one piece control valve housing is located at the front of the engine in close proximity to the AIR pump and is connected to it by a short metal pipe called an adapter. The check valves prevent exhaust gas backflow into the pump in the event of a backfire in the exhaust system or AIR pump drivebelt failure.

5   On MPFI models, a basically similar device combines the diverter valve with a solenoid controlled vent. In its energized state, the air is diverted to the switching valve; when de-energized, air is diverted to the silencer, a short metal pipe with a small muffler attached to its end. The silencer assembly is bolted to the right side of the water pump. Aside from where diverted air is dumped, the differences between the two AIR control valves are minor. Later units are actually two valves connected by a rubber hose instead of a one-piece housing, are located a little farther from the pump and are connected to the pump with a rubber hose instead of a metal pipe.

6   The air switching valve directs air to the exhaust ports during cold engine operation or whenever the system is operating in open loop mode. It sends air to the catalytic converter when the system is in closed loop.

7   When it's necessary to divert air, the ECM turns off the solenoid in the diverter valve, routing air to the air cleaner or silencer.

8   Air is diverted to the air cleaner or silencer under all of the following conditions:

If the engine is running too rich.
Any time the ECM recognizes a problem and sets the Check Engine light.
During deceleration.
During high rpm operation when air pressure is greater than the setting for the internal relief valve.

9   If no air flow is routed to the exhaust stream at the exhaust ports or the catalytic converter pipe, the HC and CO emission levels will be too high. But if air is flowing to the exhaust ports all the time, it increases the temperature of the converter to unacceptable levels. Similarly, air flowing constantly to the catalytic converter causes overheating of the converter if the engine is running too rich. Therefore, both the diverter valve and the switching valve are operated by ECM-controlled solenoids to ensure that they respond quickly to the engine's changing needs.

10   An open in the diverter valve electrical circuit will divert all air flow to the air cleaner or silencer. If an open occurs in the switching valve electrical circuit, air will be routed to the converter at all times. Mechanical failures of the valves themselves will result in air flowing incorrectly to the exhaust ports or the converter.

Check

Air pump

11   Disconnect the adapter pipe or rubber hose from the AIR pump. Accelerate the engine to about 1500 rpm and feel the air flow from the pump. If air flow increases as the engine is accelerated, the pump is operating properly.

12   If air flow does not increase, or is not present, check the drivebelt for proper tension. If the belt is tensioned properly, listen for a leaking pressure relief valve (an air leak is audible even while the pump is operating).

13   The AIR pump is not completely silent. Under normal operating conditions, its noise rises in pitch as the engine speed increases. But a harsh, whining sound is a symptom of trouble. To determine if the AIR pump is emitting excessive noise, operate the engine with the pump drivebelt removed.

14   If the noise vanishes when the belt is removed, rotate the pump by hand to see if it has seized. Inspect the hoses, tubes and all connections for leaks and proper routing. Check for air flow from the diverter and/or switching valve(s). Check the AIR pump mounting bolts to make sure they are tight. If no irregularities exist and the AIR injection pump noise is still excessive, replace the pump.

Hoses and pipes

15   Inspect the hoses and pipes for deterioration and holes. Check the routing of all hoses and pipes - close proximity to hot or sharp engine parts may cause wear. Check all hose and pipe connections and clamps for tightness.

16   If a leak is suspected on the pressure side of the system, or if a hose or pipe has been disconnected on the pressure side, the connections should be checked for leaks with a soapy water solution. With the pump running, bubbles will form if a leak exists.

Check valve

17   Inspect the check valve whenever the hose is disconnected from it or whenever check valve failure is suspected. If the pump is inoperative and an inspection reveals evidence of exhaust gases in the pump, check valve failure has occurred.

18   Blow through the check valve (toward the cylinder head), then attempt to suck back through the check valve. Flow should only be in one direction (toward the exhaust valve). Replace a valve that does not operate properly.

Air management system

Note: *If the above quick checks can't pinpoint the cause of a malfunction in the air management system, there may be an electrical problem. The following sequence of tests will help you find it.*

19   The air management system is controlled by diverter and switching valves, each with its own ECM-controlled vacuum solenoid. When a solenoid is grounded by the ECM, manifold vacuum will activate a valve and direct pump air as follows:

a) *While the diverter solenoid is ungrounded by the ECM, air is diverted to the air cleaner or silencer.*

b) *As soon as the diverter solenoid is grounded by the ECM, air is routed to the air switching valve.*

c) *When the switching solenoid is not grounded by the ECM, air is sent to the catalytic converter.*

d) *Should the switching solenoid be grounded by the ECM, air is sent to the exhaust port.*

20   To check the air management system, start the engine. With the diagnostic terminal ungrounded and the engine running at part throttle below 2000 rpm, air should be felt at the outlet to the exhaust ports during open loop operation (which takes from 6 to 120 seconds on a warm engine) and should be switched to the converter as the system goes closed loop.

21   The air management system will do one of four things when subjected to the above conditions:

It will function correctly (proceed to Step 22).
It will constantly send air to the exhaust ports (proceed to Step 24).
It will constantly send air to the divert valve (proceed to Step 25).
It will constantly send air to the catalytic converter (proceed to Step 25).

22   If the system functions correctly, disconnect the HEI bypass timing connector **(see illustration)**. This will set a Code 42. When a code is set, the ECM opens the ground to the AIR control valve and allows air to divert. Air should be diverted as soon as the CHECK ENGINE light comes on. If it does divert, the system is functioning properly. Reconnect the HEI timing connector and clear the ECM memory.

23   If the AIR control valve does not divert as soon as the CHECK ENGINE light comes on, turn the ignition to On and, with the engine stopped, disconnect the diverter valve

8.22 Disconnecting the HEI bypass timing connector (arrow) will set a Code 42, which will cause the ECM to open the ground circuit from the AIR control valve and to flash the CHECK ENGINE light - air should divert as soon as the CHECK ENGINE light comes on

8.23 If the air management system does not divert air as soon as the CHECK ENGINE light comes on, turn the ignition to On and, with the engine stopped, disconnect the diverter valve solenoid electrical connector (the lower connector) and hook up a test light between the harness connector terminals

solenoid electrical connector and connect a test light between the harness connector terminals **(see illustration)**. If the light doesn't come on, replace the control valve assembly or the diverter valve. If it does come on, check for a short to ground in the circuit from the solenoid to the ECM (refer to the wiring diagrams). If the circuit from the solenoid to the ECM is okay, check the resistance between the solenoid terminals. It should be more than 20 ohms. If it isn't, replace the control solenoid and the ECM. If it is more than 20 ohms, the ECM may be faulty. **Note:** *Before replacing the ECM, check the resistance of each ECM controlled relay and solenoid coil with an ohmmeter. Refer to the wiring diagrams for coil terminal identification for both the solenoid(s) and relay(s) to be checked. Solenoids are turned on and off by ECM internal electronic switches called "drivers." Each driver is part of a group of four called "quad-drivers." Failure of one driver can damage any other driver within the set. Solenoid coil resistance must measure more than 20 ohms. Less resistance will cause early failure of the ECM driver. Using an ohmmeter, check the resistance of the canister purge, diverter and switching solenoid coils before installing a replacement ECM. Replace any relay or solenoid if the coil resistance measures less than 20 ohms.*

24   If the system constantly sends air to the exhaust ports, turn the ignition to On and, with the engine stopped and the diagnostic terminal not grounded, disconnect the switching valve solenoid electrical connector and connect the test light between the harness connector terminals. Note whether the test light comes on. If the test light doesn't come on, replace the air switching valve. If the light comes on, check for a short to ground in the circuit from the air switching valve to the ECM (refer to the wiring diagrams). If circuit is

okay, replace the ECM. **Note:** *Before replacing the ECM, check the resistance of each ECM controlled relay and solenoid coil with an ohmmeter. Refer to the wiring diagrams for coil terminal identification for both the solenoid(s) and relay(s) to be checked. Solenoids are turned on and off by ECM internal electronic switches called "drivers." Each driver is part of a group of four called "Quad-Drivers." Failure of one can damage any other driver within the set. Solenoid coil resistance must measure more than 20 ohms. Less resistance will cause early failure of the ECM driver. Using an ohmmeter, check the resistance of the canister purge, switching and diverter solenoid coils before installing a replacement ECM. Replace any relay or solenoid if the coil resistance measures less than 20 ohms.*

25   If the system constantly sends air to the diverter valve or to the catalytic converter, check the circuits, as follows:

26   With the ignition on and the engine stopped, ground the diagnostic terminal. Disconnect the applicable solenoid wiring harness and connect the test light between the harness connector terminals. If the test light comes on, check vacuum at the valve to make sure that it is at least 10-inches while the engine is idling. If it is, either the valve connection or the valve is faulty.

27   If the test light does not come on, connect it between each connector terminal and ground. If the light comes on at one terminal, check for an open in the wire between the applicable solenoid and its ECM terminal. If there is no open, check the resistance of the applicable solenoid. If it is under 20 ohms, replace the solenoid and the ECM. If it isn't, either the ECM connector or the ECM itself is faulty. **Note:** *Before replacing the ECM, check the resistance of each ECM controlled relay and solenoid coil with an ohmmeter. Refer to the wiring diagrams for coil terminal identification for both the solenoid(s) and relay(s) to be checked. Solenoids are turned on and off by ECM internal electronic switches called "drivers." Each driver is part of a group of four called "quad-drivers." Failure of one can damage any other driver within the set. Solenoid coil resistance must measure more than 20 ohms. Less resistance will cause early failure of the ECM driver. Using an ohmmeter, check the resistance of the canister purge, switching and diverter solenoid coils before installing a replacement ECM. Replace any relay or solenoid if the coil resistance measures less than 20 ohms.*

28   If the light does not come on at all, repair the open in the circuit from the fusebox to the ignition switch. If the light comes on at both terminals, repair the short in the circuit to the ECM. Then test it again.

Component replacement

Pump centrifugal filter fan

Caution: *The centrifugal filter fan should not be cleaned with either compressed air or solvents. Nor should it be removed from the*

8.32 To replace the AIR pump filter fan, remove the pulley bolts and the pulley, then insert a pair of needle nose pliers into the vanes and pull the old filter off

pump unless it is damaged, as removal usually destroys it.

29   Disconnect the cable from the negative terminal of the battery **Caution:** *On models equipped with a Delco Loc II audio system, disable the anti-theft feature before disconnecting the battery.* Loosen the AIR pump drive pulley bolts.

30   Remove the drivebelt (Chapter 1).

31   Remove the drive pulley bolts and the pulley from the AIR pump.

32   Carefully pull the pump centrifugal filter fan from the AIR pump by inserting a pair of needle-nose pliers into the edge **(see illustration)**. **Caution:** *Do not allow any filter fragments to enter the AIR pump intake hole. Do not use a screwdriver to pry off the filter fan - it will damage the pump sealing lip. Do not drive the metal hub from the filter fan.*

33   Install the new filter fan, place the pump pulley against it and tighten the pump pulley bolts a little at a time. This will press the centrifugal filter fan into its cavity in the pump housing. **Caution:** *Do not drive the filter fan on with a hammer. A slight amount of interference with the housing bore is normal.* After a new filter fan has been installed, it may squeal upon initial operation or until the outer sealing lip has been worn in. This may require a short period of pump operation at various engine speeds.

34   The remainder of installation is the reverse of the removal procedure.

AIR injection pump

35   Hold the pump pulley from turning by compressing the drivebelt, then loosen the pump pulley bolts.

36   Remove the drivebelt (Chapter 1).

37   Remove the pulley bolts and the pulley.

38   If you are replacing the pump centrifugal filter fan, insert needlenose pliers and pull the filter fan from the hub (see Steps 32 and 33 above).

39   Disconnect the adapter pipe from the AIR pump or loosen the hose clamp.

40   Detach the hose from the AIR pump.

8.41 Typical components of the AIR pump mounting

1 AIR pump   3 Alternator bracket
2 AIR pump support bracket

8.47 To remove the control valve, unplug the electrical connectors (1) and the vacuum lines (2) from the diverter and switching valves

41   Remove the five bracket bolts (two in front and three in the rear) that mount the AIR pump (see illustration).
42   Loosen the AIR pump-to-alternator bracket bolt at the alternator and swing the bracket up and out of the way. Remove the AIR pump.
43   Installation is the reverse of removal.

AIR control valve (diverter and switching valves)

Note: On Crossfire fuel injection models, failure of either valve necessitates the replacement of the entire (one-piece) AIR control valve assembly. On newer vehicles, even though the diverter and switching valves are referred to jointly as the AIR control valve, either one may be purchased individually.

44   Disconnect the cable from the negative terminal of the battery.
45   On carbureted and TBI models, remove the air cleaner.
46   On carbureted and TBI models, remove the adapter bolts and the adapter (the pipe between the AIR pump and the control valve)

or disconnect the rubber hose that connects the AIR pump and the control valve.
47   Disconnect the electrical connectors and vacuum lines from the switching valve and the diverter valve (see illustration).
48   Loosen the hose clamps and detach all hoses from both valves.
49   Remove the bracket bolts, the brackets and the AIR control valve assembly. If you are only replacing a single valve, disconnect the two components and discard the faulty part. Connect the new component.
50   Installation is the reverse of the removal procedure.

Check valve

Note: There is a check valve for each cylinder bank. The following removal procedure applies to either valve.
51   Loosen the hose clamp and disconnect the hose from the check valve.
52   Using a backup wrench to prevent the air injection pipe from turning, loosen and

unscrew the check valve from the pipe (see illustration).
53   Installation is the reverse of the removal procedure. Be sure to coat the threads with anti-seize compound.

9   Evaporative Emission Control System (EECS)

Refer to illustration 9.2

General description

1   This system is designed to trap and store fuel vapors that evaporate from the fuel tank, throttle body and intake manifold.
2   The Evaporative Emission Control System (EECS) consists of a charcoal-filled canister and the lines connecting the canister to the fuel tank, ported vacuum and intake manifold vacuum (see illustration).
3   Fuel vapors are transferred from the fuel

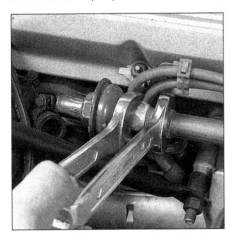

8.52 Use a backup wrench on the air injection pipe to prevent it from being twisted when the check valve is removed

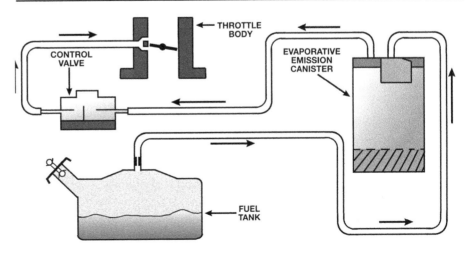

9.2 Details of a typical EVAP system

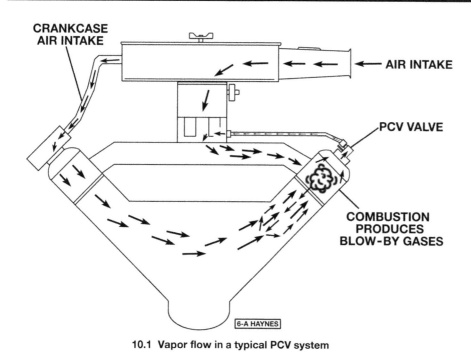

10.1 Vapor flow in a typical PCV system

tank, carburetor float bowl, throttle body and intake manifold to a canister where they are stored when the engine is not operating. When the engine is running, the fuel vapors are purged from the canister by intake air flow and consumed in the normal combustion process.

4   The ECM operates a solenoid valve which controls vacuum to the purge valve in the charcoal canister. Under cold engine or idle conditions, the solenoid is turned on by the ECM, which closes the valve and blocks vacuum to the canister purge valve. The ECM turns off the solenoid valve and allows purge when the engine is warm, after the engine has been running a specified time, above a specified vehicle speed and above a specified throttle opening.

5   The control valve opens with vacuum applied and closes with no vacuum. This prevents purge to the intake manifold under conditions of low ported vacuum, such as deceleration.

6   If the solenoid is open, or is not receiving power, the canister can purge to the intake manifold at all times. This can allow extra fuel at idle or during warm-up, which can cause rough or unstable idle, or rich operation during warm-up.

7   Poor idle, stalling and poor driveability can be caused by an inoperative purge valve, a damaged canister, split or cracked hoses or hoses connected to the wrong tubes.

8   Evidence of fuel loss or fuel odor can be caused by liquid fuel leaking from fuel lines or the TBI, a cracked or damaged canister, an inoperative bowl vent valve, an inoperative purge valve, disconnected, misrouted, kinked, deteriorated or damaged vapor or control hoses or an improperly seated air cleaner or air cleaner gasket.

Check

9   Inspect each hose attached to the canister for kinks, leaks and breaks along its entire length. Repair or replace as necessary.

10   Inspect the canister. It it's cracked or damaged, replace it.

11   Look for fuel leaking from the bottom of the canister. If fuel is leaking, replace the canister. Check the hoses and hose routing.

12   Check the filter at the bottom of the canister. If it's dirty, plugged or damaged, replace the canister.

13   Apply a short length of hose to the lower tube of the purge valve assembly and attempt to blow through it. Little or no air should pass into the canister (a small amount of air will pass because the canister has a constant purge hole).

14   With a hand vacuum pump, apply vacuum through the control vacuum signal tube to the purge valve diaphragm. If the diaphragm does not hold vacuum for at least 20 seconds, the diaphragm is leaking and the canister must be replaced.

15   If the diaphragm holds vacuum, again try to blow through the hose while vacuum is still being applied. An increased flow of air should be noted. If it isn't, replace the canister.

Component replacement

Fuel vapor canister solenoid

16   Disconnect the cable from the negative terminal of the battery.

17   Remove the solenoid cover bolts and the cover (if equipped).

18   Disconnect the wires and vacuum hoses from the solenoid.

19   Remove the solenoid.

20   Installation is the reverse of removal.

Fuel vapor canister

21   Disconnect the hoses from the canister.

22   Remove the pinch bolt and loosen the clamp.

23   Remove the canister.

24   Installation is the reverse of removal.

10   Positive Crankcase Ventilation (PCV) system

Refer to illustration 10.1

General description

1   The positive crankcase ventilation system, or PCV as it is more commonly called, reduces hydrocarbon emissions by circulating fresh air through the crankcase to pick up blow-by gases which are then rerouted through the carburetor or throttle body to be burned in the engine **(see illustration)**.

2   The main components of this simple system are vacuum hoses and a PCV valve which regulates the flow of gases according to engine speed and manifold vacuum.

Check

3   The PCV system can be checked quickly and easily for proper operation. This system should be checked regularly as carbon and gunk deposited by the blow-by gases will eventually clog the PCV valve and/or system hoses. When the flow of the PCV system is reduced or stopped, common symptoms are rough idling or reduced engine speed at idle.

4   To check for proper vacuum in the system, remove the top plate of the air cleaner and locate the small PCV filter on the inside of the air cleaner housing.

5   Disconnect the hose leading to this filter. Be careful not to break the molded fitting on the filter.

6   With the engine idling, place your thumb lightly over the end of the hose. You should feel a slight pull or vacuum. The suction may be heard as your thumb is released. This will indicate that air is being drawn all the way through the system. If a vacuum is felt, the system is functioning properly. Check that the filter inside the air cleaner housing is not clogged or dirty. If in doubt, replace the filter with a new one, an inexpensive safeguard (refer to Chapter 1).

7   If there is very little vacuum or none at all at the end of the hose, the system is clogged and must be inspected further.

8   Shut off the engine and locate the PCV valve. Carefully pull it from its rubber grommet. Shake it and listen for a clicking sound. That is the rattle of the valve's check needle. If the valve does not click freely, replace it with a new one.

9   Now start the engine and run it at idle speed with the PCV valve removed. Place your thumb over the end of the valve and feel for suction. This should be a relatively strong vacuum which will be felt immediately.

10   If little or no vacuum is felt at the PCV

valve, turn off the engine and disconnect the vacuum hose from the other end of the valve. Run the engine at idle speed and check for vacuum at the end of the hose just disconnected. No vacuum at this point indicates that the vacuum hose or inlet fitting at the engine is plugged. If it is the hose which is blocked, replace it with a new one or remove it from the engine and blow it out sufficiently with compressed air. A clogged passage at the carburetor or manifold requires that the component be removed and thoroughly cleaned to remove carbon build-up. A strong vacuum felt going into the PCV valve, but little or no vacuum coming out of the valve, indicates a failure of the PCV valve requiring replacement with a new one.

11 When purchasing a new PCV valve, make sure it is the correct one for your engine. An incorrect PCV valve may pull too little or too much vacuum, possibly leading to engine damage.

Component replacement

12 The replacement procedures for both the PCV valve and filter are covered in Chapter 1.

11 Early Fuel Evaporation (EFE) system

General description

Servo type

If your vehicle is equipped with a vacuum servo type EFE system, located between the exhaust manifold and the exhaust pipe, refer to Chapter 1 for a general description of the system and system checking procedures.

Electrically heated type

1 This unit provides rapid heat to the intake air supply on carbureted engines by means of a ceramic heater grid which is integral with the carburetor base gasket and located under the primary bore.
2 The components involved in the EFE operation include the heater grid, a relay, electrical wires and connectors and the ECM.
3 The EFE heater unit is controlled by the vehicle's Electronic Control Module (ECM) through a relay. The ECM senses the coolant temperature level and applies voltage to the heater unit only when the engine is below a pre-determined level. At normal operating temperatures, the heater unit is off.
4 If the EFE heater is not coming on, poor cold engine performance will be experienced. If the heater unit is not shutting off when the engine is warmed up, the engine will run as if it is out of tune (due to the constant flow of hot air through the carburetor).

Check

Servo type

5 To check the operation of the EFE/TVS, allow the engine temperature to fall below 80-degrees F (26-degrees C).

6 Drain the coolant from the engine until the level is below the level of the switch.
7 Disconnect and label the vacuum lines, then remove the switch.
8 Blow into either of the TVS ports. Air should flow through the valve.
9 Heat the TVS in hot water until the temperature of the valve is above 90-degrees F (32-degrees C).
10 Blow into either port of the TVS. No air should flow through the valve.
11 If the condition in either Step 9 or 10 is not met, replace the valve with a new one.
12 For other checking procedures for the servo type EFE system, refer to Chapter 1.

Electrically heated type

13 If the EFE system is suspected of malfunctioning while the engine is cold, first visually check all electrical wires and connectors to be sure they are clean, tight and in good condition.
14 With the ignition switch On, use a circuit tester or voltmeter to be sure current is reaching the relay. If not, there is a problem in the wiring leading to the relay, in the ECM thermo switch, or in the ECM itself.
15 Next, with the engine cold but the ignition switch On, disconnect the heater unit wiring connector and use a circuit tester or voltmeter to see if current is reaching the heater unit. If so, use a continuity tester to check for continuity in the wiring connector attached to the heater unit. If continuity exists, the system is operating correctly in the cold engine mode.
16 If current was not reaching the heater unit, but was reaching the relay, replace the relay.
17 To check that the system turns off at normal engine operating temperature, first allow the engine to warm up thoroughly. With the engine idling, disconnect the heater unit wiring connector and use a circuit tester or voltmeter to check for current at the heater unit.
18 If current is reaching the heater unit, a faulty ECM is indicated.
19 For confirmation of the ECM condition, refer to Section 2 or have the system checked by a dealer or automotive repair shop.

Component replacement

Servo type EFE switch

20 **Note:** *The oxygen sensor may be located near the EFE valve on some vehicles. If so, care should be taken not to damage the sensor or its connections during this procedure.*
21 Disconnect the vacuum hose from the actuator fitting.
22 Remove the exhaust pipe-to-manifold nuts, washers (if used) and tension springs (if used).
23 Lower the exhaust (crossover) pipe. In some cases, complete removal is not necessary.
24 Remove the nuts retaining the valve

assembly to the exhaust manifold.
25 Remove the valve assembly. If only the actuator assembly is to be replaced, separate it from the valve.
26 Installation is the reverse of the removal procedure.

TVS switch (V8 models equipped with carburetor only)

27 Drain the engine coolant until the level is below the level of the switch.
28 Disconnect the hoses from the TVS switch, making note of their positions for reassembly.
29 Using a wrench, remove the TVS switch.
30 Apply a soft-setting sealant uniformly to the threads of the new TVS switch. Be careful that none of the sealant gets on the sensor end of the switch.
31 Install the switch.
32 Connect the vacuum hoses to the switch in their original positions and add coolant as necessary.

Heater element

33 Remove the air cleaner.
34 Disconnect all wires, vacuum hoses and fuel lines from the carburetor.
35 Disconnect the wiring connector leading to the EFE unit.
36 Remove the carburetor, referring to Chapter 4 if necessary.
37 Lift off the heater unit.
38 Installation is the reverse of the removal procedure.
39 Following installation, start the engine and check for air and fuel leaks around the carburetor.

Heater relay

40 Disconnect the cable from the negative battery terminal.
41 Disconnect and label the relay electrical connections.
42 Remove the bolts retaining the relay and remove the relay.
43 Installation is the reverse of the removal procedure.

Solenoid

44 Disconnect the cable from the negative battery terminal.
45 If necessary to gain access to the solenoid, remove the air cleaner.
46 Disconnect and label the electrical connections and vacuum hoses at the solenoid.
47 Remove the solenoid.
48 Installation is the reverse of the removal procedure.

12 Thermostatic Air Cleaner (THERMAC)

Refer to illustration 12.2

General description

1 The thermostatic air cleaner (THERMAC) system is provided to improve engine effi-

ciency and reduce hydrocarbon emissions during the initial warm-up period by maintaining a controlled air temperature at the carburetor. This temperature control of the incoming air allows leaner carburetor and choke calibrations.

2   The system uses a damper assembly located in the snorkel of the air cleaner housing to control the ratio of cold and warm air directed into the carburetor **(see illustration)**. This damper is controlled by a vacuum motor which is, in turn, modulated by a temperature sensor in the air cleaner. On some engines, a check valve is used in the sensor, which delays the opening of the damper flap when the engine is cold and the vacuum signal is low.

3   It is during the first few miles of driving (depending on outside temperature) that this system has its greatest effect on engine performance and emissions output. When the engine is cold, the damper flap blocks off the air cleaner inlet snorkel, allowing only warm air from the exhaust manifold to enter the carburetor. Gradually, as the engine warms up, the flap opens the snorkel passage, increasing the amount of cold air allowed in. Once the engine reaches normal operating temperature, the flap opens completely, allowing only cold, fresh air to enter.

4   Because of this cold-engine-only function, it is important to periodically check this system to prevent poor engine performance when cold or overheating of the fuel mixture once the engine has reached operating temperatures. If the air cleaner valve sticks in the "no heat' position, the engine will run poorly, stall and waste gas until it has warmed up on its own. A valve sticking in the "heat' position causes the engine to run as if it is out of tune (due to the constant flow of hot air to the carburetor).

Check

5   Refer to Chapter 1 for maintenance and checking procedures for this system. If problems were encountered in the system's performance while performing the routine maintenance checks, refer to the procedures which follow.

6   If the damper door did not close off snorkel air when the cold engine was first started, disconnect the vacuum hose at the snorkel vacuum motor and place your thumb over the hose end, checking for vacuum. If there is vacuum going to the motor, check that the damper door and link are not frozen or binding within the air cleaner snorkel. Replace the vacuum motor if the hose routing is correct and the damper door moves freely.

7   If there was no vacuum going to the motor in the above test, check the hoses for cracks, crimps and proper connection. If the

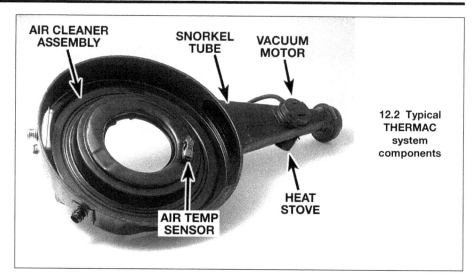

12.2 Typical THERMAC system components

hoses are clear and in good condition, replace the temperature sensor inside the air cleaner housing.

Component replacement

Note: *Check on parts availability before attempting component replacement. Some components may no longer be available for all models, requiring replacement of the complete air cleaner assembly.*

Air cleaner vacuum motor

8   Remove the air cleaner assembly from the engine and disconnect the vacuum hose from the motor.

9   Drill out the two spot welds which secure the vacuum motor retaining strap to the snorkel tube.

10   Remove the motor attaching strap.

11   Lift up the motor, cocking it to one side to unhook the motor linkage at the control damper assembly.

12   To install, drill a 7/64-inch hole in the snorkel tube at the center of the retaining strap.

13   Insert the vacuum motor linkage into the control damper assembly.

14   Using the sheet metal screw supplied with the motor service kit,
attach the motor and retaining strap to the snorkel. Make sure the sheet metal screw does not interfere with the operation of the damper door. Shorten the screw if necessary.

15   Connect the vacuum hose to the motor and install the air cleaner assembly.

Air cleaner temperature sensor

16   Remove the air cleaner from the engine and disconnect the vacuum hoses at the sensor.

17   Carefully note the position of the sensor. The new sensor must be installed in exactly the same position.

18   Pry up the tabs on the sensor retaining clip and remove the sensor and clip from the air cleaner.

19   Install the new sensor with a new gasket in the same position as the old one.

20   Press the retaining clip onto the sensor. Do not damage the control mechanism in the center of the sensor.

21   Connect the vacuum hoses and attach the air cleaner to the engine.

13   Transmission Converter Clutch (TCC)

1   Toward optimizing the efficiency of the emissions control network,
the ECM controls an electrical solenoid mounted in the automatic transmission of vehicles so equipped. When the vehicle reaches a specified speed, the ECM energizes the solenoid and allows the torque converter to mechanically couple the engine to the transmission, under which conditions emissions are at their minimum. However, because of other operating condition demands (deceleration, passing, idle, etc.), the transmission must also function in its normal, fluid-coupled mode. When such latter conditions exist, the solenoid de-energizes, returning the transmission to fluid coupling. The transmission also returns to fluid-coupling operation whenever the brake pedal is depressed.

2   Due to the requirement of special diagnostic equipment for the testing of this system, and the possible requirement for dismantling of the automatic transmission to replace components of this system, checking and replacing of the components should be handled by a dealer or automotive repair shop.

Chapter 7 Part A
Manual transmission

Contents

	Section
Clutch start switch - replacement and adjustment	9
Extension housing oil seal - replacement	7
General information	1
Shift control lever (77mm 4-speed and 5-speed) - removal and installation	4
Speedometer gear seal - replacement	8
Transmission fluid level check	See Chapter 1

	Section
Transmission (76mm and 83mm 4-speed) - removal and installation	5
Transmission (77mm 4-speed and 5-speed) - removal and installation	6
Transmission linkage (76mm and 83mm 4-speed) - removal, installation and adjustment	3
Transmission mounts - check and replacement	2

Specifications

Torque specifications

Ft-lbs (unless otherwise indicated)

76 mm 4-speed
Oil filler plug	15
Transmission-to-clutch housing bolts	55
Crossmember-to-frame bolts	35
Mount-to-crossmember bolts	35
Mount-to-transmission bolts	35
Shift lever-to-shifter shaft bolts	25

83 mm 4-speed
Oil filler plug	15
Oil drain plug	18
Transmission-to-clutch housing bolts	55
Crossmember-to-frame bolts	35
Mount-to-crossmember bolts	35
Mount-to-transmission bolts	35
Shift lever-to-shifter shaft bolts	18

77 mm 4-speed and 5-speed
Oil filler plug	20
Dust cover bolts	96 in-lbs
Transmission-to-engine bolts	55
Shift lever cover-to-case bolts	120 in-lbs
Crossmember-to-frame bolts	35
Mount-to-crossmember bolts	35
Mount-to-transmission bolts	35
Dampener-to-transmission bolts	30

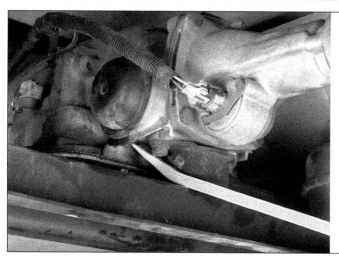

2.3 To check the transmission mount, insert a prybar between the transmission and the crossmember and pry the transmission up - if the rubber portion of the mount separates from the metal plate, replace the mount

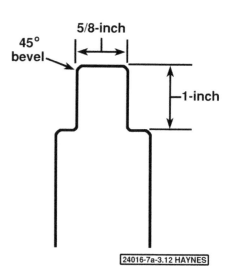

24016-7a-3.12 HAYNES

3.12 Fabricate a gauge from 5/64-inch thick steel (an old putty knife works well) to secure levers in Neutral during linkage adjustment

1   General information

Firebirds, for the years covered in this manual, are equipped with either a 4 or 5-speed manual transmission as standard equipment, depending on year, model and engine. The 5-speed is optionally available in some years for models designated to be equipped with a 4-speed as the standard transmission.

The 76mm and 83mm model 4-speed transmissions are operated through adjustable floor-mounted linkage attached to the left side plate of the transmission.

The 77mm model 4-speed and 5-speed transmissions are operated through a floor mounted gearshift lever assembly mounted atop the extension housing. The transmission case is made entirely of aluminum.

2   Transmission mounts - check and replacement

Refer to illustration 2.3

Check

1   Raise the vehicle and support it securely with jackstands.
2   Check for contact between the transmission and the crossmember. If contact is evident, replace the mount.
3   Insert a large prybar between the transmission and the crossmember and pry the transmission up **(see illustration)** . If the rubber portion of the mount separates from the metal plate, replace the mount.

Replacement

4   Raise the vehicle and support it securely with jackstands.
5   Remove the mount-to-crossmember nut.
6   Raise the rear of the transmission with a floor jack. Remove the mount-to-transmission bolts and remove the mount.
7   Installation is the reverse of removal.

3   Transmission linkage (76mm and 83mm 4-speed) - removal, installation and adjustment

Refer to illustration 3.12
1   Raise the vehicle and support it securely on jackstands.
2   Disconnect the shift rods from the control levers, labeling each one with tape to facilitate reinstallation.
3   Remove the knob from the shift lever.
4   Remove the grip from the parking brake lever.
5   Remove the retaining screws from the console cover.
6   Raise the console cover, disconnect the wire connectors and remove the console cover.
7   Remove the shift boot retaining bolts and the boot.
8   Remove the upper and lower shift control mounting bolts and remove the shifter.
9   Install the shifter by reversing the removal steps, then adjust the linkage as follows.
10   Place the transmission levers in the Neutral position.
11   Move the shift control lever to the Neutral position.
12   Align the holes in the levers beneath the shift lever with the notch in the shifter assembly and insert the gauge **(see illustration)** through the notch and holes to secure the levers in Neutral.
13   Attach the shift rod to the 3-4 shift lever at the transmission with the washer and retainer.
14   Loosely assemble the 3-4 shift rod nuts and swivel.
15   Insert the swivel into the shift control lever, install the washer and secure it with the retainer.
16   Apply rearward pressure on the 3-4 shift lever at the transmission while tightening the nuts against the swivel.
17   Tighten the nuts to 25 ft-lbs with a torque wrench.
18   Repeat Steps 14 through 18 for the 1-2

shift rod and levers and their respective nuts.
19   Repeat Steps 14 through 18 for the reverse shift rod and levers and their respective nuts.
20   Remove the gauge.
21   After the adjustments have been made, make sure the centerlines of the shift levers are aligned with each other, providing free crossover motion.

4   Shift control lever (77mm 4-speed and 5-speed) - removal and installation

Refer to illustration 4.3
1   Remove the center console.
2   With the transmission in Neutral, remove the retaining screws from the shift lever boot retainer and slide the boot up the lever.
3   Remove the shift lever retaining bolts at the transmission and remove the lever assembly by pulling it straight up and out of the transmission **(see illustration)**.
4   Installation is the reverse of the removal procedure.

5   Transmission (76mm and 83mm 4-speed) - removal and installation

1   Raise the vehicle and support it securely on jackstands.
2   Drain the lubricant from the transmission (refer to Chapter 1, if necessary).
3   Remove the torque arm from the vehicle (refer to Chapter 11).
4   Remove the driveshaft (refer to Chapter 8).

4.3 Remove the four bolts (arrows) and remove the shift lever from the transmission

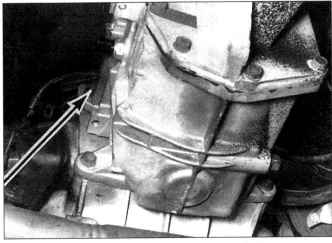

6.12 Remove the upper transmission mounting bolts (arrow) and install guide pins - other mounting bolt not visible

5   Disconnect the speedometer cable from the transmission.
6   Disconnect and label the wire connectors at the transmission.
7   Remove the exhaust brace.
8   Remove the shift linkage and shifter assembly (refer to Section 3).
9   Support the engine/transmission by placing a jack and block of wood under the engine oil pan.
10   Remove the crossmember retaining bolts.
11   Remove the transmission mount retaining bolts, then remove the crossmember and mount from the vehicle.
12   Remove the transmission-to-clutch housing upper bolts, install guide pins in the holes, then remove the lower bolts.
13   Lower the supporting jack until the transmission can be withdrawn to the rear and removed.
14   Installation is the reverse of the removal procedure. Apply a light coat of high-temperature grease to the input shaft bearing retainer and to the splined portion of the shaft to assure free movement of the clutch and transmission components during installation. Be sure to tighten all nuts and bolts to the specified torque.
15   Adjust the shift linkage (refer to Section 3).
16   Refill the transmission with the recommended lubricant.

6   Transmission (77mm 4-speed and 5-speed) - removal and installation

Refer to illustrations 6.12 and 6.14
1   Remove the shift control lever from the transmission (refer to Section 4).
2   Raise the vehicle and support it securely on jackstands.
3   Remove the torque arm from the vehicle (refer to Chapter 11).

4   Remove the driveshaft (refer to Chapter 8).
5   Disconnect the speedometer cable and wire connector at the transmission.
6   Disconnect the clutch cable at the transmission.
7   Support the engine/transmission by placing a jack and block of wood under the engine oil pan.
8   Remove the transmission mount bolts.
9   Remove the hanger supporting the catalytic converter. Take care not to damage the converter.
10   Remove the crossmember bolts, then remove the crossmember and transmission mounts.
11   Remove the dust cover bolts.
12   Remove the transmission-to-clutch housing upper bolts, install guide pins in the holes, then remove the lower bolts (see illustration). Guide pins can be fabricated from bolts of the same thread diameter and pitch as the transmission mounting bolts, but slightly longer. Cut the heads off the bolts and cut a screwdriver slot in the end with a hack-saw.
13   Lower the supporting jack until the transmission can be withdrawn to the rear

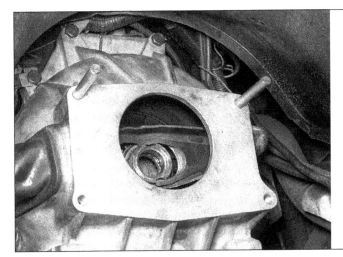

6.14 With the guide pins installed, the transmission is easier to align during installation

and removed.
14   Installation is the reverse of the removal procedure (see illustration). Apply a light coat of high-temperature grease to the main drive gear bearing retainer and to the splined portion of the transmission drive gear shaft to assure free movement of the clutch and transmission components during installation. Be sure to tighten all nuts and bolts to the specified torque.
15   Refill the transmission with the recommended lubricant.

7   Extension housing oil seal - replacement

Refer to illustrations 7.5 and 7.8
1   The extension housing oil seal can be replaced without removing the transmission from the vehicle.
2   Raise the vehicle and support it securely on jackstands.
3   Drain the lubricant from the transmission (refer to Chapter 1, if necessary).
4   Remove the driveshaft, as described in Chapter 8.

7.5 Pry the extension housing seal with a seal removal tool or large screwdriver

7.8 Install the extension housing seal with a seal driver or large socket

5   Carefully pry out the old seal **(see illustration)**.

6   Clean the counterbore and examine it for damage.

7   Lubricate the area between the lips of the new seal with transmission oil and coat the outer diameter with sealant.

8   Carefully install the seal with the lip in, until the flange seats. Use a seal driver, large socket or section of pipe of the proper diameter to drive the seal in **(see illustration)**.

9   Reinstall the driveshaft (refer to Chapter 8).

10   Refill the transmission with the recommended lubricant.

8   Speedometer gear seal - replacement

1   The speedometer gear seal can be replaced without removing the transmission from the vehicle.

2   Raise the vehicle and support it securely on jackstands.

3   Disconnect the speedometer cable, remove the lock plate-to-extension bolt and lock washer, then remove the lock plate.

4   Insert a screwdriver in the lock plate fitting and pry the fitting, gear and shaft from the extension.

5   Remove the O-ring.

6   Installation is the reverse of removal, but lubricate the new seal with transmission lubricant and hold the assembly so that the slot in the fitting is toward the lock plate boss on the extension.

9   Clutch start switch - replacement and adjustment

1   This switch is a safety device intended to prevent the vehicle from being started without the clutch pedal fully depressed. It is attached to the clutch pedal mounting bracket and is activated by a plastic shaft which is part of the switch.

2   Remove the screws retaining the hush panel and remove the panel.

3   Disconnect the wiring harness connector at the switch.

4   Pull the switch from the tubular retainer in the mounting bracket.

5   With the clutch pedal depressed, insert the new switch into the tubular retainer until the switch is completely seated. Audible clicks will be heard as the threaded portion of the switch is pushed toward the clutch pedal.

6   Allow the clutch pedal to return to its normal, fully engaged position. The switch will automatically be moved in the tubular retainer, providing adjustment.

7   Depress and release the clutch pedal once more to ensure correct adjustment.

8   Reconnect the wiring harness connector and reinstall the hush panel.

Chapter 7 Part B
Automatic transmission

Contents

	Section
Automatic transmission fluid change	See Chapter 1
Diagnosis - general	2
Extension housing oil seal - replacement	See Chapter 7A
Fluid level check	See Chapter 1
General information	1
Neutral safety and back-up light switch - adjustment	5

	Section
Shift linkage - check and adjustment	3
Speedometer gear seal - replacement	See Chapter 7A
Throttle valve (TV) cable assembly - description and adjustment	4
Transmission - removal and installation	6
Transmission mounts - check and replacement	See Chapter 7A

Specifications

Transmission type
1982	200C
1983	200C; 700-R4
1984 and later	700-R4

Torque specifications
Ft-lbs (unless otherwise indicated)

Transmission-to-engine bolts	30
Driveplate-to-converter bolts	35
Inspection cover bolts	96 in-lbs
Dipstick tube brace-to-transmission bolt	96 in-lbs
Transmission crossmember-to-frame bolts	40
Transmission mount-to-crossmember bolts	25
Catalytic converter/exhaust pipe bracket bolts	15
TV cable-to-transmission bolts/nuts	96 in-lbs
Torque arm bracket-to-transmission bolts	31
Shift cable pin-to-shift lever nut	96 in-lbs
Torque arm-to-rear axle bracket bolts	100
Driveshaft strap bolts	16

1 General information

Due to the complexity of the clutches and the hydraulic control system, and because of the special tools and expertise required to perform an automatic transmission overhaul, it should not be undertaken by the home mechanic. Therefore, the procedures in this Chapter are limited to general diagnosis, routine maintenance and adjustment and transmission removal and installation.

If the transmission requires major repair work, it should be left to a dealer service department or a reputable automotive or transmission repair shop. You can, however, remove and install the transmission yourself and save the expense, even if the repair work is done by a transmission specialist.

Adjustments that the home mechanic may perform include those involving the throttle valve cable, the shift linkage and the neutral safety switch. **Caution:** *Never tow a disabled vehicle at speeds greater than 30 mph or distances over 50 miles unless the driveshaft has been removed. Failure to observe this precaution may result in severe transmission damage caused by lack of lubrication.*

2   Diagnosis - general

Automatic transmission malfunctions may be caused by four general conditions: poor engine performance, improper adjustments, hydraulic malfunctions and mechanical malfunctions. Diagnosis of these problems should always begin with a check of the easily repaired items: fluid level and condition, shift linkage adjustment and throttle linkage adjustment. Next, perform a road test to determine if the problem has been corrected or if more diagnosis is necessary. If the problem persists after the preliminary tests and corrections are completed, additional diagnosis should be done by a dealer service department or a reputable automotive or transmission repair shop.

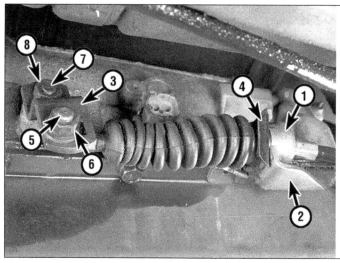

3.1 Typical shift linkage components

1   *Shift cable*
2   *Shift cable bracket*
3   *Shift lever*
4   *Shift cable retaining clip*
5   *Shift cable pin*
6   *Shift cable pin nut*
7   *Manual shift shaft*
8   *Manual shift shaft-to-shift lever nut*

3   Shift linkage - check and adjustment

Refer to illustration 3.1
Note: *Apply the parking brake and block the wheels to prevent the vehicle from rolling.*
1   Working under the vehicle, loosen the nut attaching the shift lever to the pin on the shift cable assembly **(see illustration)**.
2   Place the console shifter (inside the vehicle) in Neutral.
3   Place the manual shifting shaft on the transmission in Neutral.
4   Make sure the pin is free to move in the shift lever slot and tighten the pin-to-lever retaining nut to 96 in-lbs.
5   Make sure the engine will start in the Park and Neutral positions only.
6   If the engine can be started in any of the drive positions (as indicated by the shifter inside the vehicle), repeat the steps above or have the vehicle examined by a dealer, because improper linkage adjustment can lead to band or clutch failure and possible personal injury.

4   Throttle valve (TV) cable assembly - description and adjustment

Refer to illustrations 4.2 and 4.6
1   The throttle cable should not be thought of as just an automatic downshift cable, but rather as a cable that controls line pressure, shift feel and shift timing, as well as part throttle and detent downshifts. The function of the cable combines the functions of the vacuum modulator and the downshift (detent) cable found on other model transmissions.
2   The TV cable is attached to the link at the throttle lever and bracket assembly at the transmission and to either the throttle idler lever (L4 engines) or directly to the carburetor/TBI throttle lever on V6 and V8 engines **(see illustration)**.
3   Whenever the TV cable has been disconnected from the carburetor/TBI unit, it must be adjusted after installation.
4   The engine should be off.
5   On carbureted and TBI models, remove

the air cleaner, labeling all hoses as they are removed to simplify installation.
6   Depress and hold down the metal readjusting tab at the TV cable adjuster assembly **(see illustration)**.
7   While holding the tab down, move the cable housing into slider until it stops against the collar.
8   Release the readjustment tab.

V6 and V8 engine

9   Manually open the carburetor/TBI lever to the full throttle stop position. The cable will ratchet through the slider and automatically readjust itself.
10   Release the carburetor lever.

L4 engines

11   Rotate the idler lever to the maximum travel stop position. The cable will ratchet through the slider and automatically readjust itself.
12   Release the idler lever. **Note:** *Do not attempt to adjust the L4 engine's TV cable using the TBI lever. Be sure to use the idler lever.*

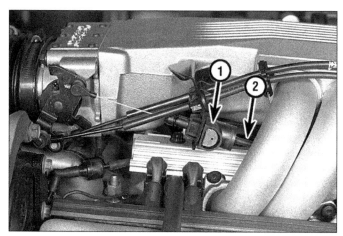

4.2 The throttle valve cable and adjuster are mounted on a bracket near the carburetor or throttle body

1   *Throttle valve cable adjuster*
2   *Throttle valve cable housing*

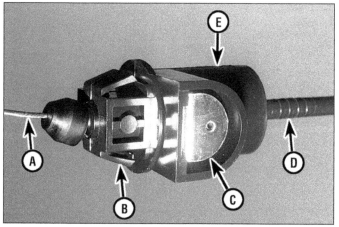

4.6 Details of the TV cable adjuster assembly

A   *TV cable*
B   *Locking lugs*
C   *Release tab*
D   *Cable housing*
E   *Slider*

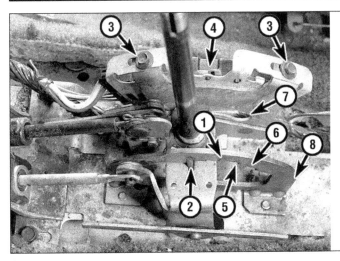

5.3 Details of the automatic transmission shifter assembly

1   Neutral notch
2   Transmission shift pin
3   Screw
4   Neutral switch assembly
5   Reverse notch
6   Park notch
7   Tang slot
8   Detent plate

All engines

13   Road test the vehicle. If delayed or only full-throttle shifts still occur, have the vehicle checked by a dealer.

5   Neutral safety and back-up light switch - adjustment

Refer to illustrations 5.3 and 5.6

1   Remove the screws retaining the shift indicator plate to the console.
2   Slide the plate up the shifter handle and secure it to the shifter knob with a rubber band.
3   Position the shifter assembly in the Neutral notch in the detent plate **(see illustration)**.
4   Loosen but do not remove the screws retaining the switch assembly to the shifter assembly.
5   Rotate the switch assembly so that the service adjustment hole on the switch is aligned with the carrier tang hole.
6   Insert a 3/32-inch drill bit as a gauge pin through the service adjustment hole to a depth of 19/32-inch **(see illustration)**.
7   Tighten the switch assembly retaining screws.
8   Remove the gauge pin.
9   Reinstall the shift indicator plate.

6   Transmission - removal and installation

1   Disconnect the cable from the negative battery terminal. **Caution:** *On models equipped with a Delco Loc II audio system, disable the anti-theft feature before disconnecting the battery.*
2   Remove the air cleaner, labeling all hoses as they are disconnected to simplify installation.
3   Disconnect the TV cable at the carburetor/TBI unit.

4   Remove the transmission fluid dipstick from the tube.
5   Remove the upper bolt retaining the dipstick tube and separate the tube from the transmission.
6   Raise the vehicle and support it securely on jackstands.
7   Remove the driveshaft (refer to Chapter 8).
8   Disconnect the speedometer cable at the transmission.
9   Disconnect the TCC wire connector at the transmission.
10   Remove the torque arm-to-transmission bracket bolts. **Caution:** *Rear spring force will cause the torque arm to move toward the floorpan when the arm is disconnected from the bracket. Carefully place a wooden block between the torque arm and floorpan when disconnecting the arm to avoid possible injury and damage to the floorpan.*
11   Loosen the rear torque arm bolts and pull the front of the arm free of the bracket, heeding the caution in Step 10.
12   Remove the catalytic converter/exhaust pipe bracket at the rear of the transmission.
13   Remove the support bracket bolts at the inspection cover.
14   Remove the bolts retaining the inspection cover and remove the cover.

15   Remove the now-exposed converter-to-driveplate bolts. It will be necessary to turn the crankshaft to bring each of the bolts into view (use a wrench on the large bolt at the front of the crankshaft). Mark the relative position of the converter and driveplate with a scribe so they can be reinstalled in the same position. Engage a large screwdriver in the teeth of the driveplate to prevent movement as the bolts are loosened.
16   Using a floor jack and a piece of wood placed between the jack and the transmission, support the transmission and remove the rear mount bolt.
17   Remove the crossmember bolts and the crossmember.
18   Lower the transmission as far as possible without causing the engine or transmission to contact the firewall, then disconnect the TV cable assembly and the oil cooler lines at the transmission.
19   Support the engine at the oil pan rail with a jack and remove the transmission-to-engine bolts. The upper bolts should be removed first.
20   Move the transmission to the rear and down. If necessary, carefully pry it free from the driveplate. Keep the rear of the transmission down at all times to prevent the converter from falling out. The converter can be held in place with a strap.
21   Installation is the reverse of the removal procedure, with the following additional instructions.
22   Before installing the driveplate-to-converter bolts, make sure that the weld nuts on the converter are flush with the driveplate and that the converter can be turned freely by hand in this position. Start the mounting bolts and tighten them finger tight, then tighten them to the specified torque. This will ensure proper alignment of the converter.
23   Tighten all nuts and bolts.
24   Adjust the shift linkage (refer to Section 3).
25   Adjust the TV cable (refer to Section 4).

5.6 Details of the Neutral safety switch adjustment

1   Service adjusment hole
2   Carrier tang

Notes

Chapter 8
Clutch and driveline

Contents

	Section
Axleshaft - removal and installation	13
Axleshaft bearing - replacement	15
Axleshaft oil seal - replacement	14
Clutch - general information	1
Clutch - removal, inspection and installation	4
Clutch adjustment	See Chapter 1
Clutch cross-shaft - removal and installation	3
Clutch pedal - removal and installation	2
Clutch pilot bearing - removal and installation	5
Driveshaft - general information	7

	Section
Driveshaft - removal and installation	10
Driveshaft out-of-balance correction	9
Hydraulic clutch assembly - removal and installation	6
Rear axle - general information and identification	12
Rear axle assembly - removal and installation	16
Rear axle oil change	See Chapter 1
Rear axle oil level check	See Chapter 1
Universal joints - disassembly, inspection and reassembly	11
Universal joints - wear check	8
Pressure plate-to-flywheel bolts	16

Specifications

Torque specifications

Ft-lbs (unless otherwise indicated)

Clutch pedal-to-mounting bracket	25
Clutch reservoir-to-mounting bracket	30 in-lbs
Clutch master cylinder-to-cowl	120 in-lbs
Clutch cross-shaft inboard ball stud	25
Clutch cross-shaft outboard ball stud nut	20
Clutch cross-shaft frame bracket-to-frame bolt	15
Clutch cross-shaft bracket-to-frame bracket bolt	35
Release cylinder heat shield-to-bellhousing	15
Release cylinder-to-bellhousing	15
Universal joint-to-rear pinion flange	15
Rear axle housing cover bolts	
1986 and earlier	30
1987 and later	20
Brake assembly-to-rear axle housing	35
Ring gear-to-differential case	80
Differential pinion shaft lock screw	20
Differential carrier cover bolts	20
Bearing cap-to-carrier	60
Filler plug	20

1 Clutch - general information

Refer to illustration 1.1

All manual transmission-equipped vehicles utilize a single dry plate, diaphragm spring-type clutch. Operation is through a foot pedal and rod linkage. The unit consists of a pressure plate assembly which contains the pressure plate, diaphragm spring and fulcrum rings. The assembly is bolted to the rear face of the flywheel **(see illustration)**. The driven plate (friction or clutch plate) is free to slide along the transmission input shaft and is held in place between the flywheel and pressure plate by the pressure exerted by the diaphragm spring. The friction lining material is riveted to the clutch plate, which incorporates a spring-cushioned hub designed to absorb driveline shocks and to assist in ensuring smooth starts.

The circular diaphragm spring assembly is mounted on shouldered pins and held in place in the cover by fulcrum rings. The spring itself is held in place by spring steel clips. On high performance vehicles, or where a heavy duty clutch is used, the diaphragm spring has bent fingers, where the standard version has flat fingers. The bent finger design is used to gain a centrifugal boost to aid rapid re-engagement of the clutch at high rotational speeds.

1982 and 1983 models

Depressing the clutch pedal pushes the release bearing forward to bear against the fingers of the diaphragm spring. This action causes the diaphragm spring outer edge to deflect and move the pressure plate to the rear to disengage the pressure plate from clutch plate.

When the clutch pedal is released, the diaphragm spring forces the pressure plate into contact with the friction linings of the clutch plate and at the same time pushes the clutch plate forward on its splines to ensure full engagement with the flywheel. The clutch plate is now firmly sandwiched between the pressure plate and the flywheel and the drive is taken up.

1984 and later models

A hydraulic clutch assembly is employed on all 1984 and later models. The assembly consists of a remote reservoir, a clutch master cylinder and a release cylinder.

The clutch master cylinder is mounted on the cowl panel in the engine compartment and the release cylinder is mounted on the engine bellhousing. The master cylinder operates directly off the clutch pedal.

When the clutch pedal is pushed in, hydraulic fluid (under pressure from the clutch master cylinder) flows into the release cylinder. Because the release cylinder is also connected to the clutch fork, the fork moves the release bearing into contact with the pressure plate release fingers, disengaging the clutch plate.

The hydraulic clutch system locates the clutch pedal and provides clutch adjustment automatically, so no adjustment of the clutch linkage or pedal position is required. **Caution:** *Prior to servicing the clutch or vehicle components that require the removal of the release cylinder (i.e. transmission and clutch housing removal), the master cylinder pushrod must be disconnected from the clutch pedal. If this is not done, permanent damage to the release cylinder will occur if the clutch pedal is depressed while the release cylinder is disconnected.*

If a malfunction of the hydraulic clutch assembly is suspected, verify it by removing the clutch housing dust shield and measuring the clutch release cylinder pushrod travel as follows.

Raise the vehicle and place it securely on jackstands. Remove the bolts retaining the clutch housing dust shield to the clutch housing and remove the shield. Note the position of the release cylinder pushrod. Have an assistant push the clutch pedal all the way to the floor. Measure the distance the pushrod travels while engaging the clutch fork. If the distance is 0.57-inch or more, the clutch hydraulic system is operating correctly. Do not remove it. If the release cylinder pushrod does not travel at least 0.57-inch, check the reservoir fluid level (the release cylinder must be in place when the fluid level is checked). **Note:** *Carefully clean the top of the reservoir before removing it to prevent contamination of the system. Fill the reservoir to the proper level, indicated by a step in the reservoir with DOT 3 brake fluid. Do not overfill the reservoir, as the upper portion must accept fluid that is displaced from the release cylinder as the clutch wears.*

If the reservoir required fluid, examine the hydraulic system components for leakage by removing the rubber boots from the master and release cylinders and checking for leakage past the pistons. A slight wetting of the surfaces is acceptable, but if excessive leakage is evident, the clutch hydraulic system must be replaced, as a whole, with a new one.

2 Clutch pedal - removal and installation

Refer to illustration 2.8

1   Disconnect the cable from the negative battery terminal. **Caution:** *On models equipped with a Delco Loc II audio system, disable the anti-theft feature before disconnecting the battery.*

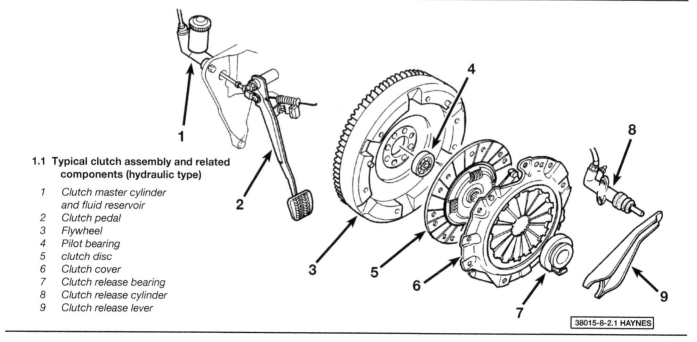

1.1 Typical clutch assembly and related components (hydraulic type)

1   Clutch master cylinder and fluid reservoir
2   Clutch pedal
3   Flywheel
4   Pilot bearing
5   clutch disc
6   Clutch cover
7   Clutch release bearing
8   Clutch release cylinder
9   Clutch release lever

38015-8-2.1 HAYNES

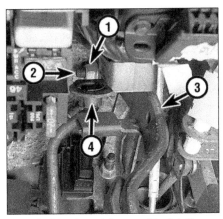

2.8 Clutch pedal assembly details

1   Mounting nut
2   Through-bolt
3   Clutch pedal
4   Clutch start switch mounting clip

2   Disconnect the clutch pedal return spring.
3   Remove the screws retaining the under-dash hush panel, then remove the hush panel.
4   If equipped with cruise control, remove the cruise control switch from the mount at the clutch pedal.
5   Disconnect and remove the clutch start switch at the clutch pedal (refer to Chapter 7A).
6   Remove the turn signal and hazard warning flasher mounting bracket screws.
7   Disconnect the clutch actuator rod (clutch master cylinder pushrod on models with hydraulic clutch) from the clutch pedal.
8   Remove the nut from the clutch pedal pivot bolt **(see illustration)**.
9   Pull the bolt out only far enough to allow removal of the clutch pedal, leaving it engaged through the brake pedal pivot and bracket.
10   Before installing the clutch pedal assembly, inspect and clean all components. Replace all components showing wear with new ones.

Note: *Do not clean the bushings with solvent; simply wipe them off with a clean rag.*
11   Installation is the reverse of the removal procedure. After installation is complete, adjust the clutch pedal free play (refer to Chapter 1).

3   Clutch cross-shaft - removal and installation

1   Remove the lever and shaft return spring.
2   Disconnect the clutch pedal and fork pushrods from their respective cross-shaft levers.
3   Loosen the outboard ball stud nut and slide the stud out of the slot in the bracket.
4   Move the cross-shaft outboard far enough to clear the inboard ball stud, then lift it out and remove it from the vehicle.
5   Check the ball stud seats on the cross-shaft, the engine bracket ball stud assembly and the anti-rattle O-ring for damage and wear; replace parts as necessary.
6   When installing, reverse the removal procedure. Lubricate the ball studs and seats with graphite grease.
7   When installation is complete, adjust the clutch pedal free play (refer to Chapter 1).

4   Clutch - removal, inspection and installation

Refer to illustrations 4.11, 4.12, 4.13 and 4.23
1   Access to the clutch is normally accomplished by removing the transmission, leaving the engine in the vehicle. If, of course, the engine is being removed for major overhaul, then the opportunity should always be taken to check the clutch assembly for wear at the same time.
2   Disconnect the clutch fork pushrod and spring, then remove the clutch housing from the engine block. On models equipped with a hydraulic clutch assembly, remove the

release cylinder heat shield and release cylinder from the clutch housing before removing the clutch housing.
3   Slide the clutch fork from the ball stud and remove the fork from the dust boot.
4   If necessary, the ball stud can be removed from the clutch housing by unscrewing it.
5   If there are no alignment marks on the clutch cover (an X-mark or white-painted letter) scribe or center-punch marks for indexing purposes during installation.
6   Unscrew the bolts securing the pressure plate and cover assembly one turn at a time in a diagonal sequence to prevent distortion of the clutch cover.
7   With all the bolts and lock washers removed, carefully lift the clutch away from the flywheel. Be careful not to drop the clutch plate.
8   DO NOT dismantle the pressure plate assembly.
9   If a new clutch plate is being installed, replace the release bearing at the same time. This will preclude having to replace it at a later date when wear on the clutch plate is still very little.
10   If the pressure plate assembly must be replaced, a rebuilt unit may be available.
11   Examine the clutch plate friction lining for wear and loose rivets, and the disc for rim distortion, cracks, broken hub springs and worn splines **(see illustration)**. The surface of the friction linings may be highly glazed, but as long as the clutch material pattern can be seen clearly, this is satisfactory. Compare the amount of lining remaining with a new clutch plate, if possible. If in doubt, replace the clutch plate with a new one.
12   Check the machined faces of the flywheel and the pressure plate **(see illustration)**. If the flywheel surface is grooved, it can be machined or replaced.
13   If the pressure plate is worn or damaged, replace it with a new one. If the condition of the diaphragm spring is suspect, the spring should be checked and replaced if necessary **(see illustration)**.

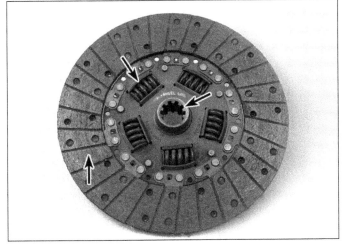

4.11 Check the clutch plate lining, spring and splines (arrows) for wear

4.12 Examine the pressure plate friction surface for score marks, cracks and evidence of overheating

NORMAL FINGER WEAR

EXCESSIVE WEAR
EXCESSIVE FINGER WEAR

BROKEN OR BENT FINGERS

4.13 Replace the pressure plate if excessive wear or damage is noted

14   Check the release bearing for smoothness of operation. There should be no harsh or loose spots in it. It should spin reasonably freely. If in doubt, replace the bearing with a new one.

15   It is important that no oil or grease gets on the clutch plate friction linings or the pressure plate and flywheel faces. Always handle the clutch with clean hands and wipe down the pressure plate and flywheel faces with a clean, dry rag before assembly begins.

16   Place the clutch plate against the flywheel, making sure that the longer splined boss faces toward the flywheel (thicker torsional spring assembly projection toward the transmission).

17   Install the pressure plate and clutch cover assembly so that the marks are in alignment. Tighten the bolts only finger tight so that the driven plate is gripped, but can still be moved sideways.

18   The clutch plate must now be aligned so that when the engine and transmission are mated, the input shaft splines will pass through the splines in the center of the clutch hub plate.

19   Alignment is accomplished by using an old transmission input shaft or alignment tool.

20   Using the alignment tool, center the clutch plate.

21   Alignment can be verified by removing the alignment tool and viewing the clutch plate hub in relation to the hole in the center of the clutch cover plate diaphragm spring. When the hub appears exactly in the center of the hole, all is correct.

22   Tighten the clutch cover bolts in a diagonal sequence to ensure that the cover plate is pulled down evenly and without distortion of the flange. Torque the bolts to Specifications.

23   Lubricate the clutch fork fingers at the release bearing end, and the ball and socket, with a high melting point grease. Also lubricate the release bearing collar and groove **(see illustration)**.

24   Install the clutch fork and dust boot in the clutch housing and attach the release bearing to the fork.

25   Install the clutch housing.

26   Install the transmission (refer to Chapter 7A).

27   Connect the fork pushrod and spring, lubricating the spring and pushrod ends. On models with hydraulic clutch assemblies, attach the release cylinder and release cylinder heat shield to the clutch housing.

28   Adjust the shift linkage if necessary (4-speed only). Refer to Chapter 7A.

29   Adjust the clutch linkage (refer to Chapter 1) on models not equipped with a hydraulic clutch assembly.

5   Clutch pilot bearing - removal and installation

Refer to illustrations 5.5 and 5.6

1   Remove the transmission (refer to Chapter 7).

2   Remove the clutch (refer to Section 4).

3   The clutch pilot bearing (bushing), which is pressed into the end of the crankshaft, supports the forward end of the transmission input shaft. It requires attention whenever the clutch is removed from the vehicle.

4   Clean the bearing thoroughly and inspect it for excessive wear and damage. If wear is noted, the bearing must be replaced with a new one.

5   Removal can be accomplished with a special puller **(see illustration)**, but an alternative method, is to remove the bearing hydraulically. First, locate a solid steel bar with a diameter that is slightly less than the inside diameter of the bearing (19/32-inch should be very close). The bar should just slip into the bearing with very little clearance. Next, pack the bearing and the area behind it (in the crankshaft recess) with heavy grease. Try to eliminate as much air as possible from the recess behind the bearing. Insert the bar into the bearing bore and rap the end of the bar with a hammer. The pressure exerted on

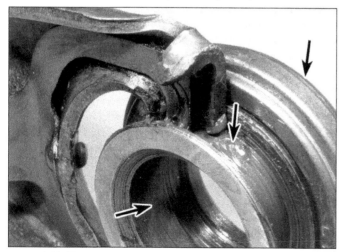
4.23 The release bearing must be lubricated at these points prior to installation

5.5 A special bearing puller can be used to remove the pilot bearing (see text for an alternate method that can be used if the puller is not available)

5.6 A bearing installer, or steel bar can be used to
install the new pilot bearing

6.4 Remove the clip and detach the master cylinder
from the clutch pedal

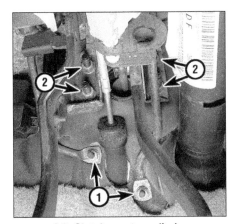

6.5 Clutch master cylinder
mounting details

1   Clutch master cylinder mounting nuts
2   Brake booster mounting nuts

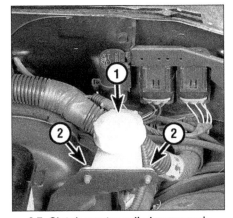

6.7 Clutch master cylinder reservoir
mounting details

1   Clutch master cylinder reservoir
2   Mounting screws

6.11 Clutch release cylinder
mounting details

1   Heat shield
2   Mounting nut
3   Clutch slave cylinder
4   Mounting bolt

the grease will be transferred to the back side of the bearing, forcing it out of the recess. Be sure to clean the grease out of the crankshaft after the bearing has been removed.

6   To install the new bearing, lubricate its outside surface with oil, then drive it into the recess with a bearing driver or steel bar and a soft-faced hammer (the radius in the bore of the bearing must face out) **(see illustration)**.

7   The remaining installation steps are the reverse of those for removal.

---

6   Hydraulic clutch assembly - removal and installation

Refer to illustrations 6.4, 6.5, 6.7 and 6.11

Note: *The hydraulic clutch assembly is serviced as a complete unit - individual components of the system cannot be serviced separately.*

1   Disconnect the cable from the negative battery terminal. **Caution:** *On models equipped with a Delco Loc II audio system, disable the anti-theft feature before disconnecting the battery.*

2   Remove the steering column trim cover.

3   Remove the underdash hush panel and disconnect the brake switch.

4   Disconnect the master cylinder pushrod from the clutch pedal **(see illustration)**.

5   Remove the clutch master cylinder-to-cowl nuts **(see illustration)**.

6   Remove the brake booster-to-cowl nuts.

7   Remove the clutch fluid reservoir from the bracket **(see illustration)**.

8   Pull the brake master cylinder forward to gain access to the clutch master cylinder. Be careful not to kink the hydraulic lines.

9   Disconnect the clutch master cylinder from the cowl.

10   Raise the vehicle and place it securely on jackstands.

11   Remove the release cylinder heat shield **(see illustration)**.

12   Remove the release cylinder from the bellhousing.

13   Remove the jackstands and lower the vehicle.

14   Remove the hydraulic clutch assembly from the engine compartment.

15   To begin installation, position the clutch master cylinder at the cowl and feed the release cylinder down to the bellhousing

mounting area. Do not install the master cylinder-to-cowl nuts at this time.

16   Feed the clutch master cylinder pushrod through the cowl and loosely attach the clutch master cylinder to the cowl with the U-bolt. Do not attach the U-bolt nuts at this time.

17   Attach the brake booster to the cowl.

18   Install the clutch master cylinder U-bolt through the clutch braces and tighten the U-bolt nuts.

19   Install and torque tighten the master cylinder-to-cowl nuts.

20   Attach the clutch master cylinder pushrod to the clutch pedal.

21   Install and adjust the brake switch (refer to Chapter 9).

22   Install the hush panel.

23   Install the steering column trim cover.

24   Attach the clutch reservoir to its bracket.

25   Raise the vehicle and place it securely on jackstands.

26   Guide the release cylinder pushrod into the clutch fork. **Note:** *When installing a new hydraulic clutch assembly, the release*

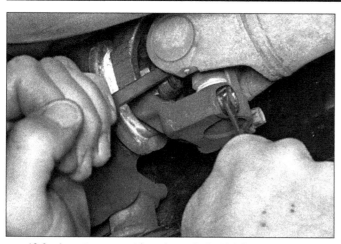

10.2a Insert a screwdriver through the U-joint to keep the driveshaft from turning and remove the retaining bolts with a wrench

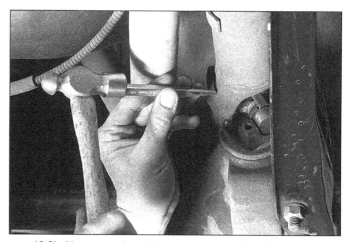

10.2b Use a punch and hammer to mark the relationship of the driveshaft to the pinion flange

cylinder contains a shipping strap that pre-positions the pushrod for proper installation and also provides a bearing insert. Following installation of the release cylinder, the first actuation of the clutch pedal will break the shipping strap and provide normal hydraulic system operation.

27   Attach the release cylinder and heat shield to the bellhousing.

28   Remove the jackstands and lower the vehicle.

29   Connect the negative battery cable.

7   Driveshaft - general information

All vehicles with which this manual is concerned are equipped with one-piece driveshafts. A splined yoke and universal joint are located at the transmission end of the shaft and a second universal joint is used where the driveshaft mates with the pinion flange at the differential.

The universal joints are lubricated for life and are not serviceable on the vehicle. If a universal joint becomes worn or noisy, a service kit containing cross and bearing assemblies is available. The kit also contains snap-rings which must be installed to substitute for the nylon injection rings installed at the factory during shaft assembly. The entire driveshaft must be removed from the vehicle whenever servicing is necessary and care should be taken when handling it to preserve the fine balance produced at the factory. Care should also be taken if the vehicle is undercoated. Never allow undercoating or any other foreign material to adhere to the driveshaft as it will disturb the factory balance.

8   Universal joints - wear check

1   Universal joint problems are usually caused by worn or damaged needle bearings. These problems are revealed as vibration in the driveline or clunking noises when the transmission is put in Drive or the clutch

is released. In extreme cases they are caused by lack of lubrication. If this happens, you will hear metallic squeaks and ultimately, grinding and shrieking sounds as the bearings are destroyed.

2   It is easy to check the needle bearings for wear and damage with the driveshaft in place on the vehicle. To check the rear universal joint, turn the driveshaft with one hand and hold the differential yoke with the other. Any movement between the two is an indication of wear. The front universal joint can be checked by holding the driveshaft with one hand and the sleeve yoke in the transmission with the other. Any movement here indicates the need for universal joint repair.

3   If they are worn or damaged, the universal joints will have to be disassembled and the bearings replaced with new ones. Read over the procedure carefully before beginning.

9   Driveshaft out-of-balance correction

1   Vibration of the driveshaft at certain speeds may be caused by any of the following:

a) *Undercoating or mud on the shaft.*
b) *Loose rear strap mounting bolts.*
c) *Worn universal joints.*
d) *Bent or dented driveshaft.*

2   Vibration which is thought to be coming from the driveshaft is sometimes caused by improper tire balance. This should be one of your first checks.

3   If the shaft is in a good, clean, undamaged condition, it is worth disconnecting the rear end mounting bolts and turning the shaft 180 degrees to see if an improvement is noticed. Be sure to mark the original position of each component before disassembly so the shaft can be returned to the same location.

4   If the vibration persists after checking for obvious causes and changing the position

of the shaft, the entire assembly should be checked out by a professional shop or replaced.

10   Driveshaft - removal and installation

Refer to illustrations 10.2a and 10.2b

1   Raise the vehicle and place it securely on jackstands.

2   Mark the relationship of the driveshaft to the pinion flange and remove the retaining bolts. Separate the driveshaft from the pinion yoke and withdraw the slip yoke from the transmission **(see illustrations)**. Cap the transmission at the output shaft to prevent fluid lose.

3   Make sure that the driveshaft is handled carefully during the removal and installation procedures, as it must maintain the factory balance to operate smoothly and quietly.

4   Make sure that the outer diameter of the transmission slip yoke is not burred, as it could damage the transmission seal when the driveshaft is installed.

5   Lubricate the slip yoke splines with engine oil, then slide the yoke onto the transmission output shaft. Do not force it or use a hammer. If resistance is met, check the splines for burrs.

6   Attach the shaft to the pinion flange. Be sure to align the marks on the shaft and pinion flange that were made before removal.

7   Install the rear bolts and tighten them to the specified torque.

8   Remove the jackstands and lower the vehicle.

11   Universal joints - disassembly, inspection and reassembly

Refer to illustrations 11.3a, 11.3b, 11.4, 11.8, 11.11, 11.13, 11.14 and 11.15

Note: *Always purchase a universal joint service kit(s) for your model vehicle before start-*

11.3a Press the bearing cup out of the front universal yoke

11.3b The cross press tool can be made from 2-1/2 x 3/4-inch wall square tubing

11.4 Remove a bearing cup from the yoke ear after it has been pressed out

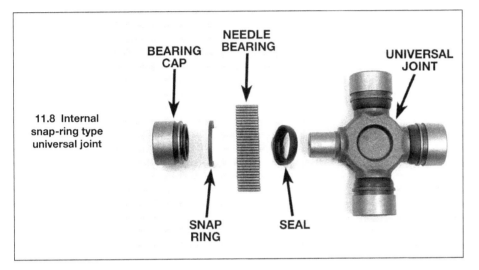

11.8 Internal snap-ring type universal joint

BEARING CAP

NEEDLE BEARING

UNIVERSAL JOINT

SNAP RING

SEAL

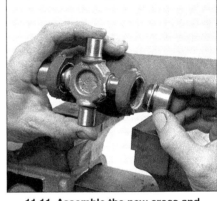

11.11 Assemble the new cross and bearing cups to the driveshaft

ing the procedure which follows. Also, read through the entire procedure before beginning work.

1   Remove the driveshaft (refer to Section 10).

2   Place the shaft on a workbench equipped with a vise.

3   Place the universal joint in the vise with a 1-1/8 inch socket against one ear of the shaft yoke and a cross press tool placed on the open horizontal bearing cups (photo). A cross press tool may be fashioned from 2-1/2 inch by 3/16-inch wall square tube or channel steel (see illustrations). Note: *Never clamp the driveshaft tubing itself in a vise, as the tube may be bent.*

4   Press the bearing cup out of the yoke ear, shearing the plastic retaining ring on the bearing (see illustration). Note: *If the cup does not come all the way out of the yoke, it may be pulled free with channel lock pliers, then removed.*

5   Turn the driveshaft 180-degrees and press the opposing bearing cup out of the yoke, again shearing the plastic retainer.

6   Disengage the cross from the yoke and remove the cross. Note: *Production universal joints cannot be reassembled because there are no bearing retaining grooves in the*

production bearing cups. All U-joint service kits contain snap-rings, which will be used when the replacement U-joints are assembled. If your driveshaft has previously had the original U-joints replaced with snap-ring types, remove the snap-rings before pressing the bearing cups out.

7   If the remaining universal joint is being replaced, press the bearing cups from the slip yoke as detailed above.

8   When reassembling the driveshaft, always install all parts included in the U-joint service kit (see illustration).

9   Remove all remnants of the plastic bearing retainers from the grooves in the yokes. Failure to do so may keep the bearing cups from being pressed into place and prevent the bearing retainers from seating properly.

10   Using multi-purpose grease to retain the needle bearings, assemble the bearings, cups and washers. Make sure the bearings do not become dislodged during the assembly and installation procedures.

11   In the vise, assemble the cross and cups in the yoke, installing the cups as far as possible by hand (see illustration).

12   Move the cross back and forth horizontally to assure alignment, then press the cups into place a little at a time, continuing to center

11.13 Install a snap-ring in the snap-ring groove

the cross to keep the proper alignment.

13   As soon as one snap-ring groove clears the inside of the yoke, stop pressing and install the snap-ring (see illustration).

14   Continue to press on the bearing cup until the opposite snap-ring can be installed. If difficulty is encountered, strike the yoke sharply with a hammer. This will spring the

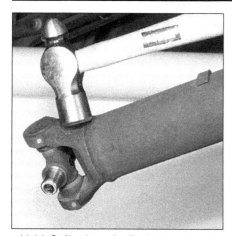

11.14 Strike the yoke firmly to seat a U-joint snap-ring - be careful not to damage the U-joint cup or the area around the cup

11.15 Install the grease fitting in a U-joint (if equipped)

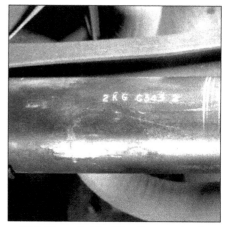

12.4 The axle identification number is stamped on the front of the right axle housing

yoke ears slightly and allow the snap-ring groove to move into position.

15   If equipped, install the grease fitting **(see illustration)**.

16   Install the driveshaft (refer to Section 10).

12   Rear axle - general information and identification

Refer to illustration 12.4

1   The rear axle is a hypoid, semi-floating type. The differential carrier is a casting with a pressed steel cover and the axle tubes are made of steel.

2   Due to the need for special gauges and equipment and the fact that individual differential components are no longer supplied (only a complete rear axle unit) the servicing and repair operations should be confined to those described in this Chapter.

3   An optional Posi-traction (limited slip) differential unit is also available. Basically, the inclusion of clutch cones or plates and springs slow the rotation of the differential case when one wheel has traction and the other does not. This slowing action applies additional force to the pinion gears and through the medium of the cone which is splined to the axleshafts, exerts equalizing rotational power to the axleshaft which is driving the wheel under traction.

4   In order to be able to undertake certain operations, particularly removal of the axleshafts, it is important to know the axle identification number. This is located on the front face of the right-hand axle tube about three inches from the differential cover **(see illustration)**. After the dirt and road grime have been brushed away, the code can be read. A typical General Motors axle code for 1982 (others similar) will read: 2 AB 0 103 2.

5   The initial number and following two letters (2 AB) indicate the axle code; the next number or letter (0) identifies the manufacturer; the next three-digit number (103) indicates the date of manufacture and the final digit (2) denotes the shift during which the axle was built.

6   By using the axle code and the accompanying chart, you can determine the gear ratio of the ring and pinion gears installed in your vehicle.

7   Any inquiries about your axle assembly to the dealer or to the factory must include the full code letters and build date numbers.

Axle code	Axle ratio	Axle type	Axle code	Axle ratio	Axle type	Axle code	Axle ratio	Axle type
1982			**1983**			**1984**		
2JE	2.73	Std	4JF	2.93	Std	4PW	2.73	Posi
2JG	3.08	Std	4JG	3.08	Std	6JG	3.08	Std
2JH	3.23	Std	4JH	3.23	Std	6JH	3.23	Std
2JJ	3.42	Std	4JJ	3.42	Std	6JJ	3.42	Std
2JR	3.42	Posi	4JK	3.73	Std	6JK	3.73	Std
2JW	2.73	Posi	4JQ	3.73	Posi	6JQ	3.73	Posi
2JX	2.93	Posi	4JR	3.42	Posi	6JR	3.42	Posi
2JY	3.08	Posi	4JX	2.93	Posi	6JY	3.08	Posi
2JZ	3.23	Posi	4JY	3.08	Posi	6JZ	3.23	Posi
2KE	2.73	Std	4JZ	3.23	Posi	6KK	3.73	Std
2KF	2.93	Std	4KJ	3.42	Std	6KY	3.08	Posi
2KG	3.08	Std	4KK	3.73	Posi	6KZ	3.23	Posi
2KH	3.23	Std	4PG	3.08	Std	6PN	3.73	Posi
2KH	3.42	Std	4PQ	3.73	Posi	6PQ	3.73	Posi
2PE	2.73	Std	4PW	2.73	Posi	6PR	3.42	Posi
2PW	2.73	Posi	4PX	2.93	Posi	6PY	3.08	Posi
2PX	2.93	Posi	4PY	3.08	Posi	6PZ	3.23	Posi
2PY	3.08	Posi	4PZ	3.23	Posi			
2PZ	3.23	Posi						

Axle code	Axle ratio	Axle type
1985		
4PW	2.73	Posi
6JG	3.08	Std
6JH	3.23	Std
6JJ	3.42	Std
6JK	3.73	Std
6JQ	3.73	Posi
6JR	3.42	Posi
6JY	3.08	Posi
6JZ	3.23	Posi
6KK	3.73	Std
6KY	3.08	Posi
6KZ	3.23	Posi
6PN	3.73	Posi
6PQ	3.73	Posi
6PR	3.42	Posi
6PY	3.08	Posi
6PZ	3.23	Posi
1985 (Borg-Warner)		
10032243	3.08	Std
10032244	3.08	Posi
10032245	3.27	Std
10032246	3.27	Posi
10032248	3.45	Posi
10032250	3.70	Posi
10032268	3.08	Posi
10032270	3.27	Posi
10032272	3.45	Posi
10032274	3.70	Posi
14085324	3.42	Std
14085325	3.42	Std
14085326	3.75	Std
14085327	3.73	Std

Axle code	Axle ratio	Axle type
1986		
2EA	2.77	Std
2EN	2.77	Posi
2ET	2.77	Posi
8JC	3.42	Std
8JD	3.73	Std
8KA	3.42	Std
8KB	3.73	Std
8XA	2.92	Std
8XB	3.08	Std
8XC	3.27	Std
8XF	3.70	Posi
8XN	2.92	Posi
8XO	3.08	Posi
8XP	3.27	Posi
8XQ	3.45	Posi
8XU	3.08	Posi
8XV	3.27	Posi
8XW	3.45Posi	
8XX	3.70	Posi
8XY	2.92	Posi
1986 (Borg-Warner)		
10032270	3.27	Posi
10032272	3.45	Posi
10032274	3.70	Posi
10040660	2.77	Posi
1987		
2HA	3.23	Std
2HL	3.42	Std
2HP	2.73	Std
2HR	3.42	Posi
2HQ	3.23	Posi

Axle code	Axle ratio	Axle type
2HT	2.73	Posi
4ET*	2.77	Posi
4EU	3.27	Posi
4EW*	3.45	Posi
4HB	3.08	Posi
4HR	3.42	Posi
4HV	2.73	Posi
4HY	3.23	Posi
1988 and 1989		
4EW*	3.45	Posi
4EU*	3.27	Posi
4ET*	2.77	Posi
6HT, 6HE	2.73	Posi
6HP	2.73	Std
6HB, 6HF	3.08	Posi
6HK	3.08	Std
6HL	3.42	Std
6HQ, 6HJ	3.23	Posi
*Denotes Borg-Warner axle		
1990 through 1992		
8HP	2.73	Std
8HT	2.73	Posi
8HE	2.73	Posi
8HK	3.08	Std
8HF	3.08	Posi
8HB	3.08	Posi
8HJ	3.23	Std
2PM	3.23	Posi
2PN	3.23	Posi
8HC	3.42	Std
2PN	3.42	Posi

13   Axleshaft - removal and installation

Refer to illustrations 13.3 and 13.4

Note: *On models equipped with a Borg-Warner rear axle assembly, service procedures should be limited to routine maintenance only. Repairs pertaining to the differential assembly, axleshafts, axle bearings and seals require special tools and expertise and should be performed by a professional mechanic. To determine if your vehicle is equipped with a Borg-Warner rear axle, match the axle number (which is stamped on the right side axle housing) with the axle codes in Section 12.*

1   Raise the rear of the vehicle, support it securely and remove the wheel and brake drum (refer to Chapter 9).

2   Unscrew and remove the pressed steel cover from the differential carrier and allow the oil to drain into a container.

3   Remove the lock bolt and pull the pinion shaft out of the differential far enough to clear the inner end of the axleshaft (**see illustration**).

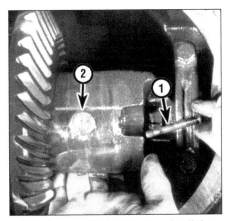

13.3 Remove the lock bolt and pull the pinion shaft out of the differential far enough to clear the inner end of the axleshaft

1   *Lock bolt*
2   *Pinion shaft*

4   Push the outer (flanged) end of the axleshaft in and remove the C-lock from the inner end of the shaft (**see illustration**).

13.4 Push the axle shaft in and remove the C-lock

5   Withdraw the axleshaft, taking care not to damage the oil seal in the end of the axle housing as the splined end of the axleshaft passes through it.

6   Installation is the reverse of removal. Tighten the lock screw to the specified torque.

15.3 Use a bearing puller to remove the rear axleshaft bearing

15.4 Using a bearing driver, drive the axleshaft bearing in the housing until it's seated

7   Always use a new cover gasket and tighten the cover bolts to the specified torque.

8   Refill the axle with the correct quantity and grade of lubricant (Chapter 1).

14   Axleshaft oil seal - replacement

Note: *On models equipped with a Borg-Warner rear axle assembly, service procedures should be limited to routine maintenance only. Repairs pertaining to the differential assembly, axleshafts, axle bearings and seals require special tools and expertise and should be performed by a professional mechanic. To determine if your vehicle is equipped with a Borg-Warner rear axle, match the axle number (which is stamped on the right side axle housing) with the axle codes in Section 12.*

1   Remove the axleshaft as described in the preceding Section.

2   Pry out the old oil seal from the end of the axle housing, using a large screwdriver or the inner end of the axleshaft itself as a lever.

3   Apply high melting point grease to the oil seal recess and tap the seal into position so that the lips are facing in and the metal face is visible from the end of the axle housing. When correctly installed, the face of the oil seal should be flush with the end of the axle housing.

4   Installation of the axleshaft is described in the preceding Section.

15   Axleshaft bearing - replacement

Refer to illustrations 15.3 and 15.4

Note: *On models equipped with a Borg-Warner rear axle assembly, service proce-*

dures should be limited to routine maintenance only. Repairs pertaining to the differential assembly, axleshafts, axle bearings and seals require special tools and expertise and should be performed by a professional mechanic. To determine if your vehicle is equipped with a Borg-Warner rear axle, match the axle number (which is stamped on the right side axle housing) with the axle codes in Section 12.

1   Remove the axleshaft (refer to Section 13) and the oil seal (refer to Section 14).

2   A bearing puller will be required or a tool which will engage behind the bearing will have to be fabricated.

3   Attach a slide hammer and pull the bearing from the axle housing **(see illustration)**.

4   Lubricate the new bearing with gear lubricant. Clean out the bearing recess and drive in the new bearing using a bearing driver or section of pipe applied against the outer bearing race **(see illustration)**. Make sure that the bearing is tapped into the full depth of its recess and that the numbers on the bearing are visible from the outer end of the housing.

5   Discard the old oil seal and install a new one, then install the axleshaft.

16   Rear axle assembly - removal and installation

Note: *On models equipped with a Borg-Warner rear axle assembly, service procedures should be limited to routine maintenance only. Repairs pertaining to the differential assembly, axleshafts, axle bearings and seals require special tools and expertise and should be performed by a professional mechanic. To determine if your vehicle is equipped with a Borg-Warner rear axle,*

match the axle number (which is stamped on the right side axle housing) with the axle codes in Section 12.

1   Raise the rear of the vehicle and support it securely on jackstands placed under the frame rails.

2   Position an adjustable floor jack under the differential housing. Raise the jack just enough to take up the weight, but not far enough to take the weight of the vehicle off the jackstands.

3   Disconnect both shock absorbers.

4   Remove the bolt that attaches the left side of the track bar to the axle.

5   Remove the brake line junction block bolt at the axle housing, then disconnect the brake lines at the junction block.

6   Lower the jack under the differential housing enough to remove the springs.

7   Remove the rear wheels and brake drums (refer to Chapter 9).

8   Remove the cover from the rear axle assembly and drain the lubricant into a container.

9   Remove the axleshafts (refer to Section 13).

10   Disconnect the brake lines from the axle housing clips.

11   Remove the brake backing plates.

12   Disconnect the lower control arms from the axle housing.

13   Disconnect the torque arm at the axle housing.

14   Mark the driveshaft and companion flange, unbolt the driveshaft and support it out of the way on a wire hanger.

15   Remove the rear axle assembly from under the vehicle.

16   Installation is the reverse of the removal procedure. When installation is complete, fill the differential with the recommended oil and bleed the brake system.

Chapter 9 Brakes

Contents

	Section
Brake check.. See Chapter 1	
Brake pedal - removal and installation...	17
Combination valve - check and replacement	13
Disc brake rotor - inspection, removal and installation................	8
Fluid level check .... See Chapter 1	
Front disc brake caliper - overhaul	4
Front disc brake caliper - removal and installation	3
Front disc brake pads - replacement...	2
General information...	1
Hydraulic brake hoses and lines - inspection and replacement	12
Hydraulic system - bleeding	16
Master cylinder - removal, overhaul and installation	11
Parking brake - adjustment..	14
Power brake booster - inspection, removal and installation.........	15
Rear disc brake caliper - overhaul	7
Rear disc brake caliper - removal and installation	5
Rear disc brake pads - replacement..	6
Rear drum brake shoes - replacement	9
Stop light switch - removal, installation and adjustment	18
Wheel cylinder (drum brakes) - removal, overhaul and installation...	10

Specifications

Disc brakes
Rotor thickness after resurfacing (minimum)..	0.980 inch
Discard thickness	
Front..	0.965 inch
Rear	
1985 through 1988..	0.965 inch
1989 and later	0.724 inch
Disc runout (maximum)..	0.005 inch
Disc thickness variation (maximum)	0.0005 inch

Rear drum brakes
Drum diameter	
Standard..	9.50 inches
Service limit...	9.59 inches
Drum taper (maximum)...	0.003 inch
Out-of-round (maximum)..	0.002 inch
Wheel cylinder bore diameter	0.748 inch

Master cylinder
Disc/drum piston diameter	0.945 inch
4-wheel disc piston diameter	1.000 inch

Torque specifications

	Ft-lbs (unless otherwise indicated)
Banjo fitting retaining bolt (front caliper)	18 to 30
Brake line-to-brake hose nut	18
Brake line-to-combination valve nut	17
Brake line-to-master cylinder nut	17
Brake pedal pivot nut	22
Brake pedal bracket-to-dash screw	18
Caliper bleeder valve	11
Caliper mount-to-steering knuckle bolt	35
Combination valve-to-master cylinder screw	15
Front brake hose-to-caliper bolt	32
Front caliper bracket bolts (dual-piston caliper)	137
Front caliper mounting bolts (single-piston caliper)	35
Front flexible hose bracket-to-frame	8
Junction block-to-axle housing screw	20
Lug nuts	
Standard wheel	80
Aluminum wheel	105
Parking brake control assembly-to-console screw	9
Parking brake lever retaining nut	30 to 40
Parking brake lever mounting bracket bolt	24 to 38
Power booster housing-to-master cylinder nut	20
Power booster studs-to-dash attaching nuts	15
Rear caliper bracket bolts (1989 and later)	70
Rear caliper mounting bolts (1988 and earlier)	35
Rear caliper guide pin bolts (1989 and later)	
Upper	26
Lower	16
Rear flexible hose bracket-to-frame	8
Splash shield-to-steering knuckle bolt	10

1   General information

All vehicles covered by this manual are equipped with hydraulically operated front and rear brake systems. All front brake systems are disc type, while rear brake systems are drum brakes on some models and disc type on others. A quick visual inspection will tell you what type of brakes you have at the rear of your vehicle.

The hydraulic system consists of two separate front and rear circuits. The master cylinder has separate reservoirs for the two circuits and in the event of a leak or failure in one hydraulic circuit, the other circuit will remain operative. A visual warning of circuit failure or air in the system is given by a warning light activated by displacement of the piston in the brake distribution (pressure differential warning) switch from its normal "in balance' position.

The parking brake mechanically operates the rear brakes only. It is activated by a pull-handle in the center console between the front seats.

A combination valve, located in the engine compartment, consists of three sections providing the following functions: The metering section limits pressure to the front brakes until a predetermined front input pressure is reached and until the rear brakes are activated. There is no restriction at inlet pressures below 3 psi, allowing pressure equalization during non-braking periods. The proportioning section proportions outlet pressure to the rear brakes after a predetermined rear input pressure has been reached, preventing early rear wheel lock-up under heavy brake loads. The valve is also designed to assure full pressure to the rear brakes should the front brakes fail, and vice versa.

The pressure differential warning switch is designed to continuously compare the front and rear brake pressure from the master cylinder and energize the dash warning light in the event of either front or rear brake system failure. The design of the switch and valve are such that the switch will stay in the "warning' position once a failure has occurred. The only way to turn the light off is to repair the cause of the failure and apply a brake pedal force of 450 psi.

The power brake booster, utilizing engine manifold vacuum and atmospheric pressure to provide assistance to the hydraulically operated brakes, is located in the engine compartment, adjacent to the cowl. All brakes are self-adjusting.

After completing any operation involving the disassembly of any part of the brake system, always test drive the vehicle to check for proper braking performance before resuming normal driving. When testing the brakes, perform the tests on a clean, dry, flat surface. Conditions other than these can lead to inaccurate test results. Test the brakes at various speeds with both light and heavy pedal pressure. The vehicle should stop evenly without pulling to one side or the other. Avoid locking the brakes because this slides the tires and diminishes braking efficiency and control.

Tires, vehicle load and front-end alignment are factors which also affect braking performance.

Torque values given in the Specifications section are for dry, unlubricated fasteners.

2   Front disc brake pads - replacement

Warning: *Dust created by the brake system may contain asbestos, which is harmful to your health. Never blow it out with compressed air and don 't inhale it. An approved respirator should be worn when working on the brakes. Do not, under any circumstances, use petroleum based solvents to clean brake parts. Use brake system cleaner or denatured alcohol only!*
Note: *Disc brake pads should be replaced on both wheels at the same time.*
1   Whenever you are working on the brake system, be aware that asbestos dust is present and be careful not to inhale any of it, because it has been proven to be harmful to your health.

2.6/1 Use an Allen wrench to remove the two caliper mounting bolts (check the bolt threads for damage)

2.6/2 Remove the brake line retaining clip

2.6/3 Separate the brake line from the retainer, then remove the inside pad and plate assembly from the caliper

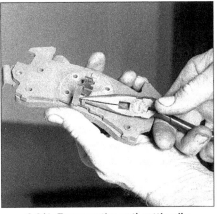

2.6/4 Remove the anti-rattle clip with pliers

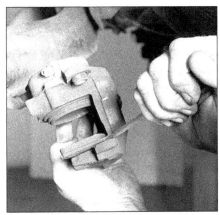

2.6/5 Break the outer pad and plate assembly loose with a screwdriver

2   Remove the cover from the brake fluid reservoir and siphon off about two ounces of the fluid into a container and discard it.
3   Raise the front of the vehicle and place it securely on jackstands.
4   Remove the front wheel, then reinstall two wheel lugs (flat side toward the rotor) to hold the rotor in place. Work on one brake

assembly at a time, using the assembled brake for reference if necessary.

Single-piston caliper

Refer to illustrations 2.6/1 through 2.6/13
5   Push the piston back into its bore. If necessary, a C-clamp can be used, but a flat bar will usually do the job. As the piston is

depressed to the bottom of the caliper bore, the fluid in the master cylinder will rise. Make sure that it does not overflow. If necessary, siphon off more of the fluid as directed in Step 2.
6   Refer to the accompanying photographs and perform the procedure illustrated, beginning with photograph 2.6/1 **(see illustrations)**.

2.6/6 Separate the outer pad and plate assembly from the caliper

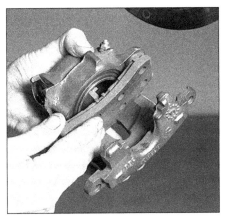

2.6/7 After cleaning the caliper, attach the inner pad to the caliper after making sure the new anti-rattle clip is in place

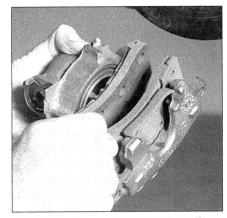

2.6/8 Attach the outer pad to the caliper

2.6/9 Apply multi-purpose grease to the ends of the caliper mounting bracket surfaces

2.6/10 Install the caliper on the mounting bracket (do not get any grease on the brake pad linings)

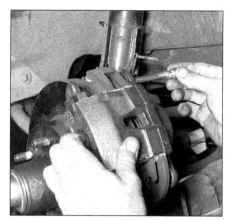

2.6/11 Install the mounting bolts and tighten them securely

2.6/12 Attach the brake hose to the bracket, then install the clip with a hammer, but be careful not to damage the hose

2.6/13 After refilling the master cylinder with fluid and pumping the brake pedal to seat the pad and plate assemblies, use channel lock pliers to bend the upper ears of the outboard plate assembly until the ears are flush with the caliper housing, with no radial clearance

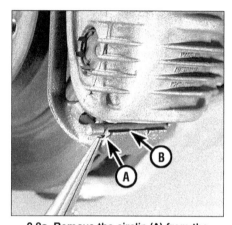

2.8a Remove the circlip (A) from the retainer pin (B)

Dual-piston caliper

Refer to illustrations 2.8a, 2.8b, 2.9, 2.10, 2.11 and 2.13

7　　Refer to Steps 1 through 4. Before pulling off the wheel, mark the wheel and hub so the wheel can be reinstalled in the same relative position.

8　　Remove the circlip, then pull out the retainer pin with a pair of pliers **(see illustra-**

tions). Discard the circlip and retainer pin and use new ones when the caliper is reinstalled.

9　　Pivot the lower end of the caliper up and disengage the upper end from the bracket **(see illustration)**. Suspend the caliper from a suspension component with a piece of stiff wire - DO NOT let it hang by the rubber brake hose!

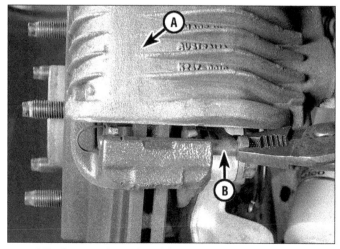

2.8b Support the caliper (A) and pull the retainer pin (B) out with pliers

2.9 Lift the caliper (A) off the disc (B) and bracket (C)

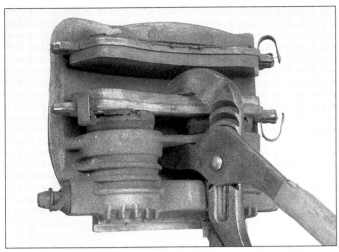

2.10 Press the pistons into the caliper core to provide clearance for the new pads

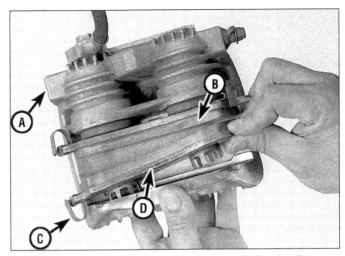

2.11 Dual piston caliper brake pad installation details

A	Caliper	C	Bias spring
B	Inner brake pad	D	Outer brake pad

10   Before removing the pads, use a large pair of channel-lock pliers or a C-clamp to push the pistons into the caliper bores to provide room for the thick linings on the new pads **(see illustration)**.

11   Note how they're installed, then remove the old pads from the caliper **(see illustration)**.

12   Use aerosol brake system cleaner to remove all residue from the caliper and bracket, particularly where the pads slide in the caliper.

13   Install the new pads in the caliper. The outer pad with the bias spring is attached to the caliper housing **(see illustration)**, while the inner pad with the wear sensor should be pressed into the pistons. Make sure the pads are seated completely and don't get oil or grease on the lining material.

14   Slip the caliper and pad assembly over the rotor and position it in the bracket, then push down on the caliper to compress the bias spring on the outer pad and insert the new retainer pin. Attach the new circlip to the end of the retainer pin.

15   Replace the pads in the remaining front caliper. After the pads on both sides have been replaced, the wheels installed and the vehicle lowered, start the engine and pump the brake pedal slowly and firmly three times to seat the pads against the rotors. Don't forget to add brake fluid to the reservoir.

3   Front disc brake caliper - removal and installation

Refer to illustrations 3.6a and 3.6b

Warning: *Dust created by the brake system may contain asbestos, which is harmful to your health. Never blow it out with compressed air and don 't inhale it. An approved respirator should be worn when working on the brakes. Do not, under any circumstances, use petroleum based solvents to clean brake parts. Use brake system cleaner or denatured alcohol only!*

1   Whenever you are working on the brake system, be aware that asbestos dust is present and be careful not to inhale any of it because it has been proven to be harmful to your health.

2   Remove the cover from the brake fluid reservoir and siphon off two-thirds of the fluid into a container and discard it.

3   Raise the front of the vehicle and place it securely on jackstands.

4   Remove the front wheel.

5   Push the caliper piston back into its bore. If necessary, a C-clamp can be used, but a flat bar will usually do the job. As the piston is depressed to the bottom of the caliper bore, the fluid in the master cylinder will rise. Make sure that it does not overflow.

6   Remove the banjo fitting bolt holding the brake hose **(see illustrations)**, then remove and discard the copper seal rings found on either side of the banjo fitting. Always use new copper seals when reinstalling the brake hose.

7   Refer to Section 2 and remove the caliper from the spindle or mounting bracket.

8   Installation is the reverse of removal. Lubricate the ends of the caliper mounting bracket surfaces before attaching the caliper to the bracket. Tighten all bolts to the specified torque.

2.13 Install the bias springs onto the new brake pads

3.6a Remove the banjo fitting bolt

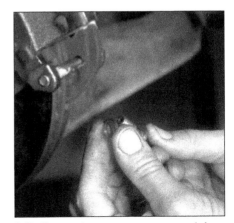

3.6b Be sure to use new copper seal rings when installing the brake hose

4.4 Compressed air is used to move the piston out of the caliper bore (note use of wooden block as a cushion)

4.5 Pry the dust boot out of the caliper bore

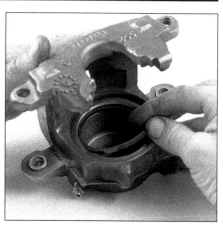

4.6 Remove the piston seal from the caliper groove

4 Front disc brake caliper - overhaul

Warning: *Dust created by the brake system may contain asbestos, which is harmful to your health. Never blow it out with compressed air and don 't inhale it. An approved respirator should be worn when working on the brakes. Do not, under any circumstances, use petroleum based solvents to clean brake parts. Use brake system cleaner or denatured alcohol only!*

Note: *Purchase a brake caliper overhaul kit for your particular vehicle before beginning this procedure.*

Single-piston caliper

Refer to illustrations 4.4, 4.5, 4.6, 4.10, 4.11, 4.12 and 4.13

1   Refer to Section 3 and remove the caliper.
2   Refer to Section 2 and remove the brake pads.
3   Clean the exterior of the brake caliper with brake fluid (never use gasoline, kerosene or cleaning solvents), then place the caliper on a clean workbench.
4   Place a wooden block or shop rag in the caliper as a cushion, then use compressed air to remove the piston from the caliper **(see illustration)**. Use only enough air pressure to ease the piston out of the bore. If the piston is blown out, even with the cushion in place, it may be damaged. **Caution:** *Never place your fingers in front of the piston in an attempt to catch or protect it when applying compressed air - serious injury could occur.*
5   Carefully pry the dust boot out of the caliper bore **(see illustration)**.
6   Using a wood or plastic tool, remove the piston seal from the groove in the caliper bore **(see illustration)** . Metal tools may cause bore damage.
7   Remove the caliper bleeder valve, then remove and discard the sleeves and bushings from the caliper ears. Also discard all rubber parts.
8   Clean the remaining parts with brake fluid. Allow them to drain and then shake them vigorously to remove as much fluid as possible.
9   Carefully examine the piston for nicks and burrs and loss of plating. If surface defects are present, parts must be replaced. Check the caliper bore in a similar way, but light polishing with crocus cloth is permissible to remove light corrosion and stains. Discard the mounting bolts if they are corroded

or damaged.
10   When assembling, lubricate the piston bores and seal with clean brake fluid; position the seal in the caliper bore groove **(see illustration)**.
11   Lubricate the piston with clean brake fluid, then install a new boot in the piston groove with the fold toward the open end of the piston **(see illustration)**.
12   Insert the piston squarely into the caliper bore, then apply force to bottom the piston in the bore **(see illustration)**.
13   Position the dust boot in the caliper counterbore, then use a drift to drive it into position **(see illustration)**. Make sure that the boot is evenly installed below the caliper face.
14   Install the bleeder valve.
15   The remainder of the installation procedure is the reverse of the removal procedure. Always use new copper gaskets when connecting the brake hose and bleed the system as described in Section 16.

Dual-piston caliper

16   The overhaul procedure for the dual-piston caliper is similar to the single-piston caliper previously described, but note that the new style caliper is made of aluminum and can be easily damaged. Also, two pistons and bores are involved, so, when

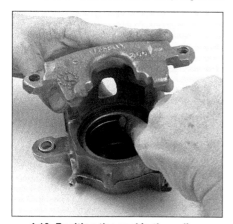

4.10 Position the seal in the caliper bore groove

4.11 Install a new dust boot in the piston groove (note that folds are toward the open end of the piston)

4.12 Square the piston in the bore

4.13 Use a hammer and driver to seat the boot in the caliper housing bore

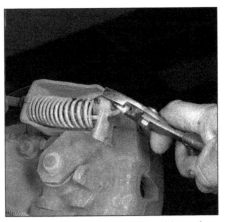

5.6a Pull the parking brake cable to the rear and disengage it from the parking brake lever

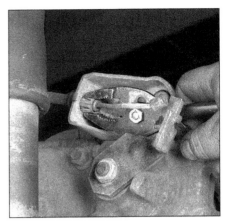

5.6b Remove the spring, squeeze the tangs on the parking brake cable ferrule and slide the cable out of the bracket

removing the pistons with compressed air, if the pistons don't come out of the caliper at the same time, remove the boot from the empty bore and reinstall the piston far enough to seal it off. Position a spacer between the caliper housing and the piston to hold it in place as compressed air is used to remove the remaining piston.

17   All inspection procedures must be repeated for both pistons and bores and all related components.

18   Do not use a tool to seat the piston boots in the caliper bores - use your fingers only.

5   Rear disc brake caliper - removal and installation

Warning: *Dust created by the brake system may contain asbestos, which is harmful to your health. Never blow it out with compressed air and don 't inhale it. An approved respirator should be worn when working on the brakes. Do not, under any circumstances, use petroleum based solvents to clean brake parts. Use brake system cleaner or denatured alcohol only!*

1   Whenever you are working on the brake system, be aware that asbestos dust

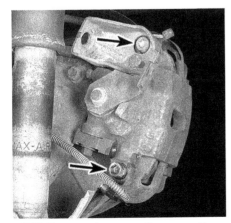

5.13 Remove the rear caliper mounting bolts (arrows)

is present and be careful not to inhale any of it, as it has been proven to be harmful to your health.

2   Remove the cover from the brake fluid reservoir and siphon off two-thirds of the fluid into a container and discard it.

3   Raise the rear of the vehicle and place it securely on jackstands, then release the parking brake handle.

4   Remove the wheel, then reinstall one wheel lug (flat side toward the rotor) to hold the rotor in place. Work on one brake assembly at a time, using the assembled brake for reference, if necessary.

1988 and earlier

Refer to illustrations 5.6a, 5.6b and 5.13

5   Loosen the tension on the parking brake cable by loosening the nuts at the equalizer.

6   Remove the cable, damper and spring from the cable lever **(see illustrations)**.

7   Hold the lever and remove the nut, then remove the lever, lever seal and anti-friction washer from the caliper actuator screw.

8   Position a C-clamp on the caliper housing. **Note:** *Make sure the C-clamp does not contact the actuator screw.*

9   Tighten the C-clamp until the piston bottoms in the cylinder bore, then remove the clamp. **Caution:** *The parking brake lever must be removed before attempting to depress the piston. If the piston does not move with moderate pressure, DO NOT force it or damage to the caliper will occur.*

10   Reinstall the anti-friction washer, lever seal (with the sealing bead against the housing), lever and nut on the actuating screw.

11   Loosen the brake line nut, then disconnect the brake line from the caliper. **Note:** *If the brake line is seized, remove the banjo fitting bolt, banjo fitting and copper washers in order to free the line.*

12   To prevent fluid loss and contamination, plug the openings in the caliper and brake line.

13   Remove the caliper mounting bolts **(see illustration)**. **Note:** *If you are working on the right brake assembly, it will be necessary to*

remove the bolt from the rear of the lower control arm and move the lower control arm to gain access to the lower caliper mounting bolt (refer to Chapter 11 if necessary).

14   Remove the caliper from the vehicle.

15   Installation is the reverse of removal. Always use a new lever seal and lubricate it with silicone brake lube before installing it. Also use a new anti-friction washer. If the banjo fitting was removed, use new copper washers when reinstalling it.

16   When installing the parking brake lever on the actuator screw hex, install the lever pointing down, then rotate the lever toward the front of the vehicle and hold it while installing the retaining nut. Tighten the nut to the specified torque, then rotate the lever back against the stop on the caliper.

17   After connecting the parking brake cable, tighten it at the equalizer until the lever starts to move off the caliper stop, then loosen the adjustment until the lever just moves back against the stop.

18   Fill the master cylinder and bleed the hydraulic system (see Section 16).

1989 and later

19   Loosen the tension on the parking brake cable by loosening the nuts at the equalizer.

20   Unbolt the parking brake cable bracket from the caliper and unhook the cable eye from the hooked end of the lever.

21   Remove the bolt and detach the brake hose from the caliper. Plug the hose to prevent the loss of fluid and the entry of dirt. Discard the brake hose fitting washers and use new ones when reinstalling the bolt.

22   Remove the caliper as described in the rear brake pad replacement procedure, but remove both guide pin bolts.

23   Installation is the reverse of removal. If the banjo fitting was removed, use new copper washers when reinstalling it.

24   Connect the parking brake cable, tighten the equalizer nut and adjust the parking brake (see Section 14).

25   Fill the master cylinder and bleed the hydraulic system (see Section 16).

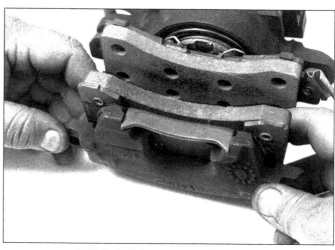

6.9 Place the inner pad against the piston, making sure the post on the pad engages with the D-shaped cutout on the piston and place the ends of the dampening spring onto the backing plate as shown

6.10 Install the new outer brake pad

6   Rear disc brake pads - replacement

Warning: *Dust created by the brake system may contain asbestos, which is harmful to your health. Never blow it out with compressed air and don't inhale it. An approved respirator should be worn when working on the brakes. Do not, under any circumstances, use petroleum based solvents to clean brake parts. Use brake system cleaner or denatured alcohol only!*

Note: *Disc brake pads should be replaced on both wheels at the same time.*

1   Whenever you are working on the brake system, be aware that asbestos dust is present and be careful not to inhale any of it, as it has been proven to be harmful to your health.

2   Remove the cover from the brake fluid reservoir and siphon off two-thirds of the fluid into a container and discard it.

3   Raise the rear of the vehicle and place it securely on jackstands, then release the parking brake handle.

4   Remove the wheel, then reinstall one wheel lug (flat side toward the rotor) to hold the rotor in place. Work on one brake assembly at a time, using the assembled brake for reference, if necessary.

1988 and earlier

Refer to illustrations 6.9 and 6.10

1   Remove the brake caliper (refer to Section 5).

2   Remove the brake pad from the caliper, using a screwdriver if necessary.

3   Remove the sleeves and bushings from the caliper ears.

4   Using a small screwdriver, remove the flexible two-way check valve from the end of the caliper piston.

5   If leakage is noted at the caliper assembly, refer to Section 7 and overhaul the caliper.

6   Lubricate the new sleeves and bushings with silicone grease and install them in the caliper ears.

7   Lubricate a new two-way check valve with silicone grease and install it in the caliper piston end.

8   Position the inboard pad in the caliper. Make sure that the D-shaped tab on the shoe engages the D-shaped notch in the piston. If the tab and notch do not line up, use a spanner wrench to turn the piston until they do.

9   With the tab and notch properly aligned, install the inboard pad in the caliper, with the wear sensor positioned so that it is at the leading edge. Slide the edge of the backing plate under the ends of the dampening spring and snap the assembly into place against the piston, making sure that the backing plate lies flat against the piston **(see illustration)**.

Note: *If the assembly fails to lie flat, recheck the alignment of the D-shaped notch and tab.*

10   Install the outboard pad in the caliper **(see illustration)**.

11   Reinstall the caliper (refer to Section 5).

12   Seat the caliper by applying the brake pedal firmly several times. Also ratchet the parking brake on and off several times.

13   Using channel lock pliers, bend the ears on the upper edge of the backing plate until they are flush with the caliper housing, with no radial clearance.

1989 and later

Refer to illustrations 6.21a and 6.21b

14   Position a C-clamp on the caliper housing. **Note:** *Position one end of the C-clamp against the inlet fitting and the other end against the outboard pad backing plate.* Tighten the C-clamp until the piston bottoms in the cylinder bore, then remove the clamp.

15   Remove the caliper upper guide pin bolt and discard it. Use a new one when the caliper is reinstalled.

16   Rotate the caliper down so the pads can be removed (be careful not to strain the brake hose or the parking brake cable housing).

17   Lift the pads out of the caliper bracket.

18   Use aerosol brake system cleaner to remove all residue from the caliper and bracket, particularly where the pads slide in the bracket. **Warning:** *Dust created by the brake system contains asbestos, which is harmful to your health. Never blow it out with compressed air and don't inhale it. An approved respirator should be worn when working on the brakes. Do not, under any circumstances, use petroleum based solvents to clean brake parts. Use brake system cleaner or denatured alcohol only!*

19   Check the guide pins for corrosion and damage-they must be free to slide in the bracket. If they're damaged and/or corroded, replace them with new ones and install new dust boots as well.

20   The outer brake pad assembly is the one with the insulator on the backing plate. The inner pad has the wear sensor.

21   Install the outer pad into the caliper housing, while the inner pad with the wear sensor should be pressed into the piston **(see illustrations)**. Make sure the pads are seated completely and don't get oil or grease on the lining material. The wear sensor on the inner pad must be at the top of the pad. If it isn't, use the other new inner pad in the set.

22   Rotate the caliper into position and make sure the springs on the pads aren't sticking through the inspection hole in the caliper. If they are, rotate the caliper down and reposition the pads.

23   Align the holes and install a new caliper upper guide pin bolt. Tighten it to the specified torque.

24   Repeat the above procedure to replace the pads in the remaining rear caliper.

25   Check the parking brake levers on the calipers to make sure they're against the stops. If not, see if the rear parking brake cables are binding. Free and lubricate the cables, if necessary.

26   If the cables are free, turn the adjusting nut at the cable equalizer until both parking

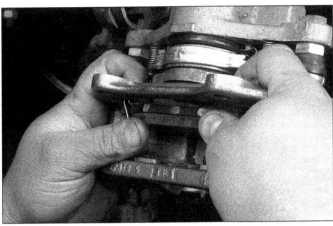

6.21a Install the inner pad with the wear sensor in the trailing position (as the wheel rotates in the forward direction)

6.21b Install the outer pad with the insulator against the caliper

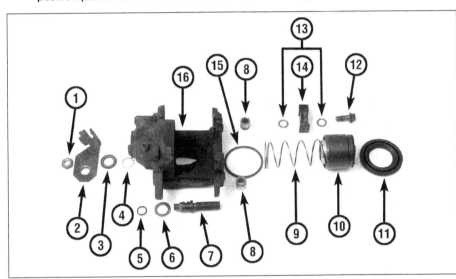

7.1 Exploded view of the rear brake caliper assembly (1988 and earlier models)

1 Parking brake lever retaining nut	9 Piston return spring
2 Parking brake lever	10 Piston
3 Parking brake lever seal	11 Piston dust boot
4 Nylon anti-friction washer	12 Brake hose fitting banjo bolt
5 Actuator screw O-ring	13 Banjo bolt sealing washers
6 Actuator screw thrust washer	14 Brake hose fitting
7 Actuator screw	15 Piston seal
8 Caliper pin bushings	16 Caliper housing

7.3 Remove the pad damping spring

7.5 Ratchet the parking brake lever back-and-forth while pulling the piston out

brake levers just touch the stops on the calipers.

27   After the pads on both sides have been replaced, the wheels installed and the vehicle lowered, start the engine and pump the brake pedal slowly and firmly three times to seat the pads against the rotors. Don't forget to add brake fluid to the reservoir.

7   Rear disc brake caliper - overhaul

Warning: *Dust created by the brake system may contain asbestos, which is harmful to your health. Never blow it out with compressed air and don 't inhale it. An approved respirator should be worn when working on the brakes. Do not, under any circumstances, use petroleum based solvents to clean brake parts. Use brake system cleaner or denatured alcohol only!*

Note: *Purchase a brake caliper overhaul kit for your particular vehicle before beginning this procedure.*

1988 and earlier models

Refer to illustrations 7.1, 7.3, 7.5, 7.6, 7.8, 7.9, 7.10, 7.19, 7.21, 7.25, 7.26a, 7.26b, 7.26c and 7.27

1   Remove the caliper (refer to Section 5) and place it on a workbench **(see illustration)**.

2   Remove the brake pads from the caliper (refer to Section 6).

3   Remove the shoe dampening spring from the end of the caliper piston **(see illustration)**.

4   Place the caliper in a vise equipped with soft jaws and cushion the interior with a piece of wood or shop towels.

5   Move the parking brake lever back and forth to work the piston out of the caliper bore **(see illustration)**. If the piston will not come out, remove the lever retaining nut and lever and use a wrench to rotate the actuator screw in the same direction as the parking brake is applied until the piston comes out of the cylinder.

7.6 Remove the piston and balance spring from the caliper

7.8 Carefully press the actuator screw out of the caliper

7.9 Pry out the dust boot

7.10 Using a wood or plastic tool, remove the piston seal

7.19 Lubricate the piston seal with clean brake fluid and install it squarely in the piston bore groove

7.21 Install a new thrust washer and O-ring on the actuator screw

6   Remove the balance spring **(see illustration)**.

7   If not previously removed, remove the retaining nut and lever, then remove the lever seal and anti-friction washer.

8   Press on the threaded end of the actuator screw to remove it from the housing **(see illustration)**.

9   Remove the dust boot from the caliper, using a screwdriver **(see illustration)**.

10   Use a wood or plastic tool to remove the piston seal from the bore.

11   If not previously removed, remove the banjo housing retaining bolt, banjo housing and copper washers.

12   Remove the parking brake lever mounting bracket only if it is damaged.

13   Carefully check the caliper bore for score marks, nicks, corrosion and excessive wear. Light corrosion may be removed with crocus cloth; otherwise, replace the caliper housing with a new one.

14   Clean all parts not included in the caliper repair kit with denatured alcohol, clean brake fluid or brake system cleaner. Do not, under any circumstances, use petroleum-based solvents.

15   Use compressed air to dry the parts and blow out all the passages in the caliper housing and bleeder valve. **Caution:** *Always wear*

eye protection when using compressed air!

16   Install the bleeder valve and tighten it to the specified torque.

17   If removed for replacement, install the parking brake lever mounting bracket and tighten the mounting bolt to the specified torque.

18   Install the banjo housing assembly, making sure to use new copper washers. Tighten the retaining bolt to the specified torque.

19   Lubricate the new piston seal with clean brake fluid and install it in the caliper bore groove **(see illustration)**. Make sure that the seal is not twisted.

20   Attach the boot to the piston with the inside lip of the boot in the piston groove and the boot fold toward the end of the piston that contacts the inboard brake shoe.

21   Install the thrust washer on the actuator screw with the bearing surface toward the caliper housing **(see illustration)**.

22   Lubricate the shaft seal with brake fluid and install it on the actuator screw.

23   Lubricate the actuator screw with brake fluid and install it in the piston.

24   Install the balance spring in the piston recess.

25   Lubricate the piston and caliper bore with brake fluid and start the piston assembly

into the caliper bore **(see illustration)**.

26   Using a special rear disc brake piston compressor (available at most auto parts stores) or other suitable tool, push the piston in until it bottoms in the caliper bore **(see illustration)**. Before removing the tool, lubricate the anti-friction washer and lever seal

7.25 Install the parking brake lever temporarily (to hold the actuator screw straight), install the piston by placing it squarely in the bore and turn it slightly to engage the actuator screw threads, then remove the parking brake lever

7.26a Using a special rear disc brake caliper piston installation tool, press the piston into the bore; do not remove the tool until the parking brake lever is installed

7.26b Install the parking brake lever on the actuator screw, positioned against the lever stop . . .

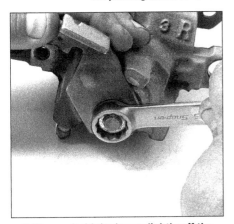

7.26c . . . hold the lever slightly off the stop while tightening the nut

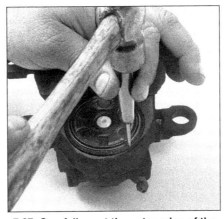

7.27 Carefully seat the outer edge of the piston boot in the caliper bore

7.31 Detach the collar return springs from the actuating collar

with silicone lubricant and install them over the end of the actuating screw, making sure that the sealing bead on the lever seal is against the housing. Install the lever over the actuating screw, then rotate the lever slightly away from the stop on the housing and hold it while installing the lever retaining nut **(see illustrations)**. Tighten the nut to the specified torque, then rotate the lever back to the stop.

27   Position the outside of the boot in the caliper recess. Using a driver and hammer, seat the boot in the recess **(see illustration)**.
28   Install the dampening spring in the groove in the piston end. **Note:** *It may be necessary to move the parking brake lever off its stop so that the piston may be extended slightly, making the spring groove accessible. If this is necessary, push the piston back into*

the bottom of the caliper bore before installing the caliper.
29   Install the caliper on the vehicle (refer to Section 5).

1989 and later models
Refer to illustrations 7.31, 7.32, 7.33, 7.41, 7.46 and 7.47
30   Remove the caliper, clean the exterior with brake system cleaner and set the caliper on a clean working surface.
31   Unhook the two collar return springs from the actuating collar **(see illustration)**.
32   Pull the actuating collar with its attached parts out of the caliper body **(see illustration)**.
33   Separate the clamp rod and bushing from the actuating collar. Discard the bushing **(see illustration)**.
34   Bend back the tabs on the boot retainers and remove them from the collar. Discard the retainers and boots.
35   Remove the piston and replace the piston seal as described in Section 4. Be sure to inspect the caliper bore and wash all parts not to be replaced in denatured alcohol, clean brake fluid or brake system cleaner. Inspect the actuating collar and the clamp

7.32 Remove the actuating collar and attached parts from the caliper

7.33 Separate the clamp rod from the actuating collar and discard the bushing

rod for corrosion and excessive wear, replacing them, if necessary.

36 Lubricate the piston bore with clean brake fluid and install the piston squarely in the bore. Press the piston in by hand until it bottoms in the bore.

37 Lightly lubricate the actuating collar with the liquid lubricant supplied in the rebuild kit. DO NOT use any other type of lubricant. Assemble the pushrod, new boots and retainers on the actuating collar and clamp the retainers firmly against the collar. Bend the tabs on the retainer to hold the assembly together.

38 Attach the preload spring to the boot retainers.

39 Lightly lubricate the clamp rod with the lubricant supplied in the kit. Insert the clamp rod into the actuator collar and push it in fully, so the boot contacts the reaction plate on the clamp rod.

40 Push a new bushing down over the end of the clamp rod.

41 Lubricate the grooved bead of the inner boot and the corresponding boot groove in the caliper housing with the lubricant furnished in the kit. Also lubricate the actuating collar around the center hole. Push the clamp rod with assembled components into the hole in the piston **(see illustration)**. Pull on the actuating collar and seat the inner boot into the groove in the caliper housing. While doing this, guide the pushrod into its hole in the caliper.

42 Install the collar return springs to the actuating collar.

43 Lubricate the caliper guide pins with high temperature grease. Inspect the guide pin boots for cracks and replace as necessary. Install the brake pads.

44 Install the caliper as described above in the brake pad replacement procedure, then bleed the brakes.

45 Anytime the caliper has been overhauled it is necessary to adjust the parking brake free travel. It is performed to ensure there is no binding between the caliper and

the parking brake mechanism. To accurately adjust the free travel, the brake pads must be new or parallel to within 0.006-inch.

46 Have an assistant press on the brake pedal lightly, just enough to take up any clearance between the caliper piston, brake pads and rotor (you should not be able to turn the rotor by hand). Apply light finger pressure to the end of the parking brake lever and measure the clearance between the lever and the caliper housing **(see illustration)**. The desired clearance should be between 0.024 and 0.028 inch.

47 If the free travel is incorrect, remove the adjusting screw, clean the screw threads, apply a thread locking compound and reinstall the screw. Turn it in far enough to obtain the desired clearance between the lever stop and the caliper body **(see illustration)**.

48 Have the assistant pump the brake pedal a few times, then re-check the free-travel.

7.41 Position the clamp rod, actuating collar and assembled components on the caliper, insert the clamp rod and bushing into the hole in the piston and guide the pushrod into the caliper

---

8   Disc brake rotor - inspection, removal and installation

Refer to illustrations 8.5 and 8.6

1 Raise the vehicle and place it securely on jackstands.

2 Remove the appropriate wheel and tire.

3 Remove the brake caliper assembly (refer to Section 3 or 5). **Note:** *On front disc brakes, it is not necessary to disconnect the brake hose. After removing the caliper mounting bolts, hang the caliper out of the way on a piece of wire. Never hang the caliper by the brake hose because damage to the hose will occur.*

4 Inspect the rotor surfaces. Light scoring or grooving is normal, but deep grooves or severe erosion is not. If pulsating has been noticed during application of the brakes, suspect disc runout.

5 Attach a dial indicator to the caliper mounting bracket, turn the rotor and note the

amount of runout. Check both inboard and outboard surfaces **(see illustration)**. If the runout is more than the maximum allowable, the rotor must be removed from the vehicle and taken to an automotive machine shop for resurfacing.

6 Using a micrometer, measure the thickness of the rotor **(see illustration)**. If it is less than the minimum specified, replace the rotor with a new one. Also measure the disc thickness at several points to determine variations in the surface. Any variation over 0.0005-inch may cause pedal pulsations during brake application. If this condition exists and the disc thickness is not below the minimum, the rotor can be removed and taken to an automotive machine shop for resurfacing.

7 To remove and install the front rotor, refer to Chapter 1, Wheel bearing check and repack.

8 The rotor on the rear wheel disc brake models can be pulled off after the caliper is removed.

7.46 Exert light pressure on the parking brake lever and measure the clearance between the lever and the caliper

7.47 Using an Allen wrench, turn the adjustment screw clockwise to increase the clearance or counterclockwise to decrease the clearance (caliper removed for clarity)

8.5 Checking the disc brake rotor pad surface
runout with a dial indicator

8.6 Using a micrometer to check rotor thickness

9   Rear drum brake shoes - replacement

Refer to illustrations 9.5/1 through 9.5/44 and 9.6
Warning: *Dust created by the brake system may contain asbestos, which is harmful to your health. Never blow it out with compressed air and don 't inhale it. An approved respirator should be worn when working on the brakes. Do not, under any circumstances, use petroleum based solvents to clean brake parts. Use brake system cleaner or denatured alcohol only!*

1   Whenever you are working on the brake system, be aware that asbestos dust is present. It has been proven to be harmful to your

health, so be careful not to inhale any of it.
2   Raise the vehicle and place it securely on jackstands.
3   Release the parking brake handle.
4   Remove the wheel and tire assembly.
Note: *All four rear shoes should be replaced at the same time, but to avoid mixing up parts, work on only one brake assembly at a time.*
5   Refer to the accompanying photographs and perform the brake shoe replacement procedure beginning with Photo 9.5/1 **(see illustrations). Note:** *If the brake drum cannot be easily pulled off the axle and shoe assembly, make sure that the parking brake is completely released, then squirt some penetrating oil around the center hub area. Allow the oil to soak in and try to pull the drum off again. If the drum still cannot be pulled off, the brake*

shoes will have to be retracted. This is accomplished by first removing the lanced cutout in the backing plate with a hammer and chisel. With this lanced area punched in, pull the lever off the adjusting screw wheel with one small screwdriver while turning the adjusting wheel with another small screwdriver, moving the shoes away from the drum. The drum may now be pulled off.
6   Before reinstalling the drum, it should be checked for cracks, score marks, deep scratches and "hard spots,' which will appear as small discolored areas. If the hard spots cannot be removed with fine emery cloth and/or if any of the other conditions listed above exist, the drum must be taken to an automotive machine shop to have it turned. If the drum will not "clean up' before the

9.5 Components of a typical drum brake assembly (left side shown, right side is the reverse)

1   Primary brake shoe
     return spring
2   Wheel cylinder
3   Shoe guide
4   Anchor pin
5   Secondary brake shoe
     return spring
6   Actuator link
7   Brake hose
8   Secondary brake shoe
9   Actuator lever pivot
10   Actuator lever
11   Lever return spring
12   Pawl
13   Adjusting screw wheel
14   Adjusting screw
15   Adjusting screw spring
16   Primary brake shoe
17   Hold-down pin
18   Hold-down spring

9.5/1 Remove the wheel mounting stud lock washers

9.5/2 Mark the relationship of the axle and brake drum

9.5/3 Remove the brake drum (if it cannot be pulled off, refer to copy)

9.5/4 Use a brake spring tool to remove the primary shoe return spring from the anchor pin pivot

9.5/5 Removing the spring after unhooking it from the primary brake shoe

9.5/6 Push on the bottom of the automatic adjuster actuator lever (bottom hand) and remove the actuator link from the top of the actuator lever

9.5/7 Remove the actuator link and secondary shoe return spring from the anchor pin pivot with a brake spring tool

9.5/8 The actuator link and secondary shoe return spring after removal

9.5/9 Remove the shoe guide from the anchor pin (the pin is removed from the rear of the backing plate)

9.5/10 Remove the primary shoe holddown spring and pin

9.5/11 The primary shoe hold-down spring and pin after removal

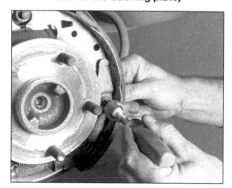

9.5/12 Remove the secondary shoe hold-down spring and pin (note that this spring is shorter than the secondary shoe hold-down spring)

9.5/13 Remove the actuator lever, pawl and lever return spring (note L mark on the actuator lever denoting left side brake assembly)

9.5/14 Turn the adjuster wheel all the way in (toward the right in this case)

9.5/15 Pivot the secondary shoe to the rear and down

9.5/16 Remove the secondary shoe from the adjusting screw spring

9.5/17 Remove the adjusting screw and spring

9.5/18 Note that the longer hook goes toward the rear

9.5/19 Pull the parking brake strut and spring assembly from behind the axle flange

9.5/20 Pull out on the primary shoe and pivot it down until it is free of the backing plate

9.5/21 Remove the spring from the parking brake lever

9.5/22 Remove the E-clip retaining the parking brake lever to the secondary shoe

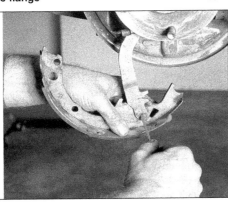

9.5/23 With the E-clip removed, the shoe may be separated from the parking brake lever
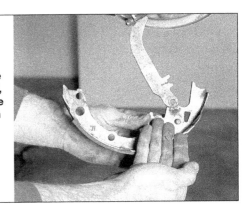

9.5/24 Check the shoe guide pads on the backing plate for wear, making sure a ridge has not formed which could hang up the brake shoe (if a sharp ridge is present, it may be removed with 150-grit emery paper). This step completes the shoe removal procedure. At this time, check all springs for tension and replace them as necessary, then check the wheel cylinder for leakage. If leakage is noted, remove the wheel cylinder for rebuilding (refer to Section 10)

9.5/25 Lubricate the shoe guide pads and wheel cylinder ends with multipurpose grease (application on the wheel cylinder ends should be light)

9.5/26 Lightly lubricate the pivot pin with multi-purpose grease

9.5/27 Attach the parking brake lever to the primary brake shoe and the parking brake lever spring to the lever (note that the primary shoe has the shorter lining)

9.5/28 Install the primary shoe in position on the backing plate and install the hold-down spring

9.5/29 Place the parking brake strut into position in the primary shoe slot

9.5/30 Lubricate both ends of the adjuster

9.5/31 Engage the short hook end of the adjusting spring in the hole in the primary shoe, with the hook inserted from the front

9.5/32 Engage the long hook end of the spring in the hole in the secondary shoe, with the hook inserted from the rear

9.5/33 With the adjuster positioned between the shoes, pivot the secondary shoe into place on the backing plate

9.5/34 Attach the pawl to the actuator lever and install the actuator assembly in place inside the secondary shoe

9.5/35 Attach the bushing to the actuator and shoe

9.5/36 Install the shoe hold-down spring

9.5/37 Install the shoe guide, with the rounded side in

9.5/38 Install the rear hook of the secondary shoe return spring in the shoe

9.5/39 Assemble the actuating link to the spring

9.5/40 Attach the spring and actuating link to the anchor pin

9.5/41 Attach the actuating link to the parking brake actuating lever

9.5/42 Attach the primary shoe return spring to the shoe . . .

9.5/43 . . . and to the anchor pin

9.5/44 With all parts installed, rock the assembly back and forth with your hands to ensure that all parts are seated

9.6 Location of the maximum diameter mark inside the brake drum

maximum drum diameter is reached in the machining operation, the drum will have to be replaced with a new one. The maximum diameter is cast into each brake drum **(see illustration)**.

7   Install the brake drum, lining up the marks made before removal if the old brake drum is used, and install the wheel stud lock washers.

8   Mount the wheel and tire, install the wheel lugs and tighten to the Specifications, then lower the vehicle.

9   Make a number of forward and reverse stops to adjust the brakes until a satisfactory pedal action is obtained.

10.2 Removing the brake line fitting from the rear of the wheel cylinder

10.3a Removing the cylinder retaining clip with needle-nose pliers

10.3b Method of removing the retaining clip using two awls (cylinder has been removed from vehicle for illustration purposes)

10 Wheel cylinder (drum brakes) - removal, overhaul and installation

Refer to illustrations 10.2, 10.3a, 10.3b, 10.5 and 10.12

Warning: *Dust created by the brake system may contain asbestos, which is harmful to your health. Never blow it out with compressed air and don't inhale it. An approved respirator should be worn when working on the brakes. Do not, under any circumstances, use petroleum based solvents to clean brake parts. Use brake system cleaner or denatured alcohol only!*

Note: *Obtain the wheel cylinder rebuild kits before beginning this procedure.*

1   Remove the brake shoes (refer to Section 9).

2   Remove the brake line fitting from the rear of the wheel cylinder **(see illustration).** Cap the brake line to prevent contamination.

3   Using curved needle-nose pliers, remove the wheel cylinder retaining clip **(see illustration)**, from the rear of the wheel cylinder. Alternatively, insert two awls into the

10.12 Place a screwdriver handle between wheel cylinder and the axle flange while the retaining clip is installed from the rear

access slots between the cylinder pilot and the retainer locking tabs and bend both tabs away at the same time **(see illustration)**.

4   Remove the cylinder and place it on a clean workbench.

5   Remove the bleeder valve, seals, pistons, boots and spring assembly from the cylinder body **(see illustration)**.

6   Clean the wheel cylinder with brake fluid, denatured alcohol or brake system cleaner. Do not, under any circumstances, use petroleum-based solvents to clean brake parts.

7   Use compressed air to remove excess

fluid from the wheel cylinder and to blow out the passages.

8   Check the cylinder bore for corrosion and scoring. Crocus cloth may be used to remove light corrosion and stains, but the cylinder must be replaced with a new one if the defects cannot be removed easily, or if the bore is scored.

9   Lubricate the new seals with clean brake fluid.

10   Assemble the brake cylinder, making sure the boots are properly seated.

11   Place the wheel cylinder in position in the backing plate.

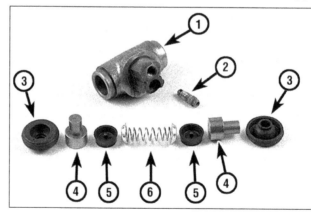

10.5 Wheel cylinder components

1   *Wheel cylinder body*
2   *Bleeder screw*
3   *Boot*
4   *Piston*
5   *Seal*
6   *Spring assembly*

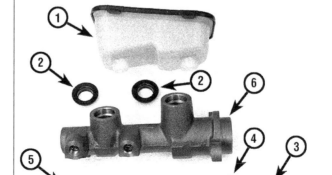

11.1 Typical master cylinder components

1   *Reservoir*
2   *Reservoir grommets*
3   *Lock-ring*
4   *Primary piston assembly*
5   *Secondary piston assembly*
6   *Master cylinder body*

11.9 Pry the plastic reservoir from the master cylinder body

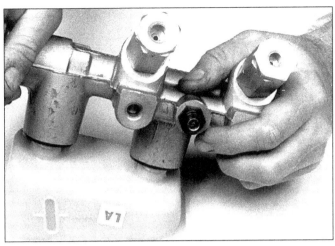

11.14 Press the master cylinder body onto the reservoir

12   Secure the cylinder in place with a new retaining clip by first starting the brake line fitting into the threads of the wheel cylinder, then placing a screwdriver handle between the wheel cylinder and the axle flange to hold the cylinder in place while the retaining clip is pushed into place with curved needle-nose pliers **(see illustration)**. Make sure the clip and cylinder are securely in place, then remove the screwdriver and tighten the brake line fitting.
13   Bleed the brake system (refer to Section 16).

11   Master cylinder - removal, overhaul and installation

Refer to illustrations 11.1, 11.9, 11.14, 11.15 and 11.18

1   A master cylinder overhaul kit should be purchased before beginning this procedure. The kit will include all the replacement parts necessary for the overhaul procedure. The rubber replacement parts, particularly the seals, are the key to fluid control within the master cylinder. As such, it's very important that they be installed securely and facing in the proper direction **(see illustration)**. Be careful during the rebuild procedure that no grease or mineral-based solvents come in contact with the rubber parts.
2   Completely cover the front fender and cowling area of the vehicle, as brake fluid can ruin painted surfaces if it is spilled.
3   Disconnect the brake line connections. Rags or newspapers should be placed under the master cylinder to soak up the fluid that will drain out.
4   Remove the two master cylinder mounting nuts, move the bracket retaining the combination valve forward slightly, taking care not to bend the hydraulic lines running to the combination valve, and remove the master cylinder from the vehicle.
5   Remove the reservoir cover and reservoir diaphragm, then discard any remaining fluid in the reservoir.

6   Remove the primary piston lock ring by depressing the piston and prying the ring out with a screwdriver.
7   Remove the primary piston assembly with a wire hook, being careful not to scratch the bore surface.
8   Remove the secondary piston assembly in the same manner.
9   Place the master cylinder in a vise and pry the reservoir from the cylinder body with a pry bar **(see illustration)**.
10   Do not attempt to remove the quick take-up valve from the cylinder body, as this valve is not serviceable.
11   Remove the reservoir grommets.
12   Inspect the cylinder bore for corrosion and damage. If any corrosion or damage is found, replace the master cylinder body with a new one, as abrasives cannot be used on the bore.
13   Lubricate the new reservoir grommets with silicone brake lube and press the grommets into the master cylinder body, making sure they are properly seated.
14   Lay the reservoir on a hard surface and press the master cylinder body onto the reservoir, using a rocking motion **(see illustration)**.
15   Remove the old seals from the secondary piston assembly and install the new seals so that the cups face out **(see**

illustration).
16   Attach the spring retainer to the secondary piston assembly.
17   Lubricate the cylinder bore with clean brake fluid and install the spring and secondary piston assembly in the cylinder.
18   Disassemble the primary piston assembly, noting the position of the parts, then lubricate the new seals with clean brake fluid and install them on the piston **(see illustration)**.
19   Install the primary piston assembly in the cylinder bore, depress it and install the lock ring.
20   Inspect the reservoir cover and diaphragm for cracks and deformation. Replace any damaged parts with new ones and attach the diaphragm to the cover.
21   **Note:** *Whenever the master cylinder is removed, the complete hydraulic system must be bled. The time required to bleed the system can be reduced if the master cylinder is filled with fluid and "bench bled" (refer to Steps 22 through 25) before the master cylinder is installed on the vehicle.*
22   Insert threaded plugs of the correct size into the cylinder outlet holes and fill the reservoirs with brake fluid (the master cylinder should be supported in such a manner that brake fluid will not spill out of it during the bench bleeding procedure).

11.15 The secondary seals must be installed with the lips facing out

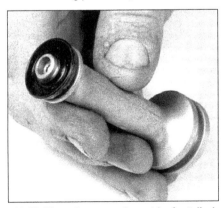

11.18 The primary seal must be installed with the lip facing away from the piston

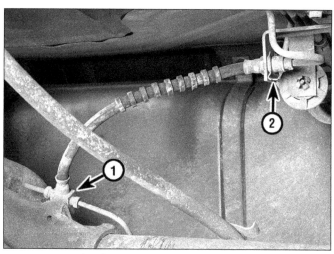

12.11 Typical junction block assembly

1　Junction block　　　　　　2　U-clip

13.1a The combination valve (A) is located just under the master cylinder, with the pressure differential switch (B) located at the center of the valve

23   Loosen one plug at a time and push the piston assembly into the bore to force air from the master cylinder. To prevent air from being drawn back into the cylinder, the appropriate plug must be tightened before allowing the piston to return to its original position.

24   Stroke the piston three or four times for each outlet to assure that all air has been expelled.

25   Refill the master cylinder reservoirs and install the diaphragm and cover assembly. **Note:** *The reservoirs should only be filled to the top of the reservoir divider to prevent overflowing when the cover is installed.*

26   Carefully install the master cylinder by reversing the removal steps, then bleed the brakes at the wheel bleed valves (refer to Section 16).

12   Hydraulic brake hoses and lines - inspection and replacement

Refer to illustration 12.11

1   About every six months, with the vehicle raised and placed securely on jackstands, the flexible hoses which connect the steel brake lines with the front and rear brake assemblies should be inspected for cracks, chafing of the outer cover, leaks, blisters and other damage. These are important and vulnerable parts of the brake system and inspection should be complete. A light and mirror will prove helpful for a thorough check. If a hose exhibits any of the above conditions, replace it with a new one as follows:

Front brake hose

2   Using a back-up wrench, disconnect the brake line from the hose fitting, being careful not to bend the frame bracket or brake line.

3   Use pliers to remove the U-clip from the female fitting at the bracket, then remove the hose from the bracket.

4   At the caliper end of the hose, remove the bolt from the fitting block, then remove the hose and the copper gaskets on either side of the fitting block.

5   When installing the hose, always use new copper gaskets on either side of the fitting block and lubricate all bolt threads with clean brake fluid before installing them.

6   With the fitting flange engaged with the caliper locating ledge, attach the hose to the caliper and tighten it to the specified torque.

7   Without twisting the hose, install the female fitting in the hose bracket (it will fit the bracket in only one position).

8   Install the U-clip retaining the female fitting to the frame bracket.

9   Using a back-up wrench, attach the brake line to the hose fitting and tighten it to the specified torque.

10   When the brake hose installation is complete, there should be no kinks in the hose. Also, make sure the hose does not contact any part of the suspension. Check this by turning the wheels to the extreme left and right positions. If the hose makes contact, remove the hose and correct the installation as necessary.

Rear brake hose

11   Locate the junction block at the rear axle and disconnect the two steel brake lines from the block **(see illustration)**.

12   Using a back-up wrench, remove the hose at the female fitting, being careful not to bend the bracket or steel lines.

13   Remove the U-clip with pliers and separate the female fitting from the bracket.

14   Note the position of the junction block carefully so that it may be reinstalled in precisely the same position.

15   Remove the bolt attaching the junction block to the axle and remove the hose from the block.

16   When installing, thread both steel line fittings into the junction block at the rear axle.

17   Bolt the junction block to the axle, then tighten the block bolt and steel lines to the

specified torque.

18   Without twisting the hose, install the female end of the hose in the frame bracket (it will fit the bracket in only one position).

19   Install the U-clip retaining the female end to the bracket.

20   Using a back-up wrench, attach the steel line fitting to the female fitting, tightening it to the specified torque. Again, be careful not to bend the bracket or steel line.

21   Check that the hose installation did not loosen the frame bracket. Retorque the bracket if necessary.

22   Fill the master cylinder reservoirs and bleed the system (refer to Section 16).

Steel brake lines

23   When it becomes necessary to replace steel lines, use only double-walled steel tubing. Never substitute copper tubing because copper is subject to fatigue cracking and corrosion. The outside diameter of the tubing is used for sizing.

24   Auto parts stores and brake supply houses carry various lengths of prefabricated brake line. Depending on the type of tubing used, these sections can either be bent by hand into the desired shape or must be bent in a tubing bender.

25   If prefabricated lengths are not available, obtain the recommended steel tubing and fittings to match the line to be replaced. Determine the correct length by measuring the old brake line, and cut the new tubing to length, leaving about 1/2-inch extra for flaring the ends.

26   Install the fittings onto the cut tubing and flare the ends using an ISO flaring tool.

27   Using a tubing bender, bend the tubing to match the shape of the old brake line.

28   Tube flaring and bending can usually be performed by a local auto parts store if the proper equipment mentioned in Steps 26 and 27 is not available.

29   When installing the brake line, leave at least 3/4-inch clearance between the line and any moving parts.

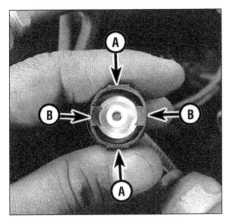

13.1b Pressure differential switch electrical connector details

A   Side lock release
B   Inside tab

13   Combination valve - check and replacement

Check

Refer to illustrations 13.1a and 13.1b

1   Disconnect the electrical connector from the pressure differential switch **(see illustrations)**. When disconnecting the connector, squeeze the connector side lock releases, moving the inside tabs away from the switch, then pull up. Pliers may be used as an aid if necessary.
2   Using a jumper wire, connect the switch wire to a good ground, such as the engine block.
3   Turn the ignition key to the On position. The warning light in the instrument panel should light up.
4   If the warning light does not light, either the bulb is burned out or the electrical circuit is defective. Replace the bulb (refer to Chapter 10) or repair the electrical circuit as necessary.
5   When the warning light functions correctly, turn the ignition switch off, disconnect the jumper wire and reconnect the wire to the switch terminal.
6   Make sure the master cylinder reservoirs are full, then attach a bleeder hose to one of the rear wheel bleeder valves and immerse the other end of the hose in a container partially filled with clean brake fluid.
7   Turn the ignition switch on.
8   Open the bleeder valve while a helper applies moderate pressure to the brake pedal. The brake warning light on the instrument panel should light.
9   Close the bleeder valve before the helper releases the brake pedal.
10   Reapply the brake pedal with moderate to heavy pressure. The brake warning light should go out.
11   Attach the bleeder hose to one of the front brake bleeder valves and repeat Steps 8 through 10. The warning light should react in the same manner as in Steps 8 and 10.

13.17 Combination valve mounting details

1   Combination valve
2   Mounting bolt
3   Mounting bracket

12   Turn the ignition switch off.
13   If the warning light did not come on in Steps 8 and 11, but does light when a jumper is connected to ground, the warning light switch portion of the combination valve is defective and the combination valve must be replaced with a new one since the components of the combination valve are not individually serviceable.

Replacement

Refer to illustration 13.17

14   Place a container under the combination valve and protect all painted surfaces with newspapers or rags.
15   Disconnect the hydraulic lines at the combination valve, then plug the lines to prevent further loss of fluid and to protect the lines from contamination.
16   Disconnect the electrical connector from the pressure differential switch (refer to Step 1 if necessary).
17   Remove the bolt holding the valve to the mounting bracket and remove the valve from the vehicle **(see illustration)**.
18   Installation is the reverse of the removal procedure.
19   Bleed the entire brake system (refer to Section 16). Do not move the vehicle until a firm brake pedal is attained.

14   Parking brake - adjustment

Refer to illustration 14.4

1   The parking brake cables may stretch over a period of time, necessitating adjustment. Also, the parking brake should be checked for proper adjustment whenever the rear brake cables have been disconnected. If the parking brake handle travel is less than 13 or more than 17 ratchet clicks, the parking brake needs adjustment.

Drum rear brakes

2   Pull the parking brake handle exactly two ratchet clicks.

14.4 Hold the brake cable stud (front) with one wrench while turning the adjusting nut with another

3   Raise the vehicle and support it securely on jackstands.
4   Locate the adjusting screw on the rear side of the equalizer bracket. To keep the brake cable stud from turning, hold it with one wrench and turn the adjusting nut with another wrench until the left rear wheel can just be turned in reverse (using two hands) but is locked when you attempt to turn it forward **(see illustration)**.
5   Release the parking brake and make sure that both rear wheels turn freely and that there is no brake drag in either direction, then remove the jackstands and lower the vehicle.

Disc rear brakes

1988 and earlier

6   Make sure the parking brake handle is completely released.
7   Raise the vehicle and place it securely on jackstands.
8   Lubricate the parking brake cables at the rubbing points on the vehicle underbody and at the equalizer hooks, then check for free movement of all cables.
9   Hold the brake cable stud with one wrench to prevent it from turning and turn the equalizer nut with another wrench until all cable slack is removed.
10   Check that the caliper levers are against the stops on the caliper housings after tightening the equalizer nut. If the levers are off the stops, loosen the cable until the levers return to the stops.
11   Operate the parking brake lever several times to check the adjustment. The parking brake handle should travel approximately 14 clicks with normal handle application effort.
12   Remove the jackstands, lower the vehicle to the ground and again check to see if the levers are on the caliper stops. If not, repeat the above steps as necessary.

1989 and later

13   Pump the brake pedal firmly three times, then apply and release the parking brake three times.
14   Loosen the rear wheel lug nuts, raise the rear of the vehicle and support it securely on

15.3 Brake master cylinder-to-power brake booster mounting details

1 *Brake master cylinder*
2 *Mounting nuts*
3 *Brake power booster*

jackstands. Mark the relationship of the wheels to the axle flanges, then remove the wheels.

15 Reinstall two lug nuts on each rotor to hold them in place. Check to see that the parking brake levers on the calipers are seated against their stops - if they aren't, loosen the cables at the equalizer until they are.

16 Turn the adjusting nut on the parking brake equalizer until the levers on the calipers just rise off their stops, then back off the adjusting nut until the levers barely touch their stops.

17 Operate the parking brake lever several times, then re-check the adjustment. The hand lever should travel no more than 16 clicks when properly adjusted. When the lever is released, the rotors should turn freely in both directions.

18 Remove the lug nuts that were installed to retain the rotors. Install the wheels and lug nuts, aligning the match marks on the wheels and axle flanges. Lower the vehicle and tighten the lug nuts to the specified torque.

15 Power brake booster - inspection, removal and installation

Refer to illustration 15.3

1 The power brake unit requires no special maintenance apart from periodic inspection of the hoses and inspection of the air filter beneath the boot at the pedal pushrod end.

2 Dismantling of the power brake unit requires special tools. If a problem develops, it is recommended that a new or factory-exchange unit be installed rather than trying to overhaul the original booster.

3 Remove the mounting nuts which hold the master cylinder to the power brake unit **(see illustration)**. Position the master cylinder out of the way, being careful not to strain the lines leading to the master cylinder. If there is any doubt as to the flexibility of the lines, disconnect them at the cylinder and plug the ends.

4 Disconnect the vacuum hose leading to

the front of the power brake booster. Cover the end of the hose.

5 Loosen the four nuts that secure the booster to the firewall. Do not remove these nuts at this time.

6 Inside the vehicle, disconnect the power brake pushrod from the brake pedal. Do not force the pushrod to the side when disconnecting it.

7 Now remove the four booster mounting nuts and carefully lift the unit out of the engine compartment.

8 When installing, loosely install the four mounting nuts and then connect the pushrod to the brake pedal. Tighten the nuts to the specified torque and reconnect the vacuum hose and master cylinder. If the hydraulic brake lines were disconnected, the entire brake system should be bled to eliminate any air which has entered the system (refer to Section 16).

16 Hydraulic system - bleeding

Refer to illustrations 16.23a and 16.23b

1 Bleeding of the hydraulic system is necessary to remove air whenever it is introduced into the brake system.

2 It may be necessary to bleed the system at all four brakes if air has entered the system due to low fluid level, or if the brake lines have been disconnected at the master cylinder.

3 If a brake line was disconnected only at a wheel, then only that wheel cylinder (or caliper) must be bled.

4 If a brake line is disconnected at a fitting located between the master cylinder and any of the brakes, that part of the system served by the disconnected line must be bled.

5 If the master cylinder has been removed from the vehicle, refer to Section 11, Step 21 before proceeding with the procedure which follows.

6 If the master cylinder is installed on the vehicle but is known to have, or is suspected of having air in the bore, the master cylinder must be bled before any wheel cylinder (or caliper) is bled. Follow Steps 7 through 16 to bleed the master cylinder while it is installed

on the vehicle.

7 Remove the vacuum reserve from the brake power booster by applying the brake several times with the engine off.

8 Remove the master cylinder reservoir cover and fill the reservoirs with brake fluid, then keep checking the fluid level often during the bleeding operation, adding fluid as necessary to keep the reservoirs full. Reinstall the cover.

9 Disconnect the forward brake line connection at the master cylinder.

10 Allow brake fluid to fill the master cylinder bore until it begins to flow from the forward line connector port (have a container and shop rags handy to catch and clean up spilled fluid).

11 Reconnect the forward brake line to the master cylinder.

12 Have an assistant depress the brake pedal very slowly (one time only) and hold it down.

13 Loosen the forward brake line at the master cylinder to purge the air from the bore, retighten the connection, then have the brake pedal released slowly.

14 Wait 15 seconds (important).

15 Repeat the sequence, including the 15 second wait, until all air is removed from the bore.

16 After the forward port has been completely purged of air, bleed the rear port in the same manner.

17 To bleed the individual wheel cylinders or calipers, first refer to Steps 7 and 8.

18 Have an assistant on hand, as well as a supply of new brake fluid, an empty clear plastic container, a length of 3/16-inch plastic, rubber or vinyl tubing to fit over the bleeder valve and a wrench to open and close the bleeder valve. The vehicle may have to be raised and placed on jackstands for clearance.

19 Beginning at the right rear wheel, loosen the bleeder valve slightly, then tighten it to a point where it is snug but can still be loosened quickly and easily.

20 Place one end of the tubing over the bleeder valve and submerge the other end in brake fluid in the container.

21 Have the assistant pump the brakes a few times to get pressure in the system, then hold the pedal firmly depressed.

22 While the pedal is held depressed, open the bleeder valve just enough to allow a flow of fluid to leave the valve. Watch for air bubbles to exit the submerged end of the tube. When the fluid flow slows after a couple of seconds, close the valve again and have your assistant release the pedal. If the pedal is released before the valve is closed again, air can be drawn back into the system.

23 Repeat Steps 21 and 22 until no more air is seen leaving the tube, then tighten the bleeder valve and proceed to the left rear wheel, the right front wheel and the left front wheel, in that order, and perform the same procedure **(see illustrations)**. Be sure to check the fluid in the master cylinder reservoir frequently.

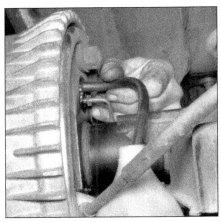

16.23a Attach a hose to the bleeder valve on a rear drum brake assembly

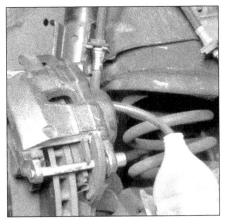

16.23b Bleeding a front disc brake caliper

24 Never use old brake fluid because it attracts moisture which will deteriorate the brake system components.
25 Refill the master cylinder with fluid at the end of the operation.
26 If any difficulty is experienced in bleeding the hydraulic system, or if an assistant is not available, a pressure bleeding kit is a worthwhile investment. If connected in accordance with the instructions, each bleeder valve can be opened in turn to allow the fluid to be pressure ejected until it is clear of air bubbles without the need to replenish the master cylinder reservoir during the process.

17 Brake pedal - removal and installation

Refer to illustration 17.4

1 Disconnect the cable from the negative battery terminal. **Caution:** *On models equipped with a Delco Loc II audio system, disable the anti-theft feature before disconnecting the battery.*
2 Disconnect the clutch pedal return spring (if equipped with a manual transmission).
3 Remove the clip retainer from the pushrod pin which travels through the pedal arm.
4 Remove the nut from the pedal shaft bolt. Slide the shaft out far enough to clear the brake pedal arm **(see illustration)**.
5 The brake pedal can now be removed, along with the spacer and bushing. The clutch pedal (if equipped) will remain in place.
6 When installing, lubricate the spacer and bushings with lightweight grease and tighten the pivot nut to the specified torque.

18 Stop light switch - removal, installation and adjustment

Refer to illustration 18.1

1 The switch is located on a flange or bracket protruding from the brake pedal support **(see illustration)**.
2 With the brake pedal in the fully released position, the plunger on the body of the switch should be completely pressed in. When the pedal is pushed in, the plunger releases and sends electrical current to the stop lights at the rear of the vehicle.
3 If the stop lights are inoperative and it has been determined that the bulbs are not burned out, push the stop light switch into the tubular clip, noting that audible clicks can be heard as the threaded portion of the switch is pushed through the clip toward the brake pedal.
4 Pull the brake pedal all the way to the rear against the pedal stop until no further clicks can be heard. This will seat the switch in the tubular clip and provide the correct adjustment.
5 Release the brake pedal and repeat Step 4 to ensure that no further clicks can be heard.
6 Make sure that the stop lights are working.
7 If the lights are not working, disconnect the electrical connectors at the stop light switch and remove the switch from the clip.
8 Install a new switch and adjust it by performing Steps 3 through 6, making sure the electrical connectors are hooked up.

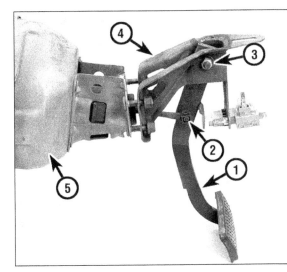

17.4 Brake pedal assembly mounting details

1 *Brake pedal*
2 *Brake booster pushrod and retainer clip*
3 *Throughbolt*
4 *Brake pedal support bracket*
5 *Brake booster*

18.1 Stop light switch location (arrow)

Notes

Chapter 10
Chassis electrical system

Contents

	Section
Airbag system - general information	28
Bulbs - replacement	13
CHECK ENGINE light	See Chapter 6
Circuit breakers - general information	5
Cluster panel instruments (except speedometer) - removal and installation	17
Clutch start switch - replacement and adjustment	See Chapter 7A
Concealed headlight actuator/motor assembly - removal and installation	10
Concealed headlight actuator switch and harness assembly - removal and installation	11
Concealed headlight assembly - general information and emergency procedures	7
Concealed headlight body assembly – removal and installation	12
Console switches - removal, servicing and installation	22
Cruise control - general information and servicing	26
Electrical troubleshooting - general information	2
Fuses - general information	3
Fusible links - general information	4
General information	1
Headlight sealed beam unit - removal and installation	8

	Section
Headlight switch - removal and installation	21
Headlights - adjustment	9
Ignition switch - replacement	See Chapter 11
Instrument cluster panel - removal, servicing and installation	20
Neutral safety and back-up light switch adjustment	See Chapter 7B
Radio - removal and installation	14
Radio power antenna - removal and installation	15
Radio speakers - removal and installation	16
Rear defogger (electric grid-type) - check and repair	27
Speedometer - removal and installation	18
Speedometer cable - replacement	19
Steering column switches - removal and installation	See Chapter 11
Stoplight switch - adjustment and replacement	See Chapter 9
Turn signals and hazard flashers - check and replacement	6
Windshield wiper arm - removal, installation and adjustment	21
Windshield wiper motor - removal and installation	24
Windshield wiper transmission - removal and installation	25

Specifications

Bulb application

	Type
Headlight	
Inner, standard	4651
Inner, Halogen	H4651
Outer	4652
Park and directional signal	
Standard Firebird	1157
Formula and Trans Am	1157NA
Side marker	194
Tail and stop	1157
Luggage compartment	561
License plate	194
Back-up	1156
Instrument cluster	161/194
Heater and A/C control panel	194
Indicators	
Headlight high beam	161
Directional signal	194
Temperature	194
Battery	194
Brake	194
Fasten seat belts	194
Check engine	194
Oil pressure	194
Choke	194
Radio (all)	194
Clock	194
Underhood	93
Automatic transmission indicator	194
Dome light	561
Courtesy light	631

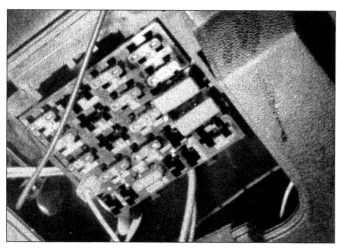

3.1 The fuse block is located under the left side of the dash (cover has been removed)

3.2 The fuses are marked to indicate the circuit they protect

1   General information

The electrical system is a 12-volt, negative ground type. Power for the lights and all electrical accessories is supplied by a lead/acid-type battery which is charged by the alternator.

This chapter covers repair and service procedures for the various electrical components not associated with the engine. Information on the battery, alternator, distributor and starter motor can be found in Chapter 5.

It should be noted that whenever portions of the electrical system are worked on, the negative battery cable should be disconnected to prevent electrical shorts and/or fires. **Caution:** *On models equipped with a Delco Loc II audio system, disable the anti-theft feature before disconnecting the battery.* **Note:** *Information concerning digital instrumentation and dash-related accessories is not included in this manual. Problems involving these components should be referred to your dealer.*

2   Electrical troubleshooting - general information

A typical electrical circuit consists of an electrical component, any switches, relays, motors, etc. related to that component and the wiring and connectors that connect the component to both the battery and the chassis. To aid in locating a problem in any electrical circuit, wiring diagrams for each model are included at the end of this Chapter.

Before tackling any troublesome electrical circuit, first study the appropriate diagrams to get a complete understanding of what makes up that individual circuit. Trouble spots, for instance, can often be narrowed down by noting if other components related to that circuit are operating properly or not. If several components or circuits fail at one time, chances are the problem lies in the fuse

or ground connection, as several circuits often are routed through the same fuse and ground connections.

Electrical problems often stem from simple causes, such as loose or corroded connections, a blown fuse or melted fusible link. Prior to any electrical troubleshooting, always visually check the condition of the fuse, wires and connections in the problem circuit.

If testing instruments are going to be utilized, use the diagrams to plan ahead of time where you will make the necessary connections in order to accurately pinpoint the trouble spot.

The basic tools needed for electrical troubleshooting include a circuit tester or voltmeter (a 12-volt bulb with a set of test leads can also be used), a continuity tester (which includes a bulb, battery and set of test leads) and a jumper wire, preferably with a circuit breaker incorporated, which can be used to bypass electrical components.

Voltage checks should be performed if a circuit is not functioning properly. Connect one lead of a circuit tester to either the negative battery terminal or a known good ground. Connect the other lead to a connector in the circuit being tested, preferably nearest to the battery or fuse. If the bulb of the tester goes on, voltage is reaching that point (which means the part of the circuit between that connector and the battery is problem free). Continue checking along the entire circuit in the same fashion. When you reach a point where no voltage is present, the problem lies between there and the last good test point. Most of the time the problem is due to a loose connection. Keep in mind that some circuits receive voltage only when the ignition key is in the Accessory or Run position.

A method of finding shorts in a circuit is to remove the fuse and connect a test light or voltmeter in its place to the fuse terminals. There should be no load in the circuit. Move the wiring harness from side-to-side while watching the test light. If the bulb goes on, there is a short to ground somewhere in that

area, probably where insulation has rubbed off of a wire. The same test can be performed on other components of the circuit, including the switch.

A ground check should be done to see if a component is grounded properly. Disconnect the battery and connect one lead of a self-powered test light such as a continuity tester to a known good ground. Connect the other lead to the wire or ground connection being tested. If the bulb goes on, the ground is good. If the bulb does not go on, the ground is not good.

A continuity check is performed to see if a circuit, section of circuit or individual component is passing electricity through it properly. Disconnect the battery and connect one lead of a self-powered test light such as a continuity tester to one end of the circuit. If the bulb goes on, there is continuity, which means the circuit is passing electricity through it properly. Switches can be checked in the same way.

Remember that all electrical circuits are composed basically of electricity running from the battery, through the wires, switches, relays, etc. to the electrical component (light bulb, motor, etc.). From there it is run to the body (ground) where it is passed back to the battery. Any electrical problem is basically an interruption in the flow of electricity to and from the battery. **Caution:** *On models equipped with a Delco Loc II audio system, disable the anti-theft feature before disconnecting the battery.*

3   Fuses - general information

Refer to illustrations 3.1 and 3.2

The electrical circuits of the vehicle are protected by a combination of fuses, circuit breakers and fusible links. The fuse block is located on the underside of the instrument panel on the driver's side **(see illustration)**.

Each of the fuses is designed to protect a specific circuit and the various circuits are identified on the fuse panel itself **(see illustration)**.

Miniaturized fuses are employed in the fuse block. These compact fuses with blade terminal design, allow fingertip removal and replacement.

If an electrical component has failed, your first check should be the fuse. A fuse which has "blown' is easily identified by inspecting the element inside the clear plastic body. Also, the blade terminal tips are exposed in the fuse body, allowing for continuity checks.

It is important that the correct fuse be installed. The different electrical circuits need varying amounts of protection, indicated by the amperage rating molded in bold, color-coded numbers on the fuse body.

At no time should the fuse be bypassed with pieces of metal or foil. Serious damage to the electrical system could result.

If the replacement fuse immediately fails, do not replace it again until the cause of the problem is isolated and corrected. In most cases, this will be a short circuit in the wiring caused by a broken or deteriorated wire.

4   Fusible links - general information

In addition to fuses, the wiring is protected by fusible links. These links are used in circuits which are not ordinarily fused, such as the ignition circuit.

Although the fusible links appear to be a heavier gauge than the wire they are protecting, the appearance is due to the thick insulation. All fusible links are four wire gauges smaller than the wire they are designed to protect.

The location of the fusible links on your particular vehicle may be determined by referring to the wiring diagram(s) at the end of this Chapter.

The fusible links cannot be repaired, but a new link of the same size wire can be put in its place. The procedure is as follows:

a) *Disconnect the battery ground cable.*
   Caution: *On models equipped with a Delco Loc II audio system, disable the anti-theft feature before disconnecting the battery.*
b) *Disconnect the fusible link from the starter solenoid.*
c) *Cut the damaged fusible link out of the wiring just behind the connector.*
d) *Strip the insulation approximately 1/2-inch.*
e) *Position the connector on the new fusible link and crimp it into place.*
f) *Use rosin core solder at each end of the new link to obtain a good solder joint.*
g) *Use plenty of electrical tape around the soldered joint. No wires should be exposed.*

7.1a View of the left side concealed headlight actutor/motor

h) *Connect the fusible link at the starter solenoid. Connect the battery ground cable. Test the circuit for proper operation.*

5   Circuit breakers - general information

A circuit breaker is used to protect the headlight wiring and is located in the light switch. An electrical overload in the system will cause the lights to go on and off, or in some cases to remain off. If this happens, check the entire headlight circuit immediately. Once the overload condition is corrected, the circuit breaker will function normally.

Circuit breakers are also used with accessories such as power windows, power door locks and rear window defogger.

The circuit breakers in your particular vehicle may be found by referring to the wiring diagram(s) at the end of this Chapter.

6   Turn signals and hazard flashers - check and replacement

1   Small canister-shaped flasher units are incorporated into the electrical circuits for the directional signals and hazard warning lights.
2   When the units are functioning properly, an audible click can be heard with the circuit in operation. If the turn signals fail on one side only and the flasher unit cannot be heard, a faulty bulb is indicated. If the flasher unit can be heard, a short in the wiring is indicated.
3   If the turn signal fails on both sides, the problem may be due to a blown fuse, faulty flasher unit or switch, or a broken or loose connection. If the fuse has blown, check the wiring for a short before installing a new fuse.
4   The hazard warning lights are checked as described in paragraph 3 above.

7.1b View of the right side concealed headlight actuator/motor

5   The hazard warning flasher and turn signal flasher are mounted either at the rear of the fuse box or at the convenience center located under the center of the dash.
6   When replacing either of the flasher units, be sure to buy a replacement of the same capacity. Compare the new flasher to the old one before installing it.

7   Concealed headlight assembly – general information and emergency procedures

Refer to illustrations 7.1a, 7.1b and 7.5
1   The electrically operated concealed headlights each use a permanent magnet/actuator motor connector to a link which raises and lowers the assembly **(see illustrations)**.
2   When the headlights are turned on at the dashboard switch, relays attached to actuators operate the motors. The motors are shut off when the headlights reach their full open positions. The motors are activated to close the headlight assemblies when the dashboard switch is pushed to the off position.
3   In emergencies, the headlights may be opened manually by one of the methods which follow, depending on the model of headlight door motor with which the vehicle is equipped.
4   On headlight door motors having a knob with a single hole in its top, insert a ballpoint pen or similarly shaped object into the hole and rotate the knob in a clockwise direction until a click is heard. Manually close the headlight assemblies by reversing the procedure. **Note:** *To use this manual procedure, there can be no voltage present at the motor terminals. To shut off the voltage, disconnect the electrical connections at the isolation relay, located on the radiator cradle on the right side of the vehicle.*
5   On headlight door motors having a knurled knob with two holes in its top, turn on

7.5 Turn the knob in the direction indicated by the arrow to manually open the concealed headlight assembly

9.2a The screw in the top adjusts the headlight beam up and down

9.2b The side screw adjusts the headlight beam left and right

9.2c The adjustment screws are located in the same positions (arrows) on concealed headlight assemblies

10.3 Removing the headlight bezel

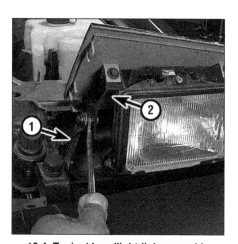

10.4 Typical headlight link assembly

    1   Link assembly
    2   Headlight assembly

the headlights and open the hood. Turn the motor knob in the direction of the arrow on the knob (clockwise) until the headlight assembly is all the way open **(see illustration)**. When the retractor motor shaft reaches the limit of its travel, the effort necessary to turn the knob will increase and a click may be heard. To manually close the headlight assembly with this type of motor, turn off the headlights and rotate the knob counterclockwise.

8 Headlight sealed beam unit - removal and installation

1 When replacing the headlight, do not turn the spring-loaded adjusting screws or the headlight aim will be changed.
2 Remove the headlight bezel screws and the decorative bezel.
3 Use a cotter pin removal tool to unhook the spring from the retaining ring.
4 Remove the two screws which secure the retaining ring and withdraw the ring. Support the light as this is done.

5 Pull the sealed beam unit out slightly and disconnect the wires from the rear of the light. Remove the light from the vehicle.
6 Position the new unit close enough to connect the wires. Make sure that the numbers molded into the lens are at the top.
7 Install the retaining ring with the mounting screws and spring.
8 Install the bezel and check for proper operation. If the adjusting screws were not turned, the new headlight should not require adjustment.

9 Headlights - adjustment

Refer to illustrations 9.2a, 9.2b and 9.2c
1 Any adjustments made by the home mechanic that affect the aim of the headlights should be considered temporary only. After adjustment, always have the beams readjusted by a facility with the proper aligning equipment as soon as possible. In some states, these facilities must be state-authorized. Check with your local motor vehicle department concerning the laws in your area.
2 Adjustment screws are provided at the front of each headlight to alter the beam

horizontally (side screw) and vertically (top screw) **(see illustrations)**. When making adjustments, be careful not to scratch the paint on the body.

10 Concealed headlight actuator/motor assembly – removal and installation

Refer to illustrations 10.3, 10.4, 10.5 and 10.7
1 Raise the headlight assembly to the open position.
2 Disconnect the cable from the negative battery terminal. **Caution:** *On models equipped with a Delco Loc II audio system, disable the anti-theft feature before disconnecting the battery.*
3 Remove the headlight bezel retaining screws and remove the bezel **(see illustration)**.
4 Using a screwdriver, pry the link assembly from the headlight body assembly **(see illustration)**.
5 Remove the actuator/motor crank arm

10.5 Actuator/motor crank arm removal

10.7 Typical actuator/motor assembly

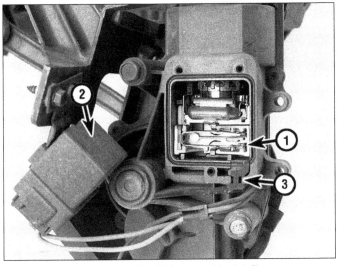

11.4 Interval switch assembly details

1   Switch assembly   3   Rubber slot filler
2   Relay

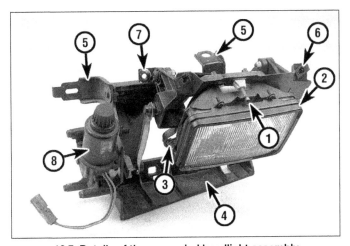

12.7 Details of the concealed headlight assembly

1   Headlight adjusting   4   Lower mounting bracket
    screws   5   Upper mounting bracket
2   Headlight capsule   6   Headlight body assembly
    assembly   7   Door mounting bracket
3   Lower retaining spring   8   Actuator/motor

attaching nut and remove the crank arm **(see illustration)**. Support the crank while it is being removed so that the actuator/motor is not damaged.

6   Remove the wiring connectors from the actuator/motor and relay.

7   Remove the actuator/motor retaining screws and remove the actuator/motor **(see illustration)**.

8   Remove the screws retaining the relay to the actuator/motor and remove the relay.

9   Installation is the reverse of the removal procedure.

11 Concealed headlight actuator switch and harness assembly – removal and installation

Refer to illustration 11.4

1   Remove the actuator/motor (see Section 10).

2   Remove the actuator/motor cover plate

retaining screws, then remove the cover, making sure not to damage the cover O-ring.

3   Remove the rubber slot filler.

4   Pull the switch assembly straight out of the cavity of the actuator/motor assembly **(see illustration)**.

5   When reassembling, install the switch assembly by snapping the brushes around the commutator. **Note:** *The switch blades must be positioned securely between the thrust block fingers.*

6   Installation is the reverse of the removal procedure.

12 Concealed headlight body assembly – removal and installation

Refer to illustration 12.7

1   With the hood raised, remove the screws retaining the rear of the headlight door.

2   Raise the headlight assembly.

3   Remove the cable from the negative battery terminal. **Caution:** *On models equipped with a Delco Loc II audio system, disable the anti-theft feature before disconnecting the battery.*

4   Remove the headlight bezel retaining screws and remove the bezel.

5   Disconnect the wiring connector at the headlight bulb.

6   Remove the lower air deflector.

7   Remove the headlight body assembly retaining screws and disconnect the wiring connectors at the actuator/motor relay **(see illustration)**.

8   Remove the actuator/motor (refer to Section 10).

9   Disengage the headlight capsule lower retaining spring and remove the capsule assembly. Do not disturb the headlight adjusting screws.

10   Remove the headlight body assembly retaining brackets and remove the body assembly.

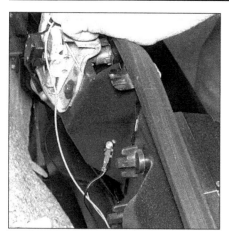

13.7 Disconnect the wire connectors at the taillight cover panel

13.8 Squeeze the tab in and rotate the socket counterclockwise to remove the bulb socket from the taillight panel

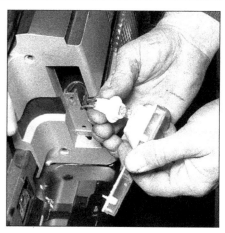

13.12 Remove the bulb socket from the license plate light lens

11   Installation is the reverse of the removal procedure.

---

13   Bulbs - replacement

Refer to illustrations 13.7, 13.8 and 13.12

Front end

1   The bulbs for the parking lights and side marker lights are accessible from the rear of each unit.
2   Locate the affected bulb housing and twist it out of the socket.
3   Remove the defective bulb from the housing, install a new bulb and twist/lock the housing back into the socket.

Rear end

4   The bulbs for the taillights, brake lights, directional signal lights and back-up lights are all contained in one assembly on each side of the rear of the vehicle.
5   From inside the vehicle, remove the plastic screws retaining the upholstery at the rear quarter panel.
6   Remove the plastic screws and metal screw retaining the upholstery panel at the rear cover panel (the metal screw is at the top of the panel), revealing the plastic wing nuts retaining the taillight cover panel.
7   Disconnect the wires attached to the panel, remove the wing nuts and pull off the taillight cover panel **(see illustration)**.
8   Remove the affected bulb socket by squeezing the lock on the bulb socket assembly and turning it counterclockwise, or by simply turning the socket out of the panel (side marker light only) **(see illustration)**.
9   Replace the burned out bulb with a new one.
10   Installation is the reverse of the removal procedure.

License plate bulb

11   Remove the screw retaining the

14.4 Disconnect the wire connector from the rear of the radio

lens/socket assembly and pull the assembly away from the taillight lens housing.
12   Twist the socket out of the lens, replace the burned out bulb with a new one and reinstall the lens/socket assembly **(see illustration)**.

Interior light

13   To replace the courtesy light, remove the translucent light panel by squeezing it on both sides and pulling it away from the fixture.
14   Replace the burned out bulb with a new one and reinstall the translucent panel by snapping it into place.

Radio light

15   Remove the radio (refer to Section 10).
16   Remove the bulb access panel and replace the bulb with a new one.
17   Installation is the reverse of the removal procedure.

Console light

18   Refer to Section 18.

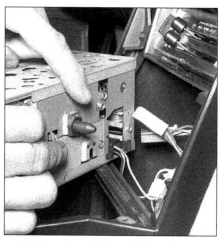

14.5 Location of the radio locating lug

---

14   Radio - removal and installation

Refer to illustrations 14.4 and 14.5
Warning: *If equipped with a Supplemental Inflatable Restraint system (SIR), more commonly known as airbags, disable the airbag system before working in the vicinity of airbag system components to avoid the possibility of accidental deployment of the airbag, which could result in personal injury (see Section 24).*

1   Disconnect the cable from the negative battery terminal. **Caution:** *On models equipped with a Delco Loc II audio system, disable the anti-theft feature before disconnecting the battery.*
2   Remove the screws retaining the console bezel and remove the bezel.
3   Remove the screws retaining the radio to the console and pull the radio out.
4   Remove all electrical connectors from the rear of the radio **(see illustration)**. To simplify reinstallation, label the connectors with pieces of numbered tape.
5   Installation is the reverse of removal. The locating lug at the rear of the radio will

position it correctly when the lug contacts the receptacle in the console frame **(see illustration)**. **Caution:** *Always connect the speaker wiring harness before turning the radio on (to prevent possible radio damage).*

15 Radio power antenna - removal and installation

1   Lower the antenna by turning off the radio or the ignition switch. **Note:** *If the mast has failed in the up position and the mast or the entire antenna assembly is being replaced, the mast can be cut off at fender level to facilitate removal.*
2   Disconnect the cable from the negative battery terminal. **Caution:** *On models equipped with a Delco Loc II audio system, disable the anti-theft feature before disconnecting the battery.*
3   Remove the screws attaching the inner fender skirt to the fender.
4   To gain access to the antenna assembly, pull down on the rear of the fender skirt and hold it away from the fender with a block of wood.
5   Remove the escutcheon nut retaining the antenna cable assembly to the fender.
6   Disconnect the electrical leads, then remove the bolts retaining the antenna assembly to the bracket and remove the assembly.
7   Installation is the reverse of the removal procedure. Make sure that the mast is in the fully retracted position before installation and that the lower bracket is secured by the fender skirt retaining screw.

16 Radio speakers - removal and installation

Refer to illustration 16.3
Warning: *If equipped with a Supplemental Inflatable Restraint system (SIR), more commonly known as airbags, disable the airbag system before working in the vicinity of airbag system components to avoid the possibility of accidental deployment of the airbag, which could result in personal injury (see Section 24).*
1   Disconnect the cable from the negative battery terminal. **Caution:** *On models equipped with a Delco Loc II audio system, disable the anti-theft feature before disconnecting the battery.*

Dash mounted speaker

2   Remove the dash pad by removing the four retaining screws underneath the lip of the pad.
3   Remove the four bolts retaining the speaker to the speaker braces **(see illustration)**.
4   Lift the speaker out and disconnect the wire connector.
5   Installation is the reverse of the removal procedure.

Rear quarter speaker

6   Remove the rear quarter trim panel.
7   Disconnect the electrical connector from the speaker.
8   Remove the screws securing the speaker to the mounting bracket and remove the speaker.
9   Installation is the reverse of removal.

Rear compartment speaker

10   Open the rear compartment.
11   Remove the screws from the speaker grill and remove the grille.
12   Remove the speaker mounting screws. Lift the speaker up, disconnect the electrical connector and remove the speaker.
13   Installation is the reverse of removal.

17 Cluster panel instruments (except speedometer) - removal and installation

Warning: *If equipped with a Supplemental Inflatable Restraint system (SIR), more commonly known as airbags, disable the airbag system before working in the vicinity of airbag system components to avoid the possibility of accidental deployment of the airbag, which could result in personal injury (see Section 24).*
1   Disconnect the cable from the negative battery terminal. **Caution:** *On models equipped with a Delco Loc II audio system, disable the anti-theft feature before disconnecting the battery.*
2   Remove the screws retaining the instrument cluster bezel and remove the bezel.
3   Remove the trip odometer reset knob by unscrewing it.
4   Remove the screws retaining the instrument cluster plastic lens.
5   The individual instruments may now be lifted out of the panel after removing the retaining screws.
6   Installation is the reverse of the removal procedure.

16.3 Remove the speaker mounting screws (arrows)

18 Speedometer - removal and installation

Warning: *If equipped with a Supplemental Inflatable Restraint system (SIR), more commonly known as airbags, disable the airbag system before working in the vicinity of airbag system components to avoid the possibility of accidental deployment of the airbag, which could result in personal injury (see Section 24).*
1   Disconnect the cable from the negative battery terminal. **Caution:** *On models equipped with a Delco Loc II audio system, disable the anti-theft feature before disconnecting the battery.*
2   Remove the dash pad by removing the four retaining screws in the defroster ducts and six retaining screws underneath the lip of the dash pad.
3   Remove the left-hand radio speaker, then disconnect the electrical connector.
4   Using a Torx screwdriver of the appropriate size, remove the screws retaining the instrument cluster outer bezel and remove the bezel.
5   Remove the cluster panel retaining screws.
6   Pull the cluster panel out just far enough to reach behind it. With one hand, reach behind the panel and locate the speedometer cable where it joins the speedometer case. Hold the cable and, with the other hand, pinch the cable release and separate the cable from the speedometer.
7   Remove the screws retaining the instrument cluster panel plastic lens.
8   From the rear of the cluster panel, remove the speedometer retaining screws and copper retaining clip, then pull the speedometer out of the panel, disconnect the VSS electrical connector with a screwdriver and remove the speedometer.
9   Installation is the reverse of the removal procedure. When reattaching the speedometer cable connector to the speedometer case, the cluster panel should be positioned so that the speedometer cable connector is visible. Hold the connector in position with one hand and "feed' the cluster panel into the connector with the other hand until the connector snaps into place.

19 Speedometer cable - replacement

Refer to illustration 19.7
Warning: *If equipped with a Supplemental Inflatable Restraint system (SIR), more commonly known as airbags, disable the airbag system before working in the vicinity of airbag system components to avoid the possibility of accidental deployment of the airbag, which could result in personal injury (see Section 24).*
1   Disconnect the cable from the negative battery terminal. **Caution:** *On models*

19.7 Typical speedometer cable routing

1   *Speedometer cable*
2   *Speedometer gear adapter*
3   *Mounting bracket*

20.3 Typical dash pad mounting screw locations (arrows)

equipped with a Delco Loc II audio system, disable the anti-theft feature before disconnecting the battery.
2   On vehicles not equipped with Cruise Control, disconnect the speedometer cable strap at the power brake booster.
3   On Cruise Control-equipped vehicles, disconnect the speedometer cable at the cruise control transducer.
4   Remove the speedometer as described in Section 14 (paragraphs 2 through 6).
5   Slide the old cable out from the upper end of the casing, or, if broken, from both ends of the casing.
6   If speedometer operation has been noisy, but the speedometer cable appears to be in good condition, take a short piece of speedometer cable with a tip to fit the speedometer and insert it in the speedometer socket. Spin the piece of cable between your fingers. If binding is noted, the speedometer is faulty and should be repaired.
7   Inspect the speedometer cable casing for sharp bends and breaks, especially at the transmission end **(see illustration)**. If breaks are noted, replace the casing with a new one.
8   When installing the cable, perform the following operations to ensure quiet operation.
9   Wipe the cable clean with a lint-free cloth.
10   Flush the bore of the casing with solvent and blow it dry with compressed air.
11   Place some speedometer cable lubricant in the palm of one hand.
12   Feed the cable through the lubricant and into the casing until lubricant has been applied to the lower two-thirds of the cable. Do not over-lubricate it.
13   Seat the upper cable tip in the speedometer and snap the retainer onto the casing.
14   The remaining installation steps are the reverse of those for removal.

20.5 Typical instrument panel lower cover screw locations (arrows)

20   Instrument cluster panel - removal, servicing and installation

Refer to illustrations 20.3, 20.5, 20.8a, 20.8b, 20.12, 20.13, 20.14a, 20.14b, 20.15a, 20.15b, 20.17, 20.19, 20.20 and 20.21
Warning: *If equipped with a Supplemental Inflatable Restraint system (SIR), more commonly known as airbags, disable the airbag system before working in the vicinity of airbag system components to avoid the possibility of accidental deployment of the airbag, which could result in personal injury (see Section 24).*
1   Disconnect the cable from the negative battery terminal. **Caution:** *On models equipped with a Delco Loc II audio system, disable the anti-theft feature before disconnecting the battery.*
2   Although it is not absolutely necessary, access to the instrument cluster panel will be easier if the steering wheel is removed at this point (refer to Chapter 11).

20.8a Remove the top retaining nuts from the instrument cluster panel

3   Remove the dash pad by removing the screws in the defroster ducts and the screws under the lip of the dash pad **(see illustration)**.
4   Remove the left-hand radio speaker and braces by removing the two retaining screws at the top front braces and the two screws from the rear braces, then disconnect the wire connector.
5   Steps 5 and 6 are not absolutely necessary, but lowering the steering column makes access to the cluster panel much easier. Remove the screws retaining the plastic trim cover below the steering column and remove the cover **(see illustration)**.
6   Support the steering column from below with a brace, so as not to damage the undercowl panel, remove the screws retaining the outer instrument cluster bezel and remove the bezel.
7   Using a Torx screwdriver of the appropriate size, remove the screws retaining the outer instrument cluster bezel and remove the bezel.
8   Remove the two nuts retaining the top

20.8b Remove the bottom retaining nuts from the instrument cluster panel

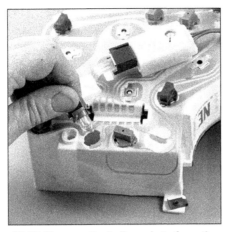

20.12 Remove the bulb sockets from the cluster panel

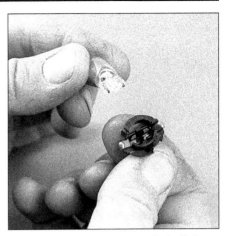

20.13 If necessary, remove the bulb from the cluster panel bulb socket

20.14a Remove the VSS printed circuit connector from the male VSS cable lead

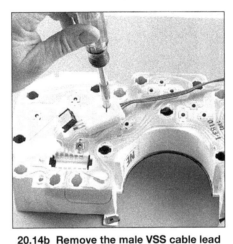

20.14b Remove the male VSS cable lead from the cluster panel

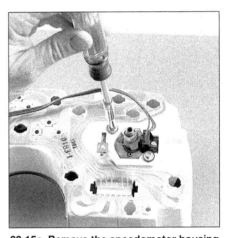

20.15a Remove the speedometer housing bolts (VSS connector is visible at right of speedometer drive housing)

of the instrument cluster panel, then remove the two nuts retaining the bottom of the cluster panel, located in the warning light housings on either side of the panel **(see illustrations)**.

9   Pull the instrument cluster panel out just far enough to reach behind it. With one hand, reach behind the panel and locate the speedometer cable where it joins the speedometer case. With the other hand, pinch the cable release and separate the cable from the speedometer.

10   Remove the cluster panel by pulling it straight out to prevent damage to the electrical contacts behind it.

11   Place the cluster panel assembly in a clean work area.

12   From the rear of the cluster panel, remove the bulb assemblies by turning the plastic retainers counterclockwise **(see illustration)**. **Note:** *To keep the bulb assemblies in order, they should be placed in the work area in a pattern simulating the back of the cluster panel. As an alternative, tag each bulb assembly and the recess from which it was removed with numbered pieces of tape.*

13   Replace burned-out bulbs with new ones by pulling them out of the plastic housings and inserting the new ones **(see illustra-**

tion). Keep the bulb assemblies in order.

14   Disconnect the vehicle speed sensor (VSS) by pinching the printed circuit connector and pulling it away from the male side of the connector, then remove the bolt retaining the male connector **(see illustrations)**.

15   Remove the bolts retaining the speedometer case and the speedometer end of the VSS cable, then remove the copper

retaining clip **(see illustrations)**.

16   From the front of the cluster assembly, remove the trip odometer reset knob by unscrewing it.

17   Remove the screws retaining the plastic face plate and remove the face plate from the cluster panel assembly **(see illustration)**.

18   Remove the screws retaining the metal

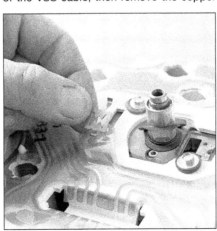

20.15b Remove the copper speedometer housing connector

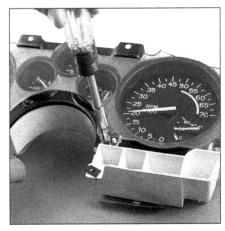

20.17 Remove the plastic face plate from the cluster panel

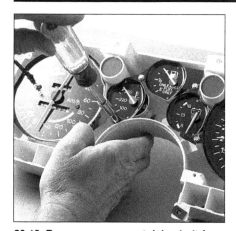

20.19 Remove a gauge retaining bolt from the cluster panel

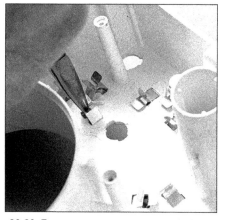

20.20 Remove a gauge contact clip from the rear of the cluster panel

20.21 If necessary, remove the printed circuit from the cluster panel

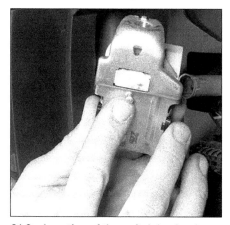

21.3a Location of the switch knob release button on the bottom of the headlight switch (shown with switch removed)

21.3b Remove the headlight switch knob from the dash

21.4 Remove the headlight switch knob panel

bezel and separate the bezel from the panel.

19   Remove the screws retaining the individual gauges and separate the gauges from the panel **(see illustration)**.

20   Remove the instrument contact clips by pinching them at the base with needle-nose pliers and pulling them out from the rear **(see illustration)**. Replace any bent or damaged clips with new ones.

21   Remove the printed circuit **(see illustration)** and replace it with a new one, if necessary.

22   Installation is the reverse of the removal procedure. When installing the speedometer, after the instrument cluster has been placed in position, pull it out just far enough so the speedometer cable connector is visible, hold the connector in position with one hand and feed the instrument cluster into the connector with the other hand until the connector snaps into place.

---

21   Headlight switch - removal and installation

Refer to illustrations 21.3a, 21.3b, 21.4, 21.5a, 21.5b, 21.6 and 21.7

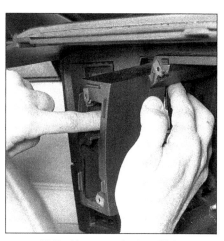

21.5a Unscrew the headlight switch retainer

21.5b Remove the headlight switch retainer

Warning: *If equipped with a Supplemental Inflatable Restraint system (SIR), more commonly known as airbags, disable the airbag system before working in the vicinity of airbag system components to avoid the possibility of accidental deployment of the airbag, which could result in personal injury (see Section 24).*

1   Disconnect the negative battery cable and remove the instrument cluster panel (refer to Section 16). **Caution:** *On models equipped with a Delco Loc II audio system, disable the anti-theft feature before disconnecting the battery.*

2   Pull the headlight switch knob to the full On position.

3   Reach behind the switch with one hand and push the release button, then pull the switch knob out of the dash with the other hand **(see illustrations)**.

21.6 Remove the headlight switch
from the dash

21.7 Disconnect the headlight switch
electrical connector

22.3 Remove the shift detent button
assembly from the shift handle

22.6 Remove the console light bulb
socket from the console cover

22.9 Pry the top half of the up/down
switch from the bottom half

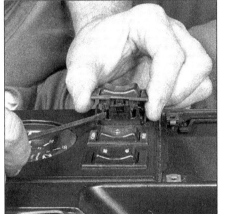

22.10 Remove the electrical connector
from the bottom half of the
up/down switch

4 Remove the screws retaining the knob panel and remove the panel **(see illustration)**.
5 Working behind the switch, hold it in position with one hand and, using the index finger of the other hand, unscrew the retainer from the front and remove it **(see illustrations)**.
6 Remove the switch from the dash **(see illustration)**. Note the angle at which it was removed to simplify installation.
7 Disconnect the wire connector **(see illustration)**.
8 Installation is the reverse of the removal procedure.

22 Console switches - removal, servicing and installation

Refer to illustrations 22.3, 22.6, 22.9 and 22.10
Warning: *If equipped with a Supplemental Inflatable Restraint system (SIR), more commonly known as airbags, disable the airbag system before working in the vicinity of airbag system components to avoid the possibility of accidental deployment of the*

airbag, which could result in personal injury (see Section 24).
1 Disconnect the cable from the negative battery terminal. **Caution:** *On models equipped with a Delco Loc II audio system, disable the anti-theft feature before disconnecting the battery.*
2 If the vehicle is equipped with a manual transmission, unscrew the shifter ball from the shifter handle.
3 If the vehicle is equipped with an automatic transmission, pull up on the shift detent button assembly and remove it from the shifter **(see illustration)**.
4 Using a screwdriver and an awl, spread the ends of the snap-ring at the bottom of the shifter detent button hole with the screwdriver and remove the snap-ring by passing the tip of the awl through one of the snap-ring eyes and pulling the snap-ring out of the hole (this method is required because the hole is too small to accept normal snap-ring pliers).
5 Pull the knob off the shifter handle.
6 Lift the console cover, twist the console light bulb socket out of the cover and remove the cover from the console **(see illustration)**.
7 To replace the left-or-right up-or-down

rocker switch, it must be removed in two pieces.
8 Raise the rear of the switch with a screwdriver while pushing on the front release lever to elevate the entire switch.
9 With the switch elevated, use a screwdriver to pry the top half of the switch from the bottom half **(see illustration)**.
10 Remove the bottom half of the switch from the underside of the console cover and separate it from the electrical connector **(see illustration)**.
11 The center open/close switch may be removed in one piece through the top of the console cover while pressing the release lever from the bottom.
12 Installation is the reverse of the removal procedure.

23 Windshield wiper arm - removal, installation and adjustment

Refer to illustrations 23.2, 23.5, 23.8 and 23.11
1 Raise the hood.
2 Lift the wiper blade off the glass and disconnect the wiper arm locking retainer

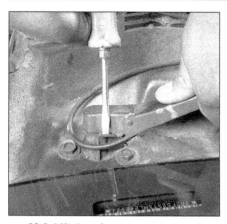

23.2 Lift the wiper arm and pry out the lock

23.5 Make sure the right wiper arm is aligned in the park position before installing the wiper arm

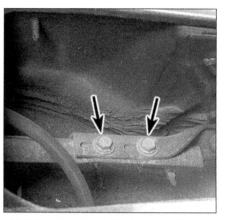

23.8 To adjust the left wiper arm, loosen the rear wiper link-to-motor crank arm attaching nuts (arrows)

from the wiper transmission shaft **(see illustration)**.

3   Pull the arm off the transmission shaft.

4   When installing the arm, the windshield wiper blade assembly release lever must be facing the base of the arm assembly for correct matching of the blade and glass contours.

5   Align the right wiper arm assembly in the parked position and attach the wiper arm to the transmission shaft **(see illustration)**.

6   If the wiper arms and blades were in the correct positions prior to wiper arm removal, no adjustment is required. However, if adjustment is required, proceed as follows.

7   To adjust the left wiper arm, loosen and raise the left side of the cowl vent screen.

8   Loosen but do not remove the rear wiper link-to-motor crank arm attaching nuts **(see illustration)**.

9   Rotate the right arm assembly to a position 3/4-inch below the ramp stops.

10   Tighten the attaching nuts on the rear wiper link(s)-to-motor crank arm to 4 to 6 ft-lbs.

11   Place the left arm and blade assembly next to the spindle and align the slot in the arm with the keyway in the spindle **(see illustration)**. Push the arm down and place the blade on the ramp stop on the vent screen.

12   Lift the right arm and blade assemblies over the stop.

13   Wet the windshield and check the wiper pattern and parked position. The distance between the top of the wiper pattern and the left edge of the windshield should be 1 to 2 inches. The overlap of the blade tips should be approximately 1-inch.

14   To adjust the right arm, remove it from the driveshaft. The correct Park position and wiper pattern dimensions are determined with the wipers operating at low speed on a wet windshield.

15   Reinstall the cowl vent assembly.

24   Windshield wiper motor - removal and installation

Refer to illustration 24.7

1   Lift the hood.

2   Remove the cowl vent screen.

3   Loosen the wiper transmission rear wiper link-to-motor crank arm retaining nuts.

4   Remove the transmission rear wiper link from the motor crank arm.

5   Disconnect the wire connectors.

6   Disconnect the washer hoses.

7   Remove the three motor mounting bolts **(see illustration)**.

8   While guiding the crank arm through the

hole in the cowl, remove the wiper motor.

9   Installation is the reverse of the removal procedure. The motor must be in the parked position before attaching the crank arm to the transmission rear wiper link.

25   Windshield wiper transmission - removal and installation

1   Lift the hood.

2   Remove the right wiper arm and the cowl vent screen.

3   Loosen but do not remove the nuts retaining the transmission rear wiper link from the motor crank arm.

4   Pull the transmission rear wiper link from the motor crank arm.

5   Remove the transmission-to-cowl bolts.

6   Remove the transmission and linkage assembly by guiding it through the plenum chamber opening.

7   To install, extend the forward motor link and rear wiper link and rotate the left-side driver housing, aligning the locator rib with the cutout in the roof panel.

8   Collapse the assembly by turning the forward motor link clockwise, the rear wiper link counterclockwise and the right driveshaft in - in effect making the assembly as short as possible.

9   Feed the assembly through the access hole in the plenum chamber from the left side.

10   Turn the right driveshaft clockwise and insert the driver through the hole so that the driver rests on the top of the roof panel.

11   Install the bolts on both the right and left driver housing driving screws and tighten them to 4 to 6 ft-lbs.

12   Install the transmission rear wiper link on the motor crank arm and tighten the bolts to 4 to 6 ft-lbs.

13   Align the transmission and attach it to the cowl.

14   Install and adjust the wiper arm and blade assemblies (refer to Section 23).

15   Check the wiper operation and wiper pattern (refer to Section 23).

16   Install the cowl vent screen.

23.11 The left wiper arm spindle keyway should be positioned as shown

24.7 Windshield wiper motor mounting bolts (arrows)

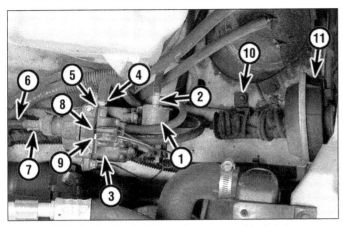

26.4 Cruise Control transducer assembly details

1 Resume solenoid valve	7 Drive cable
2 Vacuum source	8 Engage
3 Transducer assembly	9 Hold
4 Regulated bleed	10 Mounting screws
5 Orifice tube lockout	11 Servo assembly
6 Speedometer cable	

26.5 Cruise Control brake and clutch release switches and valves

1 Transmission torque converter clutch switch and vacuum valve assembly	2 Stop light and cruise control switch
	3 Retainer

26 Cruise Control - general information and servicing

Refer to illustrations 26.4 and 26.5

1 The Cruise Control with Resume system is an option which maintains a desired vehicle speed under normal driving conditions. The system also has the capability of resuming a preset speed upon driver demand after the system has been disengaged. This is accomplished by moving a slide on the Cruise Control lever handle to the Resume position. Steep grades, up or down, may cause variations in the selected speed, which is considered normal.

2 The main components of the Cruise Control system are a transducer assembly, a resume solenoid valve, a vacuum servo with linkage, an engagement switch button and an On/Off/Resume switch on the turn signal lever, and vacuum and electric release switches attached to the brake pedal (automatic) or clutch pedal (manual).

3 Because of the variations in installations, it is not possible to include all the service procedures in this manual. However, those elements of the system most often requiring service and/or adjustment are covered.

4 The transducer is calibrated in such a manner during production that overhaul operations are impractical. A defective transducer must be replaced with a new one; however, one adjustment is possible. If there is a difference between the engagement speed selected and the actual cruising speed, proceed as follows **(see illustration)**:

a) *Check all hoses for kinks and cracks. If there is still a difference between the engagement and cruising speeds, proceed to b.*

b) *If the cruising speed is lower than the engagement speed, loosen the orifice tube locknut and turn the tube out.*

c) *If the cruising speed is higher than the engagement speed, loosen the orifice tube locknut and turn the tube in.*

d) *Each 90-degree (1/4-turn) rotation will alter the engagement/cruising speed one (1) mph.*

e) *Tighten the locknut after adjustment has been made and check the system operation at 55 mph.*

5 To remove the Cruise Control switch, vacuum valve assembly/TCC switch (automatic transmission) or vacuum release valve (manual transmission), pull the component from the bracket under the dash, then disconnect the wiring and/or vacuum connector and discard the faulty component and the retainer **(see illustration)**. Install a new retainer in the bracket.

6 With the brake or clutch pedal depressed, install the new component in the retainer and make sure it is seated. Note that audible clicks can be heard as the component is pressed into the retainer.

7 Pull the brake or clutch pedal all the way up against the stop until the clicks can no longer be heard.

8 Release the clutch or brake pedal, then repeat the procedure to ensure that no more clicks can be heard. The component is now properly seated and adjusted.

9 Reconnect the wiring and/or vacuum connectors.

10 Other servicing of the Cruise Control system components should be done by your dealer.

27 Rear defogger (electric grid type) - check and repair

Refer to illustrations 27.5 and 27.11

1 This option consists of a rear window with a number of horizontal elements that are baked into the glass surface during the glass forming operation.

2 Small breaks in the element can be successfully repaired without removing the rear window.

3 To test the grids for proper operation, start the engine and turn on the system.

4 Ground one lead of a test light and carefully touch the other lead to each element line.

5 The brilliance of the test light should increase as the lead is moved across the element from right to left **(see illustration)**. If the test light glows brightly at both ends of the lines, check for a loose ground wire. All of the lines should be checked in at least two places.

6 To repair a break in a line, it is recommended that a repair kit specifically for this purpose be purchased from an auto parts store. Included in the repair kit will be a decal, a container of silver plastic and hardener, a mixing stick and instructions.

7 To repair a break, first turn off the system and allow it to de-energize for a few minutes.

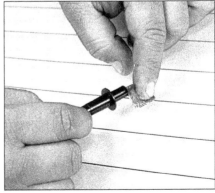

27.5 When checking the rear defogger grid, wrap a piece of aluminum foil around the probe tip and press the foil against the grid with your finger

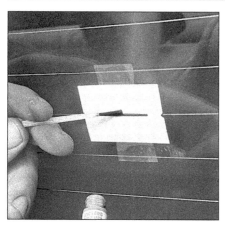

27.11 Applying repair material to a broken rear window defogger grid

8   Lightly buff the element area with fine steel wool and then clean it thoroughly with alcohol.
9   Use the decal supplied in the repair kit, or apply strips of electrician's tape above and below the area to be repaired. The space between the pieces of tape should be the same width as the existing lines. This can be checked from outside the vehicle. Press the tape tightly against the glass to prevent seepage.
10   Mix the hardener and silver plastic thoroughly.
11   Using the wood spatula, apply the silver plastic mixture between the pieces of tape, overlapping the undamaged area slightly on either end **(see illustration)**.
12   Carefully remove the decal or tape and apply a constant stream of hot air directly to the repaired area. A heat gun set at 500 to 700 degrees Fahrenheit is recommended. Hold the gun about one (1) inch from the glass for one to two minutes.
13   If the new element appears off color, tincture of iodine can be used to clean the repair and bring it back to the proper color. This mixture should not remain on the repair for more than 30 seconds.
14   Although the defogger is now fully operational, the repaired area should not be disturbed for at least 24 hours.

28   Airbag system - general information

Warning: *1990 and later models are equipped with an airbag. Airbag system components are located in the steering wheel, steering column, instrument panel and center console. The airbag(s) could accidentally deploy if any of the system components or wiring harnesses are disturbed, so be extremely careful when working in these areas and don't disturb any airbag system components or wiring. You could be injured if an airbag accidentally deploys, or the airbag might not deploy correctly in a collision if any components or wiring in the system have been disturbed. The yellow wires and connectors routed through the instrument panel and center console are for this system. Do not use electrical test equipment on these yellow wires or tamper with them in any way while working in their vicinity.*
Caution: *On models equipped with a Delco Loc II audio system, disable the anti-theft feature before disconnecting the battery.*

Description

1   1990 and late models are equipped with a Supplemental Inflatable Restraint (SIR) system, more commonly known as an airbag system. The SIR system is designed to protect the driver and passenger from serious injury in the event of a head-on or frontal collision.
2   The SIR system consists of: an airbag located in the center of the steering wheel; two impact sensors, one located in the instrument panel and another located just in front of the radiator; an arming sensor located under the center console; and a diagnostic/energy reserve module located at the right end of the instrument panel.

Sensors

3   The system has three separate sensors; two impact sensors and an arming sensor. The sensors are basically pressure sensitive switches that complete an electrical circuit during an impact of sufficient G force. The electrical signal from the crash sensors is sent to the diagnostic module, that then completes circuit and inflates the airbags.

Diagnostic/energy reserve module

4   The diagnostic/energy reserve module contains an on-board microprocessor which monitors the operation of the system. It performs a diagnostic check of the system every time the vehicle is started. If the system is operating properly, the AIRBAG warning light will blink on and off seven times. If there is a fault in the system, the light will remain on and the airbag control module will store fault codes indicating the nature of the fault.
5   If the AIRBAG warning light remains on after staring, or comes on while driving, the vehicle should be taken to your dealer immediately for service. The diagnostic/energy reserve module also contains a back-up power supply to deploy the airbags in the event battery power is lost during a collision.

Operation

6   For the airbag(s) to deploy, an impact of sufficient G force must occur within 30-degrees of the vehicle centerline. When this condition occurs, the circuit to the airbag inflator is closed and the airbag inflates. If the battery is destroyed by the impact, or is too low to power the inflator, a back-up power supply inside the diagnostic/energy reserve module supplies current to the airbag.

Self-diagnosis system

7   A self-diagnosis circuit in the module displays a light when the ignition switch is turned to the On position. If the system is operating normally, the light should go out after seven flashes. If the light doesn't come on, or doesn't go out after seven flashes, or if it comes on while you're driving the vehicle, there's a malfunction in the SIR system. Have it inspected and repaired as soon as possible. Do not attempt to troubleshoot or service the SIR system yourself. Even a small mistake could cause the SIR system to malfunction when you need it.

Servicing components near the SIR system

8   Nevertheless, there are times when you need to remove the steering wheel, radio or service other components on or near the instrument panel. At these times, you'll be working around components and wiring harnesses for the SIR system. SIR system wiring is easy to identify; they're all covered by a bright yellow conduit. Do not unplug the connectors for the SIR system wiring, except to disable the system. And do not use electrical test equipment on the SIR system wiring. *ALWAYS DISABLE THE SIR SYSTEM BEFORE WORKING NEAR THE SIR SYSTEM COMPONENTS OR RELATED WIRING.*

Disabling the SIR system

Refer to illustration 28.11
9   Turn the steering wheel to the straight ahead position, place the ignition switch in Lock and remove the key. Remove the airbag fuse from the fuse block (see Section 3).
10   Remove the panel under the steering column.
11   Unplug the yellow Connector Position Assurance (CPA) connectors at the base of the steering column **(see illustration)**.
Enabling the SIR system
12   After you've disabled the airbag and performed the necessary service, plug in the steering column CPA connector. Reinstall the lower panel.
13   Install the airbag fuse.

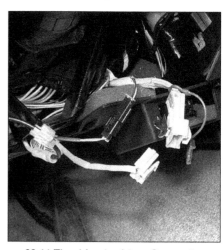

28.11 The driver's airbag Connector Position Assurance (CPA connector is located at the base of the steering column)

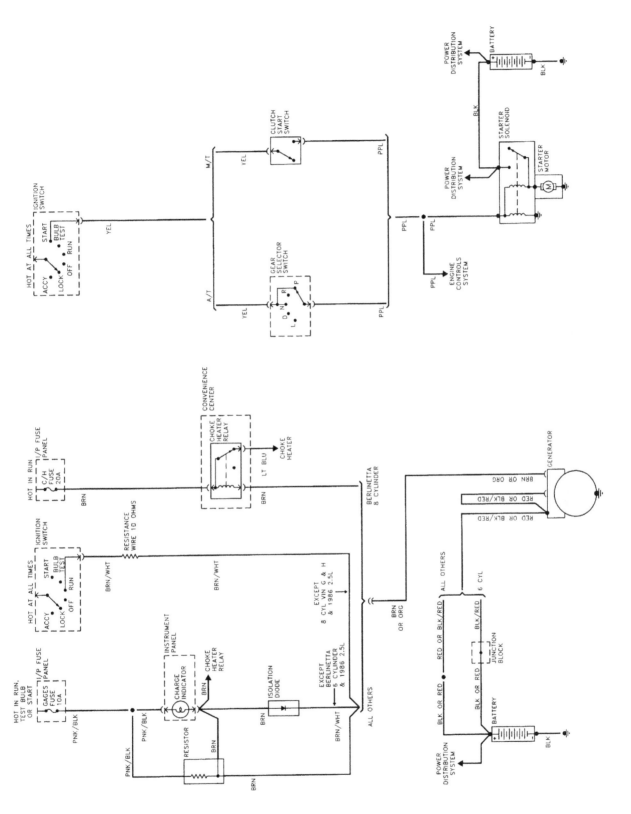

Starting and charging systems - 1984 and earlier models

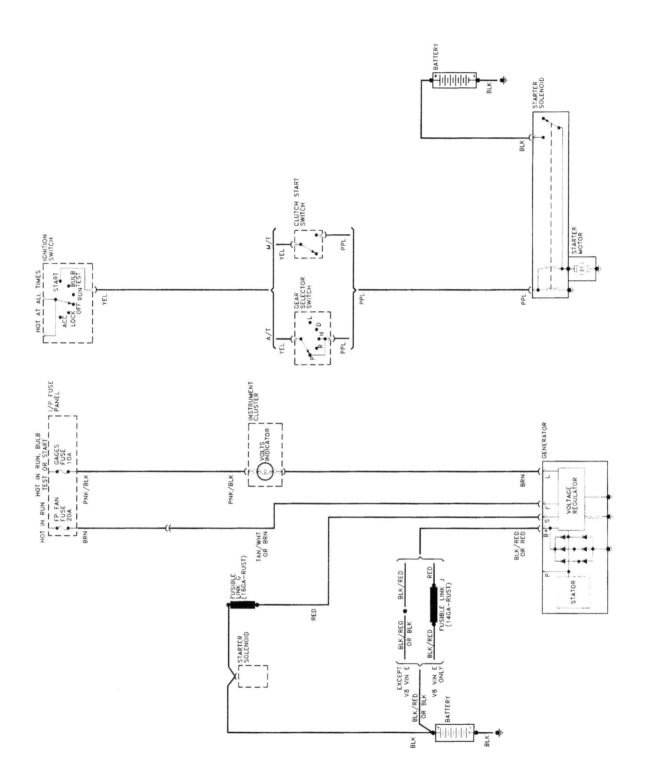

Starting and charging systems - 1985 through 1989 models

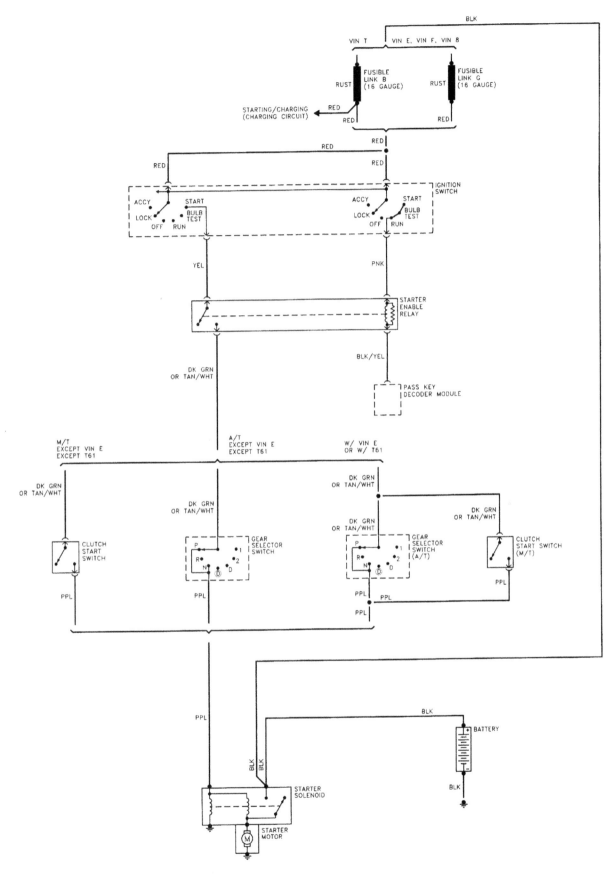

Starting system - 1990 and later models

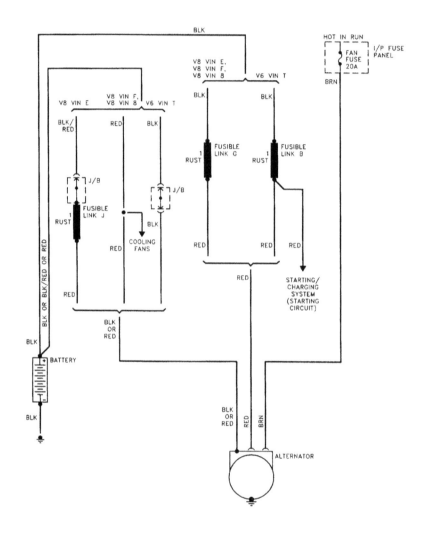

Charging system - 1990 and later models

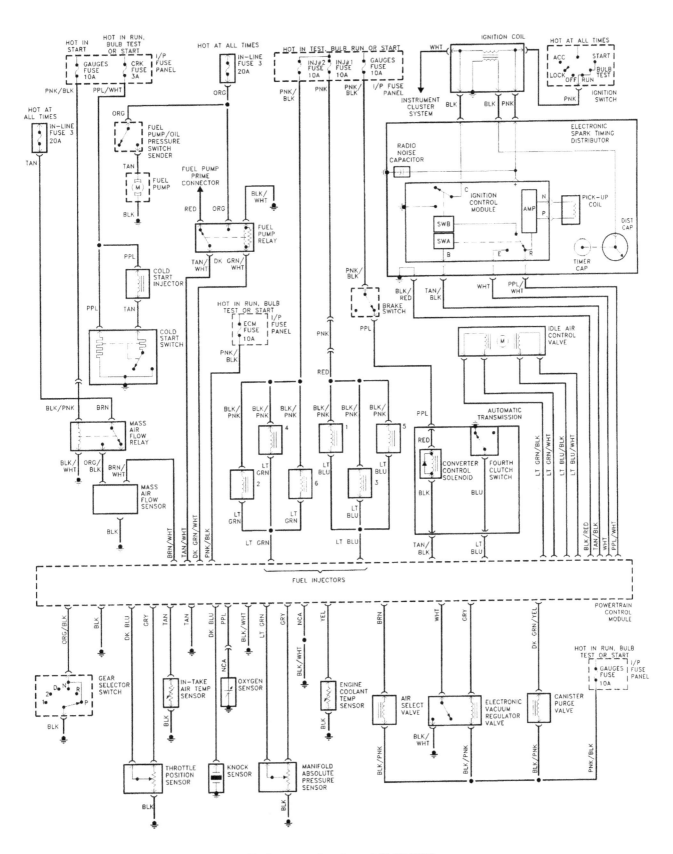

Engine control system - 2.8L V6 MPFI

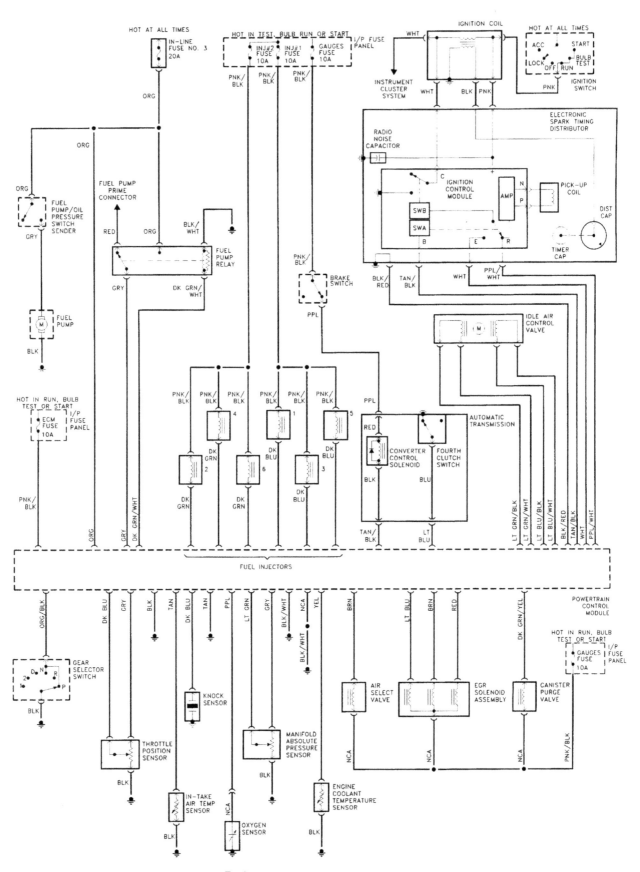

Engine control system - 3.1L V6 MPFI

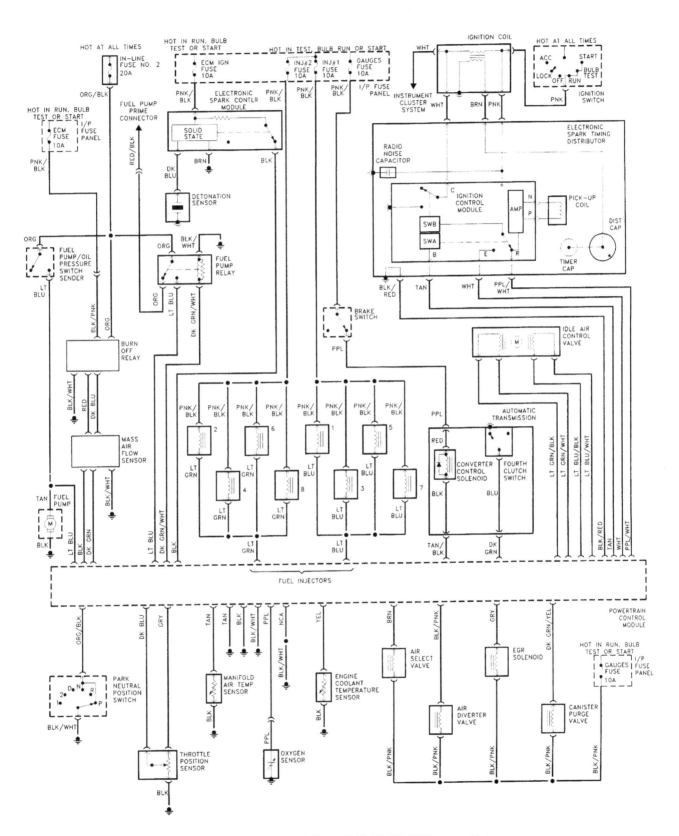

Engine control system - 5.0L and 5.7L V8 TPI (1989 and earlier)

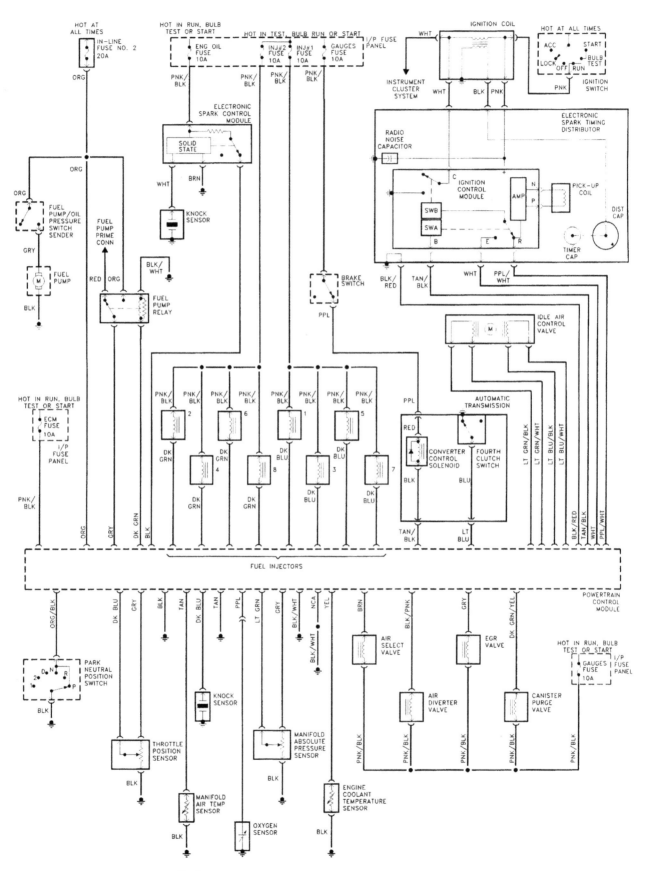

Engine control system - 5.0L and 5.7L V8 TPI (1990 and later)

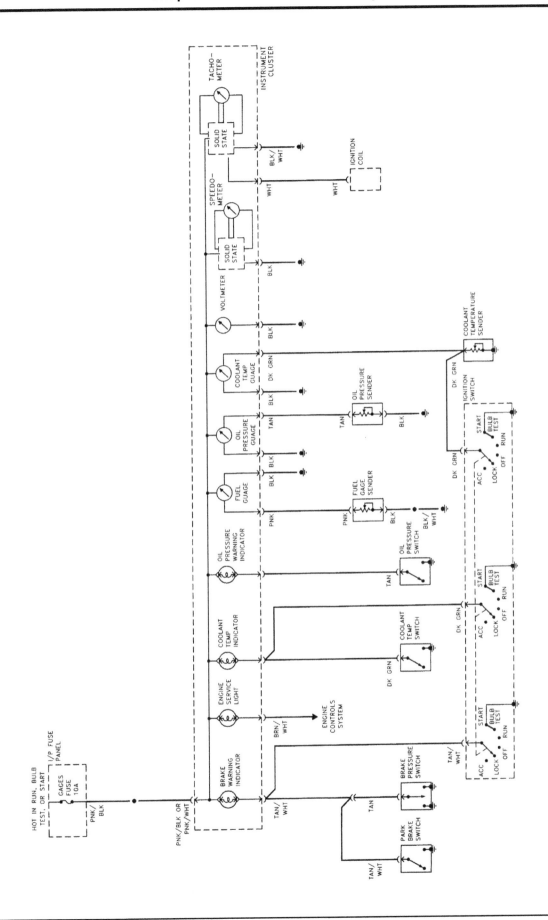

Instrument panel gauges and warning lights system

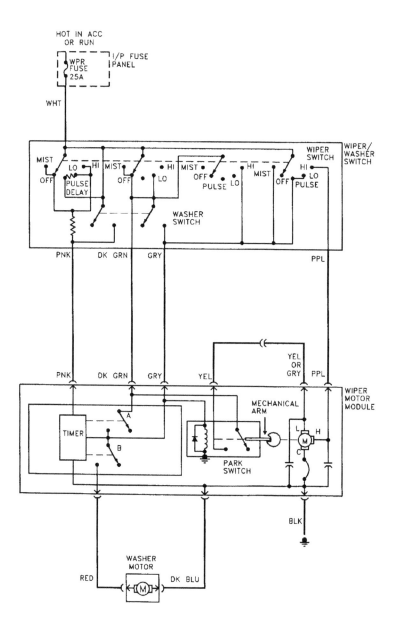

Windshield wiper and washer system

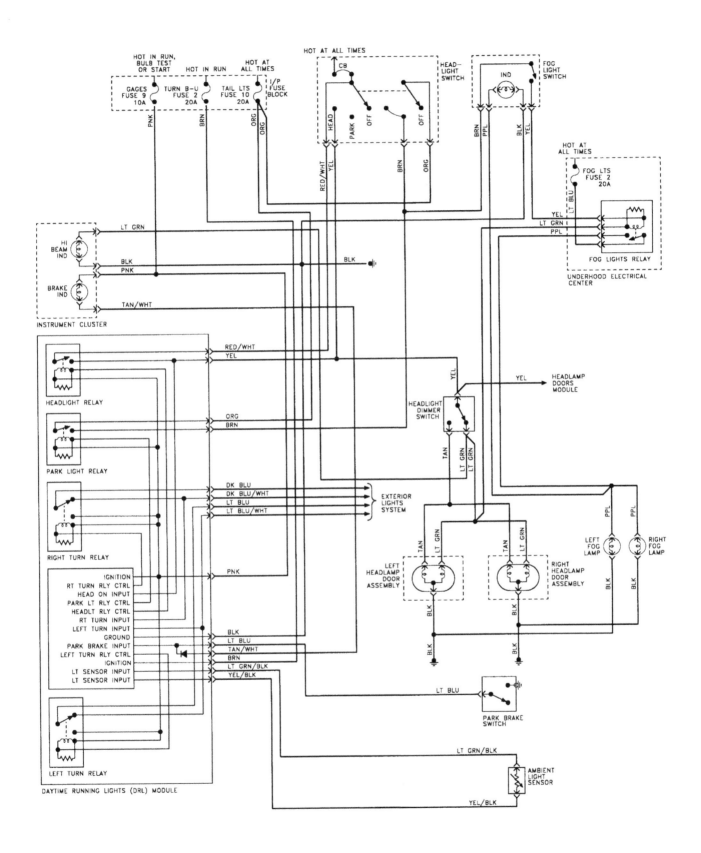

Exterior lighting system - with daytime running lights (1990 through 1992)

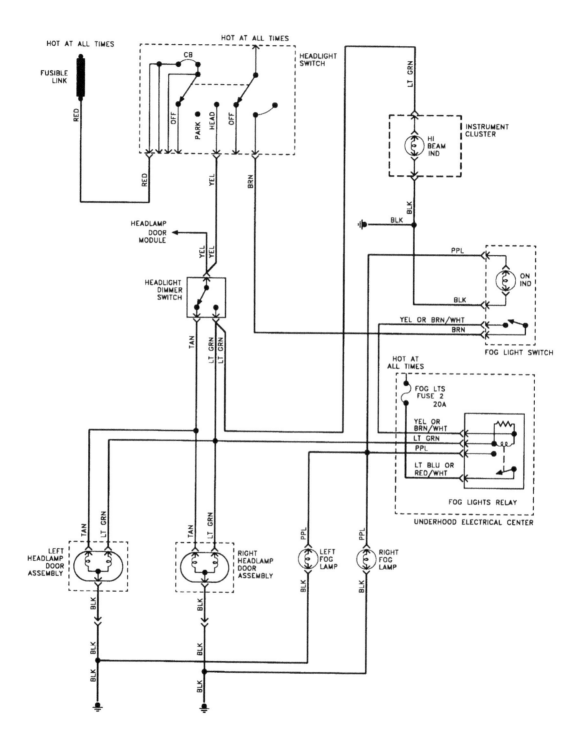

Exterior lighting system – without daytime running lights (1982 through 1992)

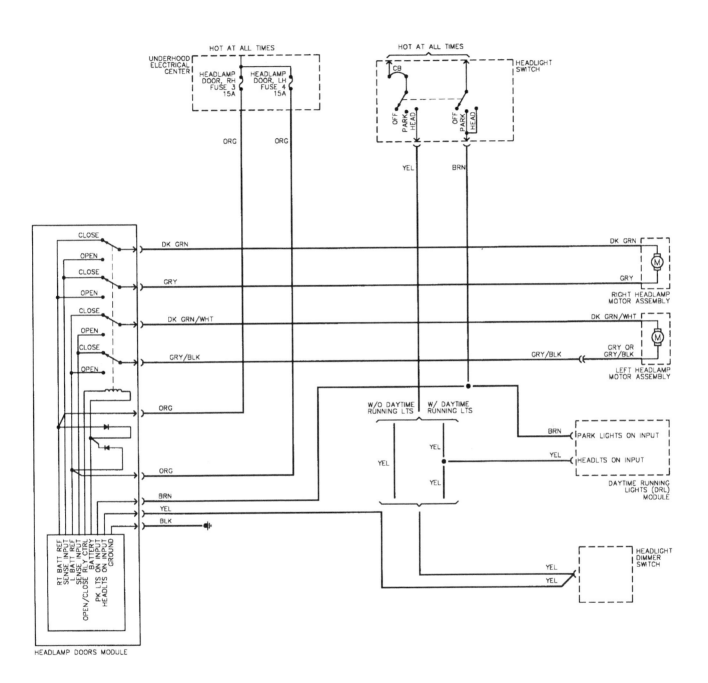

Headlamp door circuit (1990 through 1992)

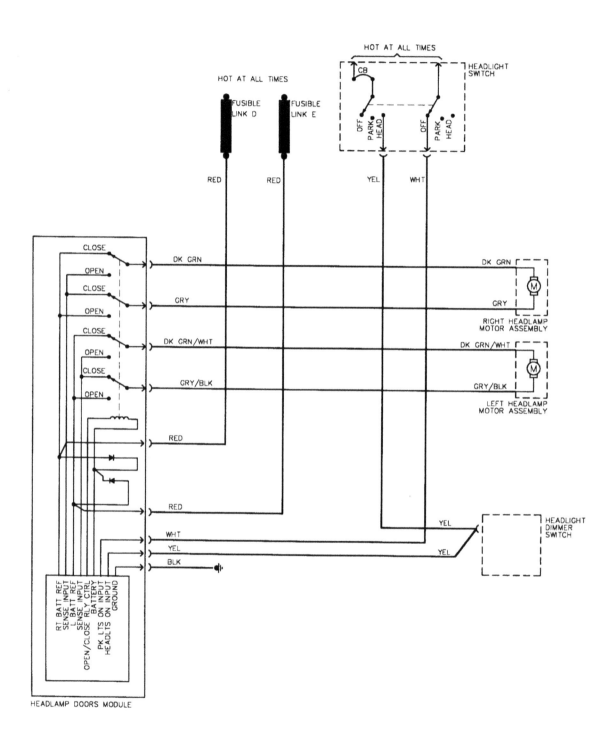

Headlamp door circuit (1987 through 1989)

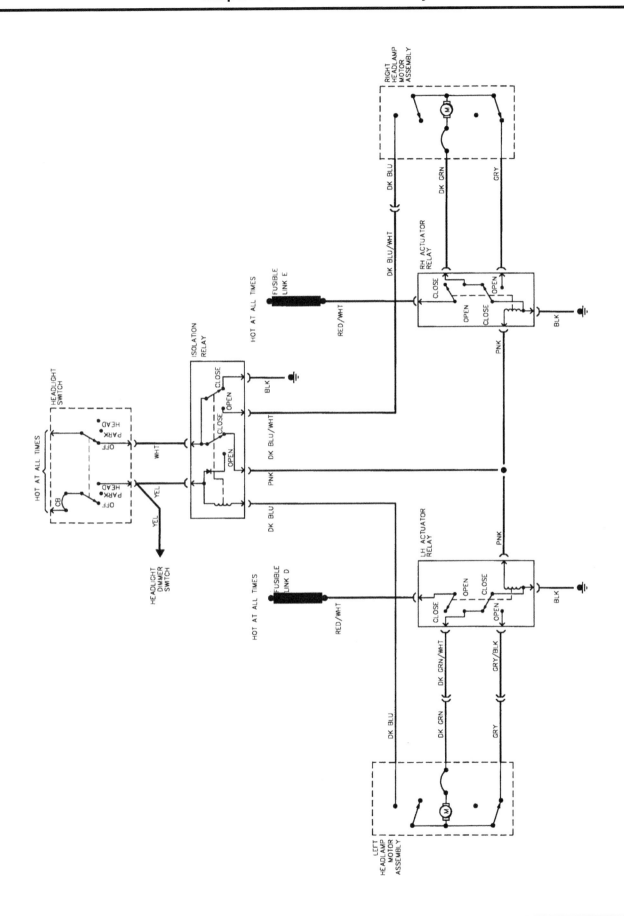

Headlamp door circuit (1982 through 1986)

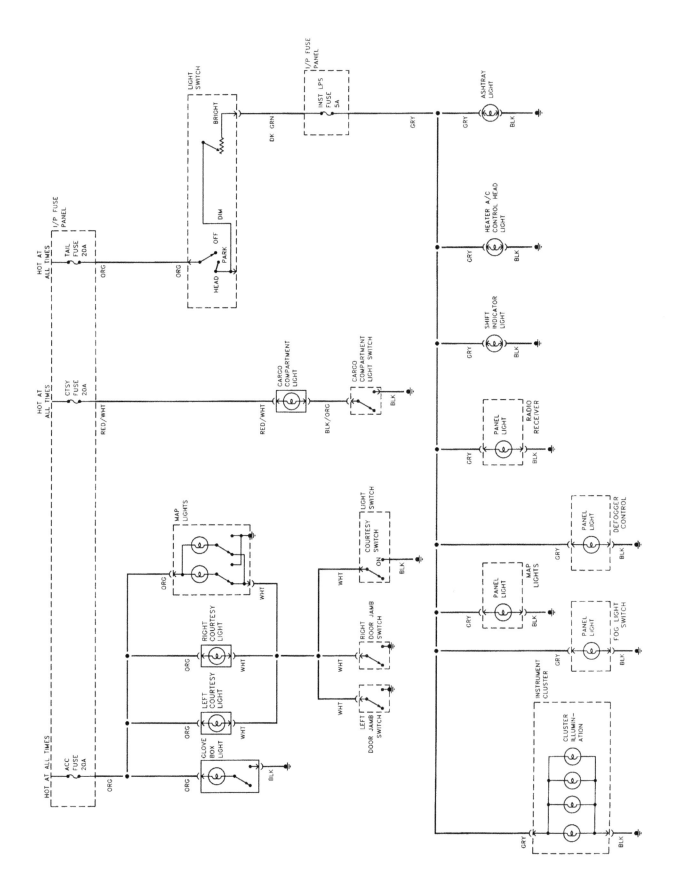

Interior lighting system

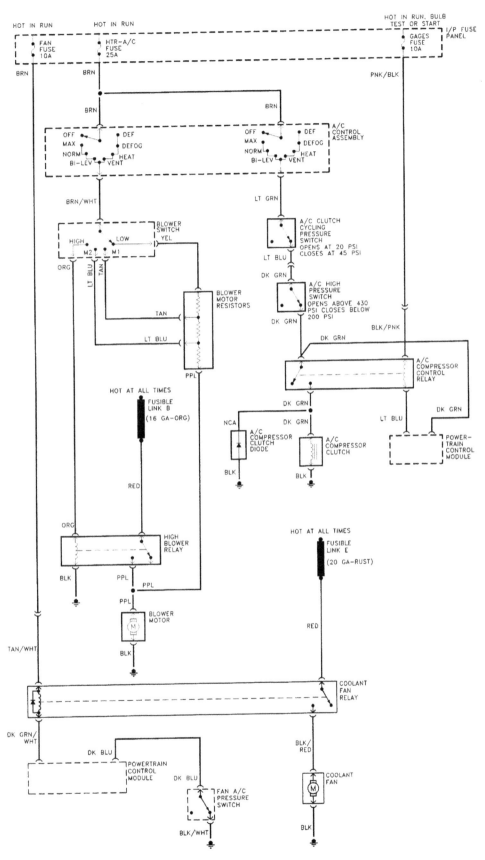

Heating and air conditioning system (includes engine cooling fan system) - 2.8L V6 engine

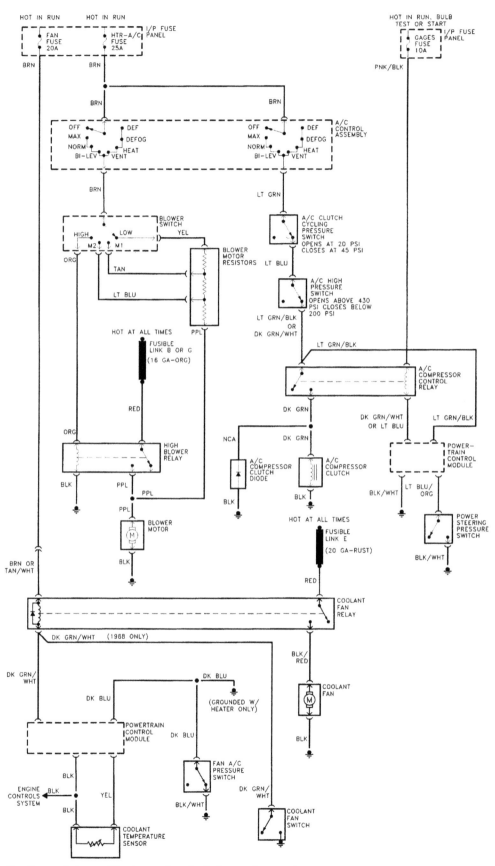

Heating and air conditioning system (includes engine cooling fan system) - 3.1L V6 engine

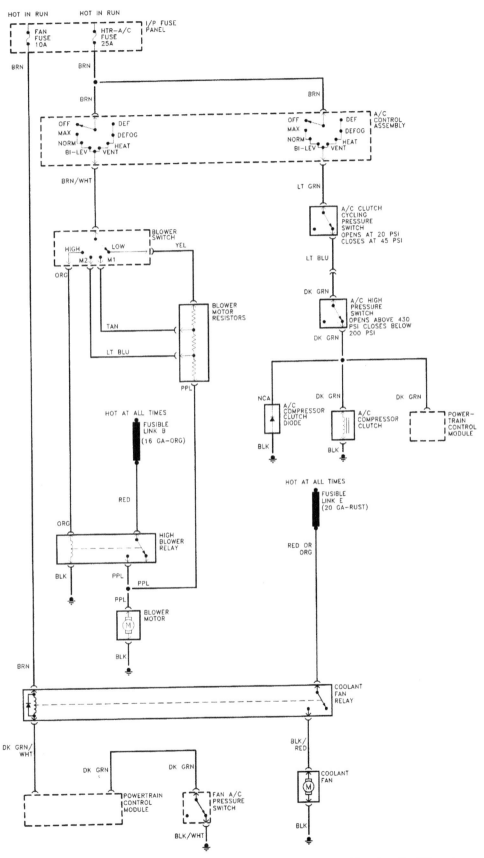

Heating and air conditioning system (includes engine cooling fan system) - 1987 and earlier V8 engine

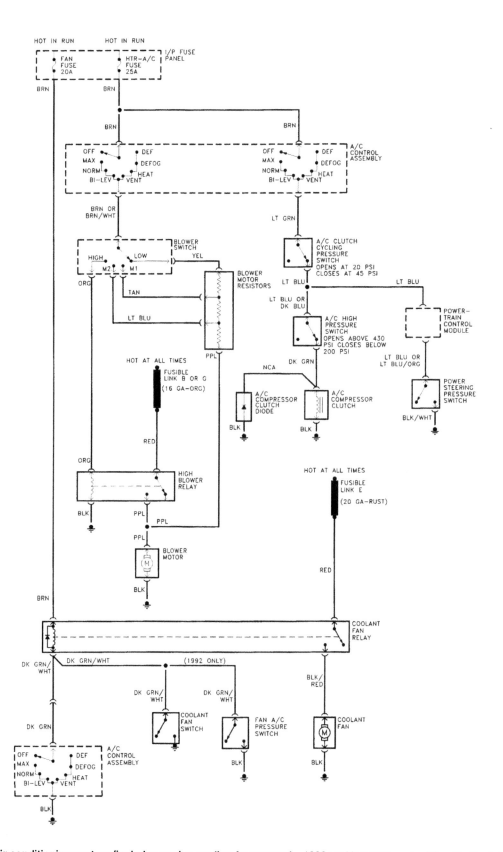

Heating and air conditioning system (includes engine cooling fan system) - 1988 and later V8 engine Throttle Body Injection

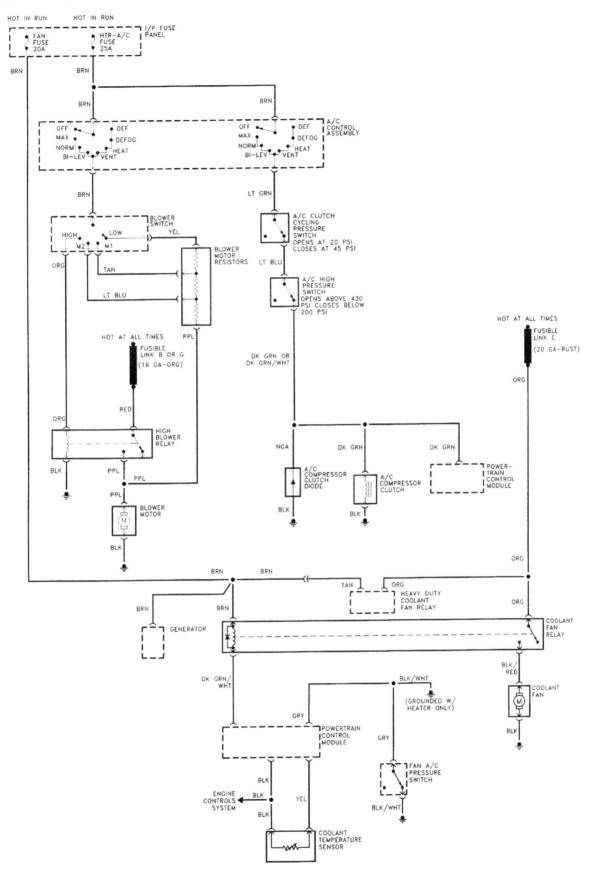

Heating and air conditioning system (includes engine cooling fan system) - 1988 and later V8 engine Tuned Port Injection (without Heavy Duty Cooling)

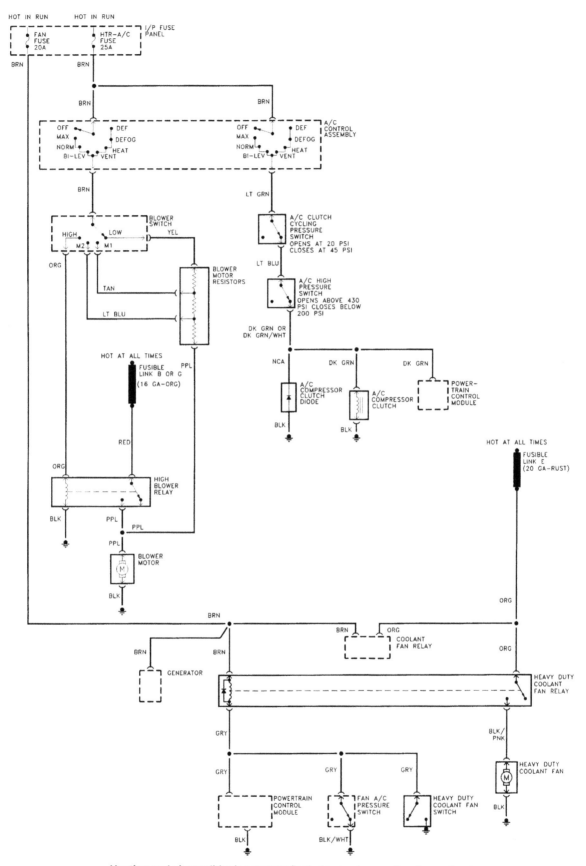

Heating and air conditioning system (includes engine cooling fan system) -
1988 and later V8 engine Tuned Port Injection (with Heavy Duty Cooling)

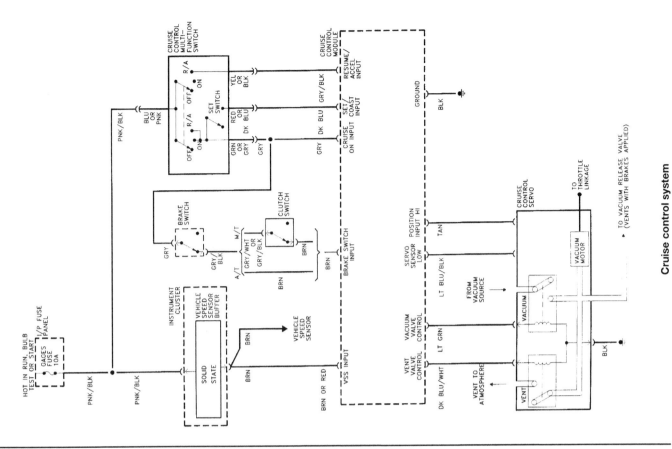

Cruise control system

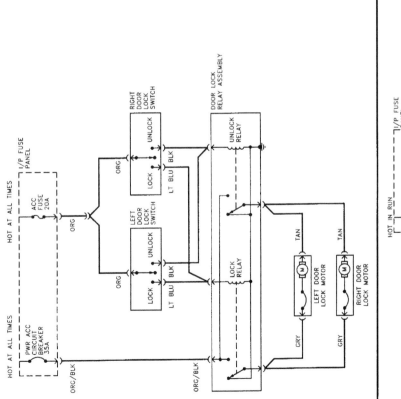

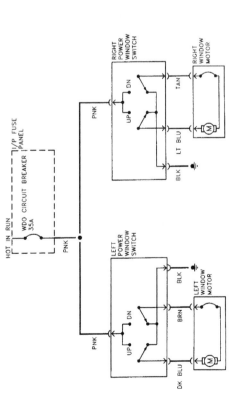

Power door lock and power window systems

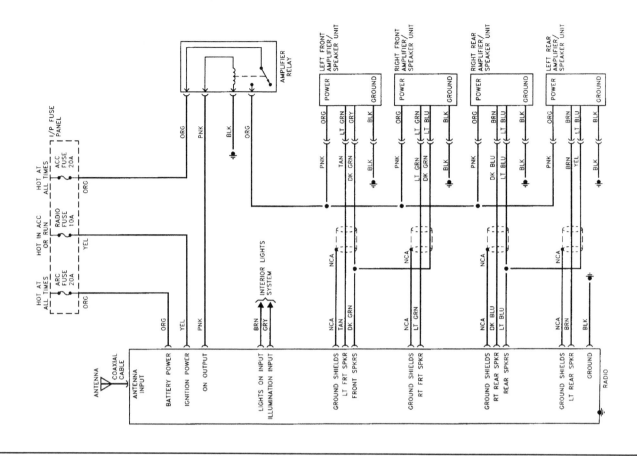

Delco-Bose stereo system

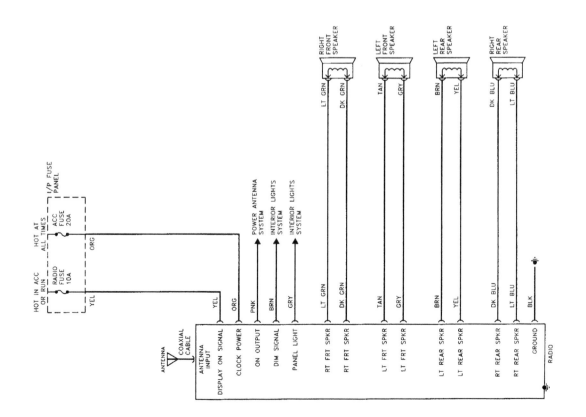

Typical base model stereo system

Chapter 11
Suspension and steering systems

Contents

	Section
Balljoints - replacement	4
Front coil springs - removal and installation	5
Front lower control arms - removal and installation	6
Front stabilizer bar - removal and installation	2
Front suspension - general information	1
Ignition lock cylinder - removal and installation	24
Power steering gear - removal and installation	22
Power steering pump - removal and installation	23
Power steering system - bleeding	21
Power steering system - general information	20
Rear coil springs and insulators - removal and installation	9
Rear lower control arm - removal and installation	12
Rear lower control arm bushing - replacement	13

	Section
Rear shock absorbers - removal and installation	8
Rear stabilizer bar - removal and installation	15
Rear suspension - general information	7
Steering column switches - removal and installation	24
Steering knuckle - removal and installation	18
Steering system - general information	16
Steering wheel - removal and installation	19
Strut assembly - removal and installation	3
Tie-rods - removal and installation	17
Torque arm - removal and installation	14
Track bar - removal and installation	10
Track bar brace - removal and installation	11

Specifications

Front wheel alignment

1986 and earlier	
Caster	3.00-degrees
Camber	1.00-degree
Toe-in	3/64-inch
1987 through 1990	
Caster	4.7-degrees
Camber	0.30-degree
Toe-in	0-inch
1991 and 1992	
Caster	4.80-degrees
Camber	0.30-degree
Toe-in	0-inch

Torque specifications

Ft-lbs (unless otherwise indicated)

Steering linkage

Steering knuckle-to-tie-rod end nut	35
Tie-rod clamp nuts	168 in-lbs
Tie-rod-to-intermediate rod nut	35
Pitman arm-to-intermediate rod nut	40
Pitman arm-to-steering gear nut	180
Idler arm-to-intermediate rod nut	35
Idler arm-to-frame nut	50

Torque specifications (continued)

Ft-lbs (unless otherwise indicated)

Power steering pump
Reservoir bolt 35
Flow control fitting 35
Pressure hose fitting 20

Power steering gear
Gear-to-frame bolts 70
High pressure line fitting at gear 20
Oil return line fitting at gear 20
Coupling flange bolt (if equipped) 30
Coupling flange nut (if equipped) 20

Steering wheel and column
Steering wheel-to-shaft nut 30
Turn signal switch attaching screws .... 35 in-lbs
Ignition switch attaching screws .... 35 in-lbs
Bracket-to-steering column support nuts 25
Toe-pan-to-dash screws .... 45 in-lbs
Toe-pan clamp screws .... 60 in-lbs
Bracket-to-steering column bolt 30
Cover-to-housing screws .... 100 in-lbs
Clamp-to-steering shaft nut 55
Support-to-lock plate screws .... 60 in-lbs

Front suspension
Strut assembly
    Strut-to-knuckle bolts 195
    Strut-to-upper mount nut 50
    Upper mount-to-wheelhouse tower 20
Lower control arm
    Pivot bolt nuts 65
    Bumper-to-lower control arm 20
    Lower control arm balljoint-to-knuckle 90
Stabilizer bar
    Link bolt/nut .... 168 in-lbs
    Insulator bracket bolt 25

Rear suspension
Shock absorber-to-upper mount nut .... 156 in-lbs
Shock absorber-to-axle nut 70
Track bar-to-axle nut
    1987 and earlier 93
    1988 and later 59
Track bar-to-body bracket
    1987 an earlier 58
    1988 and later 78
Track bar brace-to-body bracket nut 58
Track bar brace-to-body brace bracket screws 34
Control arm-to-rear axle bolt
    1987 and earlier 68
    1988 and later 85
Control arm-to-underbody bolt
    1987 and earlier 68
    1988 and later 85
Torque arm front bracket-to-transmission bolt
    With 200C automatic transmission 20
    All except 200C transmission 31
Torque arm-to-rear axle nut 100
Stabilizer shaft-to-body bracket link bolt .... 144 in-lbs
Stabilizer bracket-to-body screw 35
Stabilizer shaft clamp-to-U-bolt nut 20
Axle bumper bracket-to-underbody bolt 20

1.1 Suspension and steering components

1   *Lower control arm*
2   *Strut assembly*
3   *Upper strut mount*
4   *Steering knuckle*
5   *Hub and brake disc assembly*
6   *Coil spring*
7   *Crossmember*

2.2 Remove the link bolt from the front stabilizer bar

1 Front suspension - general information

Refer to illustration 1.1

The front suspension is independent, allowing each wheel to compensate for road surface changes without appreciably affecting the opposite wheel **(see illustration)**.

Each wheel is connected independently to the frame by a steering knuckle, strut assembly, balljoint and lower control arm. The front wheels are held in proper relationship to each other by tie-rods, which are connected to the steering knuckles, and to a relay rod assembly.

Coil springs are mounted between spring housings located on the front frame crossmember and the lower control arms. Shock absorption is provided by double, direct-acting strut assemblies, with the upper portion of each strut extending through the fender well and attaching to the upper mount assembly with a nut.

Front suspension side roll is controlled by a steel stabilizer shaft mounted in rubber bushings which are held to the frame side rails by brackets. The stabilizer ends are connected to the lower control arms by link bolts

isolated by rubber grommets.

Pressed-in bushings are located in the inner ends of the lower control arms. Bolts passing through the bushings attach the arm to the suspension crossmember. The lower balljoint assembly is press-fit in the arm and attaches to the steering knuckle with a torque-prevailing nut.

To keep dirt and moisture from entering the joints and damaging bearing surfaces, rubber grease seals are provided at the ball socket assemblies.

2 Front stabilizer bar - removal and installation

Refer to illustrations 2.2, 2.3 and 2.7
1 Raise the vehicle and support it securely on jackstands.
2 Remove the link bolt, nut, grommets, spacer and retainers from each side of the stabilizer bar, noting their positions relative to each other to simplify installation **(see illustration)**.
3 Remove the insulator and bracket from each side of the stabilizer bar, noting their positions to simplify installation **(see illustration)**.
4 Remove the stabilizer bar.

5 Installation is the reverse of the removal procedure, noting the following.
6 Apply multi-purpose grease to the insulator mounting points on the stabilizer bar before installing the insulators and brackets.
7 Make sure the slits in the insulators are toward the front of the vehicle **(see illustration)**. Also make sure the locator on top of the insulator seats properly on the frame pad.
8 Make sure the stabilizer is centered in the brackets before tightening the mounting bolts.
9 Tighten all bolts to the specified torque.

3 Strut assembly - removal and installation

Refer to illustrations 3.3, 3.4a, 3.4b, 3.8, 3.9, 3.13a and 3.13b
1 Raise the front of the vehicle and support it securely on jackstands, then remove the wheel and tire.
2 Place a hydraulic jack under the lower control arm to support the control arm.
3 Remove the brake hose retaining clip from the strut mount and remove the hose from the mount **(see illustration)**.

2.3 Remove the bracket and insulator bolts from the stabilizer bar

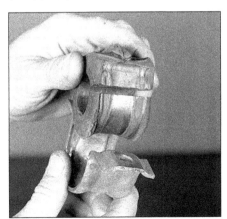

2.7 When reinstalling the insulator, make sure the slit opening is toward the front of the vehicle

3.3 Remove the brake hose from the strut by removing the clip attaching it to the bracket

3.4a Remove the caliper mounting bolts

3.4b Hang the caliper by a piece of wire, never by the brake hose

4   Remove the brake caliper mounting bolts and hang the caliper out of the way with wire **(see illustrations)**. **Caution:** *Do not allow the caliper to hang by the brake hose.*
5   Mark the relationship of the strut to the steering knuckle with white paint.
6   Loosen the top strut-to-steering knuckle bolt.
7   Loosen and remove the bottom strut-to-knuckle bolt, then remove the top bolt.
8   From under the hood, remove the cover from the upper end of the strut **(see illustration)**.
9   Hold the top of the strut piston with one wrench while removing the upper mount retaining nut with another **(see illustration)**.
10   Remove the strut.
11   Check the bearing in the upper mount by turning it with your finger.
12   If the upper mount must be removed, carefully scribe around it before removing the retaining bolts.
13   Remove the mount and retaining bracket **(see illustrations)**.

14   If the bearing is defective, the entire top strut mount assembly must be replaced because the bearing alone is not replaceable.
15   Installation is the reverse of the removal procedure. To assist in installing the upper end of the strut, carefully lower the jack supporting the lower control arm enough to get a hand on the piston, then feed it through the upper mount and attach the retaining nut.

4   Balljoints - replacement

Refer to illustration 4.3
Note: *Special tools are required to replace a balljoint. Read through the procedure and obtain the tools before beginning.*
1   Raise the front of the vehicle and support it securely on jackstands.
2   Remove the wheel and tire.
3   Place a floor jack under the lower control arm spring seat **(see illustration)**. **Warning:** *The jack must remain under the spring seat during removal and installation of the balljoint to hold the spring and control arm in position.*
4   Remove the cotter pin from the top of the balljoint and loosen, but do not remove, the castellated nut.
5   Using a balljoint separator tool, break

3.8 Remove the cover from the upper end of the strut

3.9 Remove the strut upper retaining nut

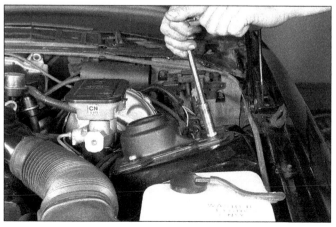

3.13a Remove the upper strut mount bolts

3.13b Remove the upper strut mount and retaining bracket

4.3 Support the lower control arm with a floor jack positioned under the lower control arm spring seat (the farther out the arm you place the jack, the greater the leverage ratio on the arm)

the balljoint loose from the steering knuckle. Remove the tool and the castellated nut.

6   Separate the steering knuckle from the balljoint. Be careful not to bend the brake disc splash shield, if necessary, us a putty knife or similar tool to guide the lower control arm out of the opening in the splash shield.

7   Remove the grease fitting from the balljoint.

8   Use a C-clamp type balljoint removal/installation tool and the proper adapters to press the balljoint out of the lower control arm.

9   Inspect the tapered hole in the steering knuckle, removing any accumulated dirt first. If out-of-roundness, deformation or other damage is noted, the knuckle must be replaced with a new one (refer to Section 18).

10   Position the balljoint in the lower control arm and press it in until it bottoms. **Note:** *The grease purge seal must face in.*

11   Position the balljoint stud in the steering knuckle, install the ball stud nut and tighten it to the specified torque.

12   If the cotter pin hole in the stud does not line up with an opening in the castellated nut, tighten the nut further to allow installation of the cotter pin.

13   Install the balljoint grease fitting and lubricate it until grease appears at the seal.

14   Install the wheel and tire and lower the vehicle.

15   Have the front end alignment checked by an alignment shop.

5   Front coil springs - removal and installation

Refer to illustration 5.5

Note: *Special tools are required to replace a coil. Read through the procedure and obtain the tools before beginning.*

1   Raise the vehicle and support it on jackstands placed under the outside frame rails.

2   Remove the wheel and tire.

3   Disconnect the stabilizer bar at the lower control arm, keeping the bolt, nut, grommet spacer and retainers in order to simplify installation.

4   Install a universal coil spring compressor and compress the spring according to the tool manufacturers instructions.

5   Remove the nuts from the lower control arm-to-front crossmember pivot bolts, but do not remove the pivot bolts at this time **(see illustration)**.

6   Place a floor jack under the lower control arm at the inner edge and centered between the control arm-to-crossmember pivot points. Raise the jack just enough to remove both pivot bolts.

7   Carefully lower the jack and guide the

lower control arm bushings out of the crossmember.

8   Release the coil spring compressor and remove the spring and insulator. The spring position should be carefully noted as it is removed.

9   Installation is the reverse of the removal procedure.

10   Note that after assembly, the end of the bottom coil of the spring must cover all or part of one inspection/drain hole. The other hole must be partly or completely uncovered.

6   Front lower control arms - removal and installation

1   Remove the coil spring (refer to Section 5).

2   Separate the balljoint from the steering knuckle (refer to Section 4).

3   Check the bushings in the lower control arm before installation. If they are worn or damaged they can be replaced with the proper tools.

4   Installation is the reverse of the removal procedure, noting the following.

5   When attaching the lower control arm to the front crossmember, install the front leg first, then the rear leg.

6   When the bolts are final tightened, the vehicle must be at normal ride height.

7   Tighten all bolts to the specified torque.

7   Rear suspension - general information

Refer to illustration 7.1

The rear axle assembly is attached to the vehicle through a link-type suspension system **(see illustration)**. A track bar and two

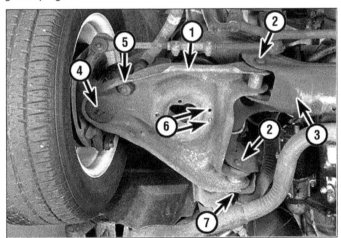

5.5 Lower control arm-to-crossmember installation details

1	Lower control arm	5	Stabilizer bar link bolt
2	Pivot bolts	6	Coil spring
3	Crossmember		inspection/drain hole
4	Balljoint	7	Nut

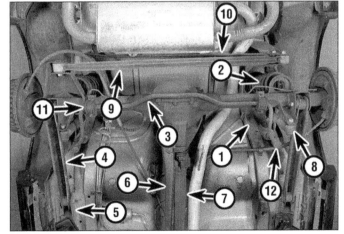

7.1 Rear axle and suspension components

1	Frame rail	8	Shock absorber
2	Coil spring	9	Track bar
3	Stabilizer bar	10	Track bar brace
4	Control arm	11	Stabilizer bar mounting
5	Vehicle underbody		bushing
6	Torque arm	12	Stabilizer bar link
7	Drive shaft		assembly

lower control arms attach the axle housing to the body. In place of upper control arms, a single torque arm is employed, rigidly mounted to the rear axle housing and, through a rubber bushing, to the transmission at the front. Coil springs support the weight of the vehicle and ride control is provided by shock absorbers mounted on both sides of the rear of the axle housing. Some models are equipped with a rear stabilizer bar.

Both standard and heavy-duty shock absorbers are sealed during production. They contain a measured amount of fluid and cannot be adjusted, refilled or disassembled. If they are damaged or worn out, they must be replaced with new units.

8   Rear shock absorbers - removal and installation

Refer to illustrations 8.3 and 8.4

1   Raise the rear of the vehicle and support it on jackstands.
2   With the rear seat in the folded-down position, carefully lift the front edge of the rear panel carpeting and fold it back to expose the rear shock upper mounting nut.
3   Hold the top of the shock piston with one wrench while removing the upper mounting nut with another wrench **(see illustration)**.
4   Remove the lower shock absorber mounting nut **(see illustration)** and remove the shock.
5   Installation is the reverse of the removal procedure.
6   Tighten all nuts and bolts to the specified torque.

9   Rear coil springs and insulators - removal and installation

Refer to illustration 9.8

1   Raise the rear of the vehicle and support it on jackstands. Support the rear axle assembly on a hydraulic jack.
2   Remove the track bar mounting bolt at the axle assembly (left side).
3   Loosen the track bar bolt at the body brace (right side).
4   Disconnect the rear brake hose clip from the vehicle underbody to allow additional axle drop. **Note:** *Never suspend the axle by the brake hose, as damage will occur.*
5   Remove the lower shock absorber mounting nut from both rear shocks.
6   On vehicles equipped with a 4-cylinder engine, remove the driveshaft (refer to Chapter 8).
7   Carefully lower the jack under the rear axle assembly until the spring and/or insulator can be removed.
8   To install, position the spring and insulator in the spring seat and raise the hydraulic jack until the spring is normally compressed **(see illustration)**.
9   Attach the shocks to the rear axle and

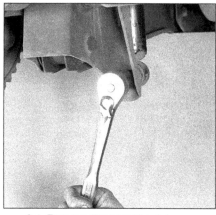

8.3 Remove the rear shock upper mounting nut

8.4 Remove the rear shock lower mounting nut

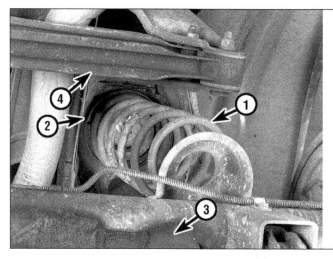

9.8 Rear coil spring installation details

1   Coil spring
2   Coil spring insulator
3   Axle housing
4   Track bar

tighten the nuts to the specified torque.
10   Install the track bar mounting bolt at the axle and tighten the nut to the specified torque.
11   Tighten the track bar-to-body bracket nut to the specified torque.
12   Attach the brake hose clip to the underbody.
13   On 4-cylinder vehicles, install the driveshaft.
14   Remove the jackstands and lower the vehicle.

10   Track bar - removal and installation

Refer to illustrations 10.2 and 10.3

1   Raise the rear of the vehicle and support it with jackstands under the axle housings.
2   Remove the track bar mounting bolt and nut at the rear axle **(see illustration)**.
3   Remove the track bar mounting bolt and nut at the body bracket and remove the track bar **(see illustration)**.

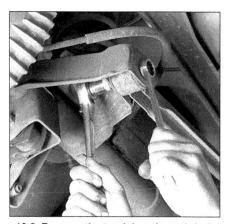

10.2 Remove the track bar through-bolt at the rear axle mount

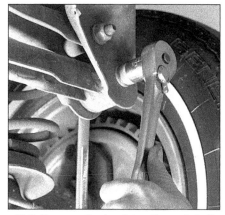

10.3 Remove the track bar through-bolt at the body bracket

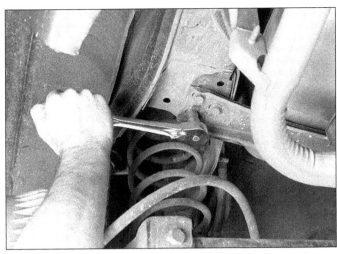

11.3 Remove the bolts retaining the track bar brace at the left body brace

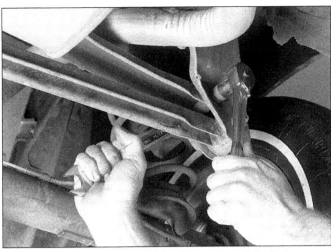

11.4 Remove the track bar brace through-bolt at the right body bracket

4   Installation is the reverse of the removal procedure. Tighten all fasteners to the specified torque.

11   Track bar brace - removal and installation

Refer to illustrations 11.3 and 11.4

1   Raise the rear of the vehicle and support it with jackstands placed under the rear axle housings.
2   If equipped, remove the heat shield screws from the track bar brace.
3   Remove the three track bar brace-to-body brace bolts on the left side of the vehicle **(see illustration)**.
4   Remove the nut and bolt attaching the track bar brace to the body bracket (right side) and remove the track bar brace **(see illustration)**.
5   Installation is the reverse of the removal procedure. Tighten all fasteners to the specified torque.

12   Rear lower control arm - removal and installation

Refer to illustrations 12.2 and 12.3
Note: *If both lower control arms are to be removed, remove and install them one at a time to prevent the axle assembly from rolling or slipping sideways, complicating the procedure.*

1   Raise the rear of the vehicle and place the rear axle housings securely on jackstands.
2   Remove the lower control arm-to-axle housing retaining bolt and nut **(see illustration)**.
3   Remove the lower control arm-to-body brace bolt and nut and remove the control arm **(see illustration)**.
4   Installation is the reverse of the removal procedure. Tighten all bolts to the specified torque.

13   Rear lower control arm bushing - replacement

Note: *This procedure requires the use of an arbor press. If you do not have access to one, take the lower control arm to a machine shop to have the bushing replaced.*
1   Refer to Section 12 and remove the control arm.
2   Mount the control arm in the press and using the proper adapters, press out the bushing. The driver used must contact the outside diameter of the bushing outer sleeve.
3   To install the bushing, press the bushing into place.
4   Refer to Section 12 and install the lower control arm.

14   Torque arm - removal and installation

Refer to illustrations 14.2 and 14.3
1   Remove both rear coil springs (refer to

12.2 Remove the lower control arm through-bolt from the axle housing bracket

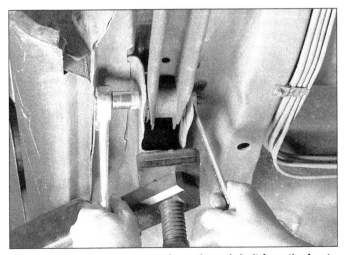

12.3 Remove the lower control arm through-bolt from the front body brace

Section 9). **Note:** *The coil springs must be removed before the torque arm is removed to avoid rear axle forward twist, which could lead to vehicle damage.*

2   Remove the torque arm rear mounting bolts and nuts **(see illustration)**.

3   Remove the torque arm front outer bracket bolts and remove the torque arm **(see illustration)**.

4   To install, position the torque arm and loosely install the rear bolts.

5   Install the front torque arm bracket and tighten the bolts to the specified torque.

6   Tighten the rear torque arm bolts to the specified torque. **Caution:** *Do not overtighten these bolts, as it could lead to distortion of the differential case and subsequent pinion bearing damage.*

7   The remaining installation steps are the reverse of those for removal. Tighten all bolts to the specified torque.

15   Rear stabilizer bar - removal and installation

Refer to illustrations 15.2 and 15.3
Note: *The rear stabilizer is an option and is not installed on all vehicles.*

1   Raise the vehicle and support it securely on jackstands.

2   Remove the link bolt, nut, grommets, spacer and retainers from each side of the stabilizer bar, noting their positions relative to each other to simplify installation **(see illustration)**.

3   Remove the insulator and bracket from each side of the stabilizer bar, noting their positions to simplify installation **(see illustration)**.

4   Remove the stabilizer bar.

5   Installation is the reverse of the removal procedure, noting the following.

6   Apply multi-purpose grease to the insulator mounting points on the stabilizer bar before installing the insulators and brackets.

7   Make sure the slits in the insulators are toward the front of the vehicle.

8   Be sure the stabilizer is centered in the brackets before tightening the mounting bolts.

9   Tighten all bolts to the specified torque.

16   Steering system - general information

The steering linkage connects both front wheels to the steering gear through a Pitman arm. The right and left tie-rods are attached to the steering knuckles (arms) and to the relay rod by balljoint studs. The left end of the relay rod is supported by the Pitman arm, which is driven by the steering gear. The right end of the relay rod is supported by the idler arm, which pivots on a support bolted to the frame rail. The Pitman arm and idler arm move in symmetrical arcs and remain parallel to each other.

14.2 Remove the torque arm through-bolt from the axle housing

15.2 Remove the stabilizer bar link bolt, grommets and spacer from each end of the bar

The tie-rods and tie-rod ends may be removed and installed by carefully following the procedures set forth in the Section that follows; however, the replacement and alignment of the relay rod, Pitman arm and/or the idler arm requires the use of special tools and alignment equipment not normally in the possession of the home mechanic. These procedures, therefore, should be done by a dealer or a front-end alignment shop.

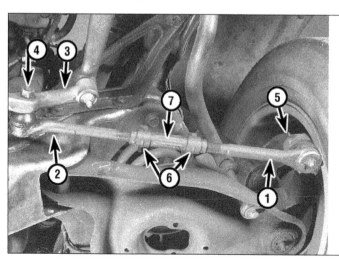

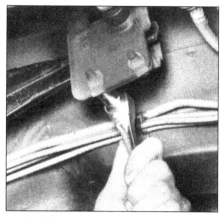

14.3 Remove the torque arm outer bracket from the transmission

15.3 Remove the U-bolt nuts and bracket from each stabilizer bar bushing

17   Tie-rods - removal and installation

Refer to illustrations 17.2a, 17.2b, 17.2c, 17.3

1   Raise the vehicle and place it securely on jackstands.

2   Remove the cotter pins and nuts from the relay rod and steering knuckle ends of the tie-rod **(see illustrations)**.

3   Using the special tool as shown in the

17.2a Tie-rod installation details

1   *Outer tie-rod*
2   *Inner tie-rod*
3   *Relay rod*
4   *Inner tie-rod-to-relay rod mounting nut*
5   *Cotter pin*
6   *Clamp*
7   *Adjuster sleeve*

17.2b Loosen, but do not remove, the nut retaining the tie-rod to the steering knuckle

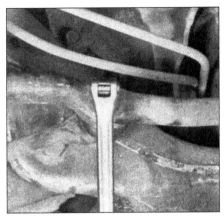

17.2c Loosen, but do not remove, the nut retaining the tie-rod to the relay rod

17.3 Use a small balljoint separator tool to separate the tie-rod end from the steering knuckle

accompanying illustration, separate the outer end of the tie-rod from the steering knuckle. **Note:** *Do not use a "pickle fork' type tool to remove the tie-rod end, as seal damage may result.*

4   Remove tool and the ballstud nut.

5   Using the same tool referred to in Step 3, separate the inner end of the tie-rod from the relay rod.

6   If the tie-rods are to be disassembled and reassembled, carefully measure the overall length of the tie-rod assembly (from the centerline of the inner ballstud to the centerline of the outer ballstud) before removing the tie-rod ends so the assemblies can be adjusted to the original length. Also measure the distance from the outer threads on the tie-rod end to the edge of the tie-rod sleeve **(see illustration).** You must perform these steps to maintain the approximate front wheel alignment on reassembly.

7   To remove the tie-rod ends, loosen the clamp nuts and unscrew the tie-rod ends.

8   If new tie-rod ends are used, lubricate the threads with multipurpose chassis grease and install the ends, making sure both ends are threaded an equal distance into the tie-rod and that the overall length is the same as recorded in Step 6. Install the clamps but do not tighten them at this time.

17.6 Measure the distance from the end of the threads to the edge of the sleeve - both must be equal when reassembling the tie-rod

9   Carefully remove all dirt from the threads of the balljoint studs and nuts. If the threads are not clean, binding may occur and the studs may turn in the tie-rod ends when attempting to tighten the nuts. Also make sure the tapered surfaces of the balljoints are clean and that the seals are installed on the studs.

10   Attach the assembled tie-rod to the steering arm and relay rod.

11   Tighten the retaining nuts to the specified torque, then, if necessary,

tighten the nuts farther to permit insertion of the cotter pins.

12   Adjust the tie-rod clamps as shown **(see illustration),** then tighten the clamps to the specified torque.

13   Lubricate the tie-rod ends with the specified grease.

14   Remove the jackstands and lower the vehicle.

15   A front-end alignment must now be performed by a dealer or an alignment shop.

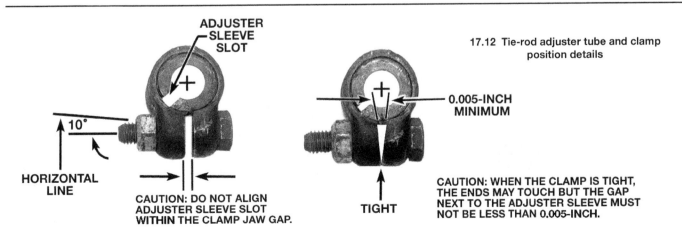

ADJUSTER SLEEVE SLOT

10°

HORIZONTAL LINE

CAUTION: DO NOT ALIGN ADJUSTER SLEEVE SLOT WITHIN THE CLAMP JAW GAP.

17.12 Tie-rod adjuster tube and clamp position details

0.005-INCH MINIMUM

TIGHT

CAUTION: WHEN THE CLAMP IS TIGHT, THE ENDS MAY TOUCH BUT THE GAP NEXT TO THE ADJUSTER SLEEVE MUST NOT BE LESS THAN 0.005-INCH.

19.1 Removing the horn pad from the
steering wheel

19.2 Remove the retaining clip from the
steering wheel nut

19.3 Remove the steering wheel nut

18 Steering knuckle - removal and installation

1   Raise the vehicle and place it securely on jackstands under the side frame rails.
2   Remove the wheel and tire.
3   Remove the brake hose bracket from the strut assembly.
4   Remove the caliper retaining bolts and hang the caliper out of the way on a piece of wire. Do not allow the caliper to hang by the brake hose.
5   Remove the hub and disc assembly (refer to the wheel bearing repack procedure in Chapter 1, if necessary).
6   Unbolt and remove the brake disc splash shield.
7   Disconnect the tie-rod from the steering knuckle (refer to Section 17).
8   Place a floor jack under the lower control arm to support the control arm **(see illustration 4.3)**. **Warning:** *The jack must remain under the spring seat during removal and installation of the steering knuckle to hold the spring and control arm in position.*
9   Disconnect the balljoint from the steering knuckle (refer to Section 4).
10   Remove the two bolts retaining the strut to the steering knuckle and remove the knuckle.

19.5 Use the proper steering wheel puller
to separate the steering wheel the
steering shaft

11   Installation is the reverse of the removal procedure. Tighten all bolts to the specified torque.

19 Steering wheel - removal and installation

Warning: *If equipped with a Supplemental Inflatable Restraint system (SIR), more commonly known as airbags, disable the airbag system before working in the vicinity of airbag system components to avoid the possibility of accidental deployment of the airbag, which could result in personal injury (see Chapter 10).*

1989 and earlier models

Refer to illustrations 19.1, 19.2, 19.3 and 19.5
1   Pull off the horn pad by pushing down on it from the top and pulling it away from the steering wheel **(see illustration)**. On models so equipped, remove the two screws securing the steering wheel shroud from the underside of the steering wheel, then lift the steering wheel shroud and horn contact lead assembly from the steering wheel.
2   Remove the retaining clip from the top of the steering wheel nut **(see illustration)**.
3   Using a socket and breaker bar, remove the steering wheel nut **(see illustration)**. Mark the relationship between the steering shaft and steering wheel.
4   Thread the anchor screws of a steering wheel puller into the threaded holes provided in the steering wheel.
5   Turn the center bolt of the puller clockwise with a wrench to separate the steering wheel from the steering shaft **(see illustration)**. Remove the tool from the steering wheel and remove steering wheel.
6   To install the wheel, make sure the turn signal lever is in the neutral position, then position the steering wheel on the shaft, making sure the flats on the wheel and shaft are aligned.
7   Install the steering wheel nut and tighten it to the specified torque.
8   The remaining installation steps are the reverse of removal.

1990 and later models

Warning: *The procedure below must be followed exactly as outlined to temporarily disable the Supplemental Inflatable Restraint (SIR) system and prevent false diagnostic codes from being set. Failure to follow this procedure could result in air bag deployment, personal injury and/or unneeded SIR system repairs.*
1   Position the front wheels pointing straight ahead, turn the ignition switch to the Lock position and remove the key.
2   Disable the airbag system (see Chapter 10).
3   Remove the screws and nuts from the rear of the steering wheel to release the SIR module. The screws will require a no. 30 Torx driver for removal and installation. Place the airbag module in a secure location. **Warning:** *When carrying a live module, make sure the bag and trim cover are pointing away from you. In case of accidental deployment, the chance of injury will be minimal. When placing a live module on a workbench or other surface, make sure the bag and trim cover face up, away from the surface. Never carry the module by the wires or connector.*
4   Refer to Steps 2 through 5 and remove steering wheel. **Caution:** *Be careful not to thread the steering wheel puller anchor bolts in too far and damage the airbag coil.*
5   Installation is the reverse of removal. When installing the airbag module, tighten the Torx screws to 25 in-lbs.
6   Enable the airbag system (see Chapter 10).

20 Power steering system - general information

With the optional power steering gear, hydraulic pressure is generated in an engine-driven vane type pump and supplied through hoses to the steering box spool valve. The valve is normally positioned in the neutral mode by a torsion bar but when the steering wheel is turned and force is applied to the steering shaft, the spool moves in relation to the body and allows oil to flow to the appropriate side of the

22.6 Separate the Pitman arm from the steering gear with a Pitman arm puller

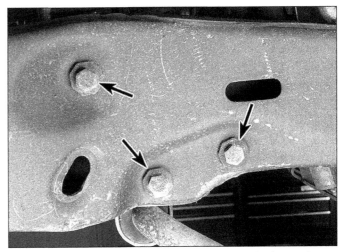

22.7 Remove the steering gear mounting bolts (arrows)

piston nut. The greater the movement of the steering wheel, the greater the hydraulic pressure which is applied and therefore the greater the power assistance given to the drive.

The Sections which follow concern procedures that can be performed by the home mechanic; however, due to the requirements for special tools and diagnostic procedures, major repairs to the power steering system should be performed by a dealer or reputable repair shop.

21 Power steering system - bleeding

1   This is not a routine operation and normally will only be required when the system has been dismantled and reassembled.
2   Fill the reservoir to the correct level with fluid of the recommended type and allow it to remain undisturbed for at least two (2) minutes.
3   Start the engine and run it for two or three seconds only. Check the reservoir and add more fluid as necessary.
4   Repeat the operations described in the preceding paragraph until the fluid level remains constant.
5   Raise the front of the vehicle until the wheels are clear of the ground.
6   Start the engine and increase the speed to about 1500 rpm. Now turn the steering wheel gently from stop-to-stop. Check the reservoir fluid level.
7   Lower the vehicle to the ground and, with the engine still running, move the vehicle forward sufficiently to obtain full right lock followed by full left lock. Recheck the fluid level. If the fluid in the reservoir is extremely foamy, allow the vehicle to stand for a few minutes with the engine switched off and then repeat the previous operations. At the same time, check the belt tightness and check for a bent or loose pulley. Check also to make sure the power steering hoses

are not touching any other part of the vehicle, especially sheet metal or the exhaust manifold.
8   The procedures above will normally remedy an extreme foam condition and/or an objectionably noisy pump (low fluid level and/or air in the power steering fluid are the leading causes of this condition). If, however, either or both conditions persist after a few trials, the power steering system will have to be checked by a dealer. Do not drive the vehicle until the condition(s) have been remedied.

22 Power steering gear - removal and installation

Refer to illustrations 22.6 and 22.7
Caution: *On models equipped with an airbag, DO NOT allow the steering shaft to rotate with the steering gear removed or damage to the airbag coil under the steering wheel may occur. To prevent the steering shaft from turning, place the ignition key in the Lock position and remove the key, or at the very least, wrap the seatbelt around the steering wheel and clip it to the buckle.*
1   Disconnect the cable from the negative battery terminal. **Caution:** *On models equipped with a Delco Loc II audio system, disable the anti-theft feature before disconnecting the battery.*
2   If equipped, remove the coupling shield.
3   Disconnect the pressure and return hoses attached to the power steering gear assembly.
4   Plug the ends of the disconnected hoses and the holes in the power steering housing to prevent contamination.
5   Remove the nuts, lock washers and bolts at the steering coupling-to-steering shaft flange.
6   Remove the Pitman arm locknut and washer. Mark the position of the Pitman arm

in relation to the shaft and disconnect the Pitman arm with a Pitman arm puller **(see illustration)**.
7   Remove the bolts securing the steering gear to the frame and separate it from the vehicle **(see illustration)**.
8   When installing the gear, place it into position so that the coupling mounts properly to the flanged end of the steering shaft. Secure the gear to the frame, install the washers and bolts, then tighten the bolts to the specified torque.
9   Secure the steering coupling to the flanged end of the column with the lock washers and nuts. Tighten the nuts.
10   Install the Pitman arm and tighten the nut.
11   Connect the negative battery cable (and the coupling shield, if applicable).

23 Power steering pump - removal and installation

1   Disconnect the hydraulic hoses from the pump and keep them in the raised position to prevent the fluid from leaking out until they can be plugged.
2   Remove the pump drivebelt by loosening the pump mounts and pushing it in toward the engine.
3   Unscrew and remove the pump mounting bolts and braces and remove the pump.
4   To remove the pump pulley, a puller will be needed.
5   Installation is the reverse of removal.
6   Prime the pump by turning the pulley in the reverse direction to that of normal rotation (counterclockwise as viewed from the front) until air bubbles cease to emerge from the fluid when observed through the reservoir filler cap.
7   Install the drivebelt and adjust it as described in Chapter 1.
8   Bleed the system as described in Section 21.

24 Steering column switches - removal and installation

Warning: *If equipped with a Supplemental Inflatable Restraint system (SIR), more commonly known as airbags, disable the airbag system before working in the vicinity of airbag system components to avoid the possibility of accidental deployment of the airbag, which could result in personal injury (see Chapter 10).*

Note: *Removal and installation of the components included in this Section are presented in a progressive manner. Those nearest the top of the steering column are covered first.*

1   Disconnect the cable from the negative battery terminal. **Caution:** *On models equipped with a Delco Loc II audio system, disable the anti-theft feature before disconnecting the battery.*

2   Position the front wheels pointing straight ahead, turn the ignition switch to the Lock position and remove the key.

3   Remove the steering wheel (refer to Section 19).

Airbag coil (1990 and later models)

Refer to illustrations 24.5 and 24.6

4   Disable the airbag system (see Chapter 10).

5   Remove the airbag coil retaining snap-ring and remove the coil assembly; let the coil hang by the wiring harness **(see illustration)**. Remove the wave washer.

6   When installing the airbag coil it, may be necessary the center the coil as follows (It will only become uncentered if the spring lock is depressed and the hub rotated with the coil off the column **(see illustration)**:

   a) *Turn the coil over and depress the spring lock.*

   b) *Rotate the hub in the direction of the arrow on the coil until it stops.*

   c) *Rotate the hub in the opposite direction 2-1/2 turns and release the spring lock.*

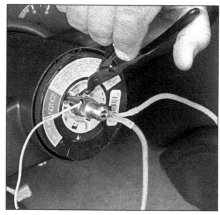

24.5 Remove the snap-ring and pull the airbag coil off the steering shaft

Canceling cam

Refer to illustrations 24.7 and 24.8

7   Remove the horn contact bushing and spring by pulling it straight out of the recess in the steering column cover plate **(see illustration)**.

8   Using a screwdriver in the slots provided, pry the plastic cover out of the steering column **(see illustration)**.

9   Remove the shaft lock from the steering column. On vehicles with standard steering columns, it is held in place with a snap-ring which fits into a groove in the steering shaft. The shaft lock must be depressed to relieve pressure on the snap-ring. A special U-shaped tool which fits on the shaft is used to depress the shaft lock as the snap-ring is removed from the groove **(see illustration)**. On models with telescopic steering wheel assemblies, remove the spacers, retracted steering shaft bumper, carrier snap-ring retainer and the shaft lock retainer.

10   Pull out and remove the canceling cam assembly and upper bearing preload spring.

11   If no other components are to be removed from the steering column, reverse the above steps for installation, making sure the turn signal switch is in the neutral position

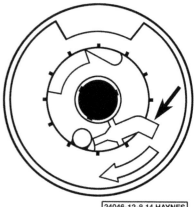

24046-12-8.14 HAYNES

24.6 To center the airbag coil, depress the spring lock (arrow); rotate the hub in the direction of the arrow on the coil until it stops; rotate the hub in the opposite direction 2-1/2 turns and release the spring lock

before installing the upper bearing preload spring and canceling cam. Always use a new snap-ring for the shaft lock.

Turn signal switch

Refer to illustration 24.14

12   Remove the turn signal lever attaching screw and withdraw the turn signal lever from the side of the column.

13   Push in on the hazard warning knob and unscrew the knob from the threaded shaft.

14   Remove the three turn signal assembly mounting screw **(see illustration)**.

15   Remove the instrument panel lower cover below the steering column and disconnect the turn signal switch wiring connector. Pull the switch wiring connector out of the bracket on the steering column jacket. Tape the connector terminals to prevent damage. Feed the wiring connector up through the column support bracket and pull the switch, wiring harness and connectors out of the top of the steering column.

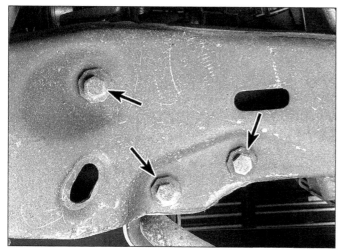

24.7 Remove the horn contact bushing and spring

24.8 Remove the plastic cover from the steering column lock plate

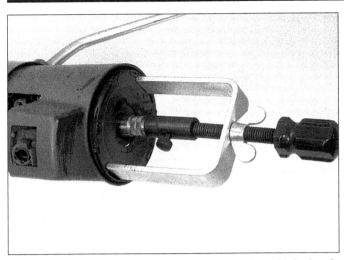

24.9 This tool is used to depress the steering column lock plate in order to remove or install the lock retaining snap-ring

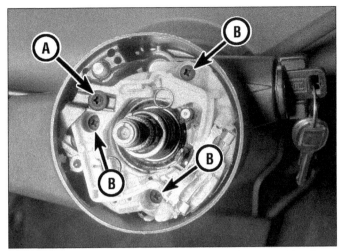

24.14 Remove the turn signal arm screw (A), followed by the switch mounting screws (B) - place the switch in the right turn position to access all the screws

16   If no other components are to be removed from the steering column, reverse the above steps for installation. Make sure the wiring harness is in the protector as it is pulled into position and that the warning knob is pulled out.

Ignition lock cylinder

Note: *If only the ignition lock cylinder and/or the ignition key warning switch is to be removed, the turn signal switch does not have to be completely removed. Instead, after removing the retaining screws, pull the switch over the end of the steering column (without having pulled the wiring out of the harness) and secure it out of the way.*

17   Place the lock cylinder in the Run position.
18   Remove the ignition key warning switch.
19   Using a magnetized screwdriver,

remove the lock retaining screw. Do not allow this screw to drop down into the column or you will have to completely disassemble the steering column to retrieve the screw.
20   Pull the lock cylinder out of the side of the steering column.
21   To install it, rotate the lock cylinder set and align the cylinder key with the keyway in the steering column housing.
22   Push the lock all the way in and install the retaining screw.
23   Install the ignition key warning switch (Steps 24 and 25).
24   Reverse the removal steps to complete the installation.

Ignition key warning switch

25   The key warning switch may be removed independently of the ignition key lock, or it may be removed at the same time if

the key lock is to be removed.
26   If the ignition lock cylinder is still intact, set the key to the Run position.
27   Use a piece of stiff wire (a paper clip will work fine) to remove the switch. Make a hook at the end of the wire and attach the hooked end into the loop of the clip at the top of the switch. Pull up on the wire and remove the clip and switch together from the steering column. Do not allow the clip to fall down into the column.
28   If the lock cylinder is still in the column, the buzzer switch actuating button (on the lock cylinder) must be depressed before the new warning switch can be installed.
29   Install the switch with the contacts toward the upper end of the steering column and with the formed end of the spring clip at the lower end of the switch. Reinstall the remaining components.

Notes

Chapter 12 Body

Contents

	Section
Body - maintenance	2
Body repair - major damage	7
Body repair - minor damage	6
Door connecting rods and locking rods - removal and installation	15
Door lock assembly - replacement	17
Door lock cylinder - removal and installation	16
Door lock knob - removal and installation	12
Door outside handle assembly - removal and installation	19
Door remote control spring clips - disengagement and engagement	14
Door trim panel - removal and installation	13
Door window glass replacement	20
Door window regulator - adjustment	21
General information	1
Hinges and locks - maintenance	5
Hood - removal and installation	8
Hood damper - replacement	9
Hood latch release cable - removal and installation	10
Power door lock actuator - removal and installation	18
Rear hatch - removal, installation and adjustment	11
Upholstery and carpets - maintenance	3
Vinyl trim - maintenance	4
Windshield replacement	22

Specifications

Torque specifications

Hinge-to-rear hatch/glass bolts	132 in-lbs
Hinge-to-body bolts	15 to 20 Ft-lbs
Gas damper-to-rear hatch assembly bolts	48 to 60 in-lbs
Door lock assembly screws	84 to 108 in-lbs
Power door lock actuator mounting bolts	96 to 120 in-lbs

1 General information

The Firebird, for the years covered by this manual, was available in three models: Firebird, Firebird Trans Am and Firebird S/E, all based on a two-door body style. Differences between the various models are noted where appropriate in the service procedures within this Chapter.

Firebird bodies are of unitized construction in which the body is designed to provide vehicle rigidity so that a separate frame is not necessary. Front and rear frame side rails integral with the body support the front end sheet metal, front and rear suspension systems and other mechanical components. Due to this type of construction, it is very important that, in the event of collision damage, the underbody be thoroughly checked by a facility with the proper equipment to do so.

Component replacement and repairs possible for the home mechanic are included in this Chapter.

2 Body - maintenance

1 The condition of your vehicle's body is very important, because it is on this that the second-hand value will mainly depend. It is much more difficult to repair a neglected or damaged body than it is to repair mechanical components. The hidden areas of the body, such as the fender wells, the frame, and the engine compartment, are equally important, although obviously do not require as frequent attention as the rest of the body.

2 Once a year, or every 12 000 miles, it is a good idea to have the underside of the body and the frame steam cleaned. All traces of dirt and oil will be removed and the underside can then be inspected carefully for rust, damaged brake lines, frayed electrical wiring, damaged cables, and other problems. The front suspension components should be greased after completion of this job.

3 At the same time, clean the engine and the engine compartment using either a steam cleaner or a water soluble degreaser.

4 The fender wells should be given particular attention, as undercoating can peel away and stones and dirt thrown up by the tires can cause the paint to chip and flake, allowing rust to set in. If rust is found, clean down to the bare metal and apply an anti-rust paint.

5 The body should be washed once a week (or when dirty). Wet the vehicle thoroughly to soften the dirt, then wash it down with a soft sponge and plenty of clean soapy water. If the surplus dirt is not washed off very carefully, it will in time wear down the paint.

6   Spots of tar or asphalt coating thrown up from the road should be removed with a cloth soaked in solvent.
7   Once every six months, give the body and chrome trim a thorough wax job. If a chrome cleaner is used to remove rust from any of the vehicle's plated parts, remember that the cleaner also removes part of the chrome, so use it sparingly.

3   Upholstery and carpets - maintenance

1   Every three months, remove the carpets or mats and clean the interior of the vehicle (more frequently if necessary). Vacuum the upholstery and carpets to remove loose dirt and dust.
2   If the upholstery is soiled, apply upholstery cleaner with a damp sponge and wipe it off with a clean, dry cloth.

4   Vinyl trim - maintenance

Vinyl trim should not be cleaned with detergents, caustic soaps or petroleum-based cleaners. Plain soap and water or a mild vinyl cleaner is best for stains. Test a small area for color fastness. Bubbles under the vinyl can be corrected by piercing them with a pin and then working the air out.

5   Hinges and locks - maintenance

Every 3000 miles or three months, the door, hood and rear hatch hinges and locks should be lubricated with a few drops of oil. The door and rear hatch striker plates should also be given a thin coat of grease to reduce wear and ensure free movement.

6   Body repair - minor damage

See photo sequence

Repair of minor scratches

If the scratch is very superficial and does not penetrate to the metal of the body, repair is very simple. Lightly rub the scratched area with a fine rubbing compound to remove loose paint and built-up wax. Rinse the area with clean water.

Apply touch-up paint to the scratch, using a small brush. Continue to apply thin layers of paint until the surface of the paint in the scratch is level with the surrounding paint. Allow the new paint at least two weeks to harden, then blend it into the surrounding paint by rubbing with a very fine rubbing compound. Finally, apply a coat of wax to the scratch area.

If the scratch has penetrated the paint and exposed the metal of the body, causing

the metal to rust, a different repair technique is required. Remove all loose rust from the bottom of the scratch with a pocket knife, then apply rust-inhibiting paint to prevent the formation of rust in the future. Using a rubber or nylon applicator, coat the scratched area with glaze-type filler. If required, the filler can be mixed with thinner to provide a very thin paste, which is ideal for filling narrow scratches. Before the glaze filler in the scratch hardens, wrap a piece of smooth cotton cloth around the tip of a finger. Dip the cloth in thinner and then quickly wipe it along the surface of the scratch. This will ensure that the surface of the filler is slightly hollow. The scratch can now be painted over as described earlier in this section.

Repair of dents

When repairing dents, the first job is to pull the dent out until the affected area is as close as possible to its original shape. There is no point in trying to restore the original shape completely as the metal in the damaged area will have stretched on impact and cannot be restored to its original contours. It is better to bring the level of the dent up to a point which is about 1/8-inch below the level of the surrounding metal. In cases where the dent is very shallow, it is not worth trying to pull it out at all.

If the back side of the dent is accessible, it can be hammered out gently from behind using a soft-faced hammer. While doing this, hold a block of wood firmly against the opposite side of the metal to absorb the hammer blows and prevent the metal from being stretched out.

If the dent is in a section of the body which has double layers, or some other factor that makes it inaccessible from behind, a different technique is required. Drill several small holes through the metal inside the damaged area, particularly in the deeper sections. Screw long, self-tapping screws into the holes just enough for them to get a good grip in the metal. Now the dent can be pulled out by pulling on the protruding heads of the screws with locking pliers.

The next stage of repair is the removal of paint from the damaged area and from an inch or so of the surrounding metal. This is easily done with a wire brush or sanding disk in a drill motor, although it can be done just as effectively by hand with sandpaper. To complete the preparation for filling, score the surface of the bare metal with a screwdriver or the tang of a file (or drill small holes in the affected area). This will provide a very good grip for the filler material. To complete the repair, see the Section on filling and painting.

Repair of rust holes or gashes

Remove all paint from the affected area and from an inch or so of the surrounding metal using a sanding disk or wire brush mounted in a drill motor. If these are not available, a few sheets of sandpaper will do the job just as effectively. With the paint

removed, you will be able to determine the severity of the corrosion and decide whether to replace the whole panel, if possible, or repair the affected area. New body panels are not as expensive as most people think and it is often quicker to install a new panel than to repair large areas of rust.

Remove all trim pieces from the affected area (except those which will act as a guide to the original shape of the damaged body, i.e. headlight shells, etc.). Then, using metal snips or a hacksaw blade, remove all loose metal and any other metal that is badly affected by rust. Hammer the edges of the hole in to create a slight depression for the filler material.

Wire brush the affected area to remove the powdery rust from the surface of the metal. If the back of the rusted area is accessible, treat it with rust-inhibiting paint.

Before filling is done, block the hole in some way. This can be done with sheet metal riveted or screwed into place, or by stuffing the hole with wire mesh.

Once the hole is blocked off, the affected area can be filled and painted (see the following sub-section on filling and painting).

Filling and painting

Many types of body fillers are available, but generally speaking, body repair kits which contain filler paste and a tube of resin hardener are best for this type of repair work. A wide, flexible plastic or nylon applicator will be necessary for imparting a smooth and contoured finish to the surface of the filler material.

Mix up a small amount of filler on a clean piece of wood or cardboard (use the hardener sparingly). Follow the manufacturer's instructions on the package, otherwise the filler will set incorrectly. Using the applicator, apply the filler paste to the prepared area. Draw the applicator across the surface of the filler to achieve the desired contour and to level the filler surface. As soon as a contour that approximates the original one is achieved, stop working the paste. If you continue, the paste will begin to stick to the applicator. Continue to add thin layers of filler paste at 20-minute intervals until the level of the filler is just above the surrounding metal.

Once the filler has hardened, the excess can be removed with a body file. From then on, progressively finer grades of sandpaper should be used, starting with a 180-grit paper and finishing with 600-grit wet-or-dry paper. Always wrap the sandpaper around a flat rubber or wooden block, otherwise the surface of the filler will not be completely flat. During the sanding of the filler surface, the wet-or-dry paper should be periodically rinsed in water. This will ensure that a very smooth finish is produced in the final stage.

At this point, the repair area should be surrounded by a ring of bare metal, which in turn should be encircled by the finely feath-

8.4 Removing the clip retaining the light wire harness to the hood

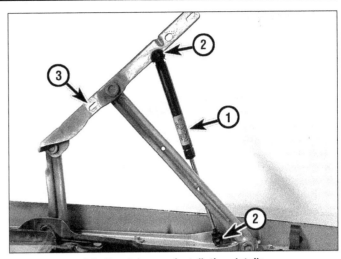

9.3 Hood damper installation details

1   Hood damper   3   Hood hinge
2   Retainer clip

ered edge of good paint. Rinse the repair area with clean water until all of the dust produced by the sanding operation is gone.

Spray the entire area with a light coat of primer. This will reveal any imperfections in the surface of the filler. Repair the imperfections with fresh filler paste or glaze filler and once more smooth the surface with sandpaper. Repeat this spray-and-repair procedure until you are satisfied that the surface of the filler and the feathered edge of the paint are perfect. Rinse the area with clean water and allow it to dry completely.

The repair area is now ready for painting. Spray painting must be carried out in a warm, dry, windless and dust-free atmosphere. These conditions can be created if you have access to a large indoor work area, but if you are forced to work in the open, you will have to pick the day very carefully. If you are working indoors, dousing the floor in the work area with water will help settle the dust which would otherwise be in the air. If the repair area is confined to one body panel, mask off the surrounding panels. This will help minimize the effects of a slight mismatch in paint color. Trim pieces such as chrome strips, door handles, etc., will also need to be masked off or removed. Use masking tape and several layers of newspaper for the masking operations.

Before spraying, shake the paint can thoroughly, then spray a test area until the spray painting technique is mastered. Cover the repair area with a thick coat of primer. The thickness should be built up using several thin layers of primer rather than one thick one. Using 600-grit wet-or-dry sandpaper, rub down the surface of the primer until it is very smooth. While doing this, the work area should be thoroughly rinsed with water and the wet-or-dry sandpaper periodically rinsed as well. Allow the primer to dry before spraying additional coats.

Spray on the top coat, again building up the thickness by using several thin layers of

paint. Begin spraying in the center of the repair area and then, using a circular motion, work out until the whole repair area and about two inches of the surrounding original paint is covered. Remove all masking material 10 to 15 minutes after spraying on the final coat of paint. Allow the new paint at least two weeks to harden, then use a very fine rubbing compound to blend the edges of the new paint into the existing paint. Finally, apply a coat of wax.

7   Body repair - major damage

1   Major damage must be repaired by an auto body shop specifically equipped to perform unibody repairs. These shops have available the specialized equipment required to do the job properly.
2   If the damage is extensive, the underbody must be checked for proper alignment or the vehicle's handling characteristics may be adversely affected and other components may wear at an accelerated rate.
3   Due to the fact that all of the major body components (hood, fenders, etc.) are separate and replaceable units, any seriously damaged components should be replaced rather than repaired. Sometimes these components can be found in a wrecking yard that specializes in used vehicle components (often at considerable savings over the cost of new parts).

8   Hood - removal and installation

Refer to illustration 8.4
1   Raise the hood.
2   Place protective pads along the edges of the engine compartment to prevent damage to the painted surfaces.
3   Disconnect the cable from the negative battery terminal. **Caution:** *On models*

equipped with a Delco Loc II audio system, disable the anti-theft feature before disconnecting the battery.
4   Remove the clip retaining the underhood light wire harness to the hood **(see illustration)**.
5   Disconnect the underhood light electrical connector and unbolt the bulb assembly from the hood.
6   Scribe lines on the underside of the hood, around the hood mounting bracket, so the hood can be installed in the same position.
7   Apply white paint around the bracket-to-hood bolts so they can be aligned quickly and accurately during installation.
8   Remove the bracket-to-hood bolts and, along with an assistant, separate the hood from the vehicle.
9   Installation is the reverse of the removal procedure.

9   Hood damper - replacement

Refer to illustration 9.3
1   Raise the hood.
2   Position a prop between the radiator support and a reinforced area at the front underside of the hood.
3   Remove the damper pin retaining clips at the top and bottom of the hood damper and remove the pins and damper **(see illustration)**.
4   Installation is the reverse of the removal procedure.

10   Hood latch release cable - removal and installation

Refer to illustrations 10.2 and 10.5
1   The hood latch cable is a one-piece assembly that includes the pull handle, control cable and housing.

These photos illustrate a method of repairing simple dents. They are intended to supplement *Body repair - minor damage* in this Chapter and should not be used as the sole instructions for body repair on these vehicles.

1 If you can't access the backside of the body panel to hammer out the dent, pull it out with a slide-hammer-type dent puller. In the deepest portion of the dent or along the crease line, drill or punch hole(s) at least one inch apart . . .

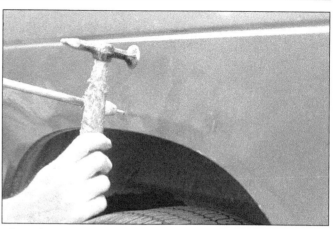

2 . . . then screw the slide-hammer into the hole and operate it. Tap with a hammer near the edge of the dent to help 'pop' the metal back to its original shape. When you're finished, the dent area should be close to its original contour and about 1/8-inch below the surface of the surrounding metal

3 Using coarse-grit sandpaper, remove the paint down to the bare metal. Hand sanding works fine, but the disc sander shown here makes the job faster. Use finer (about 320-grit) sandpaper to feather-edge the paint at least one inch around the dent area

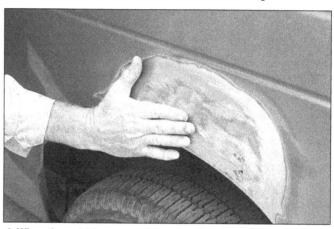

4 When the paint is removed, touch will probably be more helpful than sight for telling if the metal is straight. Hammer down the high spots or raise the low spots as necessary. Clean the repair area with wax/silicone remover

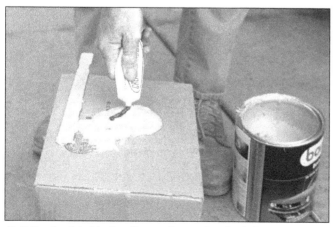

5 Following label instructions, mix up a batch of plastic filler and hardener. The ratio of filler to hardener is critical, and, if you mix it incorrectly, it will either not cure properly or cure too quickly (you won't have time to file and sand it into shape)

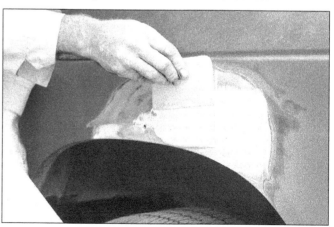

6 Working quickly so the filler doesn't harden, use a plastic applicator to press the body filler firmly into the metal, assuring it bonds completely. Work the filler until it matches the original contour and is slightly above the surrounding metal

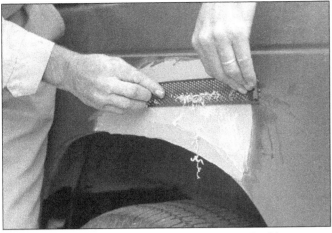

7 Let the filler harden until you can just dent it with your fingernail. Use a body file or Surform tool (shown here) to rough-shape the filler

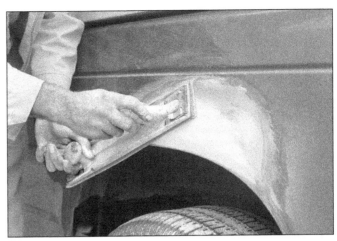

8 Use coarse-grit sandpaper and a sanding board or block to work the filler down until it's smooth and even. Work down to finer grits of sandpaper - always using a board or block - ending up with 360 or 400 grit

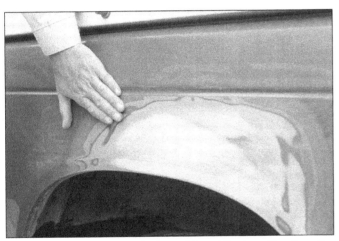

9 You shouldn't be able to feel any ridge at the transition from the filler to the bare metal or from the bare metal to the old paint. As soon as the repair is flat and uniform, remove the dust and mask off the adjacent panels or trim pieces

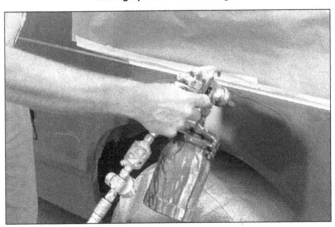

10 Apply several layers of primer to the area. Don't spray the primer on too heavy, so it sags or runs, and make sure each coat is dry before you spray on the next one. A professional-type spray gun is being used here, but aerosol spray primer is available inexpensively from auto parts stores

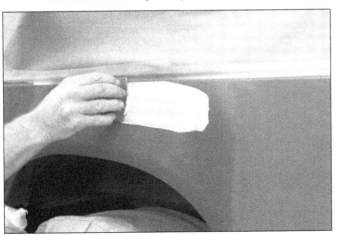

11 The primer will help reveal imperfections or scratches. Fill these with glazing compound. Follow the label instructions and sand it with 360 or 400-grit sandpaper until it's smooth. Repeat the glazing, sanding and respraying until the primer reveals a perfectly smooth surface

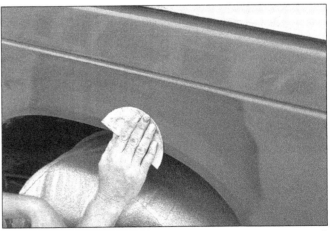

12 Finish sand the primer with very fine sandpaper (400 or 600-grit) to remove the primer overspray. Clean the area with water and allow it to dry. Use a tack rag to remove any dust, then apply the finish coat. Don't attempt to rub out or wax the repair area until the paint has dried completely (at least two weeks)

10.2 Typical hood release cable-to-hood latch installation details

1   *Hood release cable*   3   *Retainer clip*
2   *Hood release cable jacket*

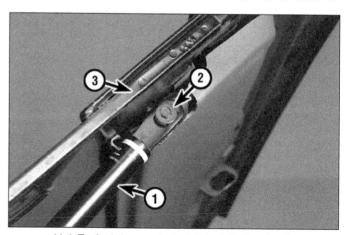

10.5 Hood release handle installation details

1   *Hood release assembly*
2   *Mounting screws*

2   To remove it, raise the hood and disengage the cable from the primary hood latch assembly **(see illustration)**. Take precautions to keep the hood from closing and locking while the cable is disconnected.

3   Unclip the cable from the body.

4   Remove the side trim panel to the left of the driver's seat.

5   Disengage the control assembly housing from the cable handle cutout in the trim panel and remove the cable assembly from the panel **(see illustration)**.

6   Withdraw the cable through the firewall.

7   Installation is the reverse of the removal procedure. When installing the cable, make sure that the sealing grommet attached to the dash panel is in place.

11   Rear hatch - removal, installation and adjustment

Refer to illustrations 11.3 and 11.4

1   Open the rear hatch and support it with a prop.

2   Place protective pads along the edges of the hatch opening to prevent damage to the painted surfaces while work is being performed.

3   Remove the nuts holding the hinges to the glass. **(see illustration)**.

4   While a helper supports the hatch, remove the bolts retaining the top of the dampers to the hatch assembly **(see illustration)**. **Caution:** *Do not try to remove or loosen the damper attachments with the hatch in any position other than completely open, as personal injury may result.*

5   Disconnect the wiring harness connector for the electric grid defogger, if so equipped.

6   With the aid of a helper, remove the hatch assembly and place it on a protected surface until it is reinstalled.

7   Installation is the reverse of the removal procedure, noting the following points.

8   When attaching the dampers to the hatch assembly, be sure to tighten the bolts to the specified torque.

9   When installing the nuts retaining the hinges to the glass, tighten the nuts to the specified torque. **Caution:** *Be sure not to overtighten the nuts, as glass breakage and possible personal injury could result.*

10   The hatch assembly height, fore-and-aft and side-to-side adjustments are made at the hinge-to-body locations.

11   Fore-and-aft and side-to-side adjustments are made by loosening the nut and bolt assemblies in the oversize holes found at these locations, moving the hatch assembly as required, then retightening the nut and bolt assemblies.

12   Hatch height may be adjusted by adding or subtracting the number of spacers at the hinge-to-body locations.

13   Tighten the hinge-to-body bolts to the specified torque when adjustments are complete.

12   Door lock knob - removal and installation

1   Insert a small flat-blade screwdriver behind the leading end of the lock knob and pry it away from the locking rod, located behind the knob.

2   After the end of the rod is free from the knob, slide the knob forward and remove it.

3   To install the knob, insert the small end

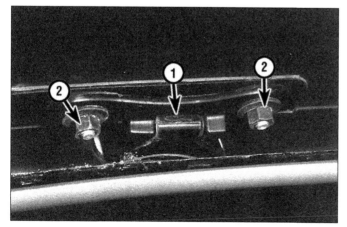

11.3 Rear hatch-to-hinge installation details

1   *Hinge*   2   *Mounting nuts*

11.4 Typical damper-to-hatch installation details

1   *Damper*   2   *Mounting bolt*   3   *Hatch*

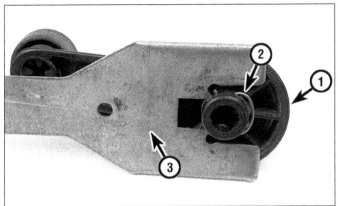

13.1 A special tool is used to remove the window crank handle retaining clip

1   *Window crank handle*   3   *Tool*
2   *Retaining clip*

13.2a Remove the front armrest assembly screw

of the knob through the hole in the escutcheon and slide the knob to the rear until the end of the locking rod engages the depression in the front end of the knob.

13   Door trim panel - removal and installation

Refer to illustrations 13.1, 13.2a, 13.2b, 13.4, 13.6, 13.7, 13.8, 13.9, 13.10, 13.11 and 13.13

1   On models with window crank handles, remove the crank handle. It is secured to the regulator shaft with a clip. The trim panel should be pushed away from the handle to expose the shaft and clip. A special spring clip removing tool is available to disengage the clip from the groove **(see illustration)**. A small hook can also be used to remove the clip if care is taken not to tear the panel.
2   Remove the three screws retaining the pull handle/armrest assembly **(see illustrations)**.
3   Separate the pull handle/armrest assembly from the trim panel by pulling it away from the retaining clips.

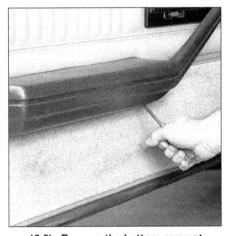

13.2b Remove the bottom armrest assembly screws

4   If so equipped, remove the two screws retaining the remote mirror control switch panel **(see illustration)**. **Note:** *On this panel and others, carefully examine the screw heads before removal because some are real and others are "dummy' heads, cast when the panel was made, which will break if a screwdriver is applied to them.*

13.4 Remove the screws retaining the remote mirror control switch panel

5   Pull out the switch panel.
6   Pry off the snap panel at the back of the switch assembly **(see illustration)**.
7   Disconnect the switch assembly from the switch panel **(see illustration)**.
8   If so equipped, remove the power door lock switch panel screws and pull out the switch assembly **(see illustration)**.

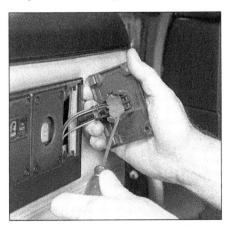

13.6 Remove the snap panel from the rear of the remote mirror control switch

13.7 Disconnect the switch assembly from the remote mirror switch panel

13.8 Remove the screws retaining the power door lock switch panel to the trim panel

13.9 Disconnect the electrical connector from the rear of the door lock switch

13.11 Remove the screws retaining the remote door handle trim panel

13.10 Remove the screws retaining the trim panel escutcheon to the trim panel

13.13 Remove the trim panel escutcheon from the trim panel

9   Disconnect the electrical connector at the rear of the switch panel **(see illustration)**.
10   Remove the screws retaining the trim panel escutcheon to the trim panel **(see illustration)**.
11   Remove the screws retaining the remote door handle trim panel **(see illustration)**.
12   Slide the trim panel escutcheon toward the rear and, using a small screwdriver inserted behind the end of the lock knob, pry the lock knob away from the lock knob rod and remove it.
13   Pull the trim panel escutcheon away from the trim panel, turn it approximately 60 degrees to clear the remote control handle and remove the escutcheon **(see illustration)**.
14   The trim panel is attached to the door with plastic retaining clips. To disengage these clips, insert a flat, blunt tool (like a screwdriver blade wrapped with tape) between the metal door skin and the trim panel. Carefully pry the door panel away from the door, keeping the tool close to the clips to prevent damage to the panel. Start at the bottom and work around the door toward the top. The top section is secured at the window channel. Once the retaining clips are pried free, lift the trim panel up and away from the door.
15   If work is to be performed on the inner door panel, remove the water deflector.
16   Installation is the reverse of the removal procedure, noting the following.

17   Before attaching the trim panel to the door, make sure that all the trim retainers are undamaged and installed tightly in the panel.
18   If a retainer is to be replaced with a new one, start the retainer flange into the 1/4-inch cutout attachment hole in the trim panel, then rotate the retainer until the flange is fully engaged.
19   When attaching the door trim panel to the door, locate the top of the panel over the upper flange of the inner door panel and press down on the trim panel to engage the upper retaining clips.
20   Position the trim panel on the inner door panel so that the panel retainers are aligned with the holes in the door panel and tap the retainers into the holes with the palm of your hand or a rubber mallet.

14 Door remote control spring clips - disengagement and engagement

1   Spring clips are used to retain remote control connecting rods and inside locking rods to door lock levers and remote handles. They are accessible after the door trim panel and water deflector are removed (refer to Section 13).
2   A slot in the clip provides for simple disengagement, allowing easy detachment of linkage rods.
3   To disengage a spring clip, use an awl

or thin-bladed screwdriver to slide the clip out of the slot in the linkage rod end.
4   To reengage the clip, return it to its original position.

15 Door connecting rods and locking rods - removal and installation

1   Connecting rods are employed to connect the door lock actuating levers with the inside and outside door handles and the inside lock knob. They are accessible after the door trim panel and water deflector have been removed (refer to Section 13).
2   To remove a door connecting or locking rod, disengage the spring clip retaining the rod to the lock lever (refer to Section 13), then disengage the rod.
3   To remove the inside remote handle connecting rod, the spring clip retaining the rod to the remote handle must also be disengaged, using the method just described.
4   To remove the inside lock knob-to-lock rod, first remove the knob (refer to Section 13).
5   To reinstall the rods, first reengage the spring clips on the lock levers, then align the rod in the installed position and push the rod end through the hole in the lock lever until it engages the spring clip.

16 Door lock cylinder - removal and installation

Refer to illustration 16.3
1   Raise the window completely.
2   Remove the door trim panel and detach the water deflector (refer to Section 13) enough to expose the access hole.
3   Disengage the lock cylinder-to-lock rod at the cylinder **(see illustration)**.
4   Using a screwdriver, slide the lock cylinder retainer forward until it is disengaged. The retainer may also be removed by hand by grasping the anti-theft shield at the top of the retainer and rotating it until the retainer disengages.
Caution: *If the lock cylinder retainer is being removed by hand, wear gloves to prevent injury from the sharp edges on the retainer.*
5   Remove the lock cylinder by pulling it out of the door from the outside.
6   Installation is the reverse of the removal procedure. Make sure the cylinder gasket is properly installed between the cylinder and the retainer. Lubricate the cylinder with WD-40 or a similar lubricant.

17 Door lock assembly - replacement

Refer to illustration 17.6
Note: *If the door lock assembly fails, it cannot be repaired. Correct door lock failures by replacing the faulty assembly.*
1   Completely raise the window glass.
2   Remove the door trim panel and the

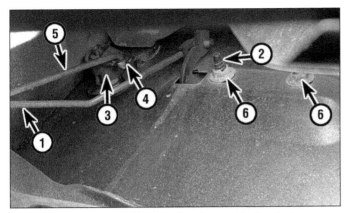

16.3 Typical outside door handles and lock cylinder installation details

1 Outside handle-to-lock rod	4 Lock cylinder retainer
2 Outside door handle	5 Lock cylinder-to-lock rod
3 Lock cylinder	6 Door handle mounting nuts

17.6 Typical door latch/lock assembly installation details

1 Latch/lock assembly	2 Mounting bolts

water deflector (refer to Section 13).

3   Disengage the inside locking rod at the lock assembly.

4   Disengage the inside handle-to-lock rod at the lock assembly.

5   Disengage the lock cylinder-to-lock rod at the lock assembly.

6   Remove the lock retaining screws, then lower the lock assembly to disengage the outside handle-to-lock rod **(see illustration)**.

7   Remove the lock assembly from the door.

8   Installation is the reverse of the removal procedure.

9   Tighten the lock screws to the specified torque.

18   Power door lock actuator - removal and installation

1   Completely raise the window glass.

2   Remove the door trim panel and the water deflector (refer to Section 13).

3   Disconnect the electrical connector at the actuator.

4   Locate the rivets retaining the actuator to the inner door panel.

5   Drive the center pins out of the rivets.

6   Using a 1/4-inch drill bit, drill out the rivets.

7   Disengage the bellcrank-to-actuator rod at the actuator.

8   Remove the actuator from the door.

9   To install the actuator, place it in position in the door and reengage the bellcrank-to-actuator rod.

10   If a hand rivet tool is available, attach the actuator to the inner door panel using 1/4 x 0.500-inch aluminum peel-type rivets.

11   If a hand rivet tool is not available, attach the actuator to the inner door panel using 1/4-20 x 1/2-inch nuts and bolts. Tighten the bolts to the specified torque.

12   Reinstall the water deflector and door trim panel.

19   Door outside handle assembly - removal and installation

1   Raise the window completely.

2   Remove the door trim panel as described in Section 13.

3   Pry back the water deflector enough to gain access to the exterior handle mounting hardware. To help prevent tools or parts from

falling down into the door cavity, place rags or newspapers in the cavity.

4   Disengage the spring clip from the exterior handle and separate the lock rod from the handle.

5   Remove the two handle mounting nuts.

6   Slide the handle to the rear and rotate it up to disengage it from the mounting hole.

7   Installation is the reverse of the removal procedure.

20   Door window glass replacement

Due to the requirements for special handling techniques, window glass should be replaced by a dealer or auto glass shop.

21   Door window regulator - adjustment

Refer to illustrations 21.2, 21.3a, 21.3b, 21.3c, 21.3d, 21.6, 21.7, 21.8, 21.9a, 21.9b and 21.10

Note: *The procedure which follows is for adjusting the front door window regulators only. The removal and installation procedure is beyond the scope of the home mechanic and must be left to a professional. One of the reasons for having experienced personnel handle this work is the extreme danger involved in handling the regulator lift arm, which is under tension from the counterbalance spring on motor-driven models. We highly emphasize the need to have all window regulator removal and installation work performed by experienced personnel.*

1   Remove the door trim panel (refer to Section 13). Remove the water deflector.

2   Some variations in door hardware design may be encountered, depending on the model, but the basic adjustment procedures are similar **(see illustration)**.

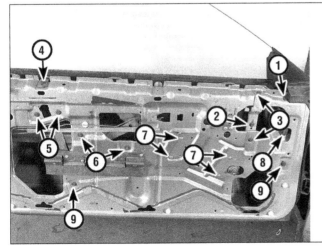

21.2 Typical door hardware installation details

1   Filler assembly -to-support bolt

2   Up-stop support bolt

3   Support-to-inner panel bolt

4   Glass stabilizer bolt

5   Rear guide channel bolt

6   Cam channel bolt

7   Regulator rivet

8   Adjusting stud and nut

9   Front run channel bolt

21.3a   Loosen the up-stop on-support bolt

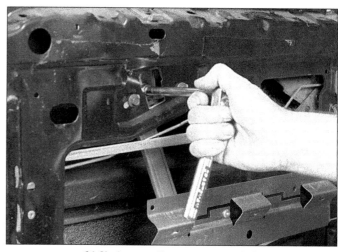

21.3b   Loosen the rear up-stop bolt

21.3c   Loosen the front cam channel bolt

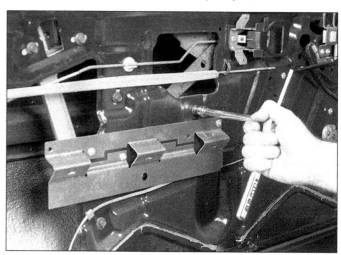

21.3d   Loosen the rear cam channel bolt

Rotation (up and down) adjustment

3   If an uneven gap shows at the top of the glass when it is rotated to the top of the channel, loosen the up-stop on-support nut/bolt, the rear up-stop nut/bolt and the cam channel nuts/bolts **(see illustrations)**.

4   Adjust the up-stops as necessary and retighten all nuts/bolts.

Channel adjustment

5   If the glass is either too tight or too loose in the channel, rotate the glass until it is half open with the door open.

6   Loosen the glass stabilizer bolts **(see illustration)**.

7   Loosen the filler assembly bolts **(see illustration)**.

8   Loosen the adjusting stud and nut-filler assembly bolt **(see illustration)**.

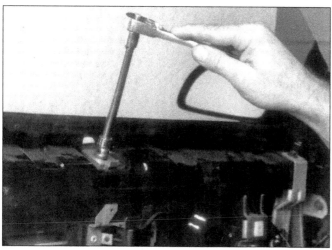

21.6   Loosen the glass stabilizer bolts

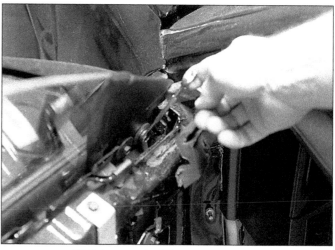

21.7   Loosen the filler assembly bolts

21.8 Loosen the adjusting stud and nut-filler assembly nut

21.9a Loosen the bottom front run channel bolt

21.9b Loosen the top front run channel bolt

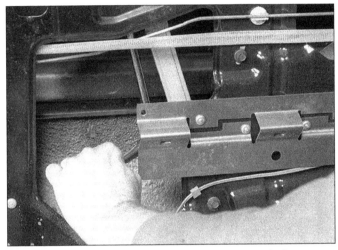

21.10 Loosen the guide channel-to-support bolts

9   Loosen the front run channel bolts **(see illustrations)**.

10   Loosen the guide channel-to-support bolts **(see illustration)**.

11   With all fasteners loose, slowly close the door, then align the glass.

12   Open the door and tighten the bolts, reversing the order in which they were loosened.

13   Make sure that the window operates correctly.

14   If misalignment still exists, repeat the above steps.

In/out adjustment

15   If the window is misaligned in (toward the passenger compartment) or out, rotate the window glass until it is half open, with the door open, and loosen the glass stabilizer bolts.

16   Loosen the filler assembly bolts.

17   Loosen the adjusting stud and nut-filler assembly bolts.

18   Loosen the front run channel bolts.

19   Loosen the rear guide channel bolts.

20   Loosen the guide channel-to-support bolts.

21   With all bolts loose, refer to Steps 9 through 12. The glass stabilizer bolts should be tightened last, with the felt bumpers just touching the glass.

22   Windshield replacement

Due to the requirements for special handling techniques, the windshield glass should be replaced by a dealer or auto glass shop.

Notes

Index

A

About this manual, 0-5
Acknowledgements, 0-2
Air conditioning
    compressor, removal and installation, 3-10
    condenser, removal and installation, 3-10
    system, servicing, 3-9
Air filter and PCV filter replacement, 1-25
Air Injection Reaction (AIR) system, 6-12
Airbag system, general information, 10-14
Alternator
    brushes and voltage regulator, replacement, 5-7
    removal and installation, 5-6
Antifreeze, general information, 3-2
Automatic transmission, 7B-1 through 7B-4
    diagnosis, general, 7B-2
    fluid change, 1-23
    fluid level check, 1-8
    general information, 7B-1
    neutral safety and back-up light switch, adjustment, 7B-3
    removal and installation, 7B-3
    shift linkage, check and adjustment, 7B-2
    Throttle valve (TV) cable assembly, description and adjustment, 7B-2
Automotive chemicals and lubricants, 0-18

Axleshaft
    bearing, replacement, 8-10
    oil seal, replacement, 8-10
    removal and installation, 8-9

B

Balljoints, replacement, 11-4
Battery
    cables, check and replacement, 5-2
    check and maintenance, 1-9
    emergency jump starting, 0-16
    removal and installation, 5-1
Body, 12-1 through 12-12
    door
        connecting rods and locking rods, removal and installation, 12-8
        outside handle assembly, removal and installation, 12-9
        remote control spring clips, disengagement and engagement, 12-8
        trim panel, removal and installation, 12-7
    door lock, 12-8
        cylinder, removal and installation, 12-8
        knob, removal and installation, 12-6

door window
    glass replacement, 12-9
    regulator, adjustment, 12-9
general information, 12-1
hinges and locks, maintenance, 12-2
hood
    damper, replacement, 12-3
    latch release cable, removal and installation, 12-3
    removal and installation, 12-3
maintenance, 12-1
power door lock actuator, removal and installation, 12-9
rear hatch, removal, installation and adjustment, 12-6
repair
    major damage, 12-3
    minor damage, 12-2
upholstery and carpets, maintenance, 12-2
vinyl trim, maintenance, 12-2
windshield replacement, 12-11
Booster battery (jump) starting, 0-16
Brake check, 1-13
Brakes, 9-1 through 9-24
brake pedal, removal and installation, 9-23
combination valve, check and replacement, 9-21
disc brake caliper
    overhaul
        front, 9-6
        rear, 9-9
    removal and installation
        front, 9-5
        rear, 9-7
disc brake pads replacement
    front, 9-2
    rear, 9-8
disc brake rotor, inspection, removal and
    installation, 9-12
general information, 9-2
hydraulic brake hoses and lines, inspection and
    replacement, 9-20
hydraulic system, bleeding, 9-22
master cylinder, removal, overhaul and installation, 9-19
parking brake, adjustment, 9-21
power brake booster, inspection, removal and
    installation, 9-22
rear drum brake shoes, replacement, 9-13
stop light switch, removal, installation and
    adjustment, 9-23
wheel cylinder (drum brakes), removal, overhaul and
    installation, 9-18
Bulbs, replacement, 10-6
Buying parts, 0-8

C

**Camshaft, lifters and bearings, inspection and bearing
    replacement, 2D-14**

Camshaft, removal and installation
L4 (four-cylinder) engine, 2A-9
V6 engine, 2B-13
V8 engine, 2C-11
Carburetor
choke check, 1-14
overhaul
    E2SE, 4A-5
    E4ME, 4A-16
removal and installation, 4A-3
**Carburetor/throttle body injection (TBI) mounting
    torque check, 1-18**
Charging system
check, 5-6
general information and precautions, 5-6
Chassis electrical system, 10-1 through 10-14
airbag system, general information, 10-14
bulbs, replacement, 10-6
circuit breakers, general information, 10-3
cluster panel instruments (except speedometer),
    removal and installation, 10-7
concealed headlight
    actuator switch and harness assembly, removal and
        installation, 10-5
    actuator/motor assembly, removal and
        installation, 10-4
    assembly, general information and emergency
        procedures, 10-3
    body assembly, removal and installation, 10-5
console switches, removal, servicing and
    installation, 10-11
cruise control, general information and
    servicing, 10-13
electrical troubleshooting, general information, 10-2
fuses, general information, 10-2
fusible links, general information, 10-3
headlight sealed beam unit, removal and
    installation, 10-4
headlight switch, removal and installation, 10-10
headlights, adjustment, 10-4
instrument cluster panel, removal, servicing and
    installation, 10-8
radio power antenna, removal and installation, 10-7
radio speakers, removal and installation, 10-7
radio, removal and installation, 10-6
rear defogger (electric grid type), check and
    repair, 10-13
speedometer cable, replacement, 10-7
speedometer, removal and installation, 10-7
turn signals and hazard flashers, check and
    replacement, 10-3
windshield wiper
    arm, removal, installation and adjustment, 10-11
    motor, removal and installation, 10-12
    transmission, removal and installation, 10-12
Chassis lubrication, 1-12
Circuit breakers, general information, 10-3
**Cluster panel instruments (except speedometer),
    removal and installation, 10-7**

Clutch
cross-shaft, removal and installation, 8-3
general information, 8-2
hydraulic clutch assembly, removal and installation, 8-5
pedal freeplay check and adjustment, 1-19
pedal, removal and installation, 8-2
pilot bearing, removal and installation, 8-4
removal, inspection and installation, 8-3
Clutch and driveline, 8-1 through 8-10
Clutch start switch, replacement and adjustment, 7A-4
Coil springs and insulators, removal and installation
front, 11-5
rear, 11-6
Combination valve, check and replacement, 9-21
Compression check, 2D-7
Computer Command Control System (CCCS), 6-3
Concealed headlight
actuator switch and harness assembly, removal and installation, 10-5
actuator/motor assembly, removal and installation, 10-4
assembly, general information and emergency procedures, 10-3
body assembly, removal and installation, 10-5
Console switches, removal, servicing and installation, 10-11
Conversion factors, 0-19
Coolant temperature sending unit, check and replacement, 3-6
Cooling system
check, 1-11
servicing (draining, flushing and refilling), 1-20
Cooling, heating and air conditioning systems, 3-1 through 3-12
air conditioning system, servicing, 3-9
antifreeze, general information, 3-2
compressor, removal and installation, 3-10
condenser, removal and installation, 3-10
coolant temperature sending unit, check and replacement, 3-6
electric cooling fan, removal and installation, 3-11
general information, 3-2
heater blower motor, removal and installation, 3-6
heater core, removal and installation, 3-6
radiator, removal and installation, 3-3
thermostat, check, 3-3
thermostat, replacement, 3-2
water pump
check, 3-4
removal and installation, 3-4
Crank signal, 6-9
Crankcase front cover, removal and installation
L4 (four-cylinder) engine, 2A-8
V6 engine, 2B-11
V8 engine, 2C-10
Crankshaft
inspection, 2D-16
installation and main bearing oil clearance check, 2D-19
removal, 2D-12

Cruise control, general information and servicing, 10-13
Cylinder head
cleaning and inspection, 2D-9
disassembly, 2D-8
reassembly, 2D-10
removal and installation
L4 (four-cylinder) engine, 2A-4
V6 engine, 2B-5
V8 engine, 2C-7

D

Disc brake caliper
overhaul
front, 9-6
rear, 9-9
removal and installation
front, 9-5
rear, 9-7
Disc brake pads, replacement
front, 9-2
rear, 9-8
Disc brake rotor, inspection, removal and installation, 9-12
Distributor reference signal, 6-9
Distributor, removal and installation, 5-2
Door
connecting rods and locking rods, removal and installation, 12-8
outside handle assembly, removal and installation, 12-9
remote control spring clips, disengagement and engagement, 12-8
trim panel, removal and installation, 12-7
Door lock
assembly, replacement, 12-8
knob, removal and installation, 12-6
Door window
glass replacement, 12-9
regulator, adjustment, 12-9
Drivebelt check and adjustment, 1-9
Driveshaft
general information, 8-6
out-of-balance correction, 8-6
removal and installation, 8-6

E

Early Fuel Evaporation (EFE) system, 1-18, 6-17
Electric cooling fan, removal and installation, 3-11
Electrical troubleshooting, general information, 10-2
Electronic Control Module/PROM, removal and installation, 6-6
Electronic feedback carburetor system, 4A-16
Electronic Spark Control (ESC) system, 6-10
Electronic Spark Timing (EST) system, 6-9

Emissions control systems
Air Injection Reaction (AIR) system, 6-12
Computer Command Control System (CCCS), 6-3
Early Fuel Evaporation (EFE) system, 6-17
Electronic Control Module/PROM, removal and installation, 6-6
Electronic Spark Control (ESC) system, 6-10
Electronic Spark Timing (EST) system, 6-9
Evaporative Emission Control System (EECS), 6-15
Exhaust Gas Recirculation (EGR) system, 6-10
information sensors, 6-7
Positive Crankcase Ventilation (PCV) system, 6-16
Thermostatic Air Cleaner (THERMAC), 6-17
Transmission Converter Clutch (TCC), 6-18
Engine coolant temperature sensor, 6-7
Engine electrical systems, 5-1 through 5-10
alternator brushes and voltage regulator, replacement, 5-7
alternator, removal and installation, 5-6
battery
cables, check and replacement, 5-2
emergency jump starting, 5-2
removal and installation, 5-1
charging system
check, 5-6
general information and precautions, 5-6
distributor, removal and installation, 5-2
ignition
coil, check and replacement, 5-5
module, replacement, 5-4
pickup coil, check and replacement, 5-3
system
check, 5-2
general information and precautions, 5-1
starter motor
brushes, replacement, 5-10
removal and installation, 5-9
testing in vehicle, 5-8
starter solenoid, replacement, 5-9
starting system, general information, 5-8
Engine idle speed check and adjustment, 1-15
Engine oil and filter change, 1-15
Engines
L4 (four-cylinder) engine, 2A-1 through 2A-12
camshaft, removal and installation, 2A-9
crankshaft pulley hub and front oil seal, removal and installation, 2A-8
cylinder head, removal and installation, 2A-4
engine mounts, replacement, 2A-11
exhaust manifold, removal and installation, 2A-4
flywheel/driveplate and rear main bearing oil seal, removal and installation, 2A-10
general information, 2A-2
hydraulic lifters, removal, inspection and installation, 2A-5
intake manifold, removal and installation, 2A-3
oil pan, removal and installation, 2A-7
oil pump driveshaft, removal and installation, 2A-6
oil pump, removal and installation, 2A-7

pushrod cover, removal and installation, 2A-2
removal and installation, 2A-11
rocker arm cover, removal and installation, 2A-2
timing gear cover, removal and installation, 2A-8
valve train components, replacement, 2A-3
V6 engine, 2B-1 through 2B-18
camshaft, removal and installation, 2B-13
crankcase front cover, removal and installation, 2B-11
cylinder heads, removal and installation, 2B-5
exhaust manifolds, removal and installation, 2B-5
front cover oil seal, replacement, 2B-11
hydraulic lifters, removal, inspection and installation, 2B-4
intake manifold, removal and installation, 2B-3
mounts, replacement with engine in vehicle, 2B-13
oil pan, removal and installation, 2B-9
oil pump, removal and installation, 2B-9
rear main bearing oil seal, replacement, 2B-9
removal and installation, 2B-14
rocker arm covers, removal and installation, 2B-1
timing chain and sprockets, inspection, removal and installation, 2B-12
valve lash, adjustment, 2B-4
valve train components, replacement, 2B-2
vibration damper, removal and installation, 2B-11
V8 engine, 2C-1 through 2C-14
camshaft, removal and installation, 2C-11
crankcase front cover, removal and installation, 2C-10
cylinder heads, removal and installation, 2C-7
engine, removal and installation, 2C-12
exhaust manifolds, removal and installation, 2C-6
front cover oil seal, replacement, 2C-10
hydraulic lifters, removal, inspection and installation, 2C-5
intake manifold, removal and installation, 2C-3
mounts, replacement, 2C-12
oil pan, removal and installation, 2C-7
oil pump, removal and installation, 2C-8
rear main bearing oil seal, replacement, 2C-8
rocker arm covers, removal and installation, 2C-2
timing chain and sprockets, removal and installation, 2C-11
valve lash, adjustment, 2C-6
valve train components, replacement, 2C-3
vibration damper, removal and installation, 2C-9
General engine overhaul procedures, 2D-1 through 2D-22
block
cleaning, 2D-12
inspection, 2D-13
camshaft, lifters and bearings, inspection and bearing replacement, 2D-14
compression check, 2D-7
crankshaft
inspection, 2D-16
installation and main bearing oil clearance check, 2D-19
removal, 2D-12

cylinder head
    cleaning and inspection, 2D-9
    disassembly, 2D-8
    reassembly, 2D-10
engine overhaul
    disassembly sequence, 2D-7
    reassembly sequence, 2D-21
engine rebuilding alternatives, 2D-6
engine removal, methods and precautions, 2D-7
initial start-up and break-in after overhaul, 2D-22
main and connecting rod bearings, inspection, 2D-16
piston rings, installation, 2D-17
piston/connecting rod assembly
    inspection, 2D-15
    installation and bearing oil clearance check, 2D-20
    removal, 2D-11
pre-oiling engine after overhaul, 2D-21
rear main oil seal, installation, 2D-18
repair operations possible with the engine in the
    vehicle, 2D-6
valves, servicing, 2D-10
Evaporative Emission Control System (EECS), 6-15
Evaporative Emissions Control System (EECS) filter
    replacement, 1-26
Exhaust Gas Recirculation (EGR)
    system, 6-10
    valve check, 1-26
Exhaust manifolds, removal and installation
    L4 (four-cylinder) engine, 2A-4
    V6 engine, 2B-5
    V8 engine, 2C-6
Exhaust system
    check, 1-13
    components, removal and installation, 4A-22

F

Fluid level checks, 1-6
    automatic transmission fluid, 1-8
    battery electrolyte, 1-7
    brake fluid, 1-7
    engine coolant, 1-6
    engine oil, 1-6
    manual transmission oil, 1-7
    power steering fluid, 1-8
    rear axle oil, 1-8
    windshield washer fluid, 1-7
Flywheel/driveplate and rear main bearing oil seal,
    removal and installation, L4 (four-cylinder)
    engine, 2A-10
Front cover oil seal, replacement
    L4 (four-cylinder) engine, 2A-8
    V6 engine, 2B-11
    V8 engine, 2C-10
Front wheel bearing check, repack and
    adjustment, 1-21

Fuel and exhaust systems, carbureted models, 4A-1
    through 4A-24
carburetor
    overhaul
        E2SE, 4A-5
        E4ME, 4A-16
    removal and installation, 4A-3
electronic feedback carburetor system, 4A-16
exhaust system components, removal and
    installation, 4A-22
fuel line, repair and replacement, 4A-2
fuel pump
    check, 4A-1
    removal and installation, 4A-2
fuel tank
    removal and installation, 4A-2
    repair, 4A-3
general information, 4A-1
Fuel and exhaust systems, fuel injected models,
    4B-1 through 4B-16
fuel injection system, check, 4B-5
fuel pressure relief, 4B-2
fuel pump
    check, 4B-3
    removal and installation, 4B-4
general information, 4B-2
minimum idle speed, adjustment, 4B-15
Multi Port Fuel Injection system, component
    replacement, 4B-10
Throttle Body Injection system, component
    replacement, 4B-5
throttle cable, removal and installation, 4B-15
Fuel filter replacement, 1-16
Fuel injection system, check, 4B-5
Fuel line, repair and replacement, 4A-2
Fuel pressure relief, 4B-2
Fuel pump
    check
        carbureted models, 4A-1
        fuel injected models, 4B-3
    removal and installation
        carbureted models, 4A-2
        fuel injected models, 4B-4
Fuel system check, 1-16
Fuses, general information, 10-2
Fusible links, general information, 10-3

G

General engine overhaul procedures,
    2D-1 through 2D-22
block
    cleaning, 2D-12
    inspection, 2D-13
camshaft, lifters and bearings, inspection and bearing
    replacement, 2D-14
compression check, 2D-7

crankshaft
    inspection, 2D-16
    installation and main bearing oil clearance
        check, 2D-19
    removal, 2D-12
cylinder head
    cleaning and inspection, 2D-9
    disassembly, 2D-8
    reassembly, 2D-10
engine overhaul
    disassembly sequence, 2D-7
    reassembly sequence, 2D-21
engine rebuilding alternatives, 2D-6
engine removal, methods and precautions, 2D-7
initial start-up and break-in after overhaul, 2D-22
main and connecting rod bearings, inspection, 2D-16
piston rings, installation, 2D-17
piston/connecting rod assembly
    inspection, 2D-15
    installation and bearing oil clearance check, 2D-20
    removal, 2D-11
pre-oiling engine after overhaul, 2D-21
rear main oil seal, installation, 2D-18
repair operations possible with the engine in the
    vehicle, 2D-6
valves, servicing, 2D-10

H

**Headlight sealed beam unit, removal and
    installation, 10-4**
Headlight switch, removal and installation, 10-10
Headlights, adjustment, 10-4
Heater blower motor, removal and installation, 3-6
Heater core, removal and installation, 3-6
Hinges and locks, maintenance, 12-2
Hood
    damper, replacement, 12-3
    latch release cable, removal and installation, 12-3
    removal and installation, 12-3
Hydraulic lifters, removal, inspection and installation
    L4 (four-cylinder) engine, 2A-5
    V6 engine, 2B-4
    V8 engine, 2C-5

I

Ignition
    coil, check and replacement, 5-5
    module, replacement, 5-4
    pickup coil, check and replacement, 5-3
Ignition system
    check, 5-2
    general information and precautions, 5-1
Ignition timing check and adjustment, 1-27

Information sensors
    air conditioning On signal, 6-9
    crank signal, 6-9
    distributor reference signal, 6-9
    engine coolant temperature sensor, 6-7
    Manifold Absolute Pressure (MAP) sensor, 6-8
    Manifold Air Temperature (MAT) sensor, 6-8
    Mass Air Flow (MAF) sensor, 6-8
    oxygen sensor, 6-8
    Park/Neutral switch, 6-9
    Throttle Position Sensor (TPS), 6-9
    vehicle speed sensor, 6-9
Initial start-up and break-in after overhaul, 2D-22
**Instrument cluster panel, removal, servicing and
    installation, 10-8**
Intake manifold, removal and installation
    L4 (four-cylinder) engine, 2A-3
    V6 engine, 2B-3
    V8 engine, 2C-3
Introduction and routine maintenance schedule, 1-4
Introduction to the Pontiac Firebird, 0-5

J

Jacking and towing, 0-17

K

Knuckle, steering, removal and installation, 11-10

L

Lower control arm, removal and installation
    front, 11-5
    rear, 11-7

M

Main and connecting rod bearings, inspection, 2D-16
Maintenance schedule, 1-4
**Maintenance techniques, tools and working
    facilities, 0-8**
Manifold Absolute Pressure (MAP) sensor, 6-8
Manifold Air Temperature (MAT) sensor, 6-8
Manual transmission, 7A-1 through 7A-4
    clutch start switch, replacement and adjustment, 7A-4
    extension housing oil seal, replacement, 7A-3
    general information, 7A-2
    mounts, check and replacement, 7A-2
    oil level check, 1-7
    shift control lever (77mm 4-speed and 5-speed),
        removal and installation, 7A-2
    speedometer gear seal, replacement, 7A-4

transmission linkage (76mm and 83mm 4-speed),
    removal, installation and adjustment, 7A-2
transmission, removal and installation
    76mm and 83mm 4-speed, 7A-2
    77mm 4-speed and 5-speed, 7A-3
Mass Air Flow (MAF) sensor, 6-8
**Master cylinder, removal, overhaul and
    installation, 9-19**
Minimum idle speed, adjustment, 4B-15
Mounts, engine, replacement
    L4 (four-cylinder) engine, 2A-11
    V6 engine, 2B-13
    V8 engine, 2C-12
**Multi Port Fuel Injection system, component
    replacement, 4B-10**

N

**Neutral safety and back-up light switch,
    adjustment, 7B-3**

O

Oil pan, removal and installation
    L4 (four-cylinder) engine, 2A-7
    V6 engine, 2B-9
    V8 engine, 2C-7
**Oil pump driveshaft, removal and installation, L4 (four-
    cylinder) engine, 2A-6**
Oil pump, removal and installation
    L4 (four-cylinder) engine, 2A-7
    V6 engine, 2B-9
    V8 engine, 2C-8
Oxygen sensor, 6-8
Oxygen sensor replacement, 1-26

P

Park/Neutral switch, 6-9
Parking brake, adjustment, 9-21
Piston rings, installation, 2D-17
Piston/connecting rod assembly
    inspection, 2D-15
    installation and bearing oil clearance check, 2D-20
    removal, 2D-11
Positive Crankcase Ventilation (PCV) system, 6-16
**Positive Crankcase Ventilation (PCV) valve
    replacement, 1-26**
**Power brake booster, inspection, removal and
    installation, 9-22**
**Power door lock actuator, removal and
    installation, 12-9**
Power steering
    fluid level check, 1-8
    gear, removal and installation, 11-11
    pump, removal and installation, 11-11

system
    bleeding, 11-11
    general information, 11-10
Pre-oiling engine after overhaul, 2D-21
**Pushrod cover, removal and installation, L4 (four-
    cylinder) engine, 2A-2**

R

Radiator, removal and installation, 3-3
Radio power antenna, removal and installation, 10-7
Radio speakers, removal and installation, 10-7
Radio, removal and installation, 10-6
Rear axle
    assembly, removal and installation, 8-10
    general information and identification, 8-8
    oil change, 1-20
**Rear defogger (electric grid type), check and
    repair, 10-13**
Rear drum brake shoes, replacement, 9-13
Rear hatch, removal, installation and adjustment, 12-6
Rear lower control arm bushing, replacement, 11-7
Rear main bearing oil seal, replacement (in vehicle)
    L4 (four-cylinder) engine, 2A-10
    V6 engine, 2B-9
    V8 engine, 2C-8
Rear main oil seal, installation (during overhaul), 2D-18
Rear shock absorbers, removal and installation, 11-6
**Repair operations possible with the engine in the
    vehicle, 2D-6**
Rocker arm covers, removal and installation
    L4 (four-cylinder) engine, 2A-2
    V6 engine, 2B-1
    V8 engine, 2C-2

S

Safety first!, 0-20
**Shift control lever (77mm 4-speed and 5-speed),
    removal and installation, 7A-2**
Shift linkage, check and adjustment, 7B-2
Shock absorbers (rear), removal and installation, 11-6
Spark plug
    replacement, 1-28
    wires, distributor cap and rotor check and
        replacement, 1-29
Speedometer
    cable, replacement, 10-7
    removal and installation, 10-7
Stabilizer bar, removal and installation
    front, 11-3
    rear, 11-8
Starter motor
    brushes, replacement, 5-10
    removal and installation, 5-9
    testing in vehicle, 5-8

Starter solenoid, replacement, 5-9
Starting system, general information, 5-8
Steering column switches, removal and
    installation, 11-12
Steering system
    general information, 11-8
    power steering
        gear, removal and installation, 11-11
        pump, removal and installation, 11-11
        system
            bleeding, 11-11
            general information, 11-10
    steering column switches, removal and
        installation, 11-12
    steering wheel, removal and installation, 11-10
    tie-rods, removal and installation, 11-8
Stop light switch, removal, installation and
    adjustment, 9-23
Strut assembly (front), removal and installation, 11-3
Suspension and steering check, 1-13
Suspension and steering systems, 11-1 through 11-14
Suspension system
    balljoints, replacement, 11-4
    coil springs and insulators, removal and installation
        front, 11-5
        rear, 11-6
    lower control arm, removal and installation
        front, 11-5
        rear, 11-7
    rear lower control arm bushing, replacement, 11-7
    rear shock absorbers, removal and installation, 11-6
    stabilizer bar, removal and installation
        front, 11-3
        rear, 11-8
    steering knuckle, removal and installation, 11-10
    strut assembly, removal and installation, 11-3
    suspension, general information
        front, 11-3
        rear, 11-5
    torque arm, removal and installation, 11-7
    track bar brace, removal and installation, 11-7
    track bar, removal and installation, 11-6

T

Tank, fuel
    removal and installation, 4A-2
    repair, 4A-3
Thermo-controlled Air Cleaner (THERMAC) check, 1-18
Thermostat
    check, 3-3
    replacement, 3-2
Thermostatic Air Cleaner (THERMAC), 6-17
Throttle Body Injection system, component
    replacement, 4B-5
Throttle cable, removal and installation, 4B-15

Throttle linkage check, 1-17
Throttle Position Sensor (TPS), 6-9
Throttle valve (TV) cable assembly, description and
    adjustment, 7B-2
Tie-rods, removal and installation, 11-8
Timing chain and sprockets, inspection, removal and
    installation
    V6 engine, 2B-12
    V8 engine, 2C-11
Timing gear cover, removal and installation, L4 (four-
    cylinder) engine, 2A-8
Tire and tire pressure checks, 1-5
Tire rotation, 1-19
Tools, 0-11
Torque arm, removal and installation, 11-7
Track bar brace, removal and installation, 11-7
Track bar, removal and installation, 11-6
Transmission Converter Clutch (TCC), 6-18
Transmission, automatic, 7B-1 through 7B-4
    diagnosis, general, 7B-2
    fluid change, 1-23
    fluid level check, 1-8
    general information, 7B-1
    neutral safety and back-up light switch,
        adjustment, 7B-3
    removal and installation, 7B-3
    shift linkage, check and adjustment, 7B-2
    Throttle valve (TV) cable assembly, description and
        adjustment, 7B-2
Transmission, manual, 7A-1 through 7A-4
    clutch start switch, replacement and adjustment, 7A-4
    extension housing oil seal, replacement, 7A-3
    general information, 7A-2
    mounts, check and replacement, 7A-2
    oil level check, 1-7
    shift control lever (77mm 4-speed and 5-speed),
        removal and installation, 7A-2
    speedometer gear seal, replacement, 7A-4
    transmission linkage (76mm and 83mm 4-speed),
        removal, installation and adjustment, 7A-2
    transmission, removal and installation
        76mm and 83mm 4-speed, 7A-2
        77mm 4-speed and 5-speed, 7A-3
Troubleshooting, 0-21
Tune-up and routine maintenance, 1-1 through 1-30
Tune-up general information, 1-5
Turn signals and hazard flashers, check and
    replacement, 10-3

U

Underhood hose check and replacement, 1-11
Universal joints
    disassembly, inspection and reassembly, 8-6
    wear check, 8-6
Upholstery and carpets, maintenance, 12-2

V

V6 engine, 2B-1 through 2B-18
camshaft, removal and installation, 2B-13
crankcase front cover, removal and installation, 2B-11
cylinder heads, removal and installation, 2B-5
exhaust manifolds, removal and installation, 2B-5
front cover oil seal, replacement, 2B-11
hydraulic lifters, removal, inspection and installation, 2B-4
intake manifold, removal and installation, 2B-3
mounts, replacement with engine in vehicle, 2B-13
oil pan, removal and installation, 2B-9
oil pump, removal and installation, 2B-9
rear main bearing oil seal, replacement, 2B-9
removal and installation, 2B-14
rocker arm covers, removal and installation, 2B-1
timing chain and sprockets, inspection, removal and installation, 2B-12
valve lash, adjustment, 2B-4
valve train components, replacement, 2B-2
vibration damper, removal and installation, 2B-11
V8 engine, 2C-1 through 2C-14
camshaft, removal and installation, 2C-11
crankcase front cover, removal and installation, 2C-10
cylinder heads, removal and installation, 2C-7
engine, removal and installation, 2C-12
exhaust manifolds, removal and installation, 2C-6
front cover oil seal, replacement, 2C-10
hydraulic lifters, removal, inspection and installation, 2C-5
intake manifold, removal and installation, 2C-3
mounts, replacement, 2C-12
oil pan, removal and installation, 2C-7
oil pump, removal and installation, 2C-8
rear main bearing oil seal, replacement, 2C-8
rocker arm covers, removal and installation, 2C-2
timing chain and sprockets, removal and installation, 2C-11
valve lash, adjustment, 2C-6
valve train components, replacement, 2C-3
vibration damper, removal and installation, 2C-9
Valve lash, adjustment
V6 engine, 2B-4
V8 engine, 2C-6
Valve train components, replacement
L4 (four-cylinder) engine, 2A-3
V6 engine, 2B-2
V8 engine, 2C-3
Valves, servicing, 2D-10
Vehicle identification numbers, 0-6
Vehicle speed sensor, 6-9
Vibration damper, removal and installation
V6 engine, 2B-11
V8 engine, 2C-9
Vinyl trim, maintenance, 12-2

W

Water pump
check, 3-4
removal and installation, 3-4
Wheel cylinder (drum brakes), removal, overhaul and installation, 9-18
Windshield replacement, 12-11
Windshield wiper
arm, removal, installation and adjustment, 10-11
blade inspection and replacement, 1-11
motor, removal and installation, 10-12
transmission, removal and installation, 10-12
Wiring diagrams, 10-15 through 10-38
Working facilities, 0-15

Haynes Automotive Manuals

ACURA
12020 **Integra** '86 thru '89 **& Legend** '86 thru '90
12021 **Integra** '90 thru '93 **& Legend** '91 thru '95
Integra '94 thru '00 - *see HONDA Civic (42025)*
MDX '01 thru '07 - *see HONDA Pilot (42037)*
12050 **Acura TL** all models '99 thru '08

AMC
14020 Concord/Hornet/Gremlin/Spirit '70 thru '83
14025 **(Renault) Alliance & Encore** '83 thru '87

AUDI
15020 **4000** all models '80 thru '87
15025 **5000** all models '77 thru '83
15026 **5000** all models '84 thru '88
Audi A4 '96 thru '01 - *see VW Passat (96023)*
15030 **Audi A4** '02 thru '08

AUSTIN-HEALEY
Sprite - *see MG Midget (66015)*

BMW
18020 **3/5 Series** '82 thru '92
18021 **3-Series** including Z3 models '92 thru '98
18022 **3-Series** including Z4 models '99 thru '05
18023 **3-Series** '06 thru '14
18025 **320i** all 4-cylinder models '75 thru '83
18050 **1500 thru 2002** except Turbo '59 thru '77

BUICK
19010 **Buick Century** '97 thru '05
Century (front-wheel drive) - *see GM (38005)*
19020 **Buick, Oldsmobile & Pontiac Full-size (Front wheel drive)** '85 thru '05
19025 **Buick, Oldsmobile & Pontiac Full-size (Rear wheel drive)** '70 thru '90
19027 **Buick LaCrosse** '05 thru '13
Regal - *see GENERAL MOTORS (38010)*
Skyhawk - *see GM (38015)*
Skylark - *see GM (38020, 38025)*
Somerset - *see GENERAL MOTORS (38025)*

CADILLAC
21015 **CTS & CTS-V** '03 thru '14
21030 **Cadillac Rear Wheel Drive** '70 thru '93
Cimarron, Eldorado & Seville - *see GM (38015, 38030, 38031)*

CHEVROLET
10305 **Chevrolet Engine Overhaul Manual**
24010 **Astro & GMC Safari** Mini-vans '85 thru '05
24015 **Camaro V8** all models '70 thru '81
24016 **Camaro** all models '82 thru '92
Cavalier - *see GM (38016)*
Celebrity - *see GM (38005)*
24017 **Camaro & Firebird** '93 thru '02
24018 **Camaro** '10 thru '15
24020 **Chevelle, Malibu, El Camino** '69 thru '87
24024 **Chevette & Pontiac T1000** '76 thru '87
Citation - *see GENERAL MOTORS (38020)*
24027 **Colorado & GMC Canyon** '04 thru '12
24032 **Corsica & Beretta** all models '87 thru '96
24040 **Corvette** all V8 models '68 thru '82
24041 **Corvette** all models '84 thru '96
24042 **Corvette** all models '97 thru '13
24044 **Cruze** '11 thru '19
24045 **Full-size Sedans** Caprice, Impala, Biscayne, Bel Air & Wagons '69 thru '90
24046 **Impala SS & Caprice and Buick Roadmaster** '91 thru '96
Impala '00 thru '05 - *see LUMINA (24048)*
24047 **Impala & Monte Carlo** all models '06 thru '11
Lumina '90 thru '94 - *see GM (38010)*
24048 **Lumina & Monte Carlo** '95 thru '05
Lumina APV - *see GM (38035)*
24050 **Luv Pick-up** all 2WD & 4WD '72 thru '82
24051 **Malibu** '13 thru '19
24055 **Monte Carlo** all models '70 thru '88
Monte Carlo '95 thru '01 - *see LUMINA (24048)*
24059 **Nova** all V8 models '69 thru '79
24060 **Nova/Geo Prizm** '85 thru '92
24064 **Pick-ups** '67 thru '87 - Chevrolet & GMC
24065 **Pick-ups** '88 thru '98 - Chevrolet & GMC
24066 **Pick-ups** '99 thru '06 - Chevrolet & GMC
24067 **Chevy Silverado & GMC Sierra** '07 thru '14
24068 **Chevy Silverado & GMC Sierra** '14 thru '19
24070 **S-10 & S-15 Pick-ups** '82 thru '93
24071 **S-10 & Sonoma Pick-ups** '94 thru '04
24072 **Chevrolet TrailBlazer, GMC Envoy & Oldsmobile Bravada** '02 thru '09
24075 **Sprint** '85 thru '88, **Geo Metro** '89 thru '01
24080 **Vans - Chevrolet & GMC** '68 thru '96
24081 **Chevrolet Express & GMC Savana** Full-size Vans '96 thru '19

CHRYSLER
10310 **Chrysler Engine Overhaul Manual**
25015 **Chrysler Cirrus, Dodge Stratus, Plymouth Breeze**, '95 thru '00
25020 **Full-size Front-Wheel Drive** '88 thru '93
K-Cars - *see DODGE Aries (30008)*
25025 **Chrysler LHS, Concorde & New Yorker, Dodge Intrepid, Eagle Vision**, '93 thru '97
25026 **Chrysler LHS, Concorde, 300M, Dodge Intrepid** '98 thru '04
25027 **Chrysler 300** '05 thru '18, **Dodge Charger** '06 thru '18, **Magnum** '05 thru '08 & **Challenger** '08 thru '18
25030 **Chrysler & Plymouth Mid-size** '82 thru '95
Rear-wheel Drive - *see DODGE (30050)*
25035 **PT Cruiser** all models '01 thru '10
25040 **Chrysler Sebring** '95 thru '06, **Dodge Stratus** '01 thru '06 & **Dodge Avenger** '95 thru '00
25041 **Chrysler Sebring** '07 thru '10, **200** '11 thru '17 **Dodge Avenger** '08 thru '14

DATSUN
28005 **200SX** all models '80 thru '83
28012 **240Z, 260Z & 280Z** Coupe '70 thru '78
28014 **280ZX** Coupe & 2+2 '79 thru '83
300ZX - *see NISSAN (72010)*
28018 **510 & PL521 Pick-up** '68 thru '73
28020 **510** all models '78 thru '81
28022 **620 Series Pick-up** all models '73 thru '79
720 Series Pick-up - *see NISSAN (72030)*

DODGE
30008 **Aries & Plymouth Reliant** '81 thru '89
30010 **Caravan & Plymouth Voyager** '84 thru '95
30011 **Caravan & Plymouth Voyager** '96 thru '02
30012 **Challenger & Plymouth Sapporo** '78 thru '83
30013 **Caravan, Chrysler Voyager & Town & Country** '03 thru '07
30014 **Grand Caravan & Chrysler Town & Country** '08 thru '18

30020 **Dakota Pick-ups** all models '87 thru '96
30021 **Durango** '98 & '99 **& Dakota** '97 thru '99
30022 **Durango** '00 thru '03 **& Dakota** '00 thru '04
30023 **Durango** '04 thru '09 **& Dakota** '05 thru '11
30025 **Dart, Challenger/Plymouth Barracuda & Valiant** 6-cylinder models '67 thru '76
30030 **Daytona & Chrysler Laser** '84 thru '89
Intrepid - *see Chrysler (25025, 25026)*
30034 **Neon** all models '95 thru '99
30035 **Omni & Plymouth Horizon** '78 thru '90
30036 **Dodge & Plymouth Neon** '00 thru '05
30040 **Pick-ups** full-size '74 thru '93
30041 **Pick-ups** full-size '94 thru '01
30042 **Pick-ups** full-size '02 thru '08
30043 **Pick-ups** full-size '09 thru '18
30045 **Ram 50/D50 Pick-ups & Raider and Plymouth Arrow Pick-ups** '79 thru '93
30050 **Dodge/Plymouth/Chrysler RWD** '71 thru '89
30055 **Shadow & Plymouth Sundance** '87 thru '94
30060 **Spirit & Plymouth Acclaim** '89 thru '95
30065 **Vans - Dodge & Plymouth** '71 thru '03

EAGLE
Talon - *see MITSUBISHI (68030, 68031)*
Vision - *see CHRYSLER (25025)*

FIAT
34010 **124 Sport Coupe & Spider** '68 thru '78
34025 **X1/9** all models '74 thru '80

FORD
10320 **Ford Engine Overhaul Manual**
10355 **Ford Automatic Transmission Overhaul**
11500 **Mustang** '64-1/2 thru '70 Restoration Guide
36004 **Aerostar** Mini-vans '86 thru '97
36006 **Contour & Mercury Mystique** '95 thru '00
36008 **Courier Pick-up** all models '72 thru '82
36012 **Crown Victoria & Mercury Grand Marquis** '88 thru '11
36016 **Escort & Mercury Lynx** '81 thru '90
36020 **Escort & Mercury Tracer** '91 thru '02
36022 **Escape** '01 thru '17, **Mazda Tribute** '01 thru '11 **& Mercury Mariner** '05 thru '11
36024 **Explorer & Mazda Navajo** '91 thru '01
36025 **Explorer & Mercury Mountaineer** '02 thru '10
36026 **Explorer** '11 thru '17
36028 **Fairmont & Mercury Zephyr** '78 thru '83
36030 **Festiva & Aspire** '88 thru '97
36032 **Fiesta** all models '77 thru '80
36034 **Focus** all models '00 thru '11
36035 **Focus** '12 thru '14
36045 **Ford Fusion** '06 thru '14 **& Mercury Milan** '06 thru '10
36048 **Mustang** V8 all models '64-1/2 thru '73
36049 **Mustang II** 4-cylinder, V6 & V8 '74 thru '78
36050 **Mustang & Mercury Capri** '79 thru '93
36051 **Mustang** all models '94 thru '04
36052 **Mustang** '05 thru '14
36054 **Pick-ups and Bronco** '73 thru '79
36058 **Pick-ups and Bronco** '80 thru '96
36059 **F-150** '97 thru '03, **Expedition** '97 thru '17, **F-250** '97 thru '99, **F-150 Heritage** '04 **& Lincoln Navigator** '98 thru '17
36060 **Super Duty Pick-up & Excursion** '99 thru '10
36061 **F-150** full-size '04 thru '14
36062 **Pinto & Mercury Bobcat** '75 thru '80
36063 **F-150** full-size '15 thru '17
36064 **Super Duty Pick-ups** '11 thru '16
36066 **Probe** all models '89 thru '92
36070 **Ranger & Bronco II** gas models '83 thru '92
36071 **Ranger** '93 thru '11 **& Mazda Pick-ups** '94 thru '09
36074 **Taurus & Mercury Sable** '86 thru '95
36075 **Taurus & Mercury Sable** '96 thru '07
36076 **Taurus** '08 thru '14, **Five Hundred** '05 thru '07, **Mercury Montego** '05 thru '07 **& Sable** '08 thru '09
36078 **Tempo & Mercury Topaz** '84 thru '94
36082 **Thunderbird & Mercury Cougar** '83 thru '88
36086 **Thunderbird & Mercury Cougar** '89 thru '97
36090 **Vans** all V8 Econoline models '69 thru '91
36094 **Vans** full size '92 thru '14
36097 **Windstar** '95 thru '03, **Freestar & Mercury Monterey** Mini-van '04 thru '07

GENERAL MOTORS
10360 **GM Automatic Transmission Overhaul**
38005 **Buick Century, Chevrolet Celebrity, Olds Cutlass Ciera & Pontiac 6000** '82 thru '96
38010 **Buick Regal, Chevrolet Lumina, Oldsmobile Cutlass Supreme & Pontiac Grand Prix** front wheel drive '88 thru '07
38015 **Buick Skyhawk, Cadillac Cimarron, Chevrolet Cavalier, Oldsmobile Firenza Pontiac J-2000 & Sunbird** '82 thru '94
38016 **Chevrolet Cavalier/Pontiac Sunfire** '95 thru '05
38017 **Chevrolet Cobalt & Pontiac G5** '05 thru '11
38020 **Buick Skylark, Chevrolet Citation, Olds Omega, Pontiac Phoenix** '80 thru '85
38025 **Buick Skylark & Somerset, Olds Achieva, Calais & Pontiac Grand Am** '85 thru '98
38026 **Chevrolet Malibu, Olds Alero & Cutlass, Pontiac Grand Am** '97 thru '03
38027 **Chevrolet Malibu** '04 thru '12
38030 **Cadillac Eldorado, Seville, Oldsmobile Toronado & Buick Riviera** '71 thru '85
38031 **Cadillac Eldorado, Seville, DeVille, Fleetwood, Oldsmobile Toronado & Buick Riviera** '86 thru '93
38032 **Cadillac DeVille** '94 thru '05, **Seville** '92 thru '04 **& Cadillac DTS** '06 thru '10
38035 **Chevrolet Lumina APV, Olds Silhouette & Pontiac Trans Sport** all models '90 thru '96
38036 **Chevrolet Venture, Olds Silhouette, Pontiac Trans Sport & Montana** '97 thru '05
38040 **Chevrolet Equinox** '05 thru '17, **GMC Terrain** '10 thru '17 **& Pontiac Torrent** '06 thru '09

GEO
Metro - *see CHEVROLET Sprint (24075)*
Prizm - '85 thru '92 see CHEVY (24060), '93 thru '02 see TOYOTA Corolla (92036)
40030 **Storm** all models '90 thru '93
Tracker - *see SUZUKI Samurai (90010)*

GMC
Vans & Pick-ups - *see CHEVROLET*

HONDA
42010 **Accord CVCC** all models '76 thru '83
42011 **Accord** all models '84 thru '89
42012 **Accord** all models '90 thru '93
42013 **Accord** all models '94 thru '97
42014 **Accord** all models '98 thru '02
42015 **Accord** '03 thru '12 **& Crosstour** '10 thru '14
42016 **Accord** '13 thru '17

42020 **Civic 1200** all models '73 thru '79
42021 **Civic 1300 & 1500 CVCC** '80 thru '83
42022 **Civic 1500 CVCC** all models '75 thru '79
42023 **Civic** all models '84 thru '91
42024 **Civic & del Sol** '92 thru '95
42025 **Civic** '96 thru '00 **& CR-V** '97 thru '01 **& Acura Integra** '94 thru '00
42026 **Civic** '01 thru '11 **& CR-V** '02 thru '11
42027 **Civic** '12 thru '15 **& CR-V** '12 thru '16
42030 **Fit** '07 thru '13
42035 **Odyssey** all models '99 thru '10
Passport - *see ISUZU Rodeo (47017)*
42037 **Honda Pilot** '03 thru '08, **Ridgeline** '06 thru '14 **& Acura MDX** '01 thru '07
42040 **Prelude CVCC** all models '79 thru '89

HYUNDAI
43010 **Elantra** all models '96 thru '19
43015 **Excel & Accent** all models '86 thru '13
43050 **Santa Fe** all models '01 thru '12
43055 **Sonata** all models '99 thru '14

INFINITI
G35 '03 thru '08 - *see NISSAN 350Z (72011)*

ISUZU
Hombre - *see CHEVROLET S-10 (24071)*
47017 **Rodeo** '91 thru '02, **Amigo** '89 thru '94 & '98 thru '02 **& Honda Passport** '95 thru '02
47020 **Trooper** '84 thru '91 **& Pick-up** '81 thru '93

JAGUAR
49010 **XJ6** all 6-cylinder models '68 thru '86
49011 **XJ6** all models '88 thru '94
49015 **XJ12 & XJS** all 12-cylinder models '72 thru '85

JEEP
50010 **Cherokee, Comanche & Wagoneer Limited** all models '84 thru '01
50011 **Cherokee** '14 thru '19
50020 **CJ** all models '49 thru '86
50025 **Grand Cherokee** all models '93 thru '04
50026 **Grand Cherokee** '05 thru '19 **& Dodge Durango** '11 thru '19
50029 **Grand Wagoneer & Pick-up** '72 thru '91
50030 **Wrangler** all models '87 thru '17
50035 **Liberty** '02 thru '12 **& Dodge Nitro** '07 thru '11
50050 **Patriot & Compass** '07 thru '17

KIA
54050 **Optima** '01 thru '10
54060 **Sedona** '02 thru '14
54070 **Sephia** '94 thru '01, **Spectra** '00 thru '09, **Sportage** '05 thru '20
54077 **Sorento** '03 thru '13

LEXUS
ES 300/330 - *see TOYOTA Camry (92007, 92008)*
RX 300/330/350 - *see TOYOTA Highlander (92095)*

LINCOLN
Navigator - *see FORD Pick-up (36059)*
59010 **Rear-Wheel Drive** Continental '70 thru '87, **Mark Series** '70 thru '92 **& Town Car** '81 thru '10

MAZDA
61010 **GLC** (rear wheel drive) '77 thru '83
61011 **GLC** (front wheel drive) '81 thru '85
61012 **Mazda3** '04 thru '11
61015 **323 & Protegé** '90 thru '03
61016 **MX-5 Miata** '90 thru '14
61020 **MPV** all models '89 thru '98
61030 **Pick-ups** '72 thru '93
Pick-ups '94 thru '09 - *see Ford (36071)*
61035 **RX-7** all models '79 thru '85
61036 **RX-7** all models '86 thru '91
61040 **626** (rear-wheel drive) all models '79 thru '82
61041 **626 & MX-6** (front-wheel drive) '83 thru '92
61042 **626** '93 thru '01 **& MX-6/Ford Probe** '93 thru '02
61043 **Mazda6** '03 thru '13

MERCEDES-BENZ
63012 **123 Series Diesel** '76 thru '85
63015 **190 Series** 4-cylinder gas models '84 thru '88
63020 **230, 250 & 280** 6-cylinder SOHC '68 thru '72
63025 **280 123 Series** gas models '77 thru '81
63030 **350 & 450** all models '71 thru '80
63040 **C-Class:** C230/C240/C280/C320/C350 '01 thru '07

MERCURY
64200 **Villager & Nissan Quest** '93 thru '01
All other titles, see FORD listing.

MG
66010 **MGB** Roadster & GT Coupe '62 thru '80
66015 **MG Midget & Austin Healey Sprite** Roadster '58 thru '80

MINI
67020 **Mini** '02 thru '13

MITSUBISHI
68020 **Cordia, Tredia, Galant, Precis & Mirage** '83 thru '93
68030 **Eclipse, Eagle Talon & Plymouth Laser** '90 thru '94
68031 **Eclipse** '95 thru '05 **& Eagle Talon** '95 thru '98
68035 **Galant** '94 thru '12
68040 **Pick-up** '83 thru '96 **& Montero** '83 thru '93

NISSAN
72010 **300ZX** all models incl. Turbo '84 thru '89
72011 **350Z & Infiniti G35** all models '03 thru '08
72015 **Altima** all models '93 thru '06
72016 **Altima** '07 thru '12
72020 **Maxima** all models '85 thru '92
72021 **Maxima** all models '93 thru '08
72025 **Murano** '03 thru '14
72030 **Pick-ups** '80 thru '97 **& Pathfinder** '87 thru '95
72031 **Frontier, Xterra & Pathfinder** '96 thru '04
72032 **Frontier & Xterra** '05 thru '14
72037 **Pathfinder** '05 thru '14
72040 **Pulsar** all models '83 thru '86
72042 **Rogue** all models '08 thru '20
72050 **Sentra** all models '82 thru '94
72051 **Sentra & 200SX** all models '95 thru '06
72060 **Stanza** all models '82 thru '90
72070 **Titan pick-ups** '04 thru '10 **& Armada** '05 thru '10
72080 **Versa** all models '07 thru '19

OLDSMOBILE
73015 **Cutlass** V6 & V8 gas models '74 thru '88
For other OLDSMOBILE titles, see BUICK, CHEVROLET or GM listings.

PLYMOUTH
For PLYMOUTH titles, see DODGE.

PONTIAC
79008 **Fiero** all models '84 thru '88
79018 **Firebird** V8 models except Turbo '70 thru '81
79019 **Firebird** all models '82 thru '92
79025 **G6** all models '05 thru '09
79040 **Mid-size Rear-wheel Drive** '70 thru '87
Vibe '03 thru '10 - *see TOYOTA Corolla (92037)*
For other PONTIAC titles, see BUICK, CHEVROLET or GM listings.

PORSCHE
80020 **911** Coupe & Targa models '65 thru '89
80025 **914** all 4-cylinder models '69 thru '76
80030 **924** all models including Turbo '76 thru '82
80035 **944** all models including Turbo '83 thru '89

RENAULT
Alliance & Encore - *see AMC (14025)*

SAAB
84010 **900** all models including Turbo '79 thru '88

SATURN
87010 **Saturn** all S-series models '91 thru '02
87020 **Saturn Ion** '03 thru '07- *see GM (38017)*
87040 **Saturn** L-series all models '00 thru '04
Saturn VUE '02 thru '09

SUBARU
89002 **1100, 1300, 1400 & 1600** '71 thru '79
89003 **1600 & 1800** 2WD & 4WD '80 thru '94
89080 **Impreza** '02 thru '11, **WRX** '02 thru '14 **& WRX STI** '04 thru '14
89100 **Legacy** all models '90 thru '99
89101 **Legacy & Forester** '00 thru '09
89102 **Legacy** '10 thru '16 **& Forester** '12 thru '16

SUZUKI
90010 **Samurai/Sidekick & Geo Tracker** '86 thru '01

TOYOTA
92005 **Camry** all models '83 thru '91
92006 **Camry** '92 thru '96 **& Avalon** '95 thru '96
92007 **Camry, Avalon, Solara, Lexus ES 300** '97 thru '01
92008 **Camry, Avalon, Lexus ES 300/330** '02 thru '06 **& Solara** '02 thru '08
92009 **Camry, Avalon & Lexus ES 350** '07 thru '17
92015 **Celica Rear-wheel Drive** '71 thru '85
92020 **Celica Front-wheel Drive** '86 thru '99
92025 **Celica Supra** all models '79 thru '92
92030 **Corolla** all models '75 thru '79
92032 **Corolla** rear-wheel drive models '80 thru '87
92035 **Corolla** front-wheel drive models '84 thru '92
92036 **Corolla & Geo/Chevrolet Prizm** '93 thru '02
92037 **Corolla** '03 thru '19, **Matrix** '03 thru '14, **& Pontiac Vibe** '03 thru '10
92040 **Corolla Tercel** all models '80 thru '82
92045 **Corona** all models '74 thru '82
92050 **Cressida** all models '78 thru '82
92055 **Land Cruiser** FJ40/43/45/55 '68 thru '82
92056 **Land Cruiser** FJ60/62/80/FZJ80 '80 thru '96
92060 **Matrix** '03 thru '11 **& Pontiac Vibe** '03 thru '10
92065 **MR2** all models '85 thru '87
92070 **Pick-up** all models '69 thru '78
92075 **Pick-up** all models '79 thru '95
92076 **Tacoma** '95 thru '04, **4Runner** '96 thru '02 **& T100** '93 thru '08
92077 **Tacoma** all models '05 thru '18
92078 **Tundra** '00 thru '06, **Sequoia** '01 thru '07
92079 **4Runner** all models '03 thru '09
92080 **Previa** all models '91 thru '95
92081 **Prius** all models '01 thru '12
92082 **RAV4** all models '96 thru '12
92085 **Tercel** all models '87 thru '94
92090 **Sienna** all models '98 thru '10
92095 **Highlander** '01 thru '19 **& Lexus RX330/330/350** '99 thru '19
92179 **Tundra** '07 thru '19 **& Sequoia** '08 thru '19

TRIUMPH
94007 **Spitfire** all models '62 thru '81
94010 **TR7** all models '75 thru '81

VW
96008 **Beetle & Karmann Ghia** '54 thru '79
96009 **New Beetle** '98 thru '10
96016 **Rabbit, Jetta, Scirocco & Pick-up** gas models '75 thru '92 **& Convertible** '80 thru '92
96017 **Golf, GTI & Jetta** '93 thru '98, **Cabrio** '95 thru '02
96018 **Golf, GTI & Jetta** '98 thru '05
96019 **Jetta, Rabbit, GLI, GTI & Golf** '05 thru '11
96020 **Rabbit, Jetta, Pick-up** diesel '77 thru '84
96021 **Jetta** '11 thru '18 **& Golf** '15 thru '19
96023 **Passat** '98 thru '05, **Audi A4** '96 thru '01
96030 **Transporter 1600** all models '68 thru '79
96035 **Transporter 1700, 1800, 2000** '72 thru '79
96040 **Type 3 1500 & 1600** '63 thru '73
96045 **Vanagon Air-Cooled** all models '80 thru '83

VOLVO
97010 **120, 130 Series & 1800 Sports** '61 thru '73
97015 **140 Series** all models '66 thru '74
97020 **240 Series** all models '76 thru '93
97040 **740 & 760 Series** all models '82 thru '88
97050 **850 Series** all models '93 thru '97

TECHBOOK MANUALS
10205 **Automotive Computer Codes**
10206 **OBD-II & Electronic Engine Management**
10210 **Automotive Emissions Control Manual**
10215 **Fuel Injection Manual** '78 thru '85
10225 **Holley Carburetor Manual**
10230 **Rochester Carburetor Manual**
10305 **Chevrolet Engine Overhaul Manual**
10320 **Ford Engine Overhaul Manual**
10330 **GM and Ford Diesel Engine Repair Manual**
10331 **Duramax Diesel Engines** '01 thru '19
10332 **Cummins Diesel Engine Performance**
10333 **GM, Ford & Chrysler Engine Performance**
10334 **GM Engine Performance**
10340 **Small Engine Repair Manual, 5 HP & Less**
10341 **Small Engine Repair Manual, 5.5 thru 20 HP**
10345 **Suspension, Steering & Driveline Manual**
10355 **Ford Automatic Transmission Overhaul**
10360 **GM Automatic Transmission Overhaul**
10405 **Automotive Body Repair & Painting**
10410 **Automotive Brake Manual**
10411 **Automotive Anti-Lock Brake (ABS) Systems**
10425 **Automotive Heating & Air Conditioning**
10435 **Automotive Tools Manual**
10445 **Welding Manual**

Over a 100 Haynes motorcycle manuals also available

10/22